Sustainable Economy and Ecotechnology

IEREK Interdisciplinary Series for Sustainable Development

The "Sustainable Economy and Ecotechnology" (SEE) book series acts as a timely initiative within the modern academic sphere, fostering a strong connection between technology, research, policy making, and a successful practical implementation for a sustainable future. In an age marked by increasingly urgent environmental challenges and by the critical importance of sustainable development. The series editor is dedicated to steering positive efforts to provide a strong platform for the dissemination of high-quality research focusing on sustainable business, technology, and ecological management and conservation. Through the contributions of experts in the field, "SEE" endeavors to create a transition toward sustainable economic models and technologically driven ecological awareness. The "SEE" series welcomes submissions of selected papers from conference proceedings, monographs, textbooks, and edited volumes, aiming to cultivate a comprehensive understanding of sustainable development by shining a spotlight on a diverse array of topics that collectively contribute to the advancement of a Sustainable Economy and Eco-technology. The scope of this series includes, but is not limited to: Sustainable business practices and management strategies and digitalization, circular economy principles and benefits, climate change impact on economy and business operation, renewable energy and its impact in driving economic growth, role of technology in ecological and environmental preservation, sustainable agriculture, sustainable economics, sustainability in digital business management, finance and investment, sustainable supply chain management: impacting marketing, finance, production, manufacturing and distribution, Implications of economic development on the environment, sustainable mobility and its economic impacts, sustainable marketing strategies and its socio-economic implications, sustainable urban and rural development impacts on economy and more topics related to business investment, economy, environment, and technology.

Klodian Dhoska · Evjola Spaho
Editors

Bridging Horizons in Artificial Intelligence, Robotics, Cybersecurity, Smart Cities, and Digital Economy

Proceedings of the 4th International Conference on Intelligence-Based Transformations of Technology and Business Trends (ICITTBT-2024)

Springer

Editors
Klodian Dhoska
Department of Mechanical Engineering
Polytechnic University of Tirana
Tirana, Albania

Evjola Spaho
Department of Electronics
and Telecommunication
Polytechnic University of Tirana
Tirana, Albania

ISSN 3005-1274 ISSN 3005-1282 (electronic)
Sustainable Economy and Ecotechnology
ISBN 978-3-031-72028-4 ISBN 978-3-031-72029-1 (eBook)
https://doi.org/10.1007/978-3-031-72029-1

This Springer imprint is published by the registered company Springer Nature Switzerland AG
The registered company address is: Gewerbestrasse 11, 6330 Cham, Switzerland

Committees

ICITTBT-2024 Organizing Committee

Prof. Dr. Ismail Kocayusufoğlu, Chair, Canadian Institute of Technology, Albania
Prof. Dr. Manuela Meçe, Member, Canadian Institute of Technology, Albania
Dr. Edmira Cakrani, Member, Canadian Institute of Technology, Albania
Dr. Wassim Ahmad, Member, Canadian Institute of Technology, Albania
Dr. Brisida Sefa, Member, Canadian Institute of Technology, Albania
Ph.D.(c). Anxhela Baraj, Member, Canadian Institute of Technology, Albania
Assoc. Prof. Klodian Dhoska, Member, Polytechnic University of Tirana, Albania
Nurgül Sarsılmaz, Secretariat of Balkan Universities Association Trakya University, Türkiye

Honorary Committee

Prof. Dr. Ismail Kocayusufoğlu—Rector—Canadian Institute of Technology, Albania
Prof. Dr. Erhan Tabakoğlu—Rector—Trakya University, Türkiye
Prof. Dr. Pavel Kozák—Rector—University of South Bohemia Ceske Budejovice, Czech Republic
Prof. Dr. Adalat Muradov—Rector—UNEC, Azerbaijan State University of Economics, Azerbaijan
Prof. Dr. Nermina Hadžigrahić—Rector—the University of Tuzla, Bosnia and Herzegovina
Prof. Dr. Fabio Pollice—Rector—University of Salento, Italy
Prof. Dr. Osman Bülent Zülfikar—Rector—Istanbul University, Türkiye
Prof. Dr. Alush Musaj—Rector—University of Mitrovica, Kosovo
Prof. Dr. Ahmed Nourreddine Helal—Rector—University of Sousse, Tunisia

International Scientific Committee

Prof. Dr. Ismail Kocayusufoğlu—Canadian Institute of Technology, Albania
Prof. Dr. Merita Xhumari—University of Tirana, Albania
Prof. Dr. Habib Hamam—University of Moncton, Canada
Prof. Dr. Suad Becirovic—International Novi Pazaru University, Serbia
Prof. Dr. Sokol Abazi—Canadian Institute of Technology, Albania
Prof. Dr. Agus Pramono—University of Sultan Ageng Tirtayasa, Jakarta, Indonesia
Prof. Dr. Hasan Korkut—Marmara University, Turkiye
Prof. Dr. Festim Halili—Director of the Fund for Innovation and Technology Development, North Macedonia
Prof. Dr. Osman Parlaktuna—Eskisehir Osmangazi University, Turkey
Prof. Dr. Manuela Meçe—Canadian Institute of Technology, Albania
Prof. Dr. Sc. Aljo Mujčić—University of Tuzla, Bosnia and Herzegovina
Prof. Dr. Hassan Ismail—Higher Institute of Business Administration, Syria
Prof. Dr. Tolga Sakallı—Trakya University, Türkiye
Assoc. Prof. Dr. Carlene Campbell—University of Wales Trinity Saint David, UK
Assoc. Prof. Vasil Qano—Canadian Institute of Technology, Albania
Assoc. Prof. Dr. Ayşegül İşcanoğlu Çekic—Trakya University, Türkiye
Prof. Dr. Sc. Almir Osmanović – University of Tuzla, Bosnia and Herzegovina
Assoc. Prof. Shkelzen Cakaj—Canadian Institute of Technology, Albania
Assoc. Prof. Nihat Adar—Canadian Institute of Technology, Albania
Assoc. Prof. Dr. Fredrick Japhet Mtenzi—Institute for Educational Development, East Africa/The Aga Khan University, Tanzania
Assoc. Prof. Albana Demi (Mosho)—University "Aleksander Moisiu" of Durres, Albania
Assoc. Prof. Altin Hoti—Canadian Institute of Technology, Albania
Prof. Dr. Sc. Edin Osmanbegović—University of Tuzla, Bosnia and Herzegovina
Assoc. Prof. Reis Mulita—Canadian Institute of Technology, Albania
Assoc. Prof. Shpresa Caslli—Polytechnic University of Tirana, Albania
Assoc. Prof. Ali Almisreb—International University of Sarajevo, Bosnia and Herzegovina
Assoc. Prof. Entele Gavoçi—Canadian Institute of Technology, Albania
Assoc. Prof. Dr. Ahmet Göçen—İbn Haldun University, Turkiye
Assoc. Prof. Artur Jaupaj—Canadian Institute of Technology, Albania
Assoc. Prof. Elena Bebi—Polytechnic University of Tirana, Albania
Assoc. Prof. Fabian Zhilla—Canadian Institute of Technology, Albania
Assoc. Prof. Evjola Spaho—Polytechnic University of Tirana, Albania
Assoc. Prof. Narasimha Rao Vajjhala—University of New York Tirana, Albania
Ph.D. Abraham Althonayan—Buckinghamshire Chilterns University College, and Brunel University, UK
Ph.D. Besim Bilali—Universitat Politècnica de Catalunya, BarcelonaTech, Spain
Ph.D. Martin Skenderas—University of Southampton, UK
Ph.D. Eshref Januzaj—Hochschule München University of Applied Sciences München, Germany

Ph.D. Ahamed Khan—University College Sedaya International, Malaysia
Ph.D. Luciana Renata De Oliveira—University of Santa Maria, Brazil
Dr. Edmira Cakrani—Canadian Institute of Technology, Albania
Dr. Nargiz Hajiyeva—Women Researchers Council/UNEC—State University of Economics, Azerbaijan
Dr. Katarzyna Strzała-Osuch—Powiślański University, Poland
Dr. Mosbeh Rim—University of Sousse, Tunisia
Dr. Wassim Ahmad—Canadian Institute of Technology, Albania
Dr. Rena Husejnova—UNEC—State University of Economics, Azerbaijan
Dr. Besarta Tafa—Canadian Institute of Technology, Albania
Dr. Katarzyna Łyp-Wrońska—AGH University of Science and Technology, Poland
Dr. Enriko Ceko—Canadian Institute of Technology, Albania
Dr. Alfons Harizaj—Canadian Institute of Technology, Albania
Dr. Eugen Musta—Canadian Institute of Technology, Albania
Dr. Blendi Shima—Canadian Institute of Technology, Albania
Dr. Ditila Ekmekciu—Canadian Institute of Technology, Albania
Dr. Zaid Shhedi—Syrian Private University, Syria
Dr. Marcis Prieditis—Riga Technical University, Latvia
Dr. Olgerta Idrizi—The Mediterranean University of Albania, Albania
Dr. Brisida Sefa—Canadian Institute of Technology, Albania
Dr. Erjona Deshati—Canadian Institute of Technology, Albania
Dr. Silva Spahija—Department of Textile and Fashion, Polytechnic University of Tirana, Albania
Dr. Albana Leti—Department of Textile and Fashion, Polytechnic University of Tirana, Albania
Dr. Elmira Dumishllari—Textile Department, Polytechnic University of Tirana, Albania
Dr. Olga Nezerenko—Logistics, Estonian Enterpreneuship University of Applied Sciences, Tallinn, Estonia
Dr. Martina Hlatka—České Budějovice, Czech Republic
Doc. Eng. Rudolf Kampf, Ph.D., MBA—České Budějovice, Czech Republic

Conference Assistants

Jona Shtini
Xheni Doksani
Hatixhe Gazidede
Zamira Greva
Renisa Berdufi
Denis Xheleku

Reviewers List

Prof. Dr. Agus Pramono
Prof. Dr. Manuela Meçe
Assoc. Prof. Abdulsalam Alkholidi
Assoc. Prof. Vasil Qano
Assoc. Prof. Shkelzen Cakaj
Assoc. Prof. Nihat Adar
Assoc. Prof. Altin Hoti
Assoc. Prof. Reis Mulita
Assoc. Prof. Shpresa Caslli
Assoc. Prof. Entele Gavoçi
Assoc. Prof. Artur Jaupaj
Assoc. Prof. Elena Bebi
Assoc. Prof. Fabian Zhilla
Assoc. Prof. Evjola Spaho
Dr. Edmira Cakrani
Dr. Wassim Ahmad
Dr. Besarta Tafa
Dr. Enriko Ceko
Dr. Alfons Harizaj
Dr. Eugen Musta
Dr. Marcis Prieditis
Dr. Brisida Sefa
Dr. Erjona Deshati
Dr. Silva Spahija
Dr. Albana Leti
Dr. Elmira Dumishllari

Preface

Welcome to the proceedings of the 4th International Conference on Innovation in Technology and Business Transformation (ICITTBT)!

It is with great pleasure that we present to you this Book, a compilation of innovative research, groundbreaking studies, and insightful analyses that reflect the dedication and hard work of our distinguished contributors.

This year's conference brings together a diverse group of experts and emerging scholars from across the globe, each contributing unique perspectives and invaluable knowledge to their respective fields.

In an era marked by dynamic changes in technology, economy, and social paradigms, the landscape of innovation is constantly evolving, challenging traditional norms and opening new frontiers for exploration and progress. The 4th ICITTBT International Conference organized by Canadian Institute of Technology (CIT), Albania aimed to transcend geographical and disciplinary boundaries, creating a global platform for the exchange of transformative ideas. As difficulties in development, employment, education, research, and business become increasingly universal, the conference sought to draw insights and experiences from diverse institutions and countries.

The papers included in this book represent the forefront of research and development, showcasing the latest advancements and future directions in Artificial Intelligence, Robotics, Cybersecurity, Smart Cities, and Digital Economy. From ICITTBT-2024 International Conference, we accepted for publication in this conference proceedings 36 papers (about 45% acceptance ratio).

We are honored to host such a distinguished assembly of minds and to provide a platform for the exchange of ideas, fostering collaboration and inspiring new lines of inquiry. The breadth and depth of topics covered in these papers highlight the dynamic and interdisciplinary nature of our field, and we are confident that the discussions and presentations at this conference will lead to significant advancements and fruitful collaborations.

We extend our deepest gratitude to authors from all around the world (Albania, Turkey, Kosovo, North Macedonia, Azerbaijan, Kuwait, South Korea, Italy, Bosnia and Herzegovina, Yemen, Lithuania, Tunisia) who have submitted their work, the reviewers who have diligently evaluated the papers, and the organizing committee whose tireless efforts have made this event possible. Your contributions are the cornerstone of this conference, and we are proud to showcase your work.

As you explore this Book, we hope you find inspiration and gain new insights that will drive innovation and progress.

Thank you for your participation and for being a part of this vibrant academic community.

We look forward to the exciting exchanges and discoveries that this conference will undoubtedly bring!

We do hope that you will enjoy the conference proceedings and readings.

Tirana, Albania

Klodian Dhoska
Evjola Spaho

Acknowledgments and Sponsor

The editors and the conference organizers would like to express their sincere gratitude to the Canadian Institute of Technology (CIT), Albania for the support and sponsorship provided in organizing the 4th International Conference on Innovation in Technology and Business Transformation (ICITTBT).

CIT
CANADIAN INSTITUTE
OF TECHNOLOGY

Keynote Speakers

1. **Professor Nicola De Filippis**, Physics Department of Politecnico di Bari, Italy
   Presentation Topic: "Machine Learning applications for fundamental science and technology"
2. **Professor Osman Parlaktuna**, Eskisehir Osmangazi University, Turkey
   Presentation Topic: "Robots in our Lives"
3. **Professor Khaldoun Dia-Eddine**, Zurich University for Applied Sciences, Switzerland
   Presentation Topic: "Education and the Smart City Life Cycle"

About This Book

This book aims to foster interdisciplinary research amongst industry and academic participants and form long-term strategic links. It provides presentation of new knowledge and development through exchange of practical experience between industry, scientific institutes and business. The carefully selected conference themes have been chosen to engender these in the fields of Engineering, Industry, Information Technology, Business, Economics and Finance, and Applied Sciences. This book aims to provide the latest research findings, innovative research results, methods and development techniques from both theoretical and practical perspectives related to the emerging areas of Artificial Intelligence, Cybersecurity, Robotics and Automation, Smart Technologies, Data Analytics and Data Science, Network and Communication, Cloud and Mobile Computing, Internet of Things, Virtual Augmented and Mixed Reality, Technology in Applied Science, Digital Economy, Management and Business, Finance and Accounting, Statistics and Econometrics, Economics and Social Sciences.

Klodian Dhoska
Evjola Spaho

Contents

Contributors

Abdulsalam Alkholidi Canadian Institute of Technology (CIT), Tirana, Albania; Faculty of Engineering, Sana'a University, Sana'a, Yemen

Uarda Allushi University College, Canadian Institute of Technology, Tirana, Albania

Said Abdala Almusawi Ministry of Education, Zahra, Kuwait

Elida Arnaut International University of Sarajevo, Bosnia and Herzegovina, Sarajevo, Bosnia-Herzegovina

Ameni Baghdadi University of Sousse, Higher Institute of Management of Sousse, Sousse, Tunisia

Anxhela Baraj Canadian Institute of Technology (CIT), Tirana, Albania

Astrit Bardhi Faculty of Electrical Engineering, Polytechnic University of Tirana, Tirana, Albania

Elena Bebi Energy Department, Polytechnic University of Tirana, Tirana, Albania

Mirza Bečirović Faculty of Mechanical Engineering, University of Tuzla, Tuzla, Bosnia and Herzegovina

Mert Çakar Department of Computer Engineering, Eskisehir Osmangazi University, Eskişehir, Turkey

Enriko Ceko Business Administration and Information Technology Department, Faculty of Economy, Canadian Institute of Technology, Tirana, Albania

Ilir Ciko Canadian Institute of Technology, Tirana, Albania

Salko Ćosić Faculty of Mechanical Engineering, University of Tuzla, Tuzla, Bosnia and Herzegovina

Klodian Dhoska Department of Production and Management, Polytechnic University of Tirana, Tirana, Albania

Ervin Domazet International Balkan University, Skopje, North Macedonia

Gentian Dume Faculty of Electrical Engineering, Polytechnic University of Tirana, Tirana, Albania

Ali Gadirov Azerbaijan State University of Economics, Baku, Azerbaijan

Eyyüp Gülbandılar Department of Computer Engineering, Eskisehir Osmangazi University, Eskişehir, Turkey

Dae Yong Han Vision Group Inc., Republic of South Korea, Buchon, South Korea

Alfons Harizaj Canadian Institute of Technology, Tirana, Albania

Miranda Harizaj The Mediterranean University of Albania, Tirana, Albania

Aldo Hasani Polytechnic University of Tirana, Tirana, Albania

Endrit Hasani Department of Mechatronics, Faculty of Mechanical Engineering, University of Prishtina, Prishtina, Kosovo

Florenc Hidri Canadian Institute of Technology, Tirana, Albania;
SEEU, South East European University, Tetovo, North Macedonia

Olgerta Idrizi The Mediterranean University of Albania, Tirana, Albania;
Polytechnic University of Tirana, Tirana, Albania

Aleksandra Janeska-Iliev Faculty of Economics-Skopje, Ss.Cyril and Methodius University in Skopje, Skopje, North Macedonia

Hiqmet Kamberaj International Balkan University, Skopje, North Macedonia

Yusuf Kartal Department of Computer Engineering, Eskisehir Osmangazi University, Eskişehir, Turkey

Mimoza Kasa-Jashari Faculty of Economics-Skopje, Ss.Cyril and Methodius University in Skopje, Skopje, North Macedonia

Najmin Khalafli Azerbaijan State University of Economics, Baku, Azerbaijan

Najada Kilic Epoka University, Tiranë, Albania

Vesna Knights Faculty of Technology and Technical Science, University "St. Kliment Ohridski", Bitola, Republic of North Macedonia

Hersi Kopani Department of Computer Science, Faculty of Technology and Business, Bedër University College, Tirana, Albania

Hafize Nagehan Koysuren Department of Energy Systems Engineering, Ankara University, Ankara, Turkey;
Department of Environmental Engineering, Kirsehir Ahi Evran University, Kirsehir, Turkey

Ozcan Koysuren Department of Energy Systems Engineering, Ankara University, Ankara, Turkey

Rigon Kqiku Department of Mechatronics, Faculty of Mechanical Engineering, University of Prishtina, Prishtina, Kosovo

Bajram Leka Faculty of Electrical Engineering, Polytechnic University of Tirana, Tirana, Albania

Rovena Llapushi Department of Mathematics "Tullio Levi-Civita", University of Padua, Padua, Italy

Fidan Mamiyeva Faculty of Economics, Sapienza University of Rome, Rome, Italy

Ragib Mammadli International Magistrate and Doctorate Center, Azerbaijan State University of Economics (UNEC), Baku, Azerbaijan

Irida Markja Department of Production and Management, Polytechnic University of Tirana, Tirana, Albania

Fariz Masimli Faculty of SABAH, Azerbaijan State University of Economics, Baku, Azerbaijan

Shefqet Meda Canadian Institute of Technology (CIT), Tirana, Albania; International Balkan University, Skopje, North Macedonia

Momeena Mehdi Canadian Institute of Technology, Tirana, Albania; Khazar University, Baku, Azerbaijan

Asemina Meraj Canadian Institute of Technology, Tirana, Albania

Vasil Mitezi Electronic and Postal Communications Authority, Tirana, Albania

Reis Mulita Canadian Institute of Technology, Tirana, Albania

Blerta Kristo Nazarko European University of Tirana, Tirana, Albania

Lobna A. Okashah Independent Consultant, Kuwait City, Kuwait

Tarik Omerćehajić Faculty of Mechanical Engineering, University of Tuzla, Tuzla, Bosnia and Herzegovina

Rufat Osmani South East European University, Tetovo, North Macedonia

Almir Osmanović Faculty of Mechanical Engineering, University of Tuzla, Tuzla, Bosnia and Herzegovina

Arbnor Pajaziti Department of Mechatronics, Faculty of Mechanical Engineering, University of Prishtina, Prishtina, Kosovo

Sidorela Pako Canadian Institute of Technology, Tirana, Albania

Olivera Petrovska Faculty of Technical Sciences, Mother Teresa University, Skopje, Republic of North Macedonia

Alfred Pjetri Faculty of Electrical Engineering, Polytechnic University of Tirana, Tirana, Albania

Marija Prchkovska Faculty of Technical Sciences, Mother Teresa University, Skopje, Republic of North Macedonia

Ornela Ramašauskaitė Mykolas Romeris University, Vilnius, Lithuania; Women Researchers Council UNEC, Baku, Azerbaijan

Mosbeh Rim University of Sousse, Higher Institute of Management of Sousse, Sousse, Tunisia

Eshev Alibek Sabirovich Karshi Engineering and Economic Institute, Baku, Azerbaijan

Shokhojaeva Zebo Safovna Karshi Engineering and Economic Institute, Baku, Azerbaijan

Zeynep Sağır International University of Sarajevo, Bosnia and Herzegovina, Sarajevo, Bosnia-Herzegovina

Dalila Salčin International University of Sarajevo, Bosnia and Herzegovina, Sarajevo, Bosnia-Herzegovina

Basri Saliu Tetovo, North Macedonia

M. S. C. Julian Saraci University College, Canadian Institute of Technology, Tirana, Albania

Brisida Sefa Canadian Institute of Technology, Tirana, Albania

Gentjan Skara Epoka University, Tiranë, Albania

Gladiola Tigno European University of Tirana, Tirana, Albania

Vilma Tomço University of Tirana, Tirana, Albania

Ensar Topanica Department of Mechatronics, Faculty of Mechanical Engineering, University of Prishtina, Prishtina, Kosovo

Bilge Tuncel Department of Energy Systems Engineering, Ankara University, Ankara, Turkey

Savas Yaglikci Department of Energy Systems Engineering, Ankara University, Ankara, Turkey

Fabrication of Composite Materials for Self-Cleaning Coatings to Enhance Solar Cell Efficiency

Klodian Dhoska, Ozcan Koysuren, Hafize Nagehan Koysuren, Irida Markja, Elena Bebi, Savas Yaglikci, and Bilge Tuncel

Abstract

Turkey and Albania are located in an advantageous position in the southern part of Europe in terms of solar energy. Both countries can receive solar light with a high intensity (~1500 kWh/m^2) when compared with most of the European countries. However, the installed photovoltaic (PV) capacity corresponds respectively to 5 GW for Turkey and 10 MW for Albania which is low compared to the European countries. The installed PV capacities of both countries can only cover 2.5% of Turkey's electricity needs and 1% of Albania's electricity needs. In addition, both countries meet the majority of their energy needs from foreign countries, which means that both countries need to increase the installed capacity of the PV systems and maintain the efficiency of the existing PV systems to be able to benefit more from the sunlight. The buildup of pollutants and contaminants on the glass covering of solar cells has a detrimental impact on the performance of PV systems, ultimately diminishing their efficiency. Our research work presents the development of self-cleaning coatings designed for PV systems that aims to mitigate the efficiency losses caused

K. Dhoska (✉) · I. Markja
Department of Production and Management, Polytechnic University of Tirana, 1001 Tirana, Albania
e-mail: kdhoska@fim.edu.al

O. Koysuren (✉) · H. N. Koysuren · S. Yaglikci · B. Tuncel
Department of Energy Systems Engineering, Ankara University, Ankara 06830, Turkey
e-mail: koysuren@ankara.edu.tr

H. N. Koysuren
Department of Environmental Engineering, Kirsehir Ahi Evran University, Kirsehir 40100, Turkey

E. Bebi
Energy Department, Polytechnic University of Tirana, 1001 Tirana, Albania

K. Dhoska and E. Spaho (eds.), *Bridging Horizons in Artificial Intelligence, Robotics, Cybersecurity, Smart Cities, and Digital Economy*, Sustainable Economy and Ecotechnology, https://doi.org/10.1007/978-3-031-72029-1_1

by contamination in existing PV systems across both countries. The composite materials SiO_2/WO_3 and $SiO_2/WO_3/ZnO$ were developed and applied onto glass substrates via a dip-coating methodology using solutions prepared from their respective compositions. The results have shown that solar cell coated with the $SiO_2/WO_3/ZnO$ film offered nearly equivalent solar cell efficiency levels compared to those of uncoated solar cells.

Keywords

Solar cell • Efficiency • Self-cleaning coatings • Composite materials

1 Introduction

In both Turkey and Albania, the adoption of photovoltaic (PV) systems has gained momentum in recent years as part of broader efforts to transition toward cleaner and more sustainable energy sources. While each country has its unique context and challenges, both share a commitment to leveraging solar energy to diversify their energy mix, enhance energy security, and mitigate environmental impacts. Both countries can receive solar light with a high intensity (~1500 kWh/m^2) when compared with most of the European countries. However, the installed PV capacity corresponds respectively 5 GW for Turkey and 10 MW for Albania which is low compared to the European countries. The installed PV capacities of both countries can only cover 2.5% of Turkey's electricity needs and 1% of Albania's electricity needs.

In addition, both countries meet the majority of their energy needs from foreign countries, which means that both countries need to increase the installed capacity of the PV systems and maintain the efficiency of the existing PV systems to be able to benefit more from the sunlight. By capitalizing on solar energy's benefits and addressing existing challenges, both countries aim to realize their vision of a cleaner, more resilient energy future driven by renewable sources like PV systems. Solar cells are composed of different layers depending on the type. Generally, they have a cover glass, a metal frame, and an encapsulant on the cell as can be seen in Fig. 1.

The efficiency of PV systems is adversely impacted by the buildup of pollution and various forms of contamination on the glass cover of solar cells. This decreases the electricity output of the PV system and diminishes its overall performance. Dust accumulation on solar panels can indirectly contribute to environmental impacts by increasing the need for water for cleaning and potentially leading to water pollution if cleaning agents are not properly managed. Furthermore, manual cleaning can be time-consuming, labor-intensive, and costly, especially for large-scale PV installations. Additionally, improper cleaning techniques or using harsh cleaning agents may damage the panels, exacerbating the problem.

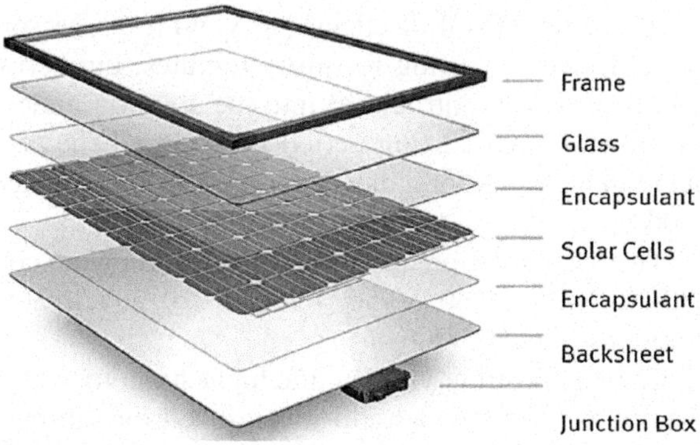

Fig. 1 Schematic illustration of a PV panel with its components (Sarkin et al. 2020)

Based on it, our research work will be focused on the development of the self-cleaning coatings designed for PV systems aiming to mitigate the efficiency losses caused by contamination in existing PV systems across both countries. The composite materials SiO_2/WO_3 and $SiO_2/WO_3/ZnO$ were developed and applied onto glass substrates via a dip-coating methodology using solutions prepared from their respective compositions. The results have shown that solar cell coated with the $SiO_2/WO_3/ZnO$ film offered nearly equivalent solar cell efficiency levels compared to those of uncoated solar cells.

2 Materials and Methods

In this section we have prepared SiO_2, WO_3, and ZnO solutions and used separately to prepare SiO_2/WO_3 and $SiO_2/WO_3/ZnO$ films. For preparing SiO_2 solution we have used Tetraethylorthoxylsilicane (TEOS) with an amount of 112 ml. Afterward, a specific volume of a mixture of distilled water and anhydrous ethanol will be prepared, with quantities corresponding to 36 ml to 1090 ml. Following this, a concentrated hydrochloric acid solution (~0.2 ml, 36%) will be introduced into the mixture to achieve a final molar ratio of TEOS: ethyl alcohol: H2O: HCl as 1: 37.5: 4: 0.004. The anticipated concentration of SiO_2 within the resulting solution is projected to be 3 wt%. It is assumed that 1 mol of TEOS can transform into 1 mol of SiO_2. The solution will undergo agitation within a sealed glass container at ambient temperature for a duration of 2 h, followed by a period of incubation in darkness for 1 day. Sodium tungstate dihydrate ($Na_2WO_4 \cdot 2H_2O$) (3 g) will be dissolved in 20 ml of distilled water and stirred at room temperature. HCl solution (20 ml, 8 M) will be added to the prepared solution. Subsequently, the solution will undergo stirring for a duration of 2 h at a temperature of 80 °C. Upon cooling to room temperature, the solution in its current state will be left to stand in darkness for a day, as outlined by Thilagavathi et al. (2021). Zinc acetate 2-hydrate ($Zn(CH_3COO)_2 \cdot 2H2O$) will serve as the precursor for zinc atoms, while absolute ethanol and diethanolamine will be employed as the solvent and solution stabilizer, respectively. To elaborate, a quantity of 0.05 mol of $Zn(CH_3COO)_2 \cdot 2H2O$ will be dissolved in 100 ml of ethanol to yield a 0.5 M solution. This solution will then be subjected to stirring for thirty minutes. Furthermore, 0.05 mol of the solution stabilizer will be introduced into the solution under continuous stirring, followed by an additional stirring period of 30 min. The resultant solution will be left to stand in darkness for a day, adhering to the methodology described by Wang et al. (2006).

The SiO_2 solution and the WO_3 solution were combined in various proportions (80/20, 70/30, 60/40, 50/50, 40/60, 30/70, and 20/80 wt./wt.) to generate composite solutions of SiO_2/WO_3.

The composition yielding the highest photocatalytic activity was identified, and the optimal SiO_2/WO_3 ratio was utilized for subsequent investigations to fabricate $SiO_2/WO_3/ZnO$ composites. Furthermore, the ZnO solution was blended with different ratios ((SiO_2/WO_3)/ZnO: 70/30, 80/20, and 90/10 wt./wt.) using the optimal SiO_2/WO_3 solution to create the $SiO_2/WO_3/ZnO$ composites. Moreover, the composite films were designated as SiO_2/WO_3(x/y) based on the weight ratio of SiO_2 to WO_3. Additionally, the composite films were labeled as $SiO_2/WO_3/ZnO$(x/y) based on the weight ratio of SiO_2/WO_3 to ZnO. Prior to the dip-coating process, the glass substrates will undergo ultrasonic cleaning. Subsequently, the prepared solutions will be applied onto the glass substrates using the dip-coating technique under ambient conditions. A consistent withdrawal speed of 100 mm/min will be maintained, and the thickness of the coated film will be optimized through additional dip-coating cycles ranging from 1 to 3. Following each dip-coating cycle, the glass substrate will be air-dried at ambient conditions for 2 h. The deposited film will undergo pre-heating at 100 °C for 1 h to eliminate any residual volatile species. Subsequently, annealing at 400 °C for 2 h will be conducted to facilitate the growth of the crystalline phase.

One of the widely adopted approaches for examining the photocatalytic self-cleaning properties of film samples is through the dye method. In this technique, a model dye is employed to simulate contamination on the film surface, and the photocatalytic breakdown of the dye molecules under UV light exposure is analyzed to unveil the self-cleaning characteristics of the film coatings, as elucidated by Khorshidi et al. (2021). According to experiments on the degradation of dyes using photocatalysis, the film samples prepared were able to break down the model dye absorbed onto themselves when exposed to UVA light. Consequently, methylene blue, serving as the model dye, underwent decomposition into water and carbon dioxide. The assessment of the photocatalytic degradation rate of methylene blue was conducted by observing the rate of water formation via contact angle measurements, as detailed in the study by Khorshidi et al. (2021). The presence of water molecules on the surface of the film resulted in a modification of surface energy, thereby reducing the water contact angle. Furthermore, the photocatalytic efficacy of the film samples was evaluated using the dye method. Contact angle measurements of water were conducted on the film samples after adsorption of methylene blue and exposure to UVA light irradiation.

3 Results

Figures 2 and 3 depicted the images of the water droplets respectively on the fabricated composite materials types SiO_2/WO_3(30/70) film and $SiO_2/WO_3/ZnO$(70/30) film with irradiation time 0 min, 60 min, 120 min, 180 min, and 240 min.

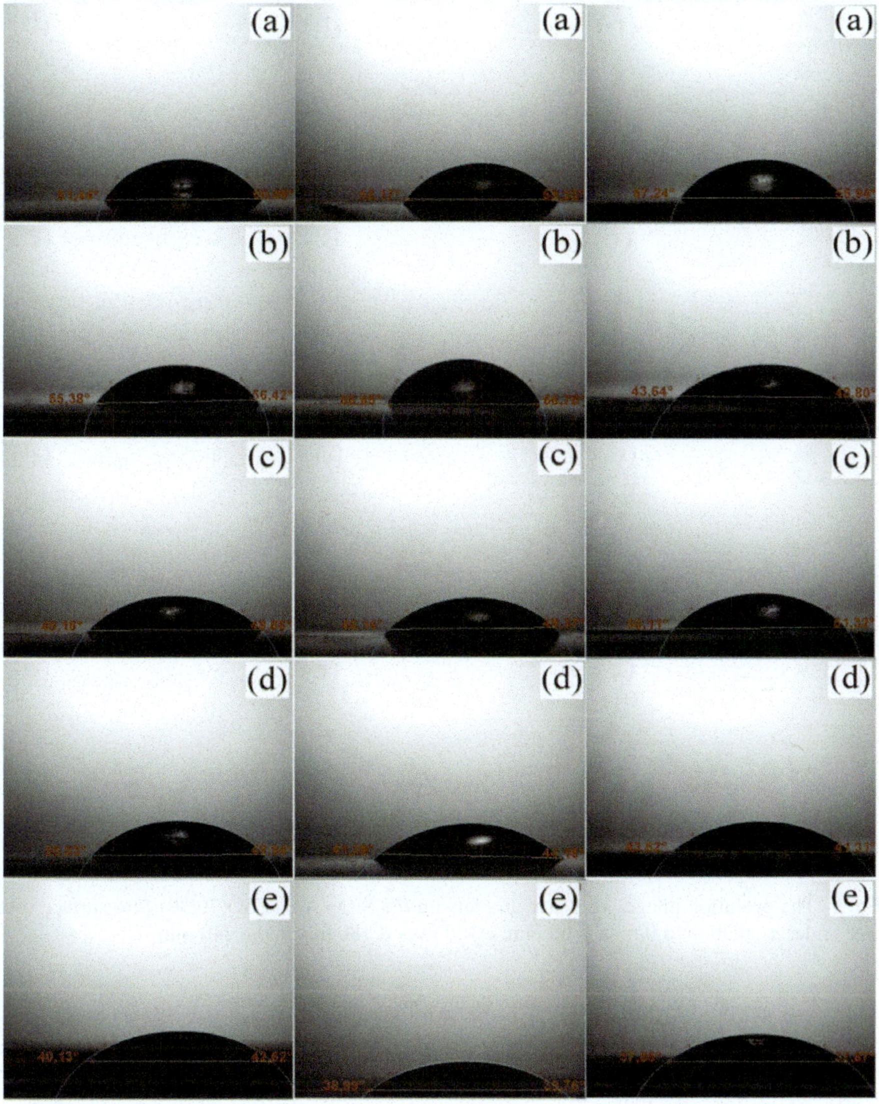

Fig. 2 The evolution images of water droplets on the SiO_2/WO_3(30/70) film surface over time under irradiation: **a** 0 min, **b** 60 min, **c** 120 min, **d** 180 min and **e** 240 min

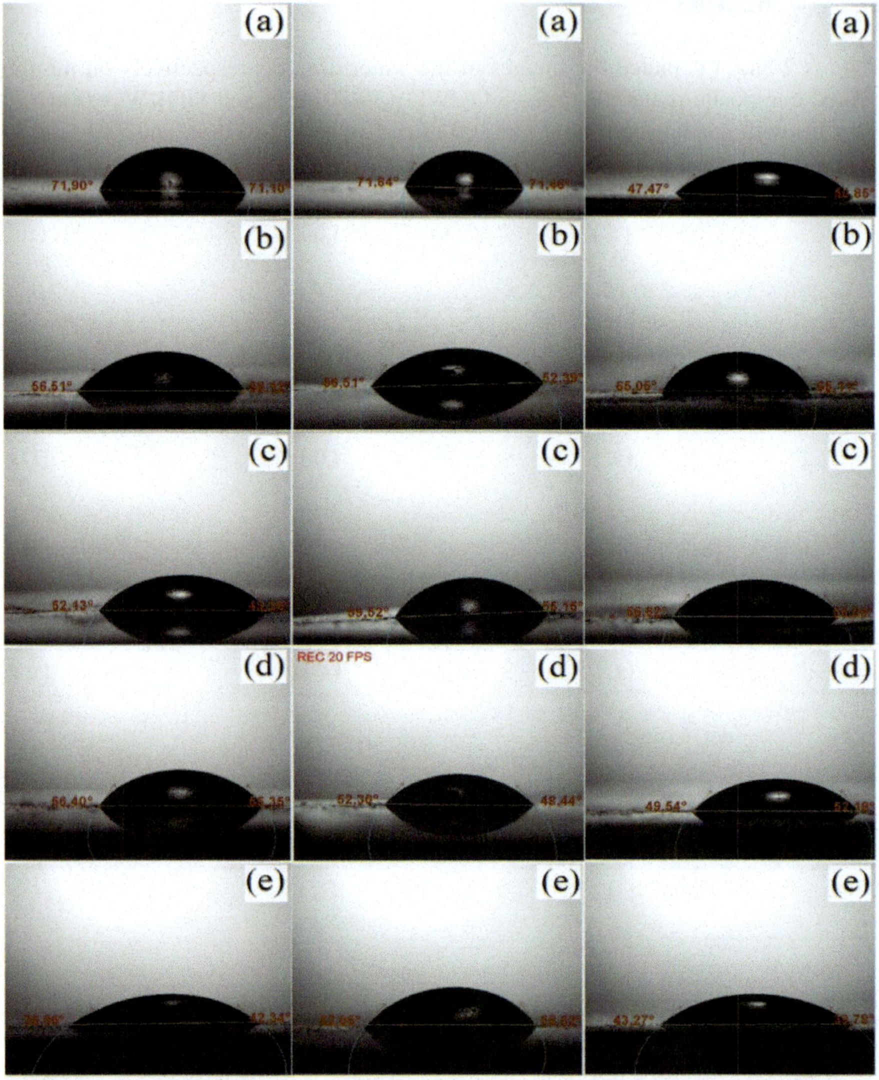

Fig. 3 The evolution images of water droplets on the $SiO_2/WO_3/ZnO$(70/30) film surface over time under irradiation: **a** 0 min, **b** 60 min, **c** 120 min, **d** 180 min and **e** 240 min

Figure 4 depicts the impact of the number of dip-coating cycles on the current–voltage characteristics of solar cells coated with the composite materials we fabricated.

Based on our findings, the water contact angle decreased as the irradiation time increased. Specifically, the water contact angles of the SiO_2/WO_3(30/70) film and the $SiO_2/WO_3/ZnO$(70/30) film decreased from 45.1° and 64.1° to 29.2° and

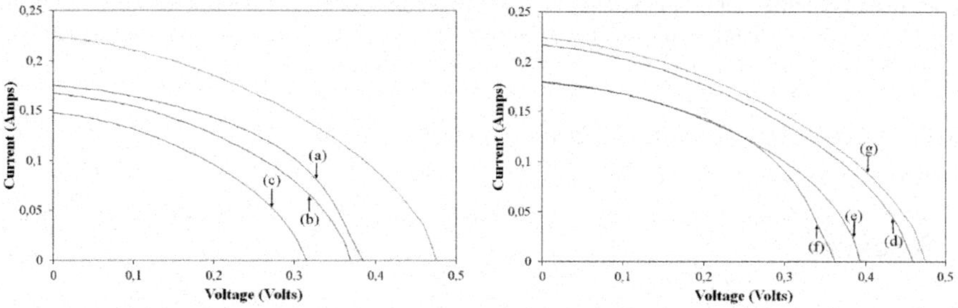

Fig. 4 The impact of the number of dip-coating cycles on the current–voltage characteristics of the solar cell coated with various compositions **a** SiO_2/WO_3(30/70) (onefold coating), **b** SiO_2/WO_3(30/70) (twofold coating), **c** SiO_2/WO_3(30/70) (threefold coating), **d** SiO2/WO3/ZnO(70/30) (onefold coating), **e** $SiO_2/WO_3/ZnO$(70/30) (twofold coating), **f** $SiO_2/WO_3/ZnO$(70/30) (threefold coating) and including **g** uncoated glass slide

45.4°, respectively, after 240 min of UVA light exposure. This observation suggests that the composite film samples exhibited the ability to degrade the model dye adsorbed on their surfaces over time, indicating their photocatalytic activity. Both the degradation of the dye and the water contact angle measurements confirmed the substantial photocatalytic activity of both the SiO_2/WO_3(30/70) and $SiO_2/WO_3/ZnO$(70/30) films. Interestingly, in comparison between the two film samples, the $SiO_2/WO_3/ZnO$(70/30) film demonstrated a higher water contact angle value (Koysuren et al., 2024). This elevation in the water contact angle could be attributed to the enhanced surface roughness introduced by the presence of ZnO.

4 Conclusions

In this research work it has been shown the fabrication of the new composite materials for self-cleaning coatings to improve solar cell efficiency. The composite materials SiO_2/WO_3 and $SiO_2/WO_3/ZnO$ were developed and applied onto glass substrates via a dip-coating methodology using solutions prepared from their respective compositions. The film coatings facilitated a self-cleaning mechanism via their photocatalytic capabilities and induced hydrophilicity upon exposure to light. The $SiO_2/WO_3/ZnO$ film coatings demonstrated notable transparency to visible light and exhibited effective photocatalytic properties when subjected to UVA light. Additionally, the self-cleaning efficacy of the developed film coatings was remarkable, with the $SiO_2/WO_3/ZnO$ film degrading over 90% of the model contaminant present on its surface within a 240-min period of UVA light exposure. The findings from this study indicate that the $SiO_2/WO_3/ZnO$ film holds promise for commercial application on the surfaces of photovoltaic (PV) panels to mitigate the buildup of various forms of organic pollutants. This application could lead to improvements in PV efficiency by reducing the adverse effects of contamination.

Acknowledgements The study is supported by National Agency of Scientific Research and Innovation (NASRI) of Albania and with Project Number 91106AB and The Scientific and Technological Research Council of Turkey (TÜBİTAK) with Project Number 122N383.

Conflict of Interests The authors declare no conflict of interest.

References

Khorshidi, Z.G., Jallab, M., Moghbelli, E., Goudarzi, A., Ghaffari, M.: Photocatalytic analysis of a hydrophilic acrylic coating/zinc oxide nanocomposite on glass substrate. Polym-Plast Technol Mater **60**(11), 1220–1232 (2021)

Koysuren, O., Dhoska, K., Koysuren, H.N., Markja, I., Yaglikci, S., Tuncel, B., Bebi, E.: SiO_2/WO_3/ZnO based self-cleaning coatings for solar cells. J. Sol-Gel Sci. Technol. (2024). https://doi.org/10.1007/s10971-024-06351-7

Sarkin, A.S., Ekren, N., Saglam, S.: A review of anti-reflection and self-cleaning coatings on photovoltaic panels. Sol. Energy **199**, 63–73 (2020)

Thilagavathi, T., Venugopal, D., Marnadu, R., Chandrasekaran, J., Alshahrani, T., Shkir, M.: An investigation on microstructural, morphological, optical, photoluminescence and photocatalytic activity of WO(3) for photocatalysis applications: An effect of annealing. J. Inorg. Organomet. Polym. Mater. **31**(3), 1217–1230 (2021)

Wang, M., Kim, E.J., Chung, J.S., Shin, E.W., Hahn, S.H., Lee, K.E., Park, C.: Influence of annealing temperature on the structural and optical properties of sol–gel prepared ZnO thin films. Phys. Status Solidi A **203**, 2418–2425 (2006)

Design and Construction of Mobile Robot for Disposal of Hazardous Materials

Arbnor Pajaziti, Ensar Topanica, Endrit Hasani, and Rigon Kqiku

Abstract

In this paper, a mobile robot prototype has been developed that in the future could serve people to remotely control and remove various hazardous and non-explosive materials. The project has been developed with low cost, indicating that in the future it is not necessarily required to have very large investments to create a mobile device for demining or just for "pick and place". The mobile platform has a robotic arm that can move accurately through hazardous areas to remove non-explosive materials. Additionally, this mobile device can overcome most obstacles encountered on the road, some of them even by passing over them. This project integrates a GPS that always informs us about its precise location. It also has two cameras that provide images of the terrain condition at any time. It is controlled by two Remote Controllers, one of which controls the robotic arm, while the other controls the robotic platform. The purpose of the robotic arm is the "pick and place" function. This robotic arm has five degrees of freedom and consists of servomotors with different power, each of which has a unique ID. The microcontroller used for this robotic arm is the CM-700. The robotic arm is controlled by a Remote Controller RC-100A, which continuously

A. Pajaziti (✉) · E. Topanica · E. Hasani · R. Kqiku
Department of Mechatronics, Faculty of Mechanical Engineering, University of Prishtina, Prishtina, Kosovo
e-mail: arbnor.pajaziti@uni-pr.edu

E. Topanica
e-mail: ensar.topanica@student.uni-pr.edu

E. Hasani
e-mail: endrit.hasani@student.uni-pr.edu

R. Kqiku
e-mail: rigon.kqiku@student.uni-pr.edu

K. Dhoska and E. Spaho (eds.), *Bridging Horizons in Artificial Intelligence, Robotics, Cybersecurity, Smart Cities, and Digital Economy*, Sustainable Economy and Ecotechnology, https://doi.org/10.1007/978-3-031-72029-1_2

sends signals to the arm controller. The mobile robot structure is entirely made of metal because it is supposed to operate in tough and wet terrains.

Keywords

Mobile robot • Robotic arm • Demining • Remote control

1 Introduction

In this work, a prototype has been developed that in the future could serve people to remotely remove various hazardous and unexploded devices. The project has been developed at low cost, indicating that in the future, creating a mobile device for demining or just object retrieval may not necessarily require significant investments.

The mobile platform features a manipulator arm capable of accurately maneuvering through hazardous areas to prevent damage from unexploded devices. Additionally, this mobile device can overcome most obstacles encountered on its path, even surpassing some of them (Trevelyan 2024; Havlik and Licko 2016).

This project integrates a GPS that always informs us of its exact location. It also has two cameras that provide real-time images of the terrain. It is controlled by two Remote Controllers, one of which controls the arm and the other controls the armless part of the platform. The purpose of the robot arm is pick and place functionality. This robotic arm has five degrees of freedom and consists of servo motors with varying power, each with its own ID. The microcontroller used for this robotic arm is the CM-700. The robot arm is controlled by an RC-100A Remote Controller, which continuously sends signals to the arm controller. Its construction is entirely made of metal, as it is assumed to operate in difficult terrains.

2 Objectives of the Research

Current demining robots often encounter difficulties in being effective in mountainous environments, which are quite challenging. This research aims to overcome this gap by designing, developing, and testing a demining robot specifically tailored to maneuver in difficult terrains, where existing machinery finds it impossible to work due to environmental conditions (Sachs 2005; Doyle et al. 1526; Manwell and McGowan 1994; UNODA 1999; United Nation (UN) (n.d.); Robots for Humanitarian Demining 2024).

The main objective of this research is to design and develop a demining robot capable of moving freely and operating efficiently in difficult terrains. Our main objectives for the development of this project have been the development of a prototype that can accomplish a great deal with very few financial resources. The electronics to realize the implementation of our project have cost very little, making our project very interesting for the future. Because today, the goal is to create technological devices that can perform tasks that have cost millions of euros in the past with very little money. The aim has been for the code written in the C++ programming language to be as optimized as possible so that in the future, it can serve as a good basis for the further development of other devices aimed at demining in difficult terrains.

3 Robotic Platform

The robotic platform is constructed with the purpose of developing and executing controlled robotic movements, as shown in Fig. 1. This platform is built from metallic and polyester materials, giving it a strong and flexible combination.

The platform dimensions are 64–39 cm with a height of 26 cm, and it is elevated 13 cm from the ground.

Weight: The total weight of the robotic platform is approximately 55 kg including all electronic and mechanical components.

Load capacity: The platform can carry a load of up to 100 kg.

The metallic structure serves as the foundation of the platform, providing support and insulation for other parts from moisture. It is constructed with strong and durable metal.

The platform is equipped with an advanced movement system that includes electric motors, wheels, and control systems. This system allows the platform to move forward, backward, and sideways accurately.

Fig. 1 Robotic platform

The platform is equipped with a range of electronic devices such as sensors, cameras, and controllers. These devices aid in gathering information and controlling the actions of the platform.

The platform can be used to explore hazardous areas and prevent damage from unexploded devices. The entire robotic platform is depicted in the figure below.

3.1 NPC-41250 Motors

The DC motor NPC-41250 is a high-performance motor designed for various applications, including robotics, automation, and industrial machinery (https://hobby.npcrobotics.com/product/npc-41250/). It offers reliable and efficient operation with precise control of speed and torque, as shown in Fig. 2.

Some of the main characteristics of the motors are presented below:

- Belongs to the group of DC (Direct Current) motors, thus operates with direct current.
- Has a weight of 3.3 kg.
- Operates with a voltage range of 12 to 24 V.
- Can draw current from 56 to 82A.
- Has an RPM range from 93 to 174 rpm.

In our project, we controlled the speed and direction of these motors using the Arduino microcontroller by utilizing PWM (Pulse Width Modulation) signals.

Fig. 2 NPC Motor within the robotic platform

Fig. 3 The power source for the electrical energy

3.2 Power Supply

The battery produced by Topla is built using high-quality materials for longevity and long-term performance. Due to the need for more electrical power in this project, an automotive battery has been used. This type of battery is characterized by high power: 12V and 65Ah. It can keep the mobile device operational for up to 1 hour.

Not all electronic devices in the project need to be supplied with 12V, which is why a voltage regulator has been used. In Fig. 3 below, the power source for the robotic platform is presented.

3.3 DC-DC Converter

In this project, it has been necessary to use converters because not all electronic devices used operate at 12 V voltage. The converter, capable of converting any voltage from 12-24 V to 5 V and 5A, was employed for this purpose. The reason the heatsink of this converter is of considerable size is due to the substantial conversion it performs. The electronic devices supplied include controllers, drivers, and routers.

Arduino Mega, Raspberry PI3, and CM700 are the controllers directly supplied by this converter. In Fig. 4, the DC-DC converter used in the project is presented.

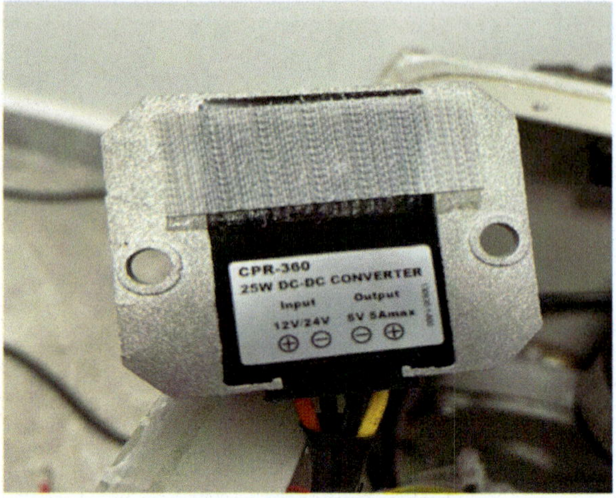

Fig. 4 DC-DC converter

3.4 Arduino Mega 2560 Microcontroller

The overall control of running on the robotic platform has been performed by Arduino Mega (https://www.techspot.com/article/2126-arduino/). The reason for using this type of Arduino is its greater number of pins, specifically PWM pins. Arduino continuously sends signals to motor drivers throughout the project to control the motors. Additionally, Arduino communicates continuously with the DX6i device, enabling remote engagement control. Arduino Mega is presented in Fig. 5.

Fig. 5 Arduino mega 2560 microcontroller

Fig. 6 BTS7960B motor driver

3.5 BTS7960b Motor-Drivers

The DC motors and stepper motors are controlled by the motor driver BTS7960B. This module is a dual-channel H-bridge motor driver module. It provides bidirectional control for two DC motors or one stepper motor, as shown in Fig. 6.

3.6 The Remote-Control Devices

The DX6i transmitter and the AR610 receiver are both products that have been used to control the robotic platform. The DX6i transmitter is designed for use with various RC models and offers advanced features for precise control, as shown in Fig. 7.

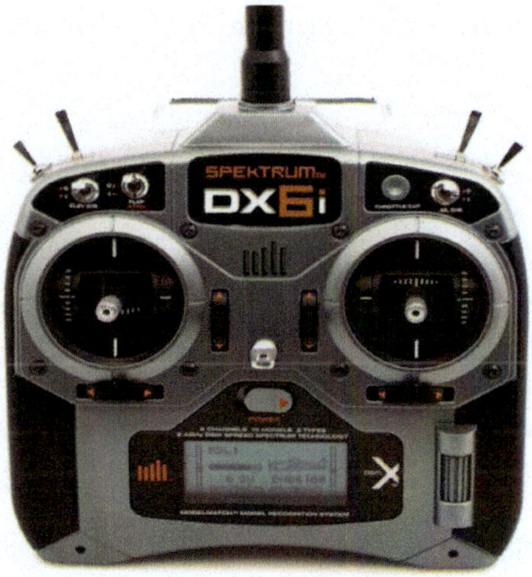

Fig. 7 DX6i transmitter

Fig. 8 AR610 receiver

The AR610 receiver is a product manufactured by Spektrum that is compatible with the DX6i transmitter, as shown in Fig. 8. It receives signals from the transmitter and transmits them to the servos and electronic speed controllers (ESCs) in your RC model.

3.7 Design of the Driver and Arduino Holders

The software used during the design process on the 3D printer is Fusion 360. The parts that were designed have been more than necessary for many reasons. The holders for the drivers have been very necessary for reasons such as the robotic platform lacking a specific place for mounting the drivers, so a holder was designed for each driver. In the space where the heatsink was mounted, a space was left for the cooler's airflow. The material used for designing these parts has been PLA (polylactic acid). Additionally, a holder was designed for mounting the Arduino Mega microcontroller onto the chassis of the robotic platform. In Fig. 9a and b, the mounted holders on the platform are depicted.

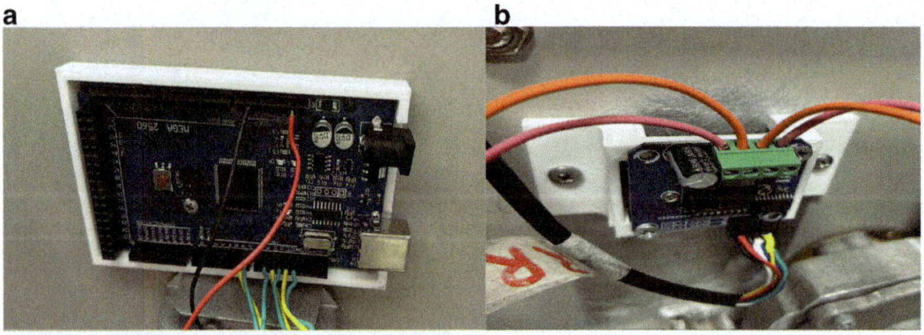

Fig. 9 **a** Holder for Arduino and **b** holder for the drivers on the robotic platform

Fig. 10 View of GPS GF-07

3.8 GPS GF-07 Tracker

The GPS tracker GF-07 is a small and lightweight device used to determine the location of unexploded ordnance during the demining process. The tracker is controlled by an application installed on a smartphone, which can read messages to report geographic coordinates, as shown in Fig. 10.

3.9 Servomotors of Robotic Arm

The servo motors used in the robotic arm are part of the intelligent actuator group, as shown in Fig. 11 (Arms and | CrustCrawler Robot Arms n.d.). These servo motors integrate: a DC motor, controller, driver, regulator, gearbox, etc. Each servo has its own unique address with which the controller communicates. The types of servo motors used in the robotic arm are AX-18A, MX-106, MX-28, AX-104, and MX-101.

3.10 CM-700 Microcontroller

The CM-700 is a robot controller developed by Robotis (robotis.com). It has been used to implement complex motions, sequences, and actions for the robotic platform. Figure 12 shows a view of the microcontroller.

Fig. 11 Robotic arm

Fig. 12 CM-700 microcontroller

Fig. 13 Remote controller RC-100A

Fig. 14 ZIG—110A receiver

3.11 The Control Device of the Robotic Arm (RC-100A) and ZIG-110A

The RC-100A is a control device developed by Robotis and has been used to control and operate the robotic platform. Figure 13 shows the RC-100A control device.

3.12 ZIG—110A Receiver

The ZIG-110A has been used to provide wireless communication based on the Zigbee protocol. Figure 14 shows the view of the Bluetooth module.

4 Image Processing with Raspberry Pi 3

In the project, Raspberry Pi has been used to control the cameras through the "MotionEye" software (GitHub). Two USB cameras are connected to this mini-computer and are also used. A router has been used on the robotic platform to generate an IP Address, which connects directly to the cameras placed, one on the side for a view of the gripper and another in front of the platform to provide a view of the front part of the platform. Figure 15 depicts the Raspberry Pi 3, which plays a crucial role in displaying the image.

Figure 16 shows the connection between the electronic devices that enable capturing images from remote distances. The "MotionEye" software allows us to virtually connect directly to the cameras using the IP address provided by the router. Raspberry Pi 3 has been used as a bridge between the router, cameras, and the laptop, which in the project only serves to find the IP address and provide the camera images in "MotionEye".

Fig. 15 Raspberry Pi 3

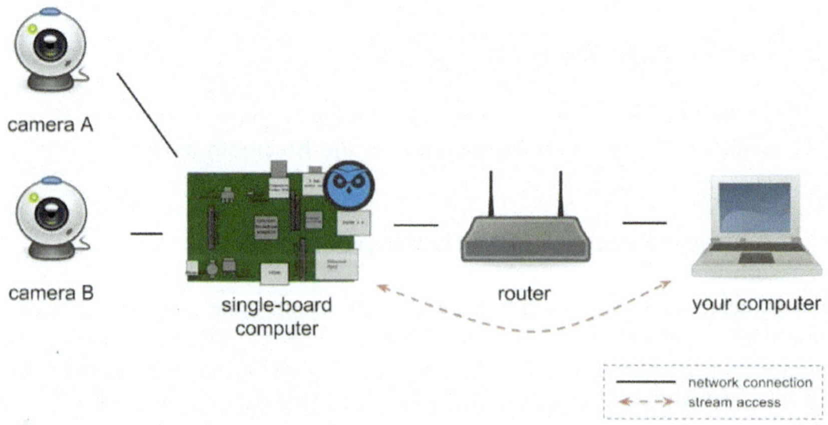

Fig. 16 Electronic device scheme for image processing ((1523) Pinterest)

5 The Process of Building the Robotic Platform and Programming

The process involves mechanical design, integration of electronic devices, software development (programming of the platform and robotic arm), and extensive testing. Testing has been conducted in laboratory conditions to evaluate the performance of the robotic platform in various scenarios. The programming is done in the C++ language.

The C++ code controls the motors of the robotic platform based on input signals from the radio receiver. It reads the control signals from the radio receiver, adjusts them for consistency, converts the input into motor power, and adjusts the motor power based on the received signals. The code demonstrates how to interface with hardware components, control motors, and work with input signals in an integrated system using the Arduino platform.

The experiment we carried out at the Laboratory of Mechatronics by performing a task by using a robotic platform for picking up the bottle by being focused on cameras. There are two IP cameras as shown in Fig. 17 and we already configured the wireless communication between cameras and computer.

There are two controllers for servomotors that move platform and robotic arm, separately. The DX6i transmitter and the AR610 receiver control the platform, and the other one, the RC-100A and ZIG—110A control the robotic arm. First, we run the platform through DX6i transmitter enabling the robot platform to approach the bottle position. When the platform comes closer to the bottle, it will stop, and the activation of the robotic arm will start, Fig. 18a. There are some pre-programed positions of the robotic arm that can be achieved by pushing into the buttons of the RC-100A, to adopt the right position of the gripper to grasp the bottle in a proper position, until we reach it. Then we close the gripper and pick up the bottle, and

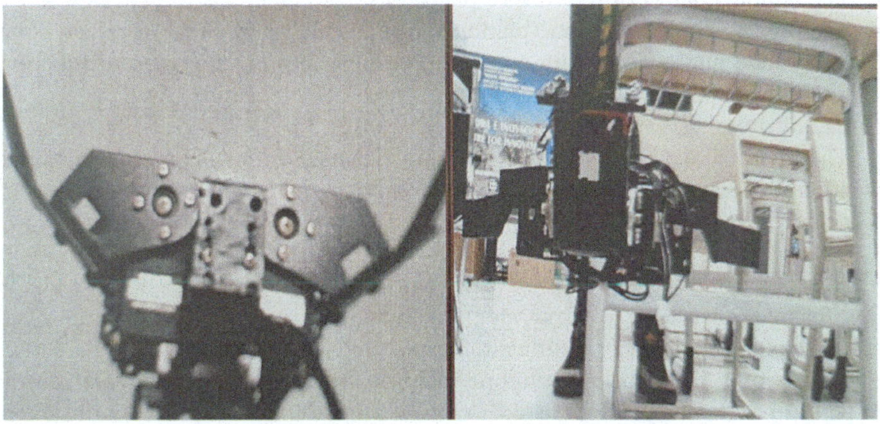

Fig. 17 View from the cameras mounted on the robotic platform

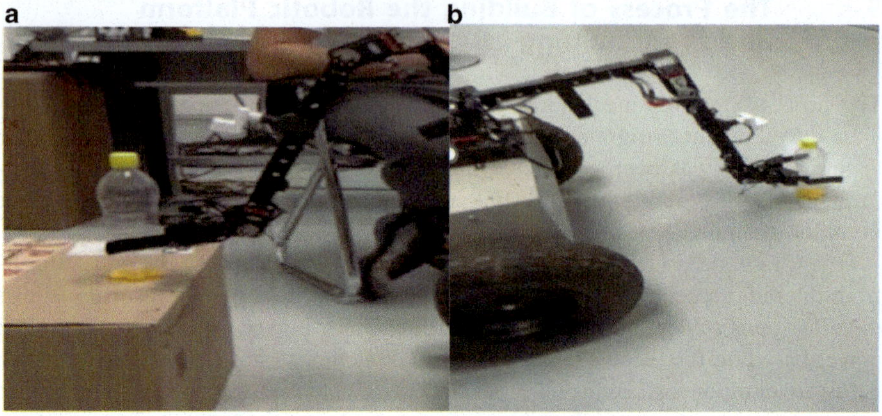

Fig. 18 **a** Pick up of the bottle **b** relocation of the bottle

we start to relocate the platform and put the bottle into another position, Fig. 18b. The platform moves backward and is ready to do another task.

6 Conclusion

Through this project, a prototype has been developed which in the future could be further developed and serve people for demining processes or simply for handling objects of certain dimensions.

Based on the design and construction process outlined in the paper, we can conclude as follows:

- The remote control of the robotic platform's movement has been achieved.
- The prototype of the robotic platform, constructed from metal parts and some 3D printed parts, along with other components, fulfills the presented function/task—the capture of unidentified objects and their relocation to safe spaces for later disposal.
- The electrical devices and electronic components used for the prototype's operation are powered by a battery, which supplies the required voltage to each device as needed.

The results achieved with this research contribute to the advancement of technology for mine clearance, offering an efficient and safe solution for clearing mines and unexploded ordnance in difficult terrain.

The robotic platform in the future could be enhanced with additional sensors to perform more complex operations, such as dismantling unexploded ordnance or placing them in secure locations to prevent explosions during transport.

References

Pro-series Robotic Arms | CrustCrawler Robot Arms [Accessed: 19 April 2024]
Doyle, M., Fuller, T.F., Newman, J.: Modeling of galvanostatic charge and discharge of the lithium/polymer/insertion cell. J. Electrochem. Soc. 6, 1526–1533.
GitHub—motioneye-project/motioneye: A web frontend for the motion daemon.
Havlík, S., Licko, P.: Humanitarian demining: The challenge for robotic research. J. Conv. Weapons Destr. 2(2) (2016)
https://hobby.npcrobotics.com/product/npc-41250/ [Accessed: 19 April 2024]
https://www.techspot.com/article/2126-arduino/ [Accessed: 19 April 2024]
Manwell, J., McGowan, J.: Extension of the kinetic battery model for wind/hybrid power systems. In: Proceedings of the fifth European Wind Energy Association Conference (EWEC '94), p. 284–289 (1994)
(1523) Pinterest [Accessed: 19 April 2024]
CM-700 (robotis.com) [Accessed: 19 April 2024]
Robots for Humanitarian Demining | IntechOpen [Accessed: 19 April 2024]
Sachs, J.: The end of poverty: How we can make it happen in our lifetime. Penguin, UK (2005)
Trevelyan, J.: Demining research [Internet]. Available from: http://staffhome.ecm.uwa.edu.au/~00006605/demining/index.html [Accessed: 10 April 2024]
United Nation (UN). Convention on the prohibition of the use, stockpiling, production, and transfer of anti-personnel mines and on their destruction, [Internet]
The UN Office for Disarmament Affairs (UNODA). Overview and Convention Text [Internet] (March 1999)

Real-Time Total Harmonic Distortion Generated from PV Power Plant Measurement and Filter Design Using LabVIEW Software

Gentian Dume, Alfred Pjetri, and Astrit Bardhi

Abstract

Nowadays, in Albania, several photovoltaic plants are in various stages of construction. The main problem encountered is the power quality, because in spite of the energy production, PV Power Plants inject into the electrical grid voltage harmonics from the conversation of DC to AC through Power Electronic devices like inverters. Except for the fundamental harmonic, the voltage and current spectrum of the inverter output contain high-order harmonics. In this work, we will present a virtual instrument built in the LabVIEW software, which is customized to measure THD and to notify in real-time for the voltage signal harmonics present in the electrical grid PoC (Point of Connection) besides the fundamental 50 Hz harmonic, generated from a small-scale PV Plant in its AC side and to allow the remote user to implement the most suitable passive filter needed to be connected to reduce the THD to the allowed value by respective standard. This virtual instrument can be upgraded easily to automatically calculate and to select a passive filter and to connect it to the electrical grid to filter a specific harmonic of the voltage and current which is detected by the same instrument. This solution can be used to reduce the voltage and current THDs generated from PV Plants and other electrical devices.

Keywords

THD • PV solar inverters • LabVIEW • Filter

G. Dume (✉) · A. Pjetri · A. Bardhi
Polyechnic University of Tirana, Tirana, Albania
e-mail: gentian.dume@gmail.com

K. Dhoska and E. Spaho (eds.), *Bridging Horizons in Artificial Intelligence, Robotics, Cybersecurity, Smart Cities, and Digital Economy*, Sustainable Economy and Ecotechnology, https://doi.org/10.1007/978-3-031-72029-1_3

1 Introduction

Last years in Albania are installed and continue to be installed a large number of Photovoltaic (PV) Plants. The high capacity in electricity production from this kind of power source has created a problem which has to be addressed and solved somehow.

The main problem encountered during energy production from inverter-based devices is the power quality because of the energy production, PV Power Plants inject into the electrical grid voltage harmonics from the conversation of DC to AC through Power Electronic devices like inverters. Except for the fundamental harmonic, the voltage and current spectrum of the inverter output contain high-order harmonics.

In a multi-parallel system, several parallel connected converters with numerous sources of power (e.g., solar plants) are utilized at the Point of Connection (PoC) to increase the quantity of injecting current in the network. The harmonics generated by power electronic converters can deteriorate the power quality across the distribution grid, due to harmonic interaction, resonance, and grid impedance (transformer and distribution power cables) and finally lead to increased losses in the grid. The power quality at the PoC is very important because it can affect the other connected equipment. Due to the employment of such devices and their non-linear behavior in power systems, power quality has emerged as a major concern for energy companies and network operators (Kroposki et al. 2010; Wang et al. 2001).

To investigate harmonic emissions in power systems, international standardization groups, e.g., the IEC and IEEE have developed many power quality standards. IEC or IEEE standards cover the harmonics limits for different frequency ranges i.e. IEEE 519–2022.

In this work, we will present a virtual instrument built in the LabVIEW software, which is customized to measure THD and to notify in real-time for the voltage signal harmonics present in the electrical grid PoC (Point of Connection) besides the fundamental 50 Hz harmonic, generated from a small-scale PV Plant in its AC side and to allow the remote user to implement the most suitable passive filter needed to be connected to reduce the THD to the allowed value by respective standard.

2 Material and Method

To monitor voltage, current, and Total Harmonics Distortion (THD) we setup a system composed by a solar inverter, a three phase voltage and current measurement module (Dume 2016), data acquisition card (USB-6008, 6009 User Guide and Specifications—National Instruments. 2023), computer with the virtual instrument running, a real filter composed by inductors and capacitors and a relay block for inserting the filter in the inverter connection to the grid path, if needed (Fig. 1).

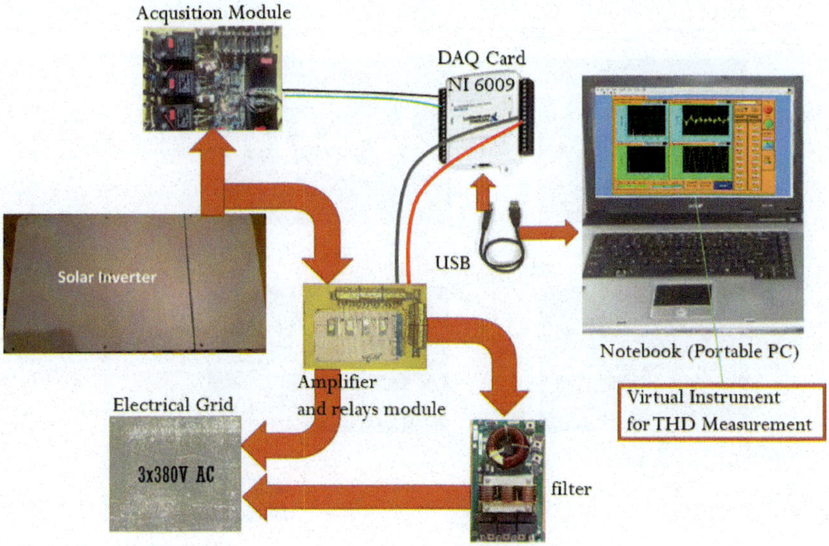

Fig. 1 Overall system for THD monitoring and reducing

First, inverter voltage and current output are isolated and lowered till allowed values from NI DAQ 6009 card. This card is configured to acquire analog inputs from the virtual instrument shown in Fig. 2. In the instrument Front Panel are inserted some ***waveform charts*** to display phase voltage, current, and respective THDs. Also, in the instrument **Front Panel** showed 10 odd multiple fundamental harmonics of the voltage and current and also Vrms, Irms, $THD_{v,}$ and THD_{i}.

From this interface we can choose also type of filter (IIR-infinite pulse response) and order till the 3rd order. Filtered signals are also displayed in the 2nd and 3rd tab of the Virtual Instrument, labelled **Phase Current** and **Phase Voltage.** In these tabs we can plot in real-time some of the filtered signals or all of them including the unfiltered voltage or current for comparison purpose (Figs. 2 and 3).

In the upper right side of the **Front Panel** are placed 2 round LED to indicate if the voltage and current THDs are inside allowed limits or not in accordance with IEEE 519–2022 standard by changing the color from red to green.

The push button labelled ***Activate real filter*** when pushed by single mouse click activate some of the digital outputs of the same DAQ 6009 card and the real filter is placed in the path between the solar inverter and the grid by the ***Relays module.***

Last, using LabVIEW G Web Server we can publish the virtual instrument online so the operator can interact and perform remote monitoring and filter design for the monitored PV plant (Archived 2024) (Fig. 4).

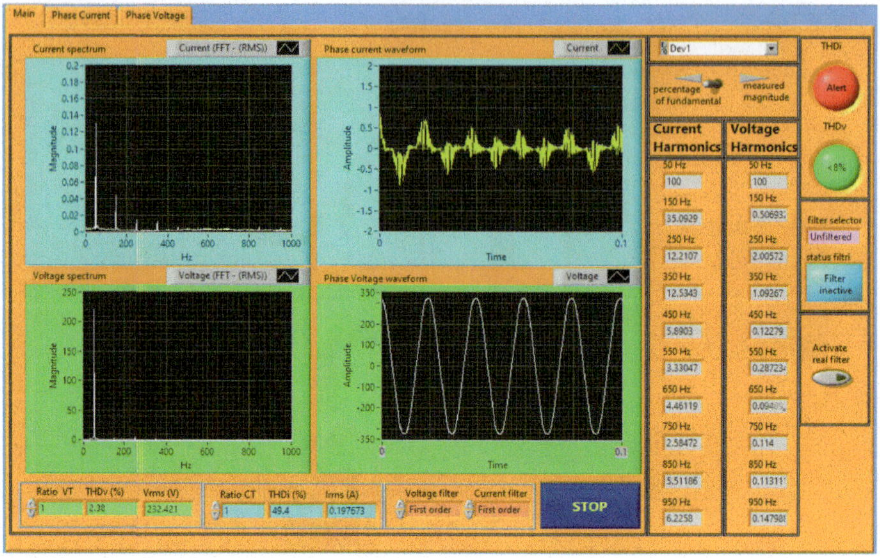

Fig. 2 Proposed virtual instrument for THD monitoring and reducing

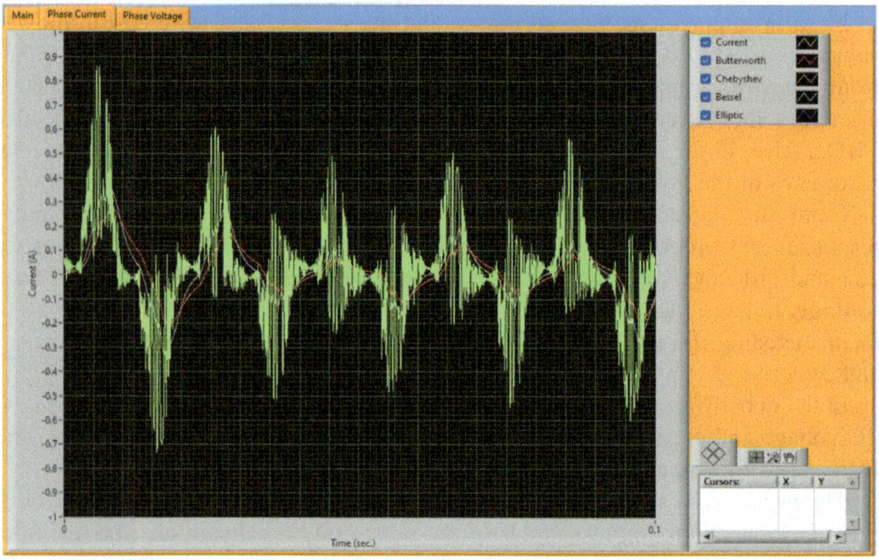

Fig. 3 Proposed virtual instrument phase current monitoring and filter design

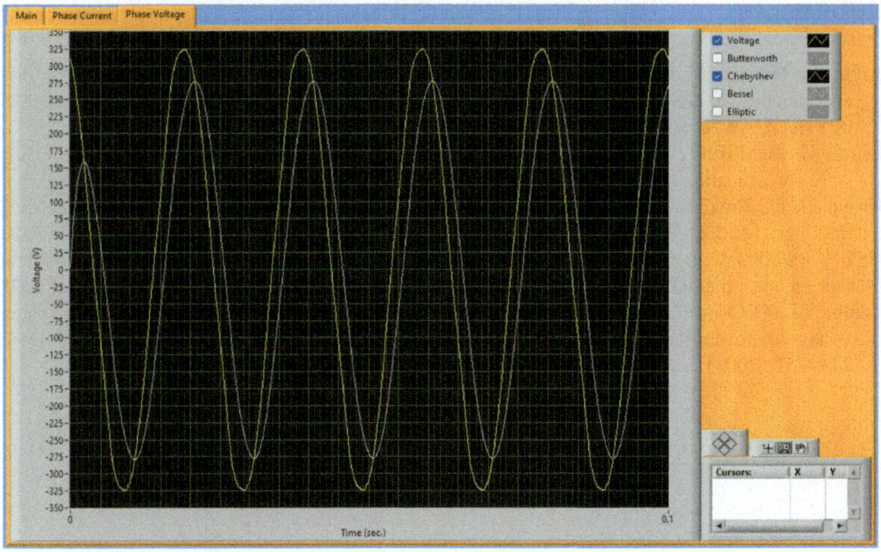

Fig. 4 Proposed virtual instrument phase voltage monitoring and filter design

3 Conclusion

Virtual instrumentation for the THD measurement can be used to detect sources that inject high-order harmonics of the voltage and current causing problems in the power system.

Online THD monitoring and improvement via the Internet is another option, which makes this Virtual System a competitive alternative to devices that perform the same task but have a fairly high maintenance cost. This can be achieved with the help of NI-PSP protocol of LabVIEW software.

Since this instrument is based on programming it is very flexible to add other functionality to it. For example, it can be used to alert the electrical grid utility if the THD of the voltage or the current passes a preset threshold value at the point where the instrument interface is connected. Besides the use of this DAQ card to acquire signals we customized this instrument by the same card (configuring it as digital output) to select a desired physical filter.

It can be configured also to interrupt any load or sources that inject high harmonics and pollute the electrical grid.

Over voltage and under voltage protection techniques can also be implemented.

References

Archived: Publish a LabVIEW VI to a Web Page With the Web Publishing Tool—NI," knowledge.ni.com. Available at https://knowledge.ni.com/KnowledgeArticleDetails?id=kA03q000000YHeGCAW&l=en-AL. Accessed on 6 May 2024

Dume, G.: Real-time control of synchronous generator in island mode based on LabVIEW software for education purpose. WSEAS Trans. Advances. Eng. Educ. 13, 50–63 (2016)

Kroposki, C., Pink, R., DeBlasio, H., Thomas, M. Simões et al.: Benefits of power electronic interfaces for distributed energy systems. IEEE Trans. Energy Convers. 25(3), 901–908 (2010)

USB-6008/6009 User Guide and Specifications—National Instruments. Available at https://courses.cit.cornell.edu/bionb442/labs/f2007/NI6008manual.pdf. Accessed on 22 November 2023

Wang, Y.J., O'Connell, R.M., Brownfield, G.: Modelling and prediction of distribution system voltage distortion caused by nonlinear residential loads. IEEE Trans. Power Delivery **16**(4), 744–751 (2001)

Methods for Determining Defects in Asynchronous Motor

Astrit Bardhi, Bajram Leka, Alfred Pjetri, and Aldo Hasani

Abstract

Induction machines, crucial in various industrial sectors and the economy, are sought after for their dependable and secure functioning. Their faults or malfunctions can cause prolonged downtimes and significant financial losses, driving the need for continuous online condition monitoring. This monitoring involves real-time measurements during operation to swiftly identify faults, thereby mitigating unexpected failures and reducing maintenance expenses. This paper commences by scrutinizing the typical operation of asynchronous motors, followed by an examination of their behaviour under abnormal conditions such as stator or rotor winding defects. Through theoretical analysis of breakdown scenarios, parameters linked to emerging defects are identified, guiding the selection of appropriate detection methods during motor operation. The alignment between theoretical analysis and experimental data from the Laboratory of Electric Machines reinforces the findings of this study.

Keywords

Induction machine • Condition monitoring • Faults • Detection methods

1 Introduction

Rotor cage induction motors are the most commonly employed in various industrial and economic sectors because they are safe to operate, free of maintenance, and have lower costs than other types of electric motors (Ahamed et al. 2012). Their role in all sectors of the economy is constantly growing as a result of

A. Bardhi (✉) · B. Leka · A. Pjetri · A. Hasani
Polytechnic University of Tirana, Tirana, Albania
e-mail: asibardhi@gmail.com

K. Dhoska and E. Spaho (eds.), *Bridging Horizons in Artificial Intelligence, Robotics, Cybersecurity, Smart Cities, and Digital Economy*, Sustainable Economy and Ecotechnology, https://doi.org/10.1007/978-3-031-72029-1_4

advances in the field of power electronics, which has made it possible to use them even in systems that require rotor speed in a wide band, torque control, and position. Research conducted by powerful companies shows that about 37% of the normal malfunction of rotor cage of asynchronous motors occurs due to damage to the stator winding. Based on extensive research conducted by the electrical corporation "General Electric" (EPRI 1982), encompassing over 5000 asynchronous motors of varying capacities, with 97% being three-phase asynchronous motors with rotor cages, the analysis of faults revealed the following breakdown: 49% of issues stemmed from rolling sphere malfunctions; 31% were attributed to faults in the stator winding; 10% were due to damage in the rotor bars; and the remaining 10% were linked to other miscellaneous problems. These defects are contingent upon the unique operational demands placed on asynchronous motors within specific industrial sectors. For example, it has been observed that in the case of using the asynchronous motor in S3 regime (on–off mode), the faults that occur in the rotor winding are more than occur in the stator winding (Thorsen and Dalva 1995). On the other hand, from the various research carried out, it has been concluded that the number of defects in asynchronous motors in environments condition is 2.5 times higher than in motors working in closed and well-maintained environments (Thorsen and Dalva 1995). Determination of defects in the asynchronous motor in recent years, many researchers have proposed different methods and techniques (Tavner and Penman 1987; Vas 1993).

In this paper we will define the methods that can be used in determining the faults of the asynchronous machine during operation. First, we will be analysed the operation of the asynchronous motor in the normal conditions. After this we have analysed operation of the asynchronous motor or induction machine in fault condition as: turn to turn faults of the same coil of stator phase windings: inter-turn short circuits between stator phase winding; inter-turn short circuits faults of the stator phase windings with magnetic core or to ground. The operation of the asynchronous motor in case of rotor bars damage will also be analysed. The theoretical examination conducted in this study led to the conclusion that defects in the induction machine can be identified through several methods: spectral analysis of the stator winding phase current, analysis of the instantaneous power of the asynchronous motor, observation of the trajectory (locus) of the space vector of the stator phase current, and measurement of both speed and electromagnetic torque.

In order to evaluate the theoretical analysis carried out in this paper, in the Laboratory of Electric Machines we have developed some experimental tests in the case of damage to the stator winding and the rotor bars cracket of induction machine during operation. The experimental data are in good agreement with the theoretical analysis carried out in this paper.

2 Working of Induction Machine in Normal Condition

If a symmetrical three-phase system of voltages is applied to the stator windings of a symmetrical three-phase asynchronous motor, then a symmetrical system of currents will pass through them. The resultant air gap magnetic field created by the currents passing through the stator winding of the asynchronous machine is distributed sinusoidal with constant amplitude and rotated at synchronous speed (Stavrou et al. 2001).

$$F_1(\alpha, t) = \frac{3}{2}F_{f1}\sin(\omega_1 t - p\theta) \tag{1}$$

The synchronous speed of the air gap magnetic field is equal to the expression:

$$\Omega_1 = 2\pi n_1 = \frac{2\pi f_1}{p} \tag{2}$$

where:

f_1 frequency of stator winding current.
p number of pole pairs of induction machine.
ω_1 synchronous speed of resultant magnetic field.

The sleep of an induction machine at a certain regime can be found as:

$$s = \frac{\Omega_1 - \Omega_r}{\Omega_1} = \frac{\omega_1 - \omega_r}{\omega_1} = \frac{n_1 - n_r}{n_1} \tag{3}$$

The frequency of rotor bar currents is $f_2 = sf_1$. The instantaneous value of the currents in the rotor bars depends on the magnetic field in the air gap and the position of the rotor bars. The instantaneous value of the currents in the rotor bars can be as the following expression:

$$i_{b(n)} = I_{mb}\sin(s\omega t - (n-1)\phi) \tag{4}$$

where:

$\phi$is phase angle between two adjacent of rotor bars.

Z_r number of rotor bars.
s sleep of induction machine.
n a number 1, 2, 3, 4….Zr.
I_{mb} is amplitude of rotor bar currents.

The outcome of the fundamental harmonic of magnetomotive force (MMF) generated by the rotor bars can be represented by the following expression:

$$F_2(\theta, t) = \frac{1}{2}\frac{Z_r}{p} I_{mb} \sin(s\omega_1 t - p\theta) \tag{5}$$

As seen the resultant *mmf* of the rotor bars rotate in the same direction as the resultant *mmf* of the stator winding. The angular speed *mmf* is created by rotor bars when the rotor rotates at an angular speed Ωr is equal to synchronous speed or rotates with the same speed as stator *mmf*. In the expression below can show:

$$\Omega_{21} = \Omega_r + \Omega_2 = [(1 - s)\Omega_1 + s\Omega_1]/p = \Omega_1 \tag{6}$$

The magnetic fields of the stator and rotor are stationary relative to each other and create a resultant flux rotating with angular velocity Ω_1. Since the air gap is not uniform due to the channels that are opened in the stator magnetic core and the rotor magnetic core, the current in the stator winding will contain, in addition to the main harmonic, an additional component due to the rotor slots. Table 1 lists the *emf* and current harmonics that exist in the stator winding of the induction machine.

As seen from Table 1, the current in the stator coil will contain the main component with frequency f_1 and a component that is a function of the rotor speed and the number of rotor bars. The total instantaneous power input to the three phases of an induction machine under normal operating conditions can be expressed as follows:

$$p(t) = 3VI \cos\varphi \tag{7}$$

Equation (10) demonstrates that the total instantaneous power of an induction machine operating under normal conditions remains constant. The current space

Table 1 The *emf* harmonic induced at stator winding

Nr	Positive wave	Harmonic range	Frequency of emf induced at stator winding
1	$F_1 \cos(w_1 t - pq)$	1	f_1
2	$F_{Zs} \cos[w_1 t - (Z_s + p)q)]$	$Z_s/p + 1$	f_1
3	$F_7 \cos(w_1 t - 7pq)$	7	f_1
4	$F_{zr} \cos[sw_1 t - (Z_r + p)(q - w_{rm} t)]$ $(Z_r + p)(q - w_{rm} t)]$	$Z_r/p + 1$	$[Z_r/p\,(1 - s) + 1] f_1$
	Valët e kundërta		
1	$F_{zs(-)} \cos[w_1 t + (Z_s - p)q]$	$Z_s/p - 1$	f_1
2	$F_5 \cos(5pq + w_1 t)$	5	f_1
4	$F_{zr(-)} \cos[sw_1 t + (Z_r - p)(q - w_m t)]$ $(Z_r - p)(q - w_m t)]$	$Z_r/p - 1$	$[Z_r/p\,(1 - s) - 1] f_1$

vector shows a constant radius circulation.

$$\bar{i}_s = \bar{I}_1 e^{j\omega_1 t} \tag{8}$$

3 Induction Machine Working Under Faulty Condition

3.1 Induction Machine Operation with Rotor Bar Breakage Fault

Damage to the rotor bars will cause asymmetry during construction of the rotor bars so the current bars are not symmetrical. To analyse the behaviour of the induction machine in the case of cutting some bars of the rotor winding, to simplify the mathematical expression in the evaluation of the *mmf*, we will accept that the currents in each bar remain constant, i.e. as in the symmetrical regime, while the current in the damaged bars is equal to zero. With the assumption made, the resultant *mmf* created by the rotor winding can be obtained from the resultant *mmf* working in the normal condition by subtracting the *mmf* of the damaged bars. So *mmf* will be equal to the following expression:

$$F_{res'}(\theta, t) = F_{res}(\theta, t) - F_{dam}(\theta, t) = \frac{1}{2}\left(\frac{Z_r}{p} - 1\right) I_{mb} \sin(s\omega_1 t - \theta) - \frac{1}{2} I_{mb} \sin\left(\theta + s\omega_1 t - \frac{2\pi p}{Z_r}\right) \tag{9}$$

As seen from the above expression the resultant *mmf* created by the rotor winding contains two components. The first component on the right-hand side of expression (3.12) represents the positive-sequence wave of *mmf* which rotates with angular speed $s\omega_1$ (rotor references) while the second component represents the negative-sequence component wave of *mmf* which rotates with the same angular speed $s\omega_1$, but in opposite direction of positive-sequence wave. The magnetic field created by the negative-sequence component of the rotor current will induce in the stator winding *emf* $e_1^{(2)}$ with frequency $f_1^{(2)} = (1 - 2s)f_1$ therefore, the current in the stator winding, in addition to the main frequency component f_1, contains a harmonic with frequency $f_1^{(2)} = (1 - 2s)f_1$. The presence of the harmonic with frequency $f_1^{(2)} = (1 - 2s)f_1$ in the stator current is an indicator of rotor asymmetry. The expression of stator current is as follows:

$$i_s(t) = I_m \cos(\omega_1 t + \phi) + I'_m \cos((1 - 2s)\omega_1 t + \phi') \tag{10}$$

The locus of the space vector of the stator current is equal to the following expression:

$$\bar{i}_s = \bar{I}_1 e^{j\omega_1 t} + \bar{I}_2 e^{-j((1-2s)\omega_1 t)} \tag{11}$$

The instantaneous power of an induction machine operating under conditions of rotor bar damage can be expressed as follows:

$$p == 3\overline{V}_1 \cdot \overline{I}_1 \cos\varphi_1 + 3\overline{V}_1 \cdot \overline{I}_1' \cos\varphi' + 3\overline{V}_1 \cdot \overline{I}_1' \cos(2s\omega_1 t - \varphi') \quad (12)$$

Thus, in the stator winding, due to the rotor speed fluctuation, additional frequency components will appear (Briz et al. 2005)

$$f_1^{(2)} = (1 \pm 2ks)f_1 \quad (13)$$

It should be noted that the amplitude of the harmonic with frequency currents $f_1^{(2)} = (1 \pm 2ks)f_1$, depends on the number of bars damaged and the motor load. From the analysis discussed above, we achieved the conclusion that due to the component that changes with frequency $f_1^{(2)} = (1 \pm 2ks)f_1$ the amplitude of the stator current will oscillate with the same frequency. The electromagnetic torque of induction machine pulsates. The existence of pulsating torque will cause noise and increased stresses on the shaft of the induction motor. Damage to the rotor bars can be determined in real time by the spectral analysis of the stator current, the instantaneous power, the dissipation flux, the rotor speed, the electromagnetic torque as well as the space vector trace of the current.

3.2 Induction Machine Working Under Faulty Condition in Case of Inter-Turn Short Circuits of Stator Winding

During the operation of the induction machine, for various reasons, inter-turn short circuits at the same coil, inter-turn short circuits at the different coil of the same winding, or between phases of the stator winding may occur. The inter-turn short circuits faults between the stator windings will cause the reduction of the number of healthy turn windings, thus the *mmf* of the air gap will decrease. The active resistance of the faulted phase will decrease depending on the inter-turn short circuits windings. On the other hand, reducing the number of turns for a phase winding will also reduce the magnetic coupling between the damaged winding with the other winding of the stator, i.e. the total flux of the damaged phase. The inter-turn short circuits windings will create an *mmf* in the opposite direction to the main component (*mmf* of the healthy part of the winding). Since the winding in such a regime is not symmetrical, therefore the phase currents of the induction machine are also asymmetrical. In this way, the analysis of the induction machine in such a regime can be realized using symmetrical components, positive-sequence components, negative-sequence components, and zero-sequence components. Since the stator winding is generally connected in an ungrounded star, the neutral component is absent. The phase current of the stator winding will be equal to the expression:

$$i_s(t) = I_m \sin(\omega t + \varphi) + I_m' \sin(\omega t + \varphi') \quad (14)$$

The instantaneous power of the induction machine in case of damage to stator winding is given by the following expression:

$$p = 3[(UI\cos\varphi_1 + U'I'\cos\varphi_2) + (UI'\cos(2\omega t + \alpha_1 + \alpha_2 + \varphi_2) + U'I\cos(2\omega t + \alpha_1 + \alpha_2 + \varphi_1)] \quad (15)$$

The space vector of currents is given by the expression:

$$\bar{i} = \bar{i}_1 + \bar{i}_2 = \bar{I}_1 e^{j\omega t} + \bar{I}_2 e^{-j\omega t} \quad (16)$$

So due to asymmetrically stator winding current system will create an elliptical magnetic field and the locus of the space vector will be an ellipse. The phase current of the stator winding expressed in symmetrical components is equal to:

$$\dot{I}_1 = \dot{I}_1^{(1)} + \dot{I}_1^{(2)} \quad (17)$$

where:

$\dot{I}_1^{(1)}$ is phasor of positive—sequence component of stator winding current.

$\dot{I}_1^{(2)}$ is phasor of negative—sequence component of stator winding current.

Symmetrical components will create corresponding magnetic fields which are distributed sinusoidally around air gap and rotating at angular speed:

$$\Omega_1^{(1)} = -\Omega_1^{(2)} = \Omega_1 = {}^{\omega_1}/_{p} \quad (18)$$

The respective fields will induce in the rotor windings the currents with frequency sf_1 (positive-sequence component) and $(2 - s)f_1$ of the negative-sequence component. These currents will create the corresponding *mmf* which rotates at the same speed as the corresponding stator field, so they are stationary relative to each other. Changing the current in one phase will affect the magnitude of the current in the other phases. In principle, the inter-turn short circuit between the phase and the ground is the same as the inter-turn short circuit between the windings of the same phase. The harmonics caused by the short connections between the windings are given by the expression (Wu 2007):

$$f_{st} = f_1\left[\frac{n}{p}(1 - s) \pm k\right] \quad (19)$$

In conclusion, we can see that in normal operation, the total resistance of each phase of the stator windings is the same and the amplitude of the phase currents are equal. Due to the fault in the stator winding, the phase resistances are not the same, so the phase currents will have different magnitudes. Also, the phase currents are not shifted by 120 degrees. Under these conditions, the third harmonic begins and becomes dominant. The detailed theoretical analysis of the impact of damage to the stator winding can be realized using (Santos et al. 2006; Syed Kamruddin Ahamed 2012): the amplitude of the phase currents, the phase angle of the currents phases, third harmonic, current spectral analysis…etc.

4 Experimental Results

To validate the theoretical analysis regarding faults in the induction machine, a sequence of experiments was conducted. These experiments utilized squirrel cage induction machines as the subjects of investigation.

First, will be studied the operation of the induction machine in normal conditions, which is feeding by a symmetrical three-phase source. In Fig. 1 a, b, c, are shown: the waveform of the stator phase currents (a), the locus of the space vector of the current (b), and the spectrum of the stator phase current (c) for the symmetrical load regime nominal on the shaft. As seen in Fig. 1, the stator phase currents are symmetrical, the locus of the space vector is a circle and the current spectrum contains only the fundamental harmonic, frequency of 50 Hz.

To simulate the inter-turn short circuit connections between the the same coil of the windings and between turns of the different windings, we have realized the extraction of several outputs of the windings of the asynchronous motor.

Figure 2 shows: the waveform of the stator phase currents (a), the space vector locus (b) and the spectral analysis of the stator phase current (c) determined the experimental data in the case of the inter-turn short circuit of coil of the windings. To limit the magnitude of the current flowing in them in series with inter-turn short circuit we have connected a resistance *R*. The asynchronous motor is loaded at nominal power. As seen in Fig. 2 (a) the stator phase currents are different. The current in the damaged phase is a slight increase specifically its value of 3.5 A. Even in the other phase, the current has undergone an increase with almost the same value. In the third phase, the current has undergone a reduction, the

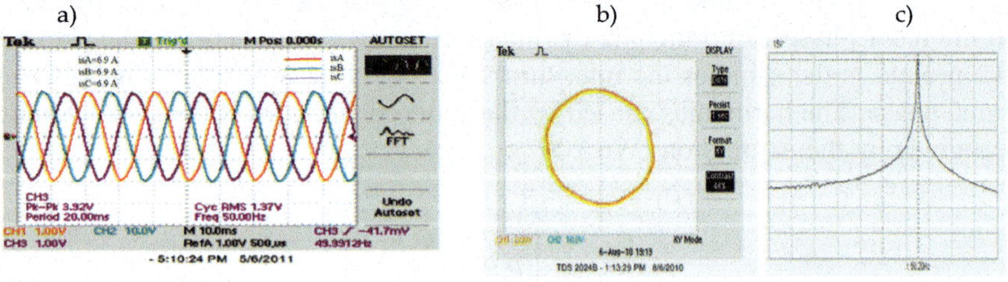

Fig. 1 The current waveforms of the induction machine during regular operation are depicted in **a**, while **b** illustrates the locus of the current space vector. Additionally, **c** displays the spectral analysis of the stator phase current during the induction machine's operation under normal conditions

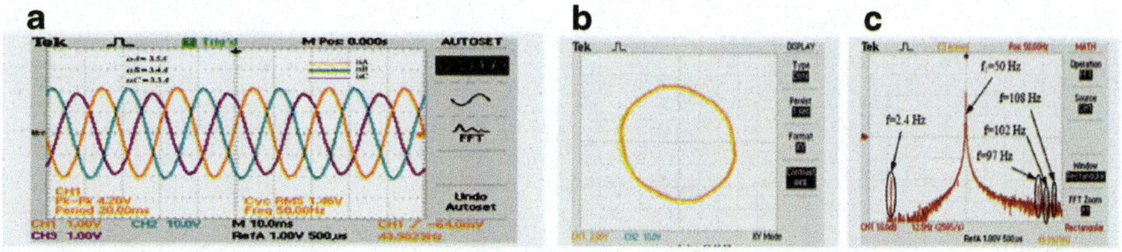

Fig. 2 Current waveforms of the induction machine during normal operation **a**, locus of current space vector **b**, and spectral analysis of stator phase current **c** during induction machine operation with inter-turn short of stator winding

Table 2 Harmonics that appear in the current spectrum due to stator winding damage

f_1	p	s	n	k	f_{st} $[Hz]$
50	2	0.048	1	1	73.4
					26.2
			2	1	97.6
					2.4
			1	5	136.9
					113.1

magnitude at a value of 3.2 A. In the spectrum of the stator current, as can be seen from Fig. 2. (c), several harmonics appear with a frequency of 2.4, 97, 102, and 108 Hz. For the given regime, the rotor rotation speed measured by the stroboscopic method was 1428 rpm, and the slip was 0.048. Table 2 shows the harmonics that appear in the spectrum of the stator phase current calculated from expression 3.48 based on the theoretical analysis carried out in about paragraph. As can be seen, the theoretically determined 2.4 and 97.6 Hz harmonics match the 2.4 and 97 Hz harmonics found experimentally for the given regime.

To confirm the theoretical analysis developed in the previous paragraphs on the damage to the rotor bars, we performed a series of experiments. In the rotor winding, the separation of the two rotor bars was realized. Initially, the induction machine was loaded at nominal power. The rotor speed of the asynchronous motor was gauged using a digital tachometer, following which we determined the slip of the induction machine using the prescribed expression $s = (n_1\text{-}n)/n_1$. Figure 3 shows the curves of the stator winding phase currents (a), the locus of space vector current (b), and the stator phase current spectrum (c). As seen in Fig. 3, the amplitude of the phase current in the case of rotor bar damage is not constant but pulsates periodically. From the spectral analysis of the stator current, additional harmonics with frequency $(1 \pm 2ks)f_1$ appear around the fundamental harmonic. This result can be observed using analog ammeters where its indicator oscillates with a certain frequency. The current space vector for the given regime as seen from graph 3 (b) is a circle whose locus is highlighted.

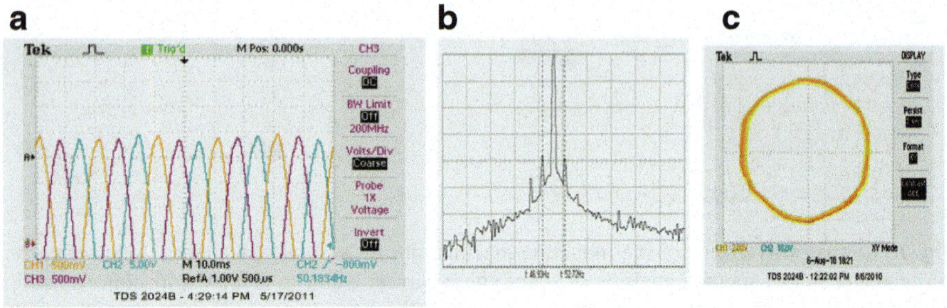

Fig. 3 Current waveforms of the induction machine **a**, locus of current space vector **b**, and spectral analysis of stator phase current **c** during induction machine rotor bars cracked operation

5 Conclusion

Inter-turn short circuits of phase windings in the asynchronous motor, cause in the spectrum of the phase current, additional harmonics with frequency $f_{st} = f_1[n/p(1-s) \pm k]$. Their appearance in the spectrum of the phase current in the asynchronous motor is a clear indication of damage to the stator winding.

The amplitude of the additional harmonics depends on both the number of inter-turn short circuits windings and the load on the motor shaft. As the number of inter-turn short circuits windings increases, as well as the load, their sizes increase. This allows us to consider the degree of damage to the stator winding.

In the case of damage to the rotor bars, additional harmonics appear around the fundamental harmonic. These harmonics indicate that the amplitude of the phase current of the asynchronous motor fluctuates rather than remaining constant, oscillating at a frequency of 2 times the fundamental frequency, denoted as $2sf_1$.

Based on a thorough theoretical analysis of the effects of stator and rotor winding damage, coupled with experimental findings, we deduce that assessing the condition of the induction machine can be accomplished through various means: monitoring phase current amplitude, phase current angle, spectral analysis of current, analysis of instantaneous power, and observing the locus of the space vector.

Using these methods is simple and easily implemented with minimal cost. The current actuator in most applications is installed at power boxes that feed the induction machine for the measurement or control system, so it is required to install a device to perform the analysis of the quantities that evaluate the condition of the induction machine.

References

Ahamed, S. K., Sarkar, A., Mitra, M., Sengupta, S.: Detection of induction motor broken bar fault through envelope analysis using start-up current, Vol. 4, pp. 646–651 (2012). https://doi.org/10.1016/j.protcy.2012.05.104

Briz F., Degner W M., Garcia P., Diez A.B.: Induction machine diagnostics using zero sequence components. In: Conference Record of the Industry Applications 2005, Fourtieth IAS Annual Meeting **1**, 2–6, 34–41 (2005)

EPRI: Improved motors for utility applications. Publication EL-2678, Vol. 1, 1763–1, final report, October, 1982 (1982)

Negrea, D. M.: Electromagnetic flux monitoring for detecting faults in electrical machines. Doctoral Dissertation, Helsinki University of Technology (2006)

Santos, P. M., Correa, M. B. R., Jacobina, C. B., et al.: A simplified induction machine model to study rotor broken bar effects and for detection. In: PESC'06, 37th IEEE, 18–22 June 2006, 1–7 (2006). https://doi.org/10.1109/PESC.2006.1712154

Stavrou, A., Sedding, H., Penman, J.: Current monitoring for detecting inter-turn short circuits in induction motors. IEEE Trans. Energy Convers. **16**(1), 32–37 (2001)

Thorsen, O.V., Dalva, M.: A survey of faults on induction motors in offshore oil industry, petrochemical industry, gas terminals and oil refineries. IEEE Trans. Ind. Appl. **31**(5), 1186–1996 (1995)

Tavner, P., Penman, J.: "Condition Monitoring of Electrical Machines", Research Studies Ltd. John Wiley & Sons, England (1987)

Vas, P.: Parameter Estimation Condition Monitoring and Diagnosis of Electrical Machines, Clarendon Press—Oxford (1993)

Wu, L.: Separating load torque oscillation and rotor faults in stator current based-induction motor condition monitoring. Doctoral Dissertation, Georgia Institute of Technology (May 2007)

Integrating Machine Learning and AI into IoT-Enabled Smart Parking

Vesna Knights, Olivera Petrovska, and Marija Prchkovska

Abstract

This study delves into the integration of machine learning (ML) and artificial intelligence (AI) techniques within Internet of Things (IoT)-enabled smart parking systems to optimize urban parking management. The purpose of this research is to enhance operational efficiency, alleviate congestion, and elevate user experience in urban environments. Methodologically, we conducted an extensive review and analysis of various data sources, ML algorithms, and AI technologies employed in smart parking infrastructures. Our investigation reveals that the convergence of IoT, ML, and AI enables intelligent data collection, analysis, and decision-making processes. By leveraging real-time and historical parking data, ML algorithms predict parking availability, optimize space utilization, and dynamically adjust pricing strategies. AI-powered systems enhance user experience through personalized recommendations, seamless navigation, and proactive maintenance. Furthermore, IoT sensors and connectivity enable remote monitoring, predictive maintenance, and efficient resource allocation, contributing to sustainable urban development. Results from case studies and simulations demonstrate significant improvements in parking efficiency, reduction in traffic congestion, and enhancement of overall urban mobility. Moreover, the integration of ML and AI fosters adaptive learning systems

V. Knights (✉)
Faculty of Technology and Technical Science, University "St. Kliment Ohridski", Bitola, Republic of North Macedonia
e-mail: vesna.knights@uklo.edu.mk

O. Petrovska · M. Prchkovska
Faculty of Technical Sciences, Mother Teresa University, Skopje, Republic of North Macedonia
e-mail: olivera.petrovska@unt.edu.mk

M. Prchkovska
e-mail: marija.prckovska@students.unt.edu.mk

K. Dhoska and E. Spaho (eds.), *Bridging Horizons in Artificial Intelligence, Robotics, Cybersecurity, Smart Cities, and Digital Economy*, Sustainable Economy and Ecotechnology, https://doi.org/10.1007/978-3-031-72029-1_5

capable of continuously optimizing parking operations based on evolving user preferences and environmental conditions. This research underscores the pivotal role of technology-driven innovations in addressing the challenges of urbanization and fostering sustainable urban development. As smart parking systems continue to evolve and integrate advanced technologies, they offer promising solutions to enhance not only parking management but also overall urban mobility, paving the way for smarter, more efficient cities of the future. Recommendations include further exploration of advanced machine learning models and the development of standardized protocols for interoperability among IoT devices. Additional data on user acceptance and system scalability are essential for comprehensive evaluation.

Keywords

Smart parking systems • Machine learning • Artificial intelligence • IoT integration • Data analysis • Predictive analytics

1 Introduction

Integrating Machine Learning and AI into IoT-Enabled Smart Parking is an innovative approach to addressing urban parking challenges. The purpose of this research is to explore how the integration of machine learning (ML) and artificial intelligence (AI) within Internet of Things (IoT)-enabled smart parking systems can optimize urban parking management, thereby enhancing operational efficiency, alleviating congestion, and improving user experience in urban environments (Raj and Shetty 2024; Singh et al. 2023a; Elfaki 2023). The study aims to provide a comprehensive analysis of the methodologies used, the results obtained, and the implications for future smart parking solutions.

Machine learning has become an essential tool in the analysis and prediction of complex data patterns for targets detection (Vink and Haan 2015) across various domains, including finance, healthcare and nutrition (Knights et al. 2023; Failed 2023b), autonomous systems (Vesna Antoska Knights 2018) and other real-world applications (Sarker 2021; Traffic Flow Prediction for Smart Traffic Lights Using Machine Learning Algorithms 2022; Boukerche et al. 2020).

In addition to these foundational works, recent studies have further expanded the application of machine learning techniques. For example, the development of guidance and control systems has been significantly enhanced by machine learning algorithms, allowing for more efficient and reliable operations in dynamic environments (Knights and Petrovska n.d; Antoska-Knights et al. 2017). Similarly, the integration for detecting and preventing vulnerabilities in the IoT, ensuring better security and functionality (Machine Learning-Enabled Internet of Things (IoT) 2022; Rafique et al. 2024; Knights and Gacovski 2024) and cyber security (Canadian institute of cybersecurity).

The advent of sophisticated algorithms has transformed how we approach problem-solving in these areas, enabling more accurate and efficient solutions.

The statistical learning theory laid out by Vapnik forms the theoretical underpinning for many machine learning algorithms, emphasizing the trade-off between model complexity and generalization capability (Vladimir 1995). Furthermore, regression analysis remains a cornerstone technique in statistical learning, as detailed by Draper and Smith, facilitating the prediction of continuous outcomes from a set of predictor variables (Norman 1998). The introduction of machine learning to regression tasks, as explained by James et al., has broadened the scope of traditional statistical methods, incorporating more flexible and powerful tools (James et al. 2013).

Among the prominent regression methods in machine learning are decision trees, random forests, and gradient boosting machines (Irmalasari and Dwiyanti 2019; Ait Naceur et al. 2024). Decision trees, introduced by Quinlan, provide a straightforward yet powerful approach for classification and regression tasks by recursively partitioning the data space based on feature values (John Ross Quinlan 1986). Building on this, Breiman's random forests combine multiple decision trees to enhance predictive accuracy and control overfitting (Breiman 2001). Similarly, gradient boosting, as described by Friedman, sequentially builds models by optimizing the residual errors of previous models, thus improving overall performance (Friedman 2001).

Performance evaluation is a critical aspect of machine learning, guiding the selection of models and tuning of hyperparameters. Common metrics such as Root Mean Square Error (RMSE) and Mean Absolute Error (MAE) offer insights into model accuracy, with ongoing debates about their relative merits (Chai and Draxler 2014; Willmott and Matsuura 2005; Hyndman and Koehler 2006). Understanding these metrics and their implications is vital for interpreting the results and effectiveness of different machine learning approaches.

In the context of this research, aim is to provide a comprehensive analysis of machine learning techniques, focusing on the application and comparison of several pivotal machine learning techniques, highlighting regressions methods (Knights and Prchkovska 2024), practical applications, and performance evaluations. By comparing these methods across various datasets, it identifies the strengths and weaknesses of each approach, offering guidance for their optimal use.

2 Material and Methods

Data Collection: The data utilized in this study was collected from four parking places located in Skopje, Republic of North Macedonia. It includes information for a period of 2019–2021 year, such as the entry and exit times of vehicles, the total duration of vehicle occupancy, and the availability of parking spaces.

The total number of vehicles is calculated as follows: $Z = i + \text{entry} - \text{exit}$, where Z indicates the total number of vehicles in the parking, i indicates the initial number of parking spaces.

Data preprocessing: To ensure data integrity and suitability for analysis, the data was subjected to several preprocessing techniques.

Outliers: Outliers were identified and handled using z-score analysis.

Missing Values: Missing values were imputed using a combination of forward-fill and backward-fill techniques.

Normalization: Data was normalized using z-score normalization to standardize the range of the features.

Z-score normalization standardizes the range of the features by subtracting the mean and dividing by the standard deviation. This process helps in ensuring that the model treats each feature equally and improves the convergence speed. This is done by transforming each feature x into a standardized value z using the following formula:

$$z = \frac{x - \mu}{\sigma}$$

where:

- x is the original feature value,
- μ is the mean of the feature,
- σ is the standard deviation of the feature.

Feature Engineering: In this configuration, lagged features are introduced to the model to capture temporal dependencies in the data.

For a given time series data $\{y_t\}$, lagged features are created as follows:

$$y_t, y_{t-1}, , y_{t-2}, \ldots y_{t-k}$$

where y_t is the value at time t, and k is the number of lagged values used as features.

By including lagged values of vehicle counts as features, the model can better understand and predict future vehicle counts based on past trends. Specifically, lagged values of vehicle counts were used as features.

Machine Learning Models: Several machine learning algorithms based on regression were employed to predict parking space availability and optimize parking operations. We utilized the RandomForestRegressor, SVM, GradientBoostingModel, and DecisionTreeModel algorithms to develop predictive models.

RandomForestRegressor: is an ensemble learning algorithm based on decision trees. It is renowned for its ability to handle nonlinear relationships and complex datasets. The algorithm constructs multiple decision trees during training and combines their predictions through averaging or voting to improve accuracy. Random Forest constructs multiple decision trees during training. The prediction $\hat{y}$ for a given input x is obtained by averaging (or voting) the predictions of individual trees (Breiman 2001).

$$\hat{y} = \frac{1}{N}\sum_{i=1}^{N} y_i$$

where N is the number of trees and y_i is the prediction of the i-th tree.

Ridge Regression is a supervised learning algorithm. It is a linear regression technique that is used when the dataset has multicollinearity, which means the independent variables are highly correlated. It is an extension of ordinary least squares (OLS) regression that introduces a penalty term to the loss function, known as the regularization parameter or the L2 penalty. In Ridge Regression, the goal is to minimize the residual sum of squares (RSS) along with a penalty term that penalizes large coefficients. This penalty term helps to reduce the variance of the estimates, making the model more robust to collinearity and less sensitive to noise in the data.

The Ridge Regression model is defined by the following function

$$\hat{y} = X\beta + b$$

where, $\hat{y}$ is predicted target variable, X is matrix of input features, β is a vector of coefficients and b is intercept term.

In Ridge Regression, the coefficients β are determined by minimizing the combined error between the observed and predicted target values, while also penalizing large coefficients to prevent overfitting. This is achieved by solving the optimization problem expressed in the equation:

$$\min\beta\left(\sum_{i=1}^{N}(y_i - \hat{y})^2 + \alpha\sum_{i=1}^{p}\beta_j^2\right)$$

Gradient Boosting Regressor: is a machine learning technique that builds models sequentially. Each new model corrects errors made by the previous ones, leading to improved accuracy over iterations. Gradient boosting is an ensemble learning method. It specifically is used for regression tasks, aiming to minimize the residual errors. Gradient Boosting builds models sequentially, where each new model is trained to correct the errors made by the previous ones. The name Boosting coming because it focuses on minimizing the residual errors of the ensemble's predictions by adding new models that target these errors. The prediction $\hat{y}$ for a given input x is obtained by summing the predictions of all models (Friedman 2001):

$$\hat{y} = \sum_{i=1}^{N} model_i(x)\; or\, \hat{y} = \sum_{i=1}^{N} f_i(x)$$

where $f_i(x)$ is the prediction from the i-th model.

Decision Tree Regressor: is a simple yet powerful algorithm for both regression and classification tasks. It recursively partitions the data into subsets based on the features, aiming to minimize the variance of the target variable within each subset. The Decision Tree Regressor predicts the target variable by averaging the values of samples in the leaf node. It recursively partitions the data into subsets based on features, aiming to minimize the variance of the target variable within each subset. The prediction for a given input x is obtained by averaging the target values of samples in the leaf node (John Ross Quinlan 1986):

$$\hat{y} = \frac{1}{N}\sum_{i=1}^{N} y_i$$

where N is the number of samples in the leaf node and $\hat{y}$ is the target value of the i-th sample.

These models were chosen based on their suitability for handling the complexity of the parking data and their performance in predicting parking space availability. Each model's performance was evaluated using metrics such as R-squared (R^2) and Root Mean Squared Error (RMSE) to assess their accuracy and effectiveness in optimizing parking operations.

Evaluation Metrics: To assess the effectiveness of the machine learning models, several evaluation metrics are utilized:

R^2 Score: A measure of the proportion of the variance in the dependent variable that is predictable from the independent variables (Norman 1998; James et al. 2013).

$$\text{R}^2 \text{ Score}: \; R^2 = 1 - \frac{\sum (y_i - \hat{y})^2}{\sum (y_i - \bar{y})^2}$$

R^2 Score: Reflects how well the model explains the variance in the data. Higher R^2 values indicate better explanatory power.

RMSE: Measures the average error in the model's predictions. Lower RMSE values signify more precise predictions.

$$\text{Mean Square Error (MSE)} \quad MSE = \frac{1}{N}\sum_{i=1}^{N} (y_i - \hat{y})^2$$

Root Mean Square Error (RMSE): Represents the square root of the mean of the squared differences between predicted and actual values, providing insight into the model's prediction accuracy (Chai and Draxler 2014; Willmott and Matsuura 2005).

$$\text{Root Mean Square Error (RMSE)} \quad RMSE = \sqrt{MSE} = \sqrt{\frac{1}{N}\sum_{i=1}^{N} (y_i - \hat{y})^2}$$

These metrics aid in evaluating the accuracy and reliability of parking space availability predictions, guiding the selection of the most suitable machine learning model.

Deployment and Integration: Upon selecting the optimal machine learning model, it is deployed into production environments to enable real-time parking space availability predictions. Integration with existing parking management systems, mobile applications, and digital signage facilitates seamless communication of parking availability information to drivers. Additionally, APIs and web services are developed to allow access to prediction results from third-party applications and platforms.

The process of methodology and the steps of machine learning to enable smart parking is shown in Fig. 1 (Raj and Shetty 2024).

3 Results

From the collected data for the three parking places, is made exploratory of data and data analysis. Figures 2, 3, 4 shows the high density of vehicles at parking places 1, parking places 2, and parking places 3. In the plots are present amounts of vehicles at parking places, by 1/4 day (6 h), half day (12 h), day (24 h), and month.

To provide a comprehensive analysis of vehicle counts in parking places, we first focus on the short-term fluctuations by plotting the data for the first 240 h. This detailed examination includes the original vehicle count, rolling mean, and rolling standard deviation on a daily basis, (Fig. 5). Original vehicle count (blue line): represents the actual recorded number of vehicles observed during each time interval, providing a raw view of the traffic patterns. Rolling mean (24 h) (orange line): displays the average vehicle count over a rolling 24-hour window, effectively smoothing short-term fluctuations to highlight daily trends and cycles. This metric helps identify patterns in average traffic volume across days. Rolling standard deviation (24 h) (green line): reflects the variability in the number of vehicles observed over a rolling 24-hour window. A higher standard deviation indicates greater fluctuations in traffic, whereas a lower standard deviation suggests more consistent traffic patterns. The combination of these visual elements enables a clearer understanding of parking usage trends and variability. For instance, periods with sharp increases or decreases in standard deviation may correspond to irregular events or sudden changes in traffic behavior.

Building upon this analysis, we extend our examination to include the autocorrelation plot (ACF) and partial autocorrelation plot (PACF) shown in Fig. 6. The blue dots are data for parking palace 1, orange dots are data for parking palace 2, and green dots are data for parking palace 3.

ACF measures the correlation between a time series and a lagged version of itself at different lag values. In the ACF plot, the x-axis represents the lag or time interval between observations, and the y-axis represents the correlation coefficient. Each point on the ACF plot represents the correlation between the original series

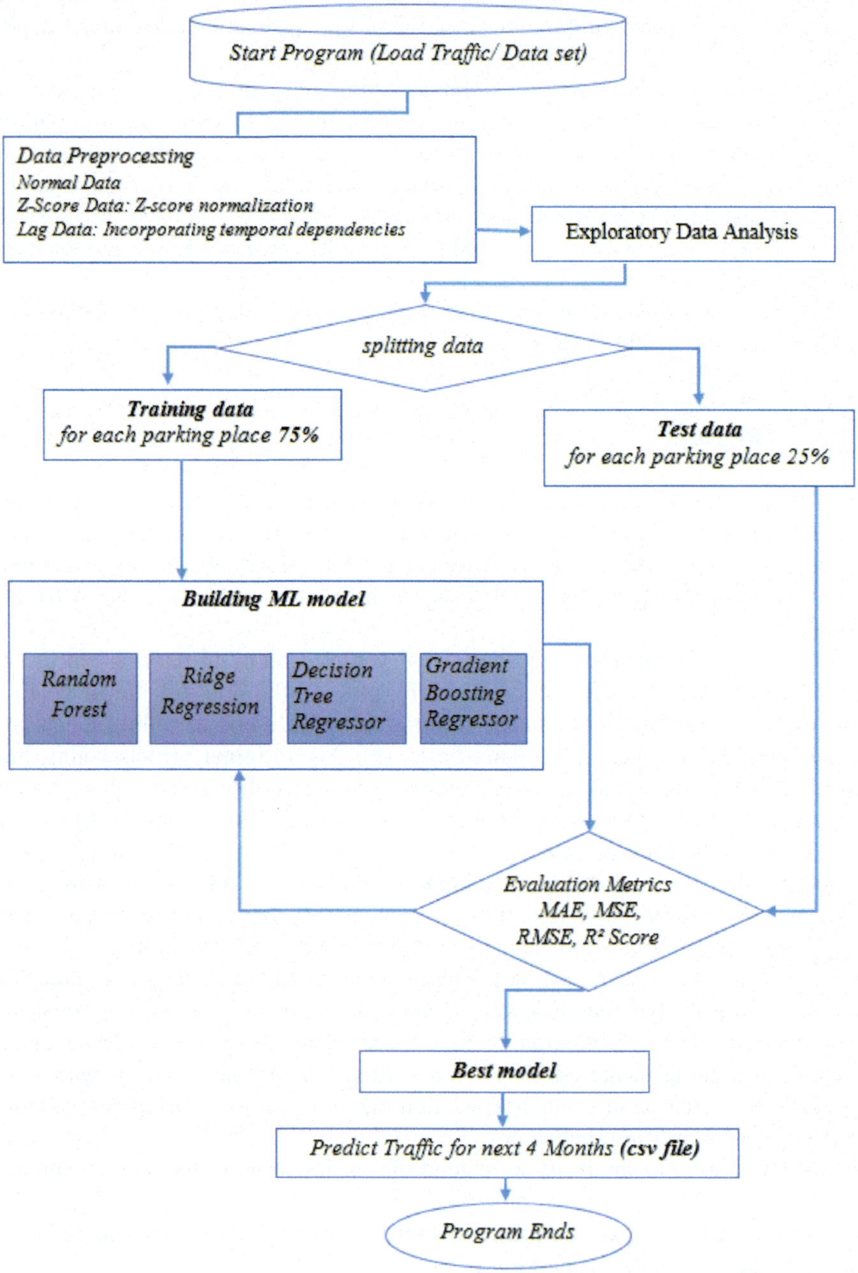

Fig. 1 Algorithm of the machine learning process for data analysis of parking place

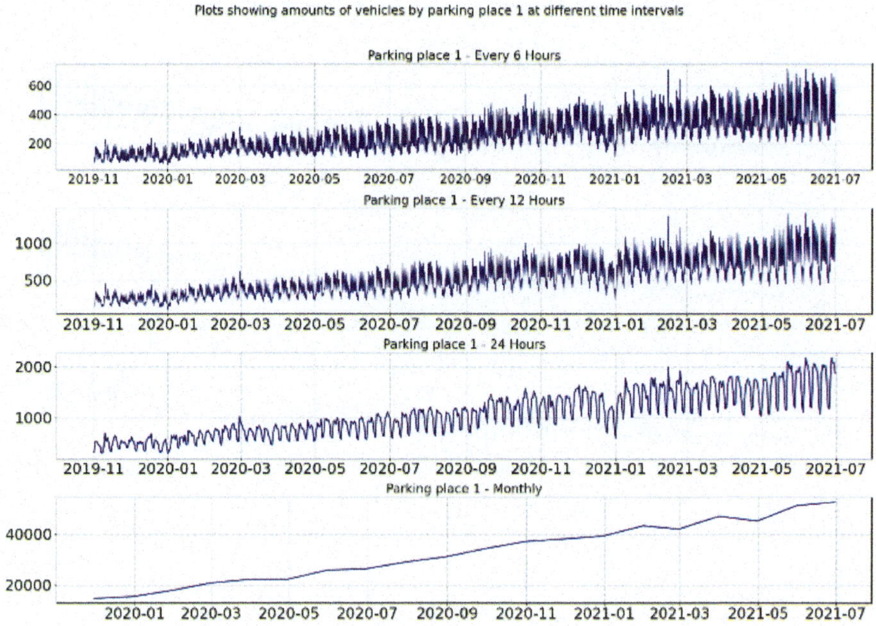

Fig. 2 Amounts of Vehicles at parking place 1, by 1/4 day (6 h), half day (12 h), day (24 h), and month

and its lagged version at a specific lag value. The blue shaded region in the plot represents the 95% confidence interval. Data points lying outside this region are considered statistically significant, indicating a strong correlation between the time series and its lagged version at that lag value.

The partial autocorrelation plot (PACF) measures the correlation between a time series and a lagged version of itself, while removing the effect of shorter lag values. In the PACF plot, similar to the ACF plot, the x-axis represents the lag or time interval between observations, and the y-axis represents the correlation coefficient. Each point on the PACF plot represents the partial correlation between the original series and its lagged version at a specific lag value, after removing the effects of shorter lags. Like the ACF plot, the blue shaded region in the PACF plot represents the 95% confidence interval. Data points lying outside this region are considered statistically significant, indicating a strong partial correlation between the time series and its lagged version at that lag value. In summary, these plots are useful for understanding the autocorrelation structure of a time series data, identifying significant lag values, and determining the appropriate lag order for autoregressive models.

An annotation is included to explain the difference between the Partial Autocorrelation Function (PACF) and the ACF. Specifically, the PACF controls for shorter lags. This means that while the ACF measures the total correlation between the series and its lags, the PACF measures the correlation between the series and its

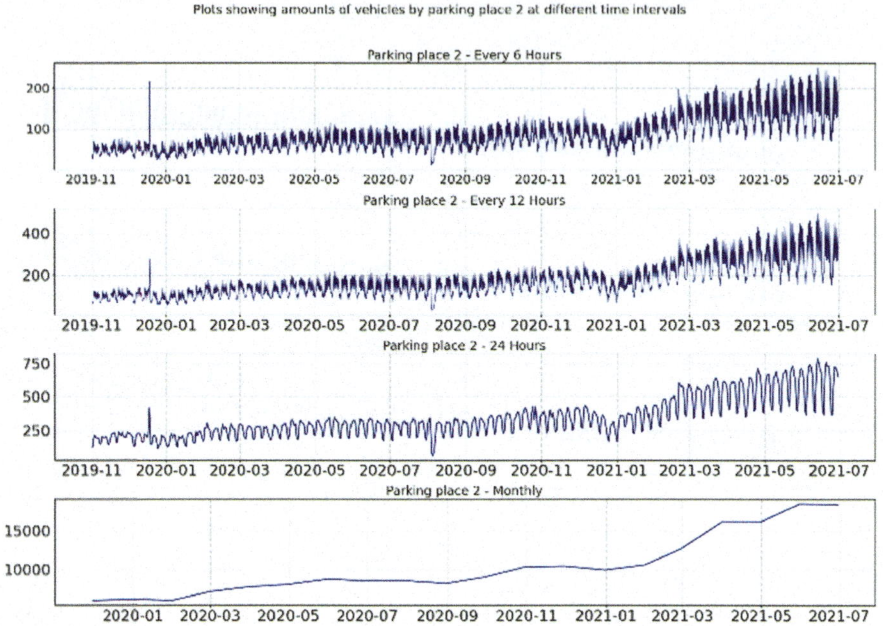

Fig. 3 Amounts of Vehicles at parking place 2, by 1/4 day (6 h), half day (12 h), day (24 h), and month

lags after removing the influence of intermediate lags. This distinction helps in identifying the direct effect of a particular lag on the time series, providing clearer insights into the underlying patterns and dependencies in the data.

Furthermore, we incorporate a heatmap analysis to visualize the correlation between various data attributes (Fig. 7).

In Fig. 7, it is shown that the variable "Month" has a negative correlation with the variable "Year." This means that as the values of "Month" increase, the values of "Year" tend to decrease, and vice versa. This indicates that different months of the year have varying frequencies of vehicle counts at the parking place, but on an annual level, there is a negative correlation.

Another observation is that the correlation between "Vehicles" and "Year" is equal to the correlation between "Vehicles" and "Hour." This indicates that the influence of the "Year" variable on "Vehicles" is similar to the influence of the "Hour" variable on "Vehicles." Based on this, it suggests that either the "Year" or the "Hour" variable could be one of them dropped from the analysis because they provide the same information.

This combined analysis through rolling statistics, autocorrelation, partial autocorrelation, and correlation heatmaps provides a comprehensive view of the traffic patterns, helping to identify both short-term and long-term trends and relationships within the data.

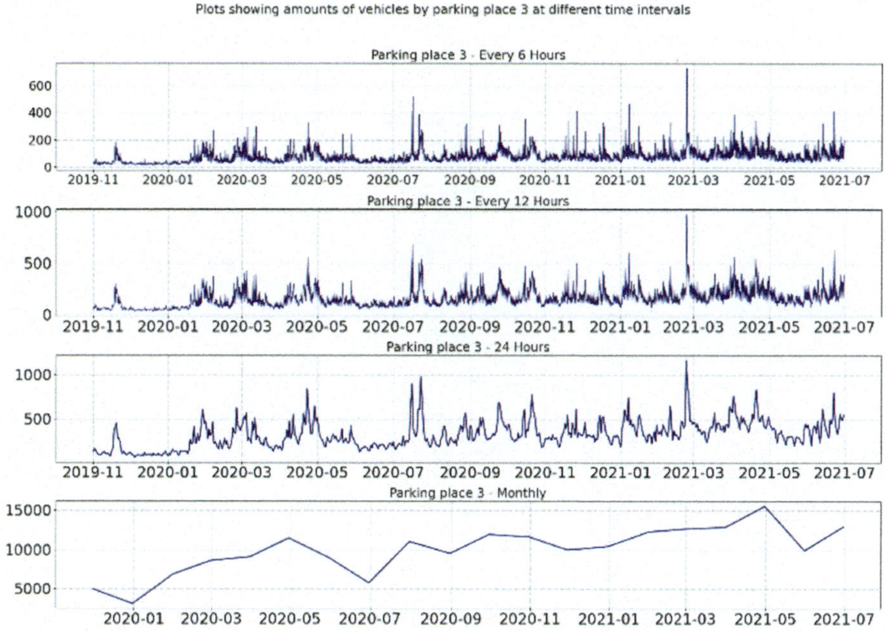

Fig. 4 Amounts of Vehicles at parking place 3, by 1/4 day (6 h), half day (12 h), day (24 h), and month

3.1 Metrics Results

Several machine learning models are employed to predict the vehicle counts for parking places, including RandomForestRegressor, Ridge Regression, GradientBoostingModel, DecisionTreeRegressor. Regression is used as a method in all of this model (Theodoridis 2020; Praba et al. 2021).

The performance of each model was evaluated using two key metrics: the coefficient of determination (R^2 Score) and the root mean squared error (RMSE). These metrics provide insights into the accuracy and precision of the models in predicting the vehicle counts.

In Table 1 the results of the Ridge Regression are presented, where 'lists `r2_scores`' and '`rmses`' are created to collect the R^2 scores and RMSE values from each model. The '`average_r2`' as the mean of the R^2 scores and '`sum_rmse`' as the sum of the RMSE values across all parking places is also calculated.

The Ridge Regression model shows moderate performance. The R^2 scores indicate that the model explains approximately 53.4%, 23.0%, and 16.9% of the variance in vehicle counts for the first, second, and third parking places, respectively. The RMSE values indicate the average prediction errors for each parking place, with higher values suggesting less precise predictions.

The GradientBoostingModel performs better than Ridge Regression, with higher R^2 scores and lower RMSE values, indicating more accurate and precise

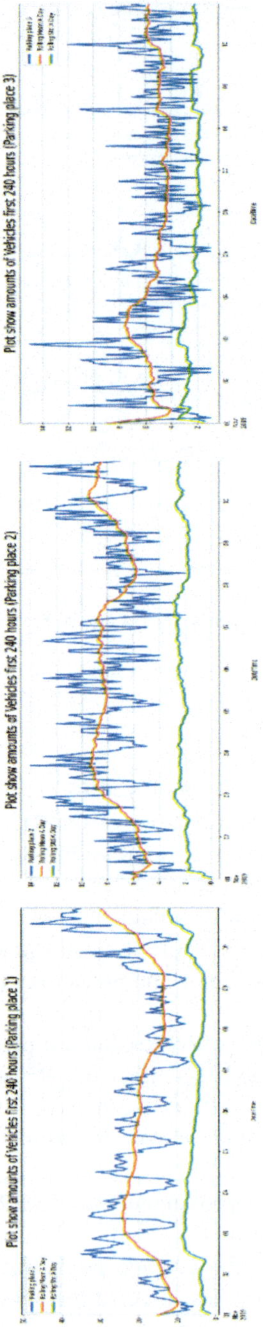

Fig. 5 Total number of vehicles in the first 240 h

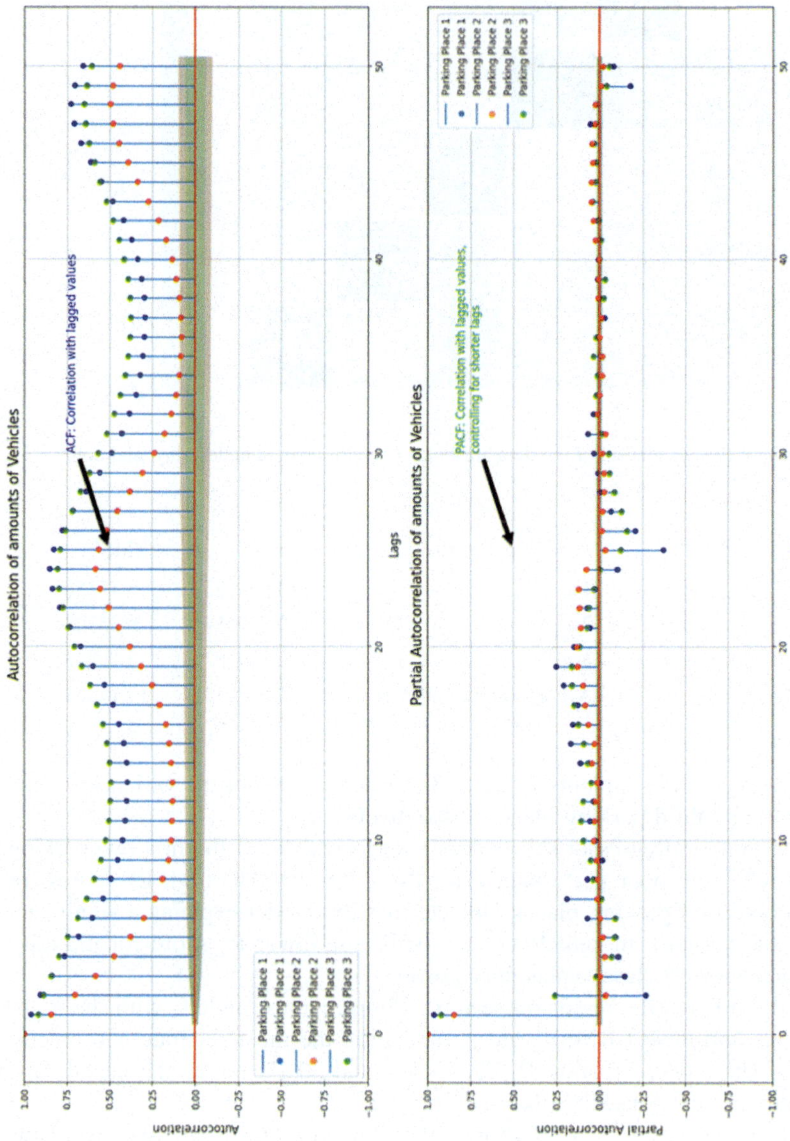

Fig. 6 Autocorrelation plot (ACF)

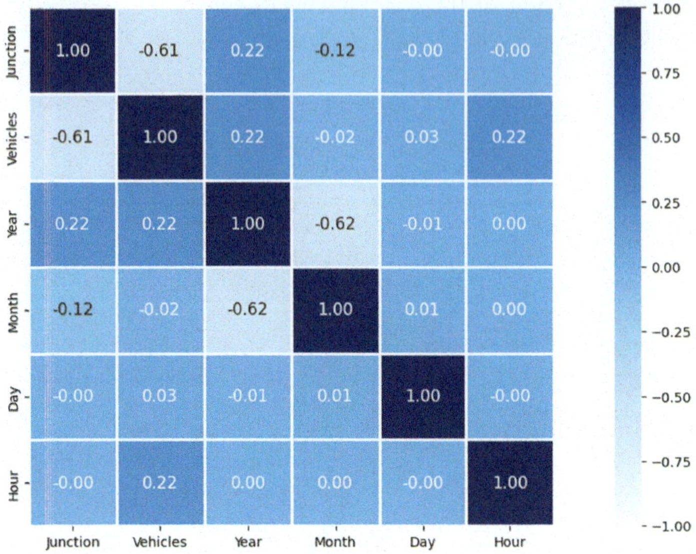

Fig. 7 Heatmap: the correlation of the data attributes (month, year, vehicles, day, and hour)

Table 1 Ridge regression

Parking place No	R^2	RMSE
1st	0.533733	5.152586
2nd	0.230414	9.35728
3rd	0.169302	3.168432
Average/sum of all data set	0.634435	14.142626

predictions. The model explains 71.3%, 43.5%, and 51.9% of the variance for the first, second, and third parking places, respectively.

The DecisionTreeRegressor shows strong performance for the first two parking places but relatively poor performance for the third one. The high R^2 scores for the first and second parking places indicate that the model explains a significant portion of the variance in vehicle counts, while the lower R^2 score for the third parking place suggests less explanatory power.

The RandomForestRegressor outperforms the previous models, with the highest R^2 scores and lowest RMSE values, indicating the most accurate and precise predictions. The model explains 85.5%, 73.4%, and 50.2% of the variance for the first, second, and third parking places, respectively.

When using Z Score Normalization, the RandomForestRegressor maintains high performance, with R^2 scores similar to those without normalization. The RMSE values are much lower due to normalization, indicating consistent and stable predictions across different scales of data.

The lag_model version of the RandomForestRegressor shows the highest R^2 scores, indicating that it explains the most variance in vehicle counts. The RMSE values are also relatively low, suggesting highly accurate predictions. This model outperforms all others, making it the best model for predicting vehicle counts for parking places.

3.2 Best Model

The analysis involves comparing multiple machine learning models to identify the best predictor for vehicle counts at parking places. By evaluating each model's R^2 and RMSE values, we gain insights into their predictive performance and reliability. Based on the results from Tables 1, 2, 3, 4, 5, 6, the RandomForestRegressor with lag_model configuration is identified as the best model for predicting vehicle counts. It demonstrates the highest R^2 scores and the lowest RMSE values, indicating superior accuracy and precision in capturing the patterns and trends in the data. This means that the lag_model configuration of RandomForestRegressor is the most reliable and effective model for forecasting vehicle counts at parking places over a given period.

Table 2 GradientBoostingModel

Parking place No	R^2	RMSE
1st	0.713491	3.880652
2nd	0.435167	7.885537
3rd	0.519839	2.309101
Average/sum of all data set	0.780945	10.798610

Table 3 DecisionTreeRegressorl

Parking place No	R^2	RMSE
1st	0.731112	3.786916
2nd	0.571639	6.846312
3rd	0.154250	3.241791
Average/sum of all data set	0.913875	6.760328

Table 4 RandomForestRegressor

Parking place No	R^2	RMSE
1st	0.854821	2.829295
2nd	0.734023	5.644854
3rd	0.501973	2.438970
Average /sum of all data set	0.940483	5.631669

Table 5 RandomForestRegressor—Z score normalization

Parking place No	R^2	RMSE
1st	0.851473	0.131294
2nd	0.737999	0.247076
3rd	0.491392	0.122755
Average/sum of all data set	0.944939	0.260053

Table 6 RandomForestRegressor—lag_ model

Parking place No	R^2	RMSE
1st	0.880554	2.543026
2nd	0.752371	4.833161
3rd	0.506155	2.597121
Average/sum of all data set	0.966429	4.223636

3.3 Prediction Vehicle Counts for Parking Places

To predict vehicle counts for parking places over a four-month period, the following steps were taken. For each parking place, the current time (start of the prediction period) and the last recorded time in the dataset were obtained. The end time for prediction was set to four months after the last recorded time. A copy of the dataset for each parking place was created to ensure that the original data remained unchanged. The features used by the prediction models were extracted from the dataset. The last row of the dataset was duplicated to facilitate the creation of lagged data necessary for prediction. Lagged data (previous time steps) were generated to use as input features for the prediction model. The values for the newly created rows were assigned based on the existing data and the lagged data structure. Using the prepared data, predictions for the number of vehicles were made for each subsequent hour within the four months. For each hour, the prediction model was applied to forecast the vehicle count. After predicting the vehicle count for the current hour, the time was incremented by one hour. The index of the new data was updated to reflect the new time range, starting from the beginning of the dataset to the end of the prediction period. The predicted data for each parking place was saved to a CSV file for further analysis and validation. The completion of predictions for each parking place was logged for tracking purposes.

Figure 8, shows the plot's visualization for each parking place's separate prediction for the next 4 months. It loads the predicted data for each junction from CSV files, plots the number of vehicles over time for each parking place, and adds labels, titles, legends, and grids to the plot for better interpretation. This visualization provides a clear overview of the predicted vehicle counts for the next four months at each parking place.

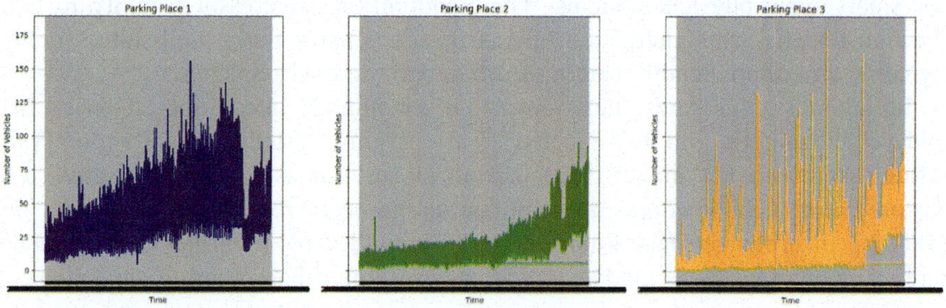

Fig. 8 Prediction and visualization for each parking place over four months period

4 Conclusion

This study demonstrates the effectiveness of integrating machine learning (ML) and artificial intelligence (AI) into IoT-enabled smart parking systems to optimize urban parking management. By utilizing models such as RandomForestRegressor, Ridge Regression, GradientBoostingModel, and DecisionTreeRegressor, we achieved significant improvements in predicting vehicle counts, thereby enhancing operational efficiency and user experience in urban environments.

The RandomForestRegressor, particularly with lag_model configuration, emerged as the best performer, achieving the highest R^2 scores and the lowest RMSE values, indicating superior accuracy and precision. This model explains a significant portion of the variance in vehicle counts and provides highly accurate predictions, aligning with findings in other studies that highlight the robustness and high performance of RandomForest in various applications (Yin et al. 2021; Shafique et al. 2020).

Comparatively, Yin et al. (2021) demonstrated the efficiency of deep learning models in traffic prediction, achieving high accuracy but often at the cost of increased computational complexity and training time. Our use of RandomForestRegressor offers a balance between performance and computational efficiency, making it a practical choice for real-time applications in smart parking systems.

Furthermore, our results are consistent with the work of Al-Garadi et al. (2020), who surveyed various machine learning methods for IoT security and found that ensemble methods like RandomForest provide robust and reliable predictions across diverse datasets and scenarios. This reinforces the suitability of RandomForest for dynamic and complex environments like urban parking management.

In contrast, traditional methods such as Ridge Regression showed moderate performance in our study, which aligns with previous research indicating that while these methods are simpler and less computationally demanding, they often lack the predictive power of more advanced ensemble techniques (Jha 2023).

The findings also echo the conclusions of Kriebel et al. (2018), who emphasized the importance of robustness and adaptability in machine learning models for smart cyber-physical systems. The RandomForestRegressor's ability to handle non-linearity and interaction effects makes it particularly well-suited for the dynamic and unpredictable nature of urban parking environments.

In summary, this study underscores the potential of integrating advanced ML and AI techniques into IoT-enabled smart parking systems. The RandomForestRegressor with lag_model configuration stands out as a highly effective tool for predicting vehicle counts, thereby facilitating better resource management and planning in urban settings. Future research should explore further enhancements and adaptations of these models to address evolving urban mobility challenges.

References

Abdelrahman Osman, E., Messoudi, W., Bushnag, A., Abuzneid, S., Alhmiedat, T.: A smart real-time parking control and monitoring system. Sensors **23**(4), 9741 (2023). https://doi.org/10.3390/s23249741

Ait Naceur, H., Abdo, H.G., Igmoullan, B., et al.: Implementation of random forest, adaptive boosting, and gradient boosting decision trees algorithms for gully erosion susceptibility mapping using remote sensing and GIS. Environ. Earth Sci. **83**, 121 (2024). https://doi.org/10.1007/s12665-024-11424-5

Al-Garadi, M. A., Mohamed, A., Al-Ali, A., Du, X., Guizani, M.: A survey of machine and deep learning methods for internet of things (IoT) security. IEEE Commun. Surv. & Tutor. **22**(3), 1646–1685 (2020). https://doi.org/10.1109/COMST.2020.2968417

Alfonso, N.-E., López-Bonilla, O. R., García-Guerrero, E. E., Tlelo-Cuautle, E., López-Mancilla D., Hernández-Mejía C., Inzunza-González E.: Traffic flow prediction for smart traffic lights using machine learning algorithms. Technologies (2022)

Antoska Knights, V., Gacovski, Z., Deskovski, S.: Guidance and control system for platoon of autonomous mobile robots. J. Electr. Eng. **6**, 281–288 (2018). https://doi.org/10.17265/2328-2223/2018.05.005. https://www.davidpublisher.com/Public/uploads/Contribute/5bd9224d81b32.pdf

Antoska-Knights, V., Gacovski, Z., Deskovski, S.: Obstacles avoidance algorithm for mobile robots, using the potential fields method. Univers. J. Electr. Electron. Eng. **5**(4), 75–84 (2017). https://doi.org/10.13189/ujeee.2017.050402. https://www.hrpub.org/download/20171230/UJEEE2-14910501.pdf

Boukerche, A., Wang, J.: Machine learning-based traffic prediction models for intelligent transportation systems. Comput. Netw. 181 (2020)

Breiman, L.: Random forests. Mach. Learn. **45**(1), 5–32 (2001)

Canadian institute of cybersecurity, university of new brunswick, iscx dataset, https://www.unb.ca/cic/datasets/index.html

Chai, T., Draxler, R. R.: Root mean square error (RMSE) or mean absolute error (MAE)? – Arguments against avoiding RMSE in the literature. Geosci. Model. Dev. **7**(3), 1247–1250 (2014). https://doi.org/10.5194/gmd-7-1247-2014. https://gmd.copernicus.org/articles/7/1247/2014/

Friedman, J.H.: Greedy function approximation: A gradient boosting machine. Ann. Stat. **29**(5), 1189–1232 (2001)

Hyndman, R.J., Koehler, A.B.: Another look at measures of forecast accuracy. Int. J. Forecast. **22**(4), 679–688 (2006)

Irmalasari, I, Dwiyanti, L.: Algorithm analysis of decision tree, gradient boosting decision tree, and random forest for classification (Case Study: West Java House of Representatives Election 2019). In: 2023 International Conference on Electrical Engineering and Informatics (ICEEI), Bandung, Indonesia, pp. 1–5 (2023). https://doi.org/10.1109/ICEEI59426.2023.10346727

Jamal, B., Alam, F., Dhafer, A., Bojović, M., Altowaijri, S. M., Khan Niazi, I., Mehmood,R.: Machine learning-enabled internet of things (IoT): Data, applications, and industry perspective. Electronics **11**(17), 2676 (2022). https://doi.org/10.3390/electronics11172676

James, G., Witten, D., Hastie, T., Tibshirani, R.: An introduction to statistical learning: with applications in R, New York, NY. Springer, USA (2013)

Jha, S.: Lightning talk: Trinity—assured neuro-symbolic model inspired by hierarchical predictive coding. In: 2023 60th ACM/IEEE Design Automation Conference (DAC), pp. 1–2 (2023). https://doi.org/10.1145/3449639.3459291

John Ross Quinlan: Induction of decision trees. Mach. Learn. **1**(1), 81–106 (1986)

Knights, V. A., Gacovski, Z.: Methods for detection and prevention of vulnerabilities in the IoT (Internet of Things) systems. Internet Things-New Insights. IntechOpen (2024). https://doi.org/10.5772/intechopen.113898

Knights, V., Petrovska, O.: Dynamic modeling and simulation of mobile robot under disturbances and obstacles in an environment. J. Appl. Math. Comput. **8**(1), 59–67. https://doi.org/10.26855/jamc.2024.03.007

Knights, V., Kolak, M., Markovikj, G., Gajdoš Kljusurić, J.: Modeling and optimization with artificial intelligence in nutrition. Appl. Sci. **13**(13), 7835 (2023). https://doi.org/10.3390/app13137835

Knights, V., Prchkovska, M.: From equations to predictions: understanding the mathematics and machine learning of multiple linear regression. J. Math. Comput. Appl. **3**(2), 1–8 (2024). https://doi.org/10.47363/jmca/2024(3)137

Kriebel, F., Rehman, S., Hanif, M. A., Khalid, F., Shafique, M.: Robustness for smart cyber physical systems and internet-of-things: from adaptive robustness methods to reliability and security for machine learning. In: 2018 IEEE Computer Society Annual Symposium on VLSI (ISVLSI), Hong Kong, China, pp. 581–586 (2018). https://doi.org/10.1109/ISVLSI.2018.00111

Norman, R.: Draper, & harry smith, applied regression analysis, 3rd edn. John Wiley & Sons, New York, NY, USA (1998)

Praba, R., Darshan, G., Roshanraj, K. T., Surya Prakash, P. B.: Study on machine learning algorithms. Int. J. Sci. Res. Comput. Sci., Eng. Inf. Technol. **7**(4), 67–72 (2021)

Purnama Muharram, A., Sajid, F.: Supervised machine learning approach for predicting cardiovascular complications risk in patients with diabetes mellitus. In: 2023 International Conference on Electrical Engineering and Informatics (ICEEI), Bandung, Indonesia, pp. 1–6 2023. https://doi.org/10.1109/ICEEI59426.2023.10346215

Raj, A., Shetty, S.D.: Smart parking systems technologies, tools, and challenges for implementing in a smart city environment: a survey based on IoT & ML perspective. Int. J. Mach. Learn. Cyber. (2024). https://doi.org/10.1007/s13042-023-02056-5

Saida Hafsa, R., Abdallah, A., Shifa Musa, N., Murugan, T.: Machine learning and deep learning techniques for internet of things network anomaly detection—current research Trends. Sensors **24**(6), 1968 (2024). https://doi.org/10.3390/s24061968

Sarker, I.H.: Machine learning: Algorithms, real-world applications and research directions. SN Comput. Sci. **2**(160), 1–21 (2021). https://doi.org/10.1007/s42979-021-00592-x

Shafique, M., Naseer, M., Theocharides, T., Kyrkou, C., Mutlu, O., Orosa, L., et al.: Robust machine learning systems: challenges, current trends, perspectives, and the road ahead. IEEE Design Test **37**(2), 30–57 (2020). https://doi.org/10.1109/MDAT.2020.2971217

Singh, T., Rathore, R., Gupta, K., Vijay, E., Harikrishnan, R.: Artificial intelligence-enabled smart parking system. In: Shaw, R.N., Siano, P., Makhilef, S., Ghosh, A., Shimi, S.L. (eds.) Innovations in Electrical and Electronic Engineering. ICEEE 2023. Lecture Notes in Electrical Engineering, vol 1115. Springer, Singapore (2024). https://doi.org/10.1007/978-981-99-8661-3_31

Theodoridis, S: Machine learning a Bayesian and optimization perspective, Elsevier Ltd (2020)

Vink, J.P., de Haan, G.: Comparison of machine learning techniques for target detection. Artif. Intell. Rev. **43**, 125–139 (2015). https://doi.org/10.1007/s10462-012-9366-7

Vladimir, N.: Vapnik, The nature of statistical learning theory, New York, NY. Springer-Verlag, USA (1995)

Willmott, C.J., Matsuura, K.: Advantages of the mean absolute error (MAE) over the root mean square error (RMSE) in assessing average model performance. Climate Res. **30**(1), 79–82 (2005)

Yin, X., Genze, Wu., Wei, J., Shen, Y., Qi, H., Yin, B.: Deep learning on traffic prediction: methods, analysis, and future directions. IEEE Trans. Intell. Transp. Syst.Intell. Transp. Syst. **23**(8), 4927–4943 (2021). https://doi.org/10.1109/TITS.2021.3049105

Space Vector Pulse Width Modulation Strategy for Three Phase Inverters

Alfred Pjetri, Astrit Bardhi, Gentian Dume, and Bajram Leka

Abstract

Adjustable speed drives for three-phase induction motors and permanent magnet synchronous motors necessitate varying both voltage and frequency. Achieving this requirement involves employing a three-phase voltage source inverter. There are two approaches to generate the inverter's output AC voltage with variable magnitude and frequency. The first method involves adjusting the input DC voltage of the inverter while maintaining a constant inverter gain. On the flip side, the second approach comes into play when the direct current input voltage of the inverter remains constant. This setup permits the manipulation of the output voltage's amplitude and frequency by adjusting the gain of the inverter. Typically, the second method is preferred due to its cost-effectiveness. Various algorithms exist for controlling the output voltage of the voltage source inverter (VSI), but they all tend to produce unwanted harmonics aside from the desired fundamental frequency. To mitigate low-order output voltage harmonics, space vector pulse width modulation (SVPWM) is employed. SVPWM is a sophisticated and highly effective technique compared to other PWM methods such as sinusoidal, trapezoidal, harmonic injected, delta, and phase-shifted PWM. It efficiently generates fundamental inverter output voltage sine waves, and helps reduce total harmonic distortion (THD) and dominant harmonics. This paper utilizes sector utilization in each switching period to implement the SVPWM algorithm strategy for generating adjustable frequency and amplitude output voltage in voltage source inverters. The SVPWM techniques have been simulated using the Matlab/Simulink environment.

A. Pjetri (✉) · A. Bardhi · G. Dume · B. Leka
Faculty of Electrical Engineering, Polytechnic University of Tirana, Tirana, Albania
e-mail: alfredpjetri@gmail.com

K. Dhoska and E. Spaho (eds.), *Bridging Horizons in Artificial Intelligence, Robotics, Cybersecurity, Smart Cities, and Digital Economy*, Sustainable Economy and Ecotechnology, https://doi.org/10.1007/978-3-031-72029-1_6

Keywords

Three-phase voltage • DC voltage • SVPWM • Total harmonic distortion

1 Introduction

Three-phase voltage source inverters are employed to transform input DC voltage into AC output voltage with adjustable magnitude and frequency. Mostly, voltage inverters are used in AC electrical drives in which speed or position control is applied. The AC output voltage of inverter can be adjusted by changing the magnitude of the DC input voltage and keeping the inverter gain unchanged. On the other hand, if the DC voltage is constant, the inverter gain must be changed to adjust the output voltage. Due to its low cost, this approach has been extensively applied in situations where there's a constant DC voltage, and frequency and magnitude of the inverter's AC output voltage is achieved through modulation of the inverter gain.

In order to have a fundamental AC sine wave that provides a higher voltage to the load, for the same value of DC voltage, and lower total harmonic distortion it is used Space Vector Pulse Width Modulation (SVPWM) strategy (Quach et al. 2012; Reney 2011; Ostadrahimi and Bifaretti 2023). SVPWM is an improved strategy to control the number of short pulses in the PWM waveform. The SVPWM provides a constant switching frequency and in this manner can be easily adjusted the inverter switching frequency. It is evident that SVPWM is more complicated than other PWM techniques but on the other hand it is easy to implement (Prasad et al. 2022; Chen 2013). This paper utilizes sector utilization in each switching period to implement the SVPWM algorithm strategy for generating adjustable frequency and amplitude output voltage in voltage source inverters (Quach et al. 2012; Reney 2011; Ostadrahimi and Bifaretti 2023; Prasad et al. 2022; Chen 2013; Hysa and Meşe 2021; Yasin 2019).

2 Strategy for Implementing SVPWM Algorithms in Three-Phase Voltage Source Inverters

Three phase voltage source inverters in which the control of the magnitude and frequency of the output voltage is realized through the SVPWM strategy is shown in Fig. 1. The ideal power switches $Ç_1$, $Ç_2$, … $Ç_6$, in practice are power electronic switches and the most widespread for inverter applications are MOSFET-s and IGBT-s transistors. These switches for the SVPWM strategy are controlled by controlling the variables a, a', b, b' and c, c'. When the upper transistors of the inverter are "ON" (state "1"), i.e. when a, b or c are "1", at the same time the corresponding transistors (of the same leg) located on the lower side of the inverter

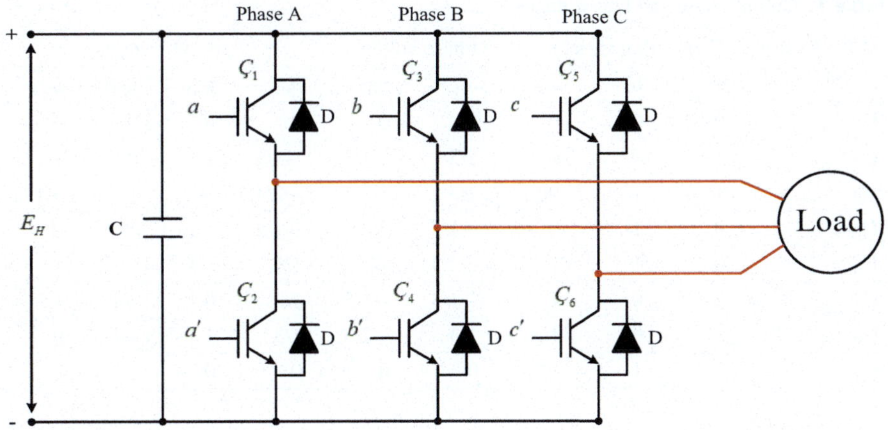

Fig. 1 Three phase voltage source inverter

are "OFF", (state "0") i.e. a', b', or c' are "0". In this way, using only the "1" or "0" states of the transistors located on the upper side, respectively $Ç_1$, $Ç_3$, and $Ç_5$, we can determine the output voltage of the inverter (Quach et al. 2012; Chen 2013).

It is clear that the relationships between the control variables (*a, b, c*) and the line voltages of Fig. 1 (u_{AB}, u_{BC}, u_{CA}) are as follows:

$$\begin{bmatrix} u_{AB} \\ u_{BC} \\ u_{CA} \end{bmatrix} = E_H \begin{bmatrix} 1 & -1 & 0 \\ 0 & 1 & -1 \\ -1 & 0 & 1 \end{bmatrix} \begin{bmatrix} a \\ b \\ c \end{bmatrix} \tag{1}$$

where E_H represents DC input voltage of the inverter.

In the same way, the relationships between the control variables (*a, b, c*) and the phase voltages (u_A, u_B, u_C) of the load (induction motor) can be expressed as:

$$\begin{bmatrix} u_A \\ u_B \\ u_C \end{bmatrix} = \frac{E_H}{3} \begin{bmatrix} 2 & -1 & -1 \\ -1 & 2 & -1 \\ -1 & -1 & 2 \end{bmatrix} \begin{bmatrix} a \\ b \\ c \end{bmatrix} \tag{2}$$

As can be seen from Fig. 1, we have eight possible combinations of the three transistors located on the upper side of the inverter. The states of the lower-side transistors are opposite to those of the upper-side and are therefore can be easily found when the positions of the upper-side transistors are known. To each of the eight combinations we put in correspondence a basic space vector from U_0–U_7. Starting from expressions (1) and (2) and taking into account the basic space vectors U_0–U_7, the phase and line voltages of the load depend from DC voltage E_H. The phase and line load voltage are given in Table 1. Calculation of phase and line

Table 1 Output phase and line voltages

Voltage space vector	Controlling variables			Phase voltages			Line voltages		
	a	b	c	u_A	u_B	u_C	u_{AB}	u_{BC}	u_{CA}
U_0	0	0	0	0	0	0	0	0	0
U_1	1	0	0	2/3	−1/3	−1/3	1	0	−1
U_2	1	1	0	1/3	1/3	−2/3	0	1	−1
U_3	0	1	0	−1/3	2/3	−1/3	−1	1	0
U_4	0	1	1	−2/3	1/3	1/3	−1	0	1
U_5	0	0	1	−1/3	−1/3	2/3	0	−1	1
U_6	1	0	1	1/3	−2/3	1/3	1	−1	0
U_7	1	1	1	0	0	0	0	0	0

voltage is performed by multiplying the corresponding voltage values of Table 1 with the DC voltage E_H.

In order to implement the pulse width modulation technique through the space vector strategy, we will need to transfer the load (induction motor) phase voltages from the phase coordinate (*A, B, C*) to an orthogonal fixed coordinate system (*D, Q, 0*). This can be done through the voltage space vector theory (Quach et al. 2012; Chen 2013). Also, the basic space vectors of the inverter voltage U_0–U_7 can be presented in the same coordinate system (D, Q, 0) as in Fig. 2.

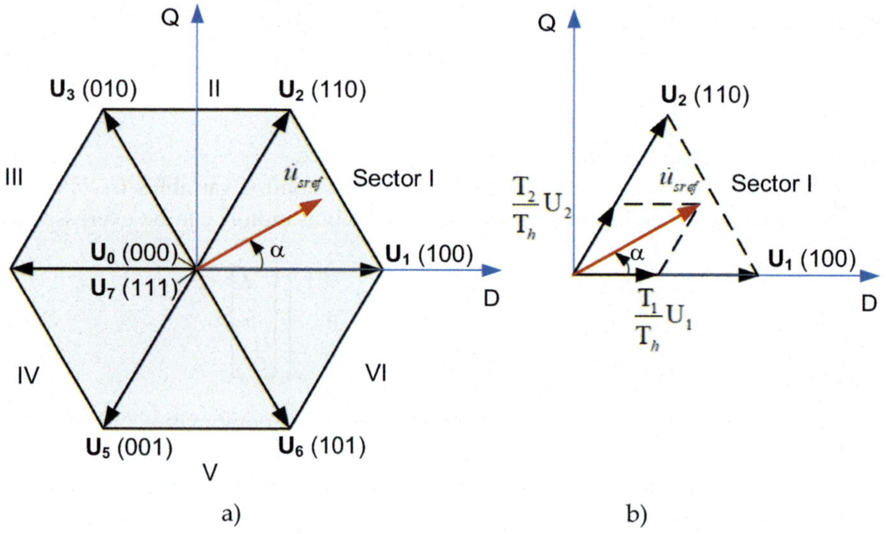

Fig. 2 The inverter's output voltages are depicted through the fundamental space vectors (a); Representation of the reference voltage space vector through two space vectors in sector I, (b)

Implementation of the SVPWM strategy for controlling the magnitude and frequency of the output voltage of the inverter is realized by following four steps (Quach et al. 2012; Chen 2013).

I. Calculation of voltage orthogonal coordinates u_{sD_ref}, u_{sQ_ref}, u_{s_ref} and space vector position angle.
II. Determining the sector in which the space vector is located.
III. Calculation of times T_1, T_2, T_0.
IV. Determining of switching times for each sector T_{m1}, T_{m2}, T_{m3}.

Step I. For a given instantaneous values of the phase voltages, we can calculate the orthogonal stator voltage components of the space vector along the D and Q axes as:

$$\begin{bmatrix} u_{sD\_ref} \\ u_{sQ\_ref} \end{bmatrix} = [T] \begin{bmatrix} u_{A\_ref} \\ u_{B\_ref} \\ u_{C\_ref} \end{bmatrix} = \frac{2}{3} \begin{bmatrix} 1 & -\frac{1}{2} & -\frac{1}{2} \\ 0 & \frac{\sqrt{3}}{2} & -\frac{\sqrt{3}}{2} \end{bmatrix} \begin{bmatrix} u_{A\_ref} \\ u_{B\_ref} \\ u_{C\_ref} \end{bmatrix} \quad (3)$$

From the voltage orthogonal components, we can calculate the modulus of the space vector.

$$\left|\dot{\mathbf{u}}_{\mathbf{s\_ref}}\right| = \sqrt{(u_{sD\_ref})^2 + (u_{sQ\_ref})^2} \quad (4)$$

While the position of the voltage space vector is calculated as:

$$\alpha = \arctan \frac{u_{sQ\_ref}}{u_{sD\_ref}} \quad (5)$$

Step II. Determination of sector in which the space vector is located is carried out as follows.

If $U_{sD\_ref} > 0$, $A = 1$; If $\sqrt{3}U_{sD\_ref} - U_{sQ\_ref} > 0$, $B = 1$; If $\sqrt{3}U_{sD\_ref} + U_{sQ\_ref} > 0$, $C = 1$.

Noting with N = A + 2B + 4C, the sector in which the space vector is located can be determined as in Table 2, (Reney 2011).

Step III. From Fig. 2b, we can calculate the switching times T1, T2 and T0 for each sector.

Table 2 Determining the sector where the space vector is located

Sector	I	II	III	IV	V	VI
N	3	1	5	4	6	2

For sector I we have.

$$\int_0^{T_h} \dot{\mathbf{u}}_{s\_ref} dt = \int_0^{T_1} U_1 dt + \int_{T_1}^{T_1+T_2} U_2 dt + \int_{T_1+T_2}^{T_h} U_0 dt \tag{6}$$

where T_h is the sampling time of the reference voltage space vector $\dot{\mathbf{u}}_{s\_ref}$, , through which can be determined the switching frequency ($f_k = 1 / T_h$) of the inverter power switches. The sampling time T_h it is constant.

From expression (6) we have:

$$T_h \dot{\mathbf{u}}_{\mathbf{s\_ref}} = (T_1 U_1 + T_2 U_2) = T_h \left|\dot{\mathbf{u}}_{\mathbf{s\_ref}}\right| \begin{bmatrix} \cos(\alpha) \\ \sin(\alpha) \end{bmatrix} = T_1 \frac{2}{3} E_H \begin{bmatrix} 1 \\ 0 \end{bmatrix} + T_2 \frac{2}{3} E_H \begin{bmatrix} \cos(\pi/3) \\ \sin(\pi/3) \end{bmatrix} \tag{7}$$

where $0 \le \alpha \le 60^0$. From the above expression (7) we can find the times T_1, T_2, and T_0 as:

$$\begin{aligned} T_1 &= T_h m \frac{\sin(\pi/3 - \alpha)}{\sin(\pi/3)} \\ T_2 &= T_h m \frac{\sin(\alpha)}{\sin(\pi/3)} \\ T_0 &= T_7 = T_h - (T_1 + T_2) \end{aligned} \tag{8}$$

where m is the modulation factor which is calcualted as:

$$m = \frac{3\left|\dot{\mathbf{u}}_{s\_ref}\right|}{2E_H} \tag{9}$$

Calculation of times for each sector it is as follow.

$$\begin{aligned} T_1 &= \frac{\sqrt{3} \cdot T_h \cdot \left|\dot{\mathbf{u}}_{\mathbf{s\_ref}}\right|}{E_H} \left(\sin\left(\frac{\pi}{3} - \alpha + \frac{n-1}{3}\pi \right) \right) \\ &= \frac{\sqrt{3} \cdot T_h \cdot \left|\dot{\mathbf{u}}_{\mathbf{s\_ref}}\right|}{E_H} \left(\sin\left(\frac{n}{3}\pi - \alpha \right) \right) \\ &= \frac{\sqrt{3} \cdot T_h \cdot \left|\dot{\mathbf{u}}_{\mathbf{s\_ref}}\right|}{E_H} \left(\sin \frac{n}{3}\pi \cos(\alpha) - \cos \frac{n}{3}\pi \sin(\alpha) \right) \\ T_2 &= \frac{\sqrt{3} \cdot T_h \cdot \left|\dot{\mathbf{u}}_{\mathbf{s\_ref}}\right|}{E_H} \left(\sin\left(\alpha - \frac{n-1}{3}\pi \right) \right) \\ &= \frac{\sqrt{3} \cdot T_h \cdot \left|\dot{\mathbf{u}}_{\mathbf{s\_ref}}\right|}{E_H} \left(-\cos(\alpha) \sin(\frac{n-1}{3}\pi) + \sin(\alpha) \cos(\frac{n-1}{3}\pi) \right) \end{aligned} \tag{10}$$

Table 3 Relative times t_1 and t_2

Sector	Time t_1	Time t_2
I	$t_1 = Z$	$t_2 = Y$
II	$t_1 = Y$	$t_2 = -X$
III	$t_1 = -Z$	$t_2 = X$
IV	$t_1 = -X$	$t_2 = Z$
V	$t_1 = X$	$t_2 = -Y$
VI	$t_1 = -Y$	$t_2 = -Z$

Table 4 Determining of switching times for each sector

N	1	2	3	4	5	6
T_{cm1}	T_b	T_a	T_a	T_c	T_c	T_b
T_{cm2}	T_a	T_c	T_b	T_b	T_a	T_c
T_{cm3}	T_c	T_b	T_c	T_a	T_b	T_a

$$T_0 = T_h - (T_1 + T_2)$$

where $n = 1, 2, ...6$ and $0 \leq \alpha \leq 60^0$

The relative times t_1 and t_2 for each sector are defined according to Table 3.

Where X, Y and Z can be calculated according to expression (11)

$$\begin{aligned} X &= \frac{\sqrt{3}}{E_H} T_h U_{sD\_ref} \\ Y &= \frac{3}{2E_H} T_h (U_{sD\_ref} + \frac{\sqrt{3}}{3} U_{sQ\_ref}) \\ Z &= \frac{3}{2E_H} T_h (\frac{\sqrt{3}}{3} U_{sQ\_ref} - U_{sD\_ref}) \end{aligned} \tag{11}$$

Step IV. The relationship between the switching times for each phase and the switching times for each sector is shown in Table 4.

Where:

$T_a = \frac{1}{4}(T_h - T_1 - T_2)$; $T_b = T_a + \frac{1}{2}T_1$; $T_c = T_b + \frac{1}{2}T_2(12)$.

Represents switching times for each phase.

3 SVPWM Model in the Matlab/Simulink Environment

Four steps described above are exactly the algorithm of the pulse width modulation method by using space vector strategy. By implementing these steps, we are able to implement the SVPWM strategy for controlling the magnitude and frequency

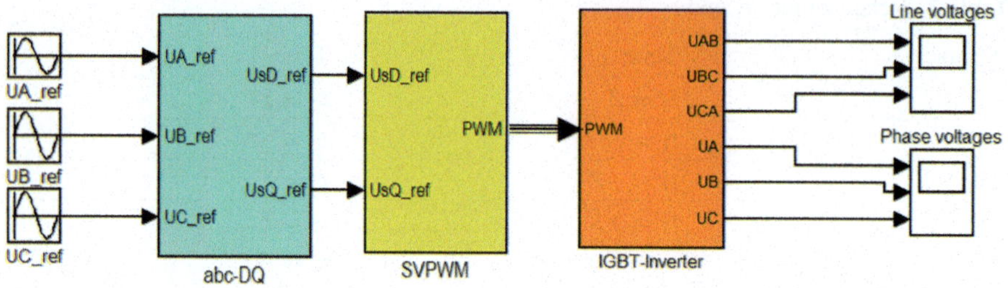

Fig. 3 General block diagram implementation of the SVPWM strategy in Matlab/Simulink

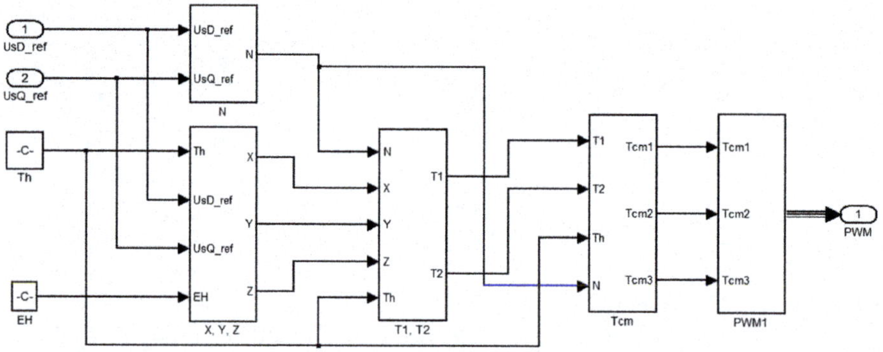

Fig. 4 Block diagram of SVPWM implementation in Matlab/Simulink

of the output voltage of the three-phase inverter. Figure 3 shows the general block diagram, which gives us the output AC voltage with adjustable magnitude and frequency by applying the SVPWM strategy. The magnitude and frequency of the first harmonic of the output voltage of the inverter are exactly the same as the reference.

Block diagram implementation in the Matlab/Simulink environment of the SVPWM strategy is shown in Fig. 4

Figure 5 shows each block implementation in order to have SVPWM algorithm strategy applied in Matlab/Simulink.

4 Simulation Results

For the simulation, the modulation factor is taken $m = 0.81$.

For the modulation factor $m = 0{,}81$ and for the reference sinusoidal voltages with an effective value of $U = 230$ V, from the calculations we have a value of the DC voltage $E_{\mathrm{H}} = 600$ VDC. The frequency of the reference three phase symmetric voltages system is taken $f = 50$ Hz. The sampling time is chosen $T_{\mathrm{h}} = 0.0001$ (s), through which the IGBT inverter switching frequency ($f_{\mathrm{k}} = 1 / T_{\mathrm{h}}) = 10$ kHz.

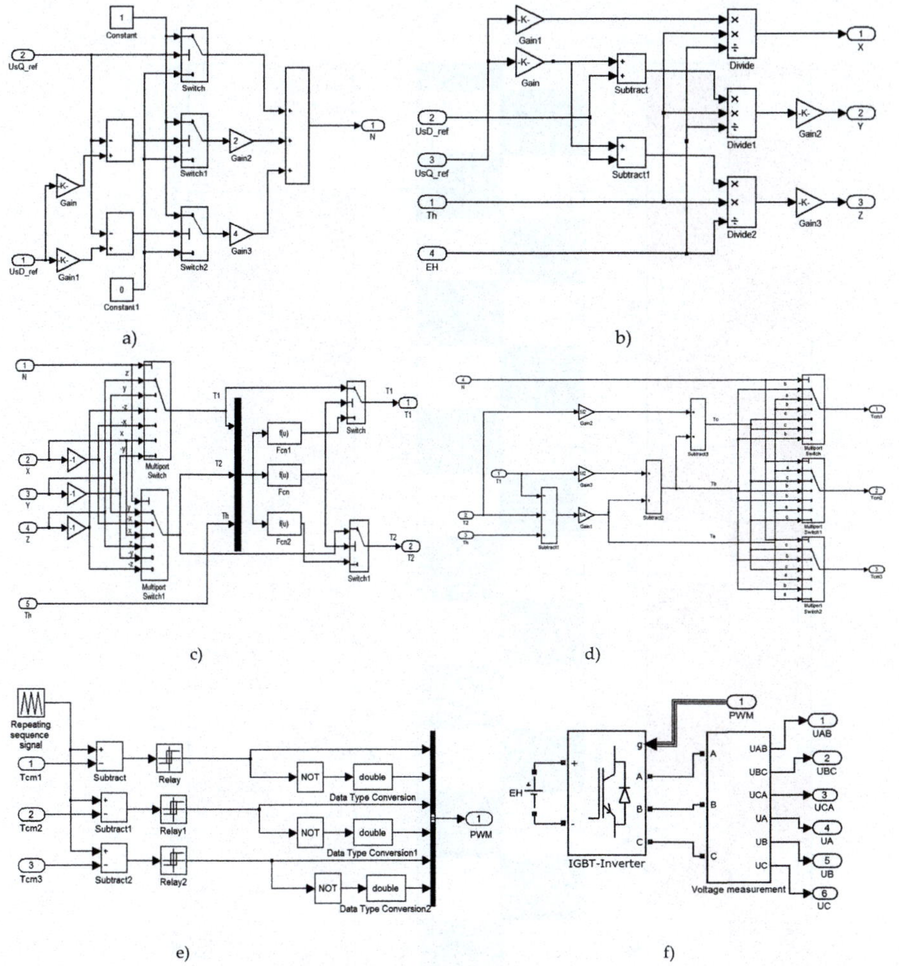

Fig. 5 Block diagrams of each SVPWM block in Matlab/Simulink

As can be seen from Fig. 5f, for the simulation we have considered a three phase IGBT inverter.

The simulation results are obtained for a simulation time of 0.2 s. Figures 6 and 7 show line to line and phase output voltages of the inverter. As it can be seen, the line voltages for a half cycle period changes in two levels, while the phase voltages change in three levels.

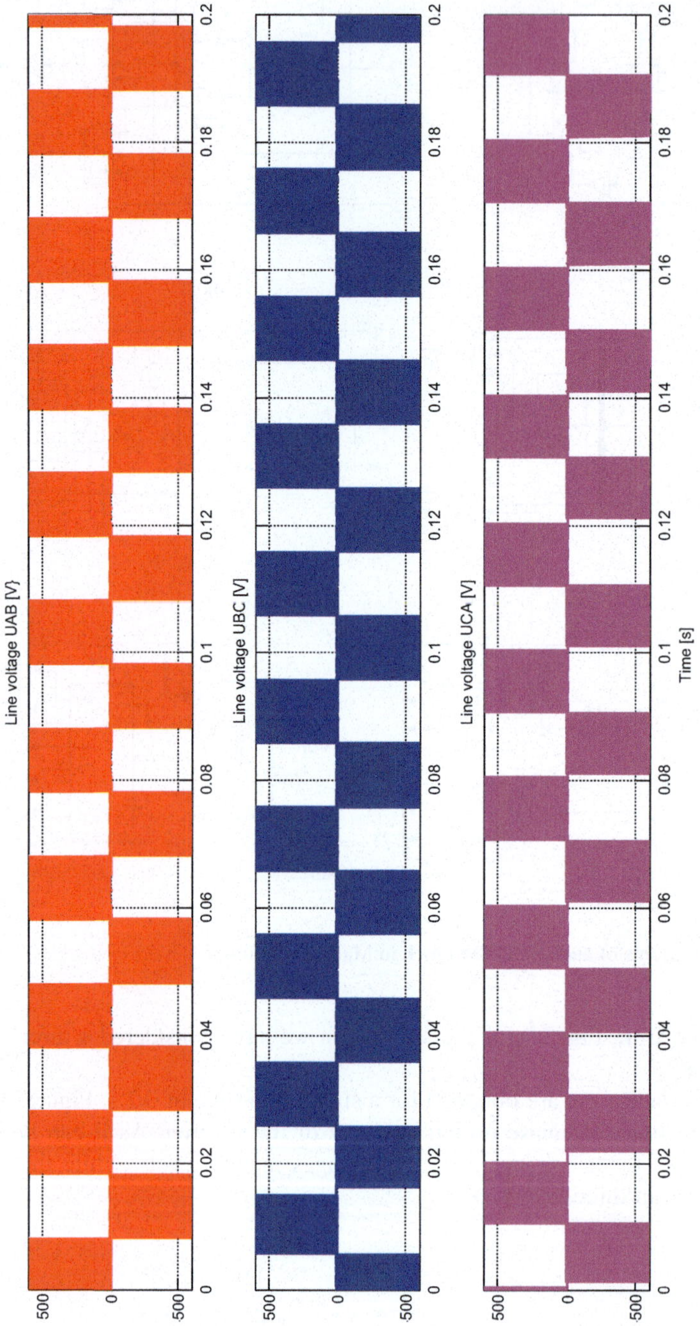

Fig. 6 Line to line inverter output voltages

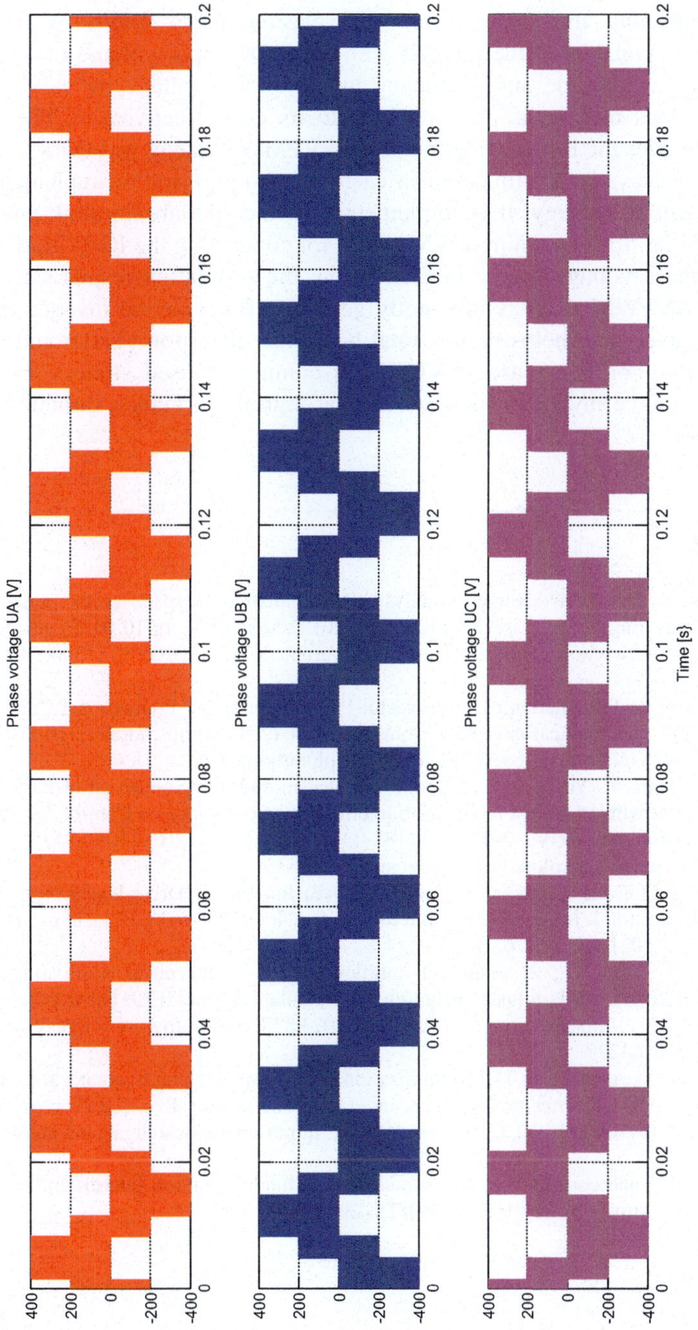

Fig. 7 Inverter output phase voltages

5 Conclusions

This paper presents the strategy of space vector pulse width modulation (SVPWM). It is considered the method with fixed DC input voltage in order to have adjustable magnitude and frequency in the output voltage by varying the inverter gain. This method is preferred due to its cost-effectiveness. The algorithm which enables the implementation of the SVPMW strategy was demonstrated step by step. It uses sector utilization in each switching period to implement the SVPWM algorithm strategy. It is implemented in the Matlab/Simulink environment the algorithm for generating PWM pulses for controlling the IGBT transistors of the three-phase voltage source inverter. From the simulation results it is clear that using of SVPWM strategy efficiently generates fundamental inverter output voltage sine waves, and helps reduce total harmonic distortion (THD) and dominant harmonics. For this reason SVPWM is commonly used strategy in three phase voltage source inverter which are used at high performance variable speed electrical drives.

References

Chen, G., Kang, J., Zhao, J.W: Numeric analysis and simulation of space vector pulse width modulation. Adv. Eng. Softw. Elsevier, pp. 60–65 (2013). https://doi.org/10.1016/j.advengsoft.2013.04.017. http://www.sciencedirect.com/science/article/abs/pii/S0965997813000562?via%3Dihub.

Dolly, R.: Modeling and simulation of space vector PWM inverter. In: International Conference on Devices and Communications (ICDeCom), pp. 1–4 (2011). https://doi.org/10.1109/ICDECOM.2011.5738466. https://ieeexplore.ieee.org/Xplore/home.jsp

Duc-Cuong, Q., Quan, Y., Yu-Feng, S., Chunjie, Z.: Design and implementation of three phase SVPWM inverter with 16-bit dsPIC. In: 12th International Conference on Control Automation Robotics & Vision (ICARCV), 2012, pp. 1181–1186, https://doi.org/10.1109/ICARCV.2012.6485354. http://www.ieeexplore.ieee.org/document/6485354

Hysa, G., Meşe, E.: On the Behavior of Five-Phase Induction Motor Drive Under Normal and Faulty Conditions. Int. J. Innov. Technol. Interdiscip. Sci. 4(3), 754–763 (2021). https://doi.org/10.15157/IJITIS.2021.4.3.754-763

Ostadrahimi, A., Bifaretti, S.: A simplified multilevel space vector pulse width modulation (SVPWM) based on boundary lines, including overmodulation zone. IEEE Open J. Ind. Appl. 4, 215–226 (2023). https://doi.org/10.1109/OJIA.2023.3289094, http://www.ieeexplore.ieee.org/document/10164229

Prasad, M., Vinesh, A., Ashish, M.: THD minimization under low switching frequency for superior two level three phase inverter performance. Int. J. Electr. Electron. Res. (IJEER), pp. 1247–1255 (2022). https://doi.org/10.37391/ijeer.100476. http://www.researchgate.net/publication/366720700

Yasin, H.: Modelling and control of hybrid vehicle. Int. J. Innov. Technol. Interdiscip. Sci. 2(3), 212–222 (2019). https://doi.org/10.15157/IJITIS.2019.2.3.212-222

A Bibliometric Analysis of Agile Enterprise

Florenc Hidri and Gladiola Tigno

Abstract

Agile methodologies are becoming very popular in software development companies due to their flexibility, adaptability, and emphasis on collaboration. However, implementing agile methodologies in larger organizations—referred to as Agile Enterprises—presents unique challenges that must be carefully addressed. A lot of efforts have been made recently to implement Agile in Large enterprises. As a results methodology such as SAFe, LeSS and others have emerged. This study employs a bibliometric analysis to assess the status of research on Agile Enterprises. By analysing research papers from the Scopus database over the past five years, the study aims to identify trends, patterns, and gaps in the literature. Keywords used in the search criteria on the Scopus database to retrieve the data is "Agile Enterprise". The findings reveal that the concept of Agile Enterprises is gaining momentum, with research focusing on areas such as project management, scaling, governance, and digital transformation. Despite this growing interest, there is a notable scarcity of empirical studies that evaluate the effectiveness of Agile Enterprise adoption and its impact on organizational performance. This lack of empirical evidence highlights a critical gap in the literature. Consequently, this study not only provides a comprehensive review of the current state of Agile Enterprise research but also underscores the need for more empirical studies. These future studies

F. Hidri (✉)
Canadian Institute of Technology, Tirana, Albania
e-mail: florenc.hidri@cit.edu.al; fh31409@seeu.edu.mk

SEEU, South East European University, Tetovo, North Macedonia

G. Tigno
European University of Tirana, Tirana, Albania
e-mail: gladiola.tigno@uet.edu.al

K. Dhoska and E. Spaho (eds.), *Bridging Horizons in Artificial Intelligence, Robotics, Cybersecurity, Smart Cities, and Digital Economy*, Sustainable Economy and Ecotechnology, https://doi.org/10.1007/978-3-031-72029-1_7

are essential to aid decision-making processes regarding Agile Enterprise adoption and to understand its broader implications. The paper highlights future research directions and stresses the need for empirical evidence to support theoretical frameworks and models. By addressing these gaps, future research can significantly contribute to the practical and theoretical understanding of Agile Enterprise implementation and its outcomes.

Keywords

Agile enterprise • Bibliometric analysis • Research landscape • Trends • Patterns • Themes

1 Introduction

The Agile Enterprise concept, focusing on an organization's capability to swiftly adjust to market changes and customer demands, has received significant recognition in business and management research (Lee and Kim 2017) Organizations are progressively embracing Agile principles and methods, derived from software development, to enhance their overall performance, innovation potential, and competitiveness in dynamic and unpredictable business settings (Ambler and Lines 2012) Agile Enterprise focuses on adaptability, cooperation, innovation, and customer-focused approach, with the goal of helping organizations react swiftly and efficiently to evolving market situations, technological progress, and customer desires. (Denning and Dunham 2010; Highsmith 2002).

With the increasing popularity of Agile Enterprise in the business world, researchers have become more intrigued to investigate different facets of this trend. Scholars from various fields like business, management, information systems, and engineering have all played a part in shaping the Agile Enterprise literature, leading to a broad and interdisciplinary research field (O'Connor 2016). To grasp the present research status on Agile Enterprise, it is crucial to carry out a bibliometric analysis, which offers a methodical and quantitative method for examining academic publications and recognizing trends, patterns, and themes within a specific area (Kundu and Iyengar 2019; Pereira et al. 2017).

In this article, we introduce a bibliometric method for comprehending the research environment of Agile Enterprise. We analyze scholarly publications on Agile Enterprise from sources like academic databases, conference proceedings, and online repositories, including journal articles and conference papers. We employ bibliometric methods such as co-citation analysis, bibliographic coupling, and keyword analysis to discover the top authors, journals, and keywords, and to uncover the trends, patterns, and themes in Agile Enterprise research (Li et al. 2020).

2 Literature Review

Agile methodologies are gaining more popularity in software development because they offer flexibility, collaboration, and adaptability. Nonetheless, the implementation of agile approaches in big corporations poses distinct obstacles that need to be thoroughly thought through. The objective of this literature review is to give a summary of the latest research on Agile Enterprise through bibliometric analysis. A research was carried out on Google Scholar using the keywords "Agile Enterprise," "Agile transformation," "Agile governance," "Agile project management," and "Agile scaling" to find pertinent publications in the academic literature. The search only included writings from the most recent ten years (2017–2023). The study of bibliometrics showed a substantial rise in research on Agile Enterprise over the past ten years. The examination pinpointed various research topics such as agile transformation, scaling, governance, and project management. The examination also showed an increasing fascination with the idea of Agile Enterprise and its possible advantages for big corporations.

Multiple research projects have centered around transitioning to agile practices throughout an entire company, a process known as agile transformation. A study conducted by O'Reilly and Tushman indicated that transitioning to agile methods necessitates a substantial change in culture, including reassessing work methods and establishing a more flexible and adaptable organizational framework. There has been significant focus in the literature on implementing agile methodologies on a large scale in enterprises. (O'Reilly and Tushman 2013) A study conducted by Larman and Vodde in 2010 introduced the Large-Scale Scrum (LeSS) framework with the aim of extending agile methodologies beyond the team level (Larman and Vodde 2010).

Agile governance has also become a crucial focus of research. A governance framework for agile projects was suggested by Kettunen et al. (2019), focusing on collaboration, transparency, and adaptability. In different research conducted by Schneider and Sunyaev, the difficulties of incorporating agile governance in big businesses were outlined, along with a suggested list of principles for effective execution (Schneider and Sunyaev 2015).

The study of Agile Enterprise also emphasizes the importance of project management. Sutherland and Schwaber (2013) suggested the Scrum framework as a way to oversee agile projects, focusing on teamwork, interaction, and input. Different research conducted by Tavares et al. (2020) indicated that moving to agile project management entails changing mindset from conventional project management approaches, focusing on customer partnership, product excellence, and ongoing enhancements.

Scaled agile methodologies like SAFe and DAD are also being investigated in research as strategies for applying agile practices in big companies. Leffingwell introduced the SAFe framework in a study, suggesting it as a way to expand agile practices to numerous teams, programs, and portfolios by offering direction on how to synchronize and organize agile initiatives on a larger scale (Leffingwell 2010). A different study carried out by Ghaisas, Ambler, and Sharanappa reviewed existing literature to examine how SAFe implementation affects large organizations, emphasizing both the advantages and difficulties linked to its adoption (Ghaisas et al. 2019). In the same way, research conducted by Moe, Nygaard, and Seim investigated how the DAD framework was utilized in a major corporation, offering insights on its efficiency in handling complicated solutions (Moe et al. 2019). These models offer a systematic method for expanding agile practices in big companies by tackling issues in communication, coordination, and alignment. However, additional research is required to assess how effective these frameworks are in actual work environments and their influence on organizational results.

Although Agile Enterprise has gained interest, there is a shortage of empirical research assessing its effectiveness in adoption and its influence on organizational performance in the literature. This disparity underscores the necessity for additional empirical research to back decision-making in the implementation of Agile Enterprise.

3 Research Questions

RQ1: What is the overall growth and trend in research on Agile Enterprise over the past 5 years?

RQ2: Who are the most influential authors and what are their contributions to Agile Enterprise literature?

RQ3: Which journals publish the most Agile Enterprise research?

RQ4: What are the most frequently used keywords in Agile Enterprise literature?

4 Methodology

This research paper utilizes bibliometric methods to determine the present status of research on Agile Enterprise. The research examines papers from top academic journals and conferences listed in the Scopus database in agile methodologies and enterprise transformation. The procedure includes the subsequent stages:

Literature search: An extensive search of academic databases like Scopus will be performed to find pertinent publications. The search was conducted using the keyword Agile Enterprise. The research will only include materials published within the most recent 5 years (2017–2023).

Data extraction: The importance of publications was determined by evaluating their titles and abstracts. The chosen publications were later examined for their bibliographic details, including author, year of publication, journal/conference, and keywords.

Bibliometric analysis: The analysis of publications was conducted using R Studio and Biblioshiny software. The examination involved network analysis, co-occurrence analysis, and bibliographic coupling. Network analysis was employed to pinpoint the most dominant authors, journals, and keywords. Co-occurrence analysis was utilized to pinpoint the keyword pairs and clusters that were most commonly seen. Bibliographic coupling was employed to pinpoint the publications with the highest citation counts and the most significant research topics.

Data interpretation: The bibliometric analysis results were interpreted to uncover trends, patterns, and gaps within the literature. The analysis concentrated on determining the most studied subjects, the most important researchers, and the most impactful publications. The analysis also identified gaps in the literature, such as the need for more empirical studies to evaluate the effects of implementing Agile Enterprise (Fig. 1).

The bibliometric data, which comprises valuable information on scholarly publications, was exported from the renowned Scopus database. To gain insights and

MAIN INFORMATION ABOUT DATA	
Timespan	2017:2023
Sources (Journals, Books, etc)	153
Documents	265
Annual Growth Rate %	-9.8
Document Average Age	2.85
Average citations per doc	6.253
References	10927
DOCUMENT CONTENTS	
Keywords Plus (ID)	1496
Author's Keywords (DE)	925
AUTHORS	
Authors	837
Authors of single-authored docs	18
AUTHORS COLLABORATION	
Single-authored docs	19
Co-Authors per Doc	3.43
International co-authorships %	20.75
DOCUMENT TYPES	
article	152
conference paper	108
editorial	1
erratum	1
review	3

Fig. 1 Main Information About Data

uncover patterns within the data, rigorous analyses were conducted using powerful tools such as BiblioShine and RStudio, known for their advanced data analysis capabilities. The extracted data was carefully examined, analysed, and processed to generate meaningful results. These results are expected to provide valuable insights and contribute to the understanding of the research landscape in the respective field of study.

5 Results

5.1 Annual Scientific Production

Agile Enterprise has become a popular subject in recent years, as companies look to adjust and succeed in today's fast-changing business environment. The yearly publication of articles on this subject has continually risen, indicating the increasing significance and popularity of Agile methodologies in contemporary corporate settings.

In 2017, 26 articles were released discussing Agile Enterprise, indicating a modest beginning to studying this concept in both academia and industry. However, in the next year, 2018, there was a rise in the number of published articles to 29, showing a higher acknowledgment of Agile methods for attaining organizational agility and resilience.

In 2019, there was a notable increase in the number of articles published on Agile methodologies, with 40 articles reflecting a growing interest in applying them across different industries. In 2020, the trend persisted with the publication of 42 articles, demonstrating an ongoing fascination and investigation into Agile Enterprise practices to help organizations adapt efficiently to quickly evolving business circumstances.

During 2021, the number of articles published on Agile Enterprise increased to 52, reflecting a consistent expansion in academic and industry writings concerning this subject. In 2022, there was a significant increase with 61 articles published, showing continued interest in Agile practices within enterprise operations.

However, the publication of articles dropped to 14 in 2023, attributed to the brief timeframe (Fig. 2).

The articles published on the topic of Agile Enterprise cover a wide range of aspects, including case studies, best practices, frameworks, tools, and empirical research, providing valuable insights and guidance for organizations looking to embrace Agile principles and practices to achieve organizational success. The growing body of literature on Agile Enterprise reflects the evolving nature of business practices and the need for organizations to continuously adapt and evolve to remain competitive in today's dynamic and rapidly changing business landscape.

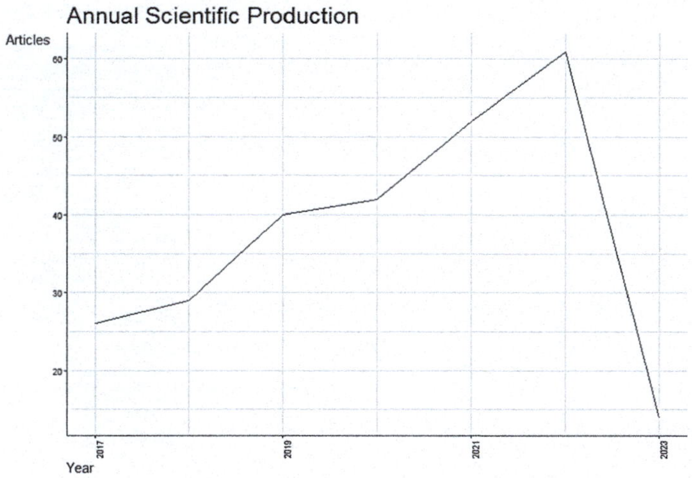

Fig. 2 Annual Scientific Production

5.2 Most Relevant Sources

According to bibliometric analyses, certain sources have been identified as highly important in the field of study. These sources have published a large number of articles, showing their significance in the literature concerning the subject. Some of the most relevant sources identified are:

- IOP Conference Series: Materials Science and Engineering—with 13 articles published, this conference series is a popular venue for research related to materials science and engineering, showcasing the latest advancements in this field.
- Procedia Computer Science—also with 13 articles published, this journal covers a wide range of topics related to computer science, including applications, methodologies, and technologies, making it a valuable source of research in this area.
- Sustainability (Switzerland)—with 13 articles published, this journal focuses on sustainability-related research, covering topics such as environmental science, sustainable development, and renewable energy, making it a prominent source for research on sustainable practices in various domains.
- Procedia CIRP—With 12 articles published, this journal focuses on research related to the field of Computer Integrated Manufacturing and provides insights into the latest developments in this area.
- IEEE Access—With 8 articles published, this open-access journal covers a wide range of topics related to electrical engineering, computer science, and information technology, making it a valuable source for interdisciplinary research in these fields (Fig. 3).

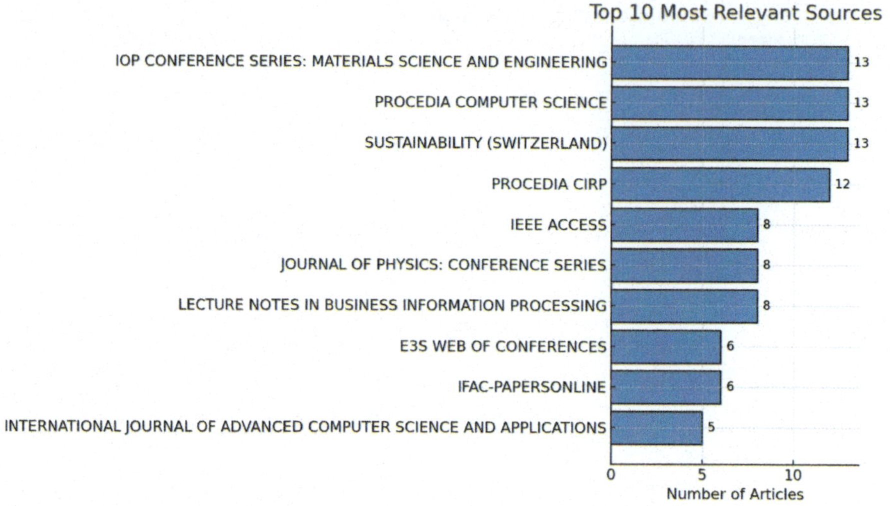

Fig. 3 Most Relevant Sourses

These sources have contributed significantly to the literature on the topic of study, and researchers may find them valuable for gaining insights, accessing the latest research, and identifying emerging trends in the field.

5.3 Sources' Local Impact

The local effects of the sources, as determined by different bibliometric measures, offer understanding into their impact and importance among the research community. Below are the specifics regarding the local consequences of the sources referenced (Fig. 4).

These local impact metrics provide a quantitative measure of the influence and relevance of these sources in the research community, with higher values indicating greater impact. Researchers can use these metrics to assess the quality and significance of research published in these sources when conducting their literature review or identifying potential sources for their own research.

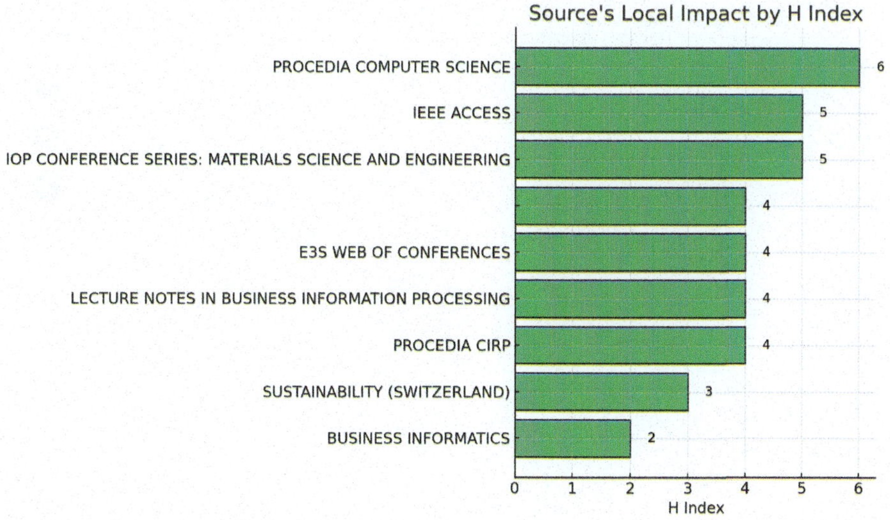

Fig. 4 Sourse' Local Impact by H Index

5.4 Most Relevant Authors

Some of the most relevant authors and their articles:

- KOTTKE M: This author has contributed 6 articles and has a fractionalized articles count of 1.916666667, indicating a significant contribution to the field.
- POTH A: This author has also contributed 6 articles with a fractionalized articles count of 1.916666667, indicating a substantial contribution to the field.
- RIEL A: Similarly, this author has contributed 6 articles with a fractionalized articles count of 1.916666667, indicating a significant contribution to the field.
- DE VRIES M: This author has contributed 4 articles with a fractionalized articles count of 2, indicating a high level of contribution to the field.
- KRENCZYK D: This author has contributed 3 articles with a fractionalized articles count of 1.5, indicating a notable contribution to the field.

In general, these writers have added greatly to the field through their research papers, as shown by their number of articles and fractionalized article counts. Their research results indicate their knowledge and active participation in their specific area of study (Fig. 5).

5.5 Most Relevant Affiliations

The most relevant affiliations and their article counts:

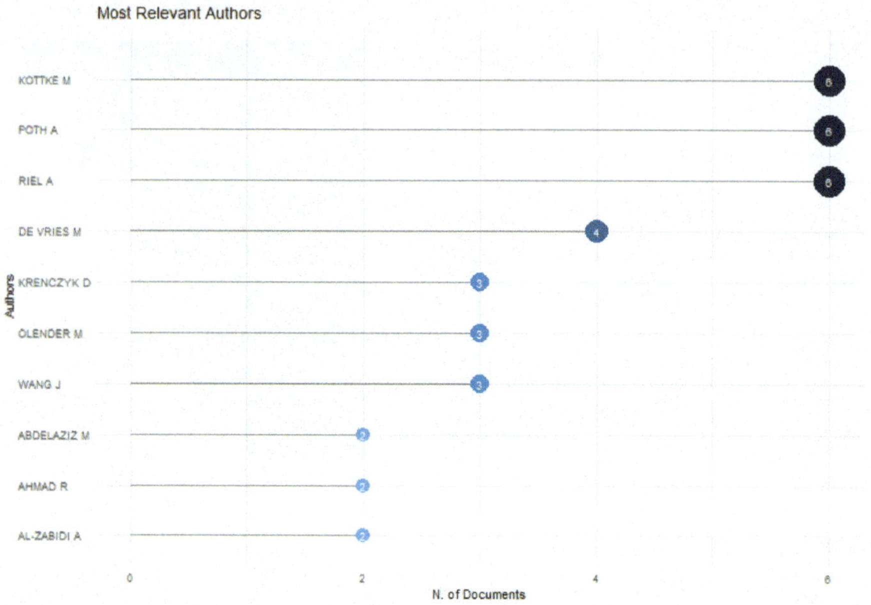

Fig. 5 Most Relevant Authors

- RWTH Aachen University: This affiliation has contributed 13 articles, indicating a significant research output in the field.
- University of Texas Southwestern Medical Center: This affiliation has contributed 12 articles, indicating a substantial research output in the field.
- Guizhou University of Finance and Economics: Similarly, this affiliation has contributed 9 articles, indicating a notable research output in the field.
- The Third Affiliated Hospital of Sun Yat-sen University: This affiliation has also contributed 9 articles, indicating a notable research output in the field.
- King Saud University: This affiliation has contributed 8 articles, indicating a significant research output in the field.
- Swiss Federal Laboratories for Materials Science and Technology: Similarly, this affiliation has contributed 8 articles, indicating a significant research output in the field.

These affiliations have displayed a notable amount of research productivity in the area of study, as seen in the number of articles they have produced. They are actively participating in research and helping advance knowledge in the field. The expertise and dedication of these affiliations are evident in their research contributions to the field of study (Fig. 6).

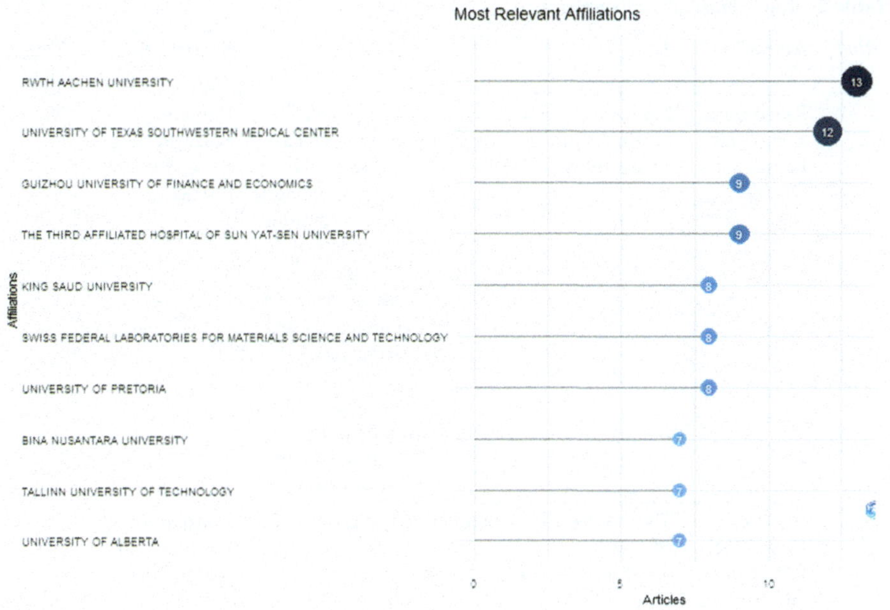

Fig. 6 Most Relevant Affiliations

5.6 Most Cited Documents

See Table 1.

5.7 Most Used Keywords

Agile Manufacturing Systems (Occurrences: 59): Agile manufacturing prioritizes flexibility, responsiveness, and adaptability to stay in line with changing customer demands and market conditions.

Lean Production (Occurrences: 28): Lean Production is a strategy focused on minimizing inefficiencies, optimizing resource utilization, and continuously improving processes to enhance the value delivered to customers.

Software Design (Occurrences: 23): The study makes 23 references to software design. It involves the steps of imagining, organizing, and developing software systems.

Small and Medium-sized Enterprise (Occurrences: 19): SMEs are referenced 19 times in the research. Small and medium-sized enterprises (SMEs) have fewer employees and generate lower profits than large corporations. Information Systems (Occurrences: 17): The study mentions information systems a total of 17 times. All elements related to collecting, storing, organizing, analyzing, and distributing information within a company make up the system (Fig. 7).

Table 1 Top 5 most cited documents

Rank	Author (s)	Title	Summary	Source title	Times cited
1	Paasivaara, Behm and Lassenius (2018)	Large-scale agile transformation at Ericsson: a case study	This paper presents a case study on the large-scale agile transformation at Ericsson, providing insights into the challenges and strategies involved in implementing agile practices within a large organization	Empire Software Eng 23, 2550–2596 (2018)	79
2	Jovovic, Draskovic, Delibasic, and Jovovic (2017)	The concept of sustainable regional development—institutional aspects, policies and prospects	The proposed paper suggests tools and a framework to guide regional development, trade-offs, and institutional cooperation for a balanced approach that meets economic and social needs while reducing environmental impact	Journal of International Studies 10(1), 255–266	66
3	Chen (2017)	Continuous Delivery: Overcoming adoption challenges	This paper explores the challenges faced during the adoption of continuous delivery practices, providing strategies and solutions to overcome these challenges	Journal of Systems and Software, 128(Complete), 72–86	65

(continued)

Table 1 (continued)

Rank	Author (s)	Title	Summary	Source title	Times cited
4	Antosz and Stadnicka (2017)	Lean philosophy implementation in smes—study results	The results of the study show that a significant number of small and medium-sized enterprises (SMEs) are willing to adopt Lean Manufacturing (LM) philosophy. These companies recognize the importance of enhancing their operations and eliminating waste	Procedia Engineering	48
5	Romero and Stahre (2021)	Towards the resilient operator 5.0: The future of work in Smart Resilient Manufacturing System	This paper discusses the concept of the resilient operator 5.0, exploring the future of work in Smart Resilient Manufacturing Systems and the implications for the manufacturing industry	Procedia CIRP, 104, pp. 1089–1094	46

6 Conclusion

In summary, this study performed an comprehensive bibliometric analysis to explore the research field of Agile Enterprise. The results show a rising number of articles about Agile Enterprise, demonstrating a growing interest in this area. The IOP Conference Series: Materials Science and Engineering was determined to be the most important resource in this study, highlighting its importance in the field of Agile Enterprise literature. This platform is a valuable resource

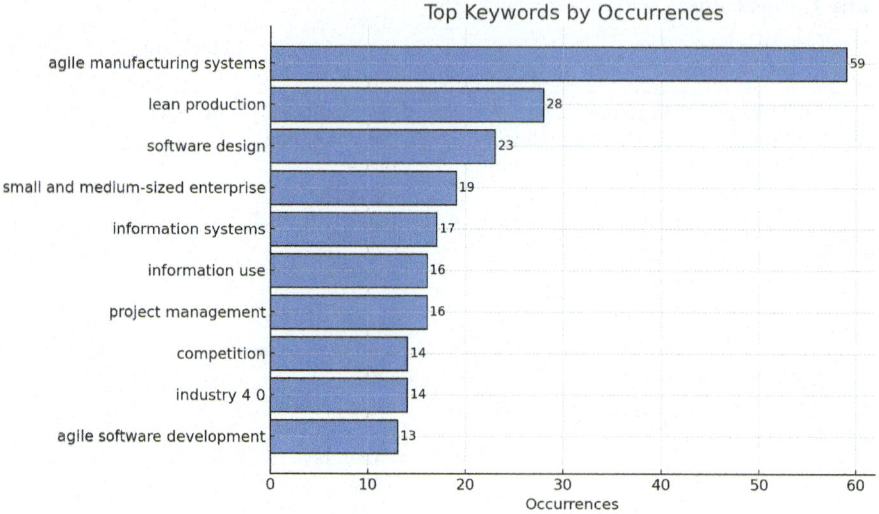

Fig. 7 Most Relevant Words

for researchers to exchange their discoveries and insights regarding Agile Enterprise. The examination also discovered that POTH A was the most productive author in this field of research, demonstrating their important contributions to the literature on Agile Enterprise. Moreover, the primary association discovered was RWTH AACHEN UNIVERSITY, indicating their significant participation in Agile Enterprise research. Poland stood out as the country with the highest number of citations, showing its significance in the academic conversations about Agile Enterprise. Furthermore, the article titled "Large-scale Agile Transformation at Ericsson: A Case Study" by Paasivaara, Behm, and Lassenius, published in Empirical Software Engineering in 2018, was the most referenced. It has received 79 citations (Paasivaara et al. 2018).

Agile manufacturing systems, lean production, software design, and small and medium-sized enterprises are the key keywords pinpointed in this analysis. These key terms reflect important topics and research priorities in the Agile Enterprise field.

This bibliometric analysis offers important information about the research field of Agile Enterprise. The findings illuminate the growing interest in this area, pinpoint important contributors and resources, showcase influential articles, and pinpoint key research areas through relevant keywords. These results add to the current understanding and can be used as a starting point for additional research and examination in the field of Agile Enterprise.

This study highlights the importance of customizing agile methods to suit particular organizational environments.

Further research, such as case studies and scientific assessments, is advised to enhance comprehension of agile customization and offer practical advice for organizations aiming to utilize agile methodologies effectively in their specific circumstances.

References

Ambler, S. W., Lines, S.: Disciplined agile delivery: A practitioner's guide to agile software delivery in the enterprise. IBM Press (2012)

Antosz, K., Stadnicka, D.: Lean philosophy implementation in SMEs—study results. Procedia Engineering **182**, 25–32 (2017)

Chen, L.: Continuous delivery: overcoming adoption challenges. J. Syst. Softw. **128**, 72–86 (2017). https://doi.org/10.1016/j.jss.2017.02.013

Denning, S., Dunham, R.: The innovator's way: Essential practices for successful innovation. MIT Press (2010)

Ghaisas, S., Ambler, S., Sharanappa, S.: Benefits and challenges of SAFe implementation: A systematic literature review. In: Proceedings of the 20th International Conference on Agile Software Development (XP 2019), pp. 71–86 (2019)

Highsmith, J.: Agile software development ecosystems. Addison-Wesley (2002)

Jovovic, R., Draskovic, M., Delibasic, M., Jovovic, M.: The concept of sustainable regional development—institutional aspects, policies and prospects. J. Int. Stud. **10**(1), 255–266 (2017). https://doi.org/10.14254/2071-8330.2017/10-1/18

Kettunen, P., Karimäki, E., Kautz, K.: Agile governance in IT projects: A framework for agile decision-making. Inf. Syst. J. **29**(5), 1000–1028 (2019)

Kundu, A., Iyengar, A.: Bibliometric analysis of Industry 4.0 research: A scientometric approach. J. Manuf. Technol. Manag. **30**(7), 1171–1194 (2019)

Larman, C., Vodde, B.: Scaling lean and agile development: Thinking and organizing in large complex projects. Addison-Wesley (2010)

Lee, J., Kim, Y.G.: How firms can effectively implement Agile project management: A bibliometric analysis. Sustainability **9**(5), 690 (2017)

Leffingwell, D.: Agile software requirements: lean requirements practices for teams, Programs, and the Enterprise. Addison-Wesley (2010)

Li, Y., Liu, Z., Chen, Y., Zhang, L., Gao, Y.: Bibliometric analysis of agile project management research: A scientometric approach. Int. J. Inf. Manage. **50**, 56–70 (2020)

Moe, N. B., Nygaard, D., Seim, E.: Scaling agile in a large organization: a case study of disciplined agile delivery. In: Proceedings of the 2019 ACM/IEEE International Symposium on Empirical Software Engineering and Measurement (ESEM), pp. 1–6, (2019)

O'Connor, R. V.: Agile maturity model for software process improvement in small and medium enterprises. Softw.: Pract. Exp. **46**(2), 149–171 (2016)

O'Reilly, C.A., Tushman, M.L.: Organizational ambidexterity: Past, present, and future. Acad. Manag. Perspect. **27**(4), 324–338 (2013)

Paasivaara, M., Behm, B., Lassenius, C., et al.: Large-scale agile transformation at Ericsson: a case study. Empir. Softw. Eng. **23**, 2550–2596 (2018)

Pereira, R., MacDonell, S.G., Silva, C.E., Sampaio, R.B.: Bibliometric analysis of Agile project management literature. Int. J. Inf. Manage. **37**(6), 746–755 (2017)

Romero, D., Stahre, J.: Towards the resilient operator 5.0: The future of work in smart resilient manufacturing systems. Procedia CIRP **104**, 1089–1094 (2021). https://doi.org/10.1016/j.procir.2021.11.183

Schneider, C., Sunyaev, A.: Implementing agile governance in IT projects: An empirical analysis. Int. J. Project Manage. **33**(4), 772–783 (2015)

Privacy-Enhancing Elliptical Curve Cryptography Techniques in Ambient Intelligence

Anxhela Baraj

Abstract

Ambient Intelligence refers to the intelligent computing and sensing equipment that comprehensively surrounds human being, providing them with personalized context-aware, adaptive, and predictive services. As ambient intelligence extends to various aspects of human life, data security and privacy have become crucial problems. In response to these pressing privacy issues, this study explores the incorporation of privacy-enhancing elliptic curve cryptography (ECC) methods into ambient intelligence systems. One of the main advantages of Elliptic Curve Cryptography is its strong security and efficient computation. This advantage can be used as a promising solution for safeguarding sensitive information. The Elliptic Curve Discrete Logarithm Problem demonstrates ECC's inherent computational challenges and is the underlying reason for its powerful security features. The study analyses the Elliptic Curve Cryptography foundations, highlighting core constructs and operations, such as key generation, encryption, decryption, and digital signatures in preserving user privacy in smart settings. These Elliptical Curve Cryptography constructs/operations can be used to efficiently handle privacy challenges of many components of ambient intelligence systems, promoting a safe and privacy-conscious environment for users and devices.

Keywords

Ambient intelligence • Elliptic curve cryptography • Privacy challenge • Privacy-enhancing techniques • Encryption • Decryption • Digital signature

A. Baraj (✉)
Canadian Institute of Technology, Tirana, Albania
e-mail: anxhela.baraj@cit.edu.al

K. Dhoska and E. Spaho (eds.), *Bridging Horizons in Artificial Intelligence, Robotics, Cybersecurity, Smart Cities, and Digital Economy*, Sustainable Economy and Ecotechnology, https://doi.org/10.1007/978-3-031-72029-1_8

1 Introduction

Significantly, the rapid development of technology during the last few years has sparked the widespread use of ambient intelligence systems indefinitely changing the way we interact with the surrounding environment. Ambient Intelligence refers to a "paradigm with computing devices and intelligent sensors pervasively embedded in the got environment providing information and services whenever and wherever needed" (Ramos 2009).

These environments need to be aware of people's needs, adapt requirements, and predict behaviors. Ambient intelligence systems are capable of creating an interconnected network that serves to collect significant amounts of data. Their ability expands even more in the processing of these data, and their subsequent use to serve human needs. Ambient intelligence promises to improve our overall living experience by providing flexibility, efficiency, and even safety for people in everyday environments such as homes, offices, meeting rooms, schools, hospitals, control centers, transports, touristic attractions, stores, sport installations, and music devices. However, with the pursuit of these extraordinary benefits come significant concerns regarding the privacy and security of personal information.

This pervasive nature of ambient intelligence systems raises serious privacy challenges. As I mentioned above, these systems continuously collect data from various smart devices, and the processing of this data leads to the generation of very sensitive information about the behaviors, preferences and daily routines of individuals, which can be used for personalized services, such as intelligent automation and predictive assistance, but also opens the door to possible misuse and unauthorized access.

One of the ways to address these privacy concerns within ambient intelligence solutions is cryptography. Ambient intelligence systems can benefit from various cryptographic techniques to strengthen user privacy. The cryptographic possibilities are varied, and some of the options are provided down below: homomorphic encryption, secure multi-party computation (SMPC), attribute-based encryption (ABE), secure hashing algorithms, elliptic curve cryptography (ECC), etc. (Cook et al. 2009). The most important thing about each cryptographic techniques is that they can help to improve data privacy in ambient intelligence systems as long as sensitive information remains confidential and free from hostile actions of malicious third parties. Elliptic curve cryptography is one of the instruments that have become possible for preserving sensitive data within ambient intelligence systems.

This paper will outline the possible uses of ECC within ambient intelligence use cases and the benefits ECC can offer to end users. Elliptic Curves is a modern family of public-key cryptosystems allowing a secure exchange of keys. ECC is based on the properties of the elliptic curves over finite fields and uses the Elliptic Curve Discrete Logarithm Problem (ECDLP) as a mathematical basis (Vliet 2016) ECC allows different privacy preservation solutions, including data encryption, digital signatures, and key exchange mechanisms. The difference is that, instead of using very large prime numbers, ECC uses the properties of the elliptic curve equation to generate keys. That is why ECC is suitable for ambient intelligence as a solution used in environments with limited resources.

2 Foundations of Elliptic Curve Cryptography (ECC)

Elliptic curve cryptography, as a cryptographic technique, was first instigated in the mid-1980s by Koblitz, and Miller (1985) as a possible substitute for cryptographic schemes reliant on the discrete logarithm problem in the multiplicative set of a finite field. Since those times, this cryptographic technique has grown in importance and use, offering efficient and powerful security measures for various areas. Elliptic curve cryptography is a type of public key cryptography that is based on the algebraic structure of elliptic curves over finite fields (Menezes et al. 1996). ECC offers stronger security and faster computation than other traditional public-key cryptosystems such as RSA and Diffie-Hellman, whose keys are generated as the product of very large prime numbers (Tutănescu et al. 2012). ECC uses smaller keys and signatures than RSA and Diffie-Hellman for the same level of security and provides very fast key generation, fast key agreement, and fast signatures.

The main strength of ECC lies in the difficulty of the Elliptic Curve Discrete Logarithm Problem (ECDLP). ECDLP states that given a point P on an elliptic curve and a public point $Q = kP$, it is computationally infeasible to determine the private scalar k even with knowledge of P and Q (Menezes 2001). The foundation of ECC's security relies on this inherent computational complexity.

2.1 Key Components of Elliptic Curve Cryptography

To enable secure cryptographic operations, ECC relies on several components that have a significant role in ensuring confidentiality, integrity, and authentication. In this part, we will present a summary of these components, including their roles and importance in elliptic curve cryptography.

1. Key generation

The first operation of elliptic curve cryptography is creating a unique pair key occurrence key for each user. This pair consists of a public key and a corresponding private key. The random private key is selected, and the corresponding public-key is computed using the elliptic curve operation (Froehlich 2022). Usually the public–private key pair is represented as (Q, d), where Q is the public key and d is the corresponding private key. The public key—Q is usually represented as a point on the elliptic curve and is derived through scalar multiplication of the generator point by the private key: $Q = d * P$, where "P" is the generator point, as represented in Fig. 1. In contrast, the private key is a scalar value and must be kept confidential as it is used for cryptographic operations. The message is encrypted

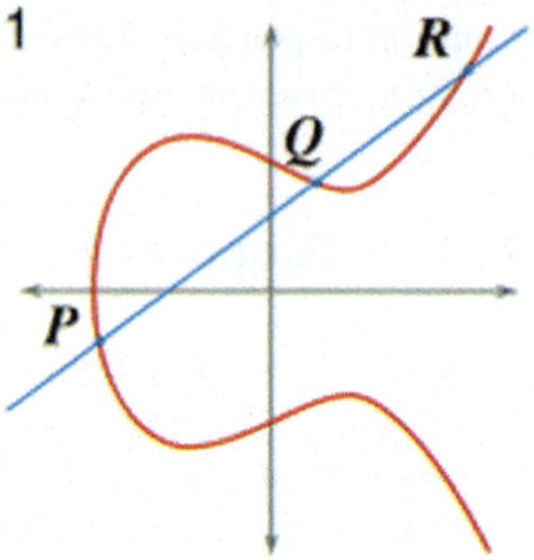

Fig. 1 Elliptic curve (Networks 2023)

by the sender with the receiver's public key, and the receiver decrypts it with his private key.

Private keys in ECC are integers (typically 256-bit integers). An example of a 256-bit ECC private key (hex encoded, 32 bytes, 64 hex digits) is:

Q = 0 × 51897b64e85c3f714bba707e867914295a1377a7463a9dae8ea6a86394 (Nakov 2018).

The public keys in ECC are EC points—pairs of integer coordinates {x, y}, located on the curve. Due to their special properties, EC points can only be compressed to a coordinate +1 bit (odd or even). Thus, the compressed public key, corresponding to a 256-bit ECC private key, is a 257-bit integer. Example ECC public key (corresponding to the above private key, encoded in Ethereum format, as hex prefix 02 or 03) is:

d = 0 × 02f54ba86dc1ccb5bed0224d23f01ed87e4a443c47fc690d747d (Nakov2018).

2. Encryption and decryption

Cryptographic processes such as encryption and decryption are used to protect communication and user data. Data can be encrypted using the recipient's public key and decrypted using the user's private key with the help of the ECC public key encryption algorithm. Decryption is the reverse process, where the shared secret is derived using the sender's public key and the receiver's private key.

Key agreement and symmetric encryption are the two main ECC-based encryption processes. Combining the recipient's public key (Q) with the sender's private key (d) to create a shared ciphertext results in key agreement (AlMajed and AlMogren 2020).

The ECC encryption and decryption processes steps are summarized in a simplified manner as follows:

Encryption (Amounas and Kinani 2012):

- The sender generates a random scalar value "k" and derives the elliptic curve point C1 = k * P to encrypt a message M for its intended recipient.
- A shared secret point S = k * Q_receiver is then calculated by the sender.

- Both the encryption key (Ke) and the message authentication code key (Kmac) are derived from the shared secret point S.
- The ciphertext C2 is created by encrypting the message M with the encryption key Ke and a symmetric encryption method (like DES, AES, etc.).
- The sender sends the encrypted message (C1, C2) to the intended receiver.

Decryption (Amounas and Kinani 2012):

- To decrypt a text encrypted with ECC, a pair of keys (Q, d) is required.
- The recipient's private key (d), which is kept secret, and the corresponding public key used during encryption (Q = d * P), are both owned by the recipient.
- The receiver determines S = d * C1, the common secret point.
- The receiver derives the encryption key (Ke) and the message authentication code key (Kmac) using the shared secret point S.
- The original message M is obtained by the receiver by decrypting the ciphertext C2 with the encryption key Ke.

3. Digital signature

ECC is also known to particularly generate digital signatures, which on the other hand is popular for ensuring data integrity, authentication, and non-repudiation (N. I. o. S. a. 2023). The digital signature of the message is created by the sender using the message's hash value and private key. On the other hand, signature's legitimacy is confirmed by the receiver using the sender's public key and the message's hash.

2.2 Advantages of Elliptic Curve Cryptography

Elliptic curve cryptography (ECC) is a relatively modern option compared to other cryptographic methods. The advantages of ECC, such as its strong security, computational efficiency, and versatility, make it a compelling choice for modern cryptographic applications, especially in contexts where resource constraints and security considerations are critical factors. The most significant advantages of ECC are introduced in this subsection.

Shorter key-size: The keys generated by ECC are much shorter than the keys used by traditional cryptographic algorithms such as RSA. However, the level of security offered by ECC is the same, or even higher compared to these algorithms (Saho and Ezin 2020).

This results in significant computational savings, reduced memory requirements, and faster encryption and decryption processes. The efficiency of ECC is particularly useful in resource-constrained environments, such as IoT devices and mobile devices, where computing power and memory are limited.

Compactness and space efficiency: Compared to other public key cryptography methods, ECC keys are significantly shorter. As a result, encryption and digital signatures have reduced data size, which reduces the amount of data transferred and stored. Compactness is very useful for systems with limited storage and limited bandwidth communication paths (Karki 2014).

Faster computational speeds: ECC procedures use direct algebraic operations on elliptic curve points. Due to their computational efficiency, these processes handle cryptographic data faster. Therefore, ECC is a good choice for real-time applications and circumstances that require fast cryptographic operations.

Reduced power consumption: Devices using ECC-based cryptographic operations have lower power consumption due to the computational efficiency of ECC. This benefit is particularly important for battery-powered devices, because extending battery life requires energy efficiency.

3 Privacy Challenges in Ambient Intelligent Systems

Ambient Intelligence (AmI) systems are otherwise known as "Smart Environments" or "Pervasive Computing". These systems are capable of creating intelligent environments with the aim of improving the user experience (Matjaza et al. 2019). At the heart of these systems is the integration of various smart devices, sensors, actuators and communication technologies to create an intelligent environment, able to understand and respond to the needs and preferences of users without problems.

The main capabilities of these AmI systems include the large amount of data collection and processing, and the seamless integration of smart devices. However, pursuing these amazing capabilities of these systems raises important privacy issues. Ensuring data protection and user privacy is essential to build trust and user acceptance. This section will summarize the key privacy challenges that arise in Ambient Intelligence systems.

Data collection and supervision

The enormous data collecting and surveillance that smart ambient systems carry out is one of their biggest privacy problems. To build context-aware environments, these systems continually collect data from a variety of sensors, smart devices, and ambient sources. Although this data is necessary for offering individualized services, it raises concerns about continuing surveillance and potential invasions of people's privacy (Streitz et al. 2019).

Behavioral profiling and tracking

Advanced algorithms are used by intelligent ambient systems to build user profiles and monitor patterns of behavior. Individuals may not want to reveal sensitive information, therefore profiling individuals based on their preferences, routines, and habits might result in privacy violations.

Data security

Significant challenges to data security are presented by the vast amounts of data that can be collected and processed in smart ambient settings. Individuals may suffer significant consequences as a result of data breaches and illegal access to sensitive information, which can result in identity theft, financial fraud, and other types of abuse (Tian et al. 2022). The lack of data security can also come as a result of insufficient encryption. The usage of inappropriate or weak encryption algorithms can allow attackers to bypass security measures and gain access to sensitive information.

Lack of user awareness and consent

It is possible that many consumers are not aware of how much data the smart home systems around them collect and process. The ability of users to give informed consent to the use of their data can be hindered by the lack of transparency and complexity of data flows, potentially resulting in privacy violations (Caire et al. 2016).

Termination of sensitive information

In smart ambient environments, gathering and analyzing apparently unimportant data points might reveal vital information. For instance, a person's lifestyle and preferences may be inferred from geographical data, behavioral patterns, and contextual information combined (Caire et al. 2016).

Compliance with privacy regulations

For businesses, organizations and/or institutions implementing ambient intelligence systems, complying with privacy requirements such as the General Data Protection Regulation (GDPR) and other local data protection legislation can be difficult. In order to maintain users' right to privacy, data processing techniques must be made to comply with legal obligations.

To overcome these privacy challenges and protect sensitive user data, it is essential to apply privacy-enhancing techniques.

4 Privacy-Enhancing Techniques Using Elliptic Curve Cryptography in Ambient Intelligence

One way to address privacy issues in smart environment systems is to use stable cryptographic algorithms. Elliptic curve cryptography provides strong security and thus increases the level of user privacy by using the intrinsic computational difficulty of the Elliptic Discrete Logarithm Curve (ECDLP). ECC is considered particularly suitable for ambient intelligence systems involving resource-constrained environments, also as a result of its efficiency in terms of key size and computational complexity.

At the core of this paper is the exploration of different ways in which ECC can be used to improve privacy and data protection in ambient intelligence systems.

4.1 Secure Data Encryption and Decryption

As explained in Sect. 2.1, one of the key components of elliptic curve cryptography is the secure encryption and decryption process, compared to other traditional public key encryption methods. This advantage of ECC can be applied in ambient intelligence systems to provide reliable and effective methods for secure data encryption and decryption (AlMajed and AlMogren 2020). ECC protects data security and privacy during data transmission and storage using the intrinsic computational complexity of the Elliptic Curve Discrete Logarithm Problem (ECDLP).

ECC-based encryption and decryption algorithms also allow secure user and smart device communication. ECC can be used by ambient intelligence systems to create a private and secure ecosystem that protects sensitive data and promotes user trust. The use of ECC-based key exchange protocols allows secure deployment of shared secrets, ensuring that only authorized parties have access to encrypted data (Roy and Khatwani 2017).

It is essential to integrate ECC with effective key management, safe key storage, and robust cryptographic protocols in order to guarantee the general security and privacy of ambient intelligence systems. ECC may successfully protect data, protect privacy, and promote confidence in the communication and data exchange processes within ambient intelligence settings by adhering to these rules.

4.2 Secure Digital Signature

A solid foundation for creating secure digital signatures in ambient intelligence systems is provided by elliptic curve cryptography (ECC). By ensuring that messages or transactions come from a trusted source and have not been compromised in transit, digital signatures are essential to confirm the validity and integrity of data. Strong security, reduced key length, and resistance to quantum attacks are some benefits of ECC-based digital signatures (Muthukuru 2016).

The digital signature component of ECC offers specific ways to enhance privacy within ambient intelligent systems, as follows:

Secure identity verification: Secure identity verification is made possible by the digital signature feature of ECC without the disclosure of sensitive personal data. Users can digitally sign communications, guaranteeing the accuracy of the material and confirming their identities without disclosing personal information.

Message signing: When a user or device in the ambient intelligence environment wants to send a message or data, they generate a digital signature using the private key (secret scalar) “d”, as explained in Sect. 2.1.

Message verification: After the message reaches the sender with the digital signature, using the sender’s public key “Q”, obtained from a trusted source (such as a public key infrastructure or a previously established trust relationship) uses the public key “Q” and signature value “s” to run an ECC-based verification procedure.

Trust and Non-repudiation: ECC digital signatures offer non-repudiation, making sure that senders cannot retract their acts. In interactions using ambient intelligence, this fosters accountability and trust.

5 Conclusion

The development of ambient intelligence systems, where connected devices are seamlessly integrated into our lives, has raised important issues regarding data privacy and security. In this study, the adaptation and incorporation of Elliptic Curve Cryptography (ECC) techniques to improve these privacy issues in ambient intelligence systems is studied.

ECC's robust security is built on the complexity of ECDLP, which makes it a great choice for ambient intelligence applications that respect user privacy.

It is shown that ECC has benefits in terms of effectiveness, quantum resistance, and compatibility for devices with limited resources. These characteristics guarantee that ECC is not only safe but also appropriate for the varied and changing ambient intelligence environment.

ECC has proven to be effective in resolving certain privacy challenges such as risks associated with data security breaches, user identity protection, and the requirement for data aggregation without sacrificing personal privacy. Because of its capacity to provide forward secrecy, even in the event of long-term key breaches, user data is always protected.

ECC techniques, including secure data encryption and decryption, secure message signing and verification, secure identity verification and trust and non-repudiation, have proven fruitful in ambient intelligent systems in terms of privacy enhancement.

References

Amounas, F., Kinani, E.H.E.: ECC encryption and decryption with a data sequence. Appl. Math. Sci. **6** (2012)

AlMajed, H., AlMogren, A.: A secure and efficient ECC-based scheme for edge computing and internet of things. MDPI Sens. (2020)

Caire, P., Moawad, A., Efthymiou, V., Bikakis, A.: Privacy challenges in ambient intelligence systems. J. Ambient Intell. Smart Environ. (2016)

Cook, D.J., Wrede, J.C.A., Jakkula, V.R.: Review: ambient intelligence: technologies, applications, and opportunities. J. Ambient Intell. Res. (2009)

Froehlich, A.: Elliptical Curve Cryptography (ECC). Andrew Froehlich (2020)

Karki, A.: Elliptic curve cryptography—a new approach to public key cryptography. Int. J. Eng. Sci. Res. (2014)

Matjaza, G., Yu-Huab, G.I., Akic, H., Andrésd, M., Vincente, T.: Artificial intelligence and ambient intelligence. J. Ambient Intell. Smart Environ. (2019)

Menezes, A., Oorschot, P.V., Vanstone, S.: Key management techniques. In: Handbook of applied cryptography. CRC Press (1996)

Miller, V.S.: Use of elliptic curves in cryptography. In: Advances in Cryptology—CRYPTO '85 Proceedings (1985)

Menezes, A.: Evaluation of Security Level of Cryptography: The Elliptic Curve Discrete Logarithm Problem. University of Waterloo (2001)
Muthukuru, S.K.M.: Secure digital signature scheme based on elliptic curves for internet of things. Int. J. Electr. Comput. Eng. (2016)
N. I. o. S. a (2023) Technology digital signature. https://csrc.nist.gov/glossary/term/digital_signature#:~:text=signer%20non%2Drepudiation.,-Sources%3A&text=An%20asymmetric%20key%20operation%20where,repudiation%2C%20but%20not%20confidentiality%20protection
Nakov, S.: Elliptic Curve Cryptography (ECC). In: Practical Cryptography. MIT License (2018)
Networks, A.: Elliptic Curve Cryptography. AVI Networks (2023)
Ramos, C.: What is ambient intelligence. In: Encyclopedia of Artificial Intelligence. IGI Global (2009)
Roy, S., Khatwani, C.: Cryptanalysis and improvement of ECC based authentication and key exchanging protocols. In: International Conference on Computational Intelligence and Communication Networks (CICN) (2017)
Saho, N.G.S., Ezin, E.C.: Comparative study on the performance of elliptic curve cryptography algorithms with cryptography through RSA algorithm. In: African Conference on Research in Computer Science and Applied Mathematics (2020)
Streitz, N., Charitos, D., Kaptein, M., Böhlen, M.: Grand challenges for ambient intelligence and implications for design contexts and smart societies. J. Ambient Intell. Smart Environ. **11**, 87–107 (2019)
Tutănescu, I., Anton, C., Ionescu, L., Caragaţă, D.: On assurance of information security using elliptic curves cryptosystems. J. Internet Technol. Secur. Trans. (JITST) (2012)
Tian, H., Jian, Y., Ge, X.: Blockchain-based AMI framework for data security and privacy protection. Sustain. Energy, Grids Netw. (2022)
Vliet, G.V.: The Use of Elliptic Curves in Cryptography. Radboud University (2016)

Position Control of Hybrid Stepper Motor

Alfred Pjetri, Astrit Bardhi, Gentian Dume, and Bajram Leka

Abstract

The stepper motor, power supply, and load mechanism comprise an electric drive where the moving part executes discrete step movements following voltage pulses generated by the control system. It's essentially an electric drive with angular position control, where each voltage impulse corresponds to a change in the rotor's position. Unlike other motors, stepper motors don't require feedback signals to determine rotor angular displacement, offering a significant advantage. This paper focuses on open-loop control for stepper motor control, emphasizing the preparation of suitable synchronous phase voltages to manage rotor angular displacement. The study includes modeling in the Matlab/Simulink environment and experimental implementation of a hybrid stepper motor position control system, confirming stepping behavior through both simulation and experimentation. The control algorithm utilizes an Arduino processor, while a PWM driver A4988 is employed to supply the hybrid stepper motor winding coils in the experimental setup. Furthermore, the ability to control the direction of rotation of the stepper motor is demonstrated. The simulation results align closely with experimental findings.

Keywords

Stepper motor position control • Stepper motor control • Hybrid stepper motor

A. Pjetri (✉) · A. Bardhi · G. Dume · B. Leka
Fakulteti Inxhinierisë ELektrike, Universiteti Politeknik i Tiranës, Tirana, Albania
e-mail: alfredpjetri@gmail.com

K. Dhoska and E. Spaho (eds.), *Bridging Horizons in Artificial Intelligence, Robotics, Cybersecurity, Smart Cities, and Digital Economy*, Sustainable Economy and Ecotechnology, https://doi.org/10.1007/978-3-031-72029-1_9

1 Introduction

Stepper motors are important in low power industrial applications as they are characterized by simplicity of construction and low cost. In stepper motors, the moving part called rotor, performs step movements (discrete movements), after each voltage pulse generated by the control system which, through the driver, supplies sequentially the phases of the stator winding according to a certain algorithm. Stepper motors are used in low-cost, and low-dynamics industrial applications where position control is required. For stepper motor operation it is not needed rotor feedback position (Teodor-Ionut et al. 2023; Pop and Maer 2021; Ionica et al. 2017). The stepper motor has several advantages, where it can mention:

- The motor step angle is proportional with the input pulses.
- They have precise positioning and repeatability of movement, since stepper motors have a position error in the steady state usually no greater than 3–5% and this error does not accumulate from step to step.
- They have good response for Start/Stop/Reverse.
- They have very robust operation due to the fact that they do not have brushes and a collector. For this reason, their life time mainly depends on the life time of the bearings and stator winding.
- The motor response against the digital pulses makes it possible to implement the control system in an open loop, which makes the system simpler in design and the cost is significantly reduced.
- The stepper motor from the operation view is a synchronous motor.
- Provides high starting torque.
- They have very low maintenance costs.

When choosing a stepper motor, the greatest care must be taken regarding the maximum torque that the motor can develop in order not to lose steps (synchronism). The mechanism torque must be lower than the maximum torque developed by the motor and also must be taking into account the safety reserve from the synchronism (Pop and Maer 2021).

The hybrid motor is the type of stepper motor that is widely used in practice, compared to the permanent magnet stepper motor or the stepper motor with variable reluctance, as it provides a small step angle and has high efficiency. This motor is a combination of a stepper motor with variable reluctance and a permanent magnet motor. The hybrid motor does not change its position even when the stator winding coil is not supplied, this is due to the permanent magnets placed in the rotor. The position control of the stepper motor consists of feeding the stator phase winding according to an algorithm which is implemented by using a driver (Teodor-Ionut et al. 2023; Ionica et al. 2017). A microcontroller or computer is needed to send command signals to the driver. Increasing the frequency of voltage pulses reduces the time it takes to move from one position to another, but it is worth to notice the fact that there is a maximum frequency at which the motor loses its step. This maximum frequency depends on the inertia of the motor-mechanism drive. Stepper motors are found to be used in practice in "pick and place" applications, CNC machines, printers, photocopiers, scanners (Iqteit et al. 2022; Revanth Kuma 2018; Poçari and Londo 2022; Vu et al. 2023; Hysa and Meşe 2021), etc.

2 Matlab/Simulink Block Diagram of the Stepper Motor Position Control System

In order to implement the motor position control system block diagram in Matlab/Simulink environment, it is needed the dynamic model of the motor. The dynamic model of the hybrid stepping motor is based on the equivalent electric circuit shown in Fig. 1 (Ionica et al. 2017; Iqteit et al. 2022).

For the equivalent circuit shown in Fig. 1, applying voltage Kirchhoff' law, we have:

$$u = iR + L\frac{di}{dt} + e(\theta) \tag{1}$$

Where R and L are specifically the resistance and inductance of the phase winding of the motor, while $e(\theta)$ represents the counter electromotive force induced in the stator winding due to the permanent magnets. Since the rotor–stator air space is large, practically the inductance of the phase does not depend on the position of the rotor for the hybrid stepping motor.

The counter electromotive force is a periodic function of the rotor position and varies according to a sinusoidal law. The counter electromotive force for phase '*a*' can be calculated as:

$$\mathrm{e}(\theta) = -p\Psi_{\mathrm{m}}\sin(p\theta)L\frac{\mathrm{d}\theta}{\mathrm{dt}} \tag{2}$$

where p and Ψ_{m} are respectively the number of pole pairs and the maximum phase flux.

At two phase hybrid stepper motor, electromagnetic torque is equal to the sum of the torque resulting from the interaction of magnetic fluxes created by the magnets with the phase currents and the detent torque T_{dm}.

$$T = -p\Psi_{\mathrm{m}}i_A\sin(p\theta) - p\Psi_{\mathrm{m}}i_B\sin\left(p\theta - \frac{\pi}{2}\right) - T_{\mathrm{dm}}\sin(2p\theta) \tag{3}$$

In order to implement the motor position control system block diagram in Matlab/Simulink environment,we used the library provided by Simulink, which are is available, among others type types of the motors,the hybrid stepper motor dynamic model (Ionica et al. 2017; Iqteit et al. 2022; Revanth Kumar 2018). The block diagram of the control system is shown in Fig. 2.

As can be seen from Fig. 2, in addition to the hybrid motor, we have the 12 VDC power supply, the driver and the signal builder, which serves to give the step

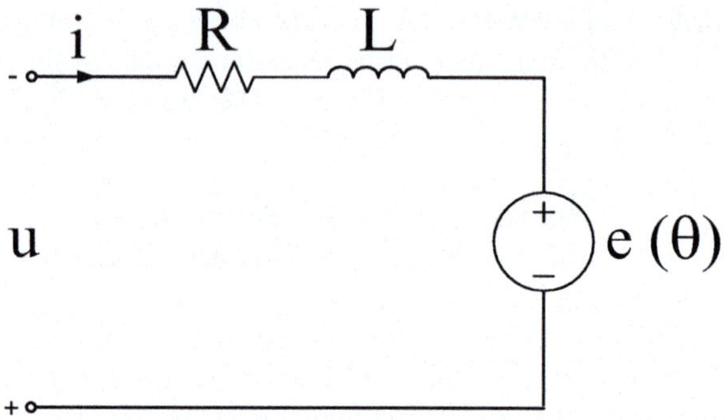

Fig. 1 Equivalent electrical circuit for one phase of the hybrid motor

pulses and clockwise or counterclockwise rotational direction of the motor. To perform the simulation, we first set the hybrid stepper motor data given in Table 1.

Simulation results are obtained by sequentially feeding the two winding phases of the motor from the 12 VDC power supply. We have specifically applied 26 V pulses clockwise and counterclockwise direction. Initially, 26 pulses are given in the clockwise direction, and for each pulse we have the rotor displacement with a step angle of 1.8°. Then for the same position of the rotor, only by changing the rotational direction in counterclockwise sensing, are given the same pulse number as before in clockwise direction. The simulation results are presented in Figs. 3 and 4.

Figure 3 shows the voltage and current waveforms of each phase of the hybrid stepper motor obtained from the simulation. The angular position of the rotor in the clockwise and counterclockwise directions is shown in Fig. 4.

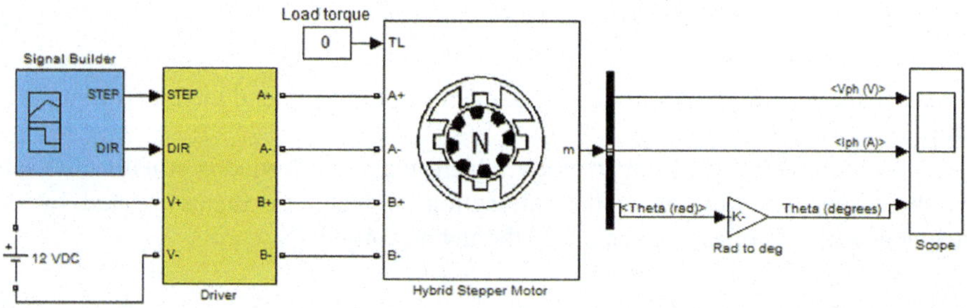

Fig. 2 Position control system for hybrid stepper motor

Table 1 Hybrid stepper motor data

Phase number	R [Ω]	L [mH]	Stepp angle [θ]	Ψ_m [Wb]	T_m [Nm]	Inertia [Kgm^2]	Friction [Kg*m/s]	I [A]	V [V]
2	8	10	1.8°	0.0015	0.0016	$34 * 10^{-9}$	10^{-4}	0.6	12

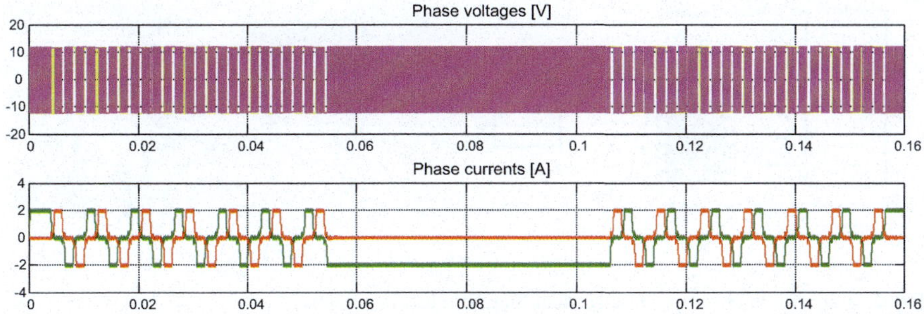

Fig. 3 Phase voltages and currents for hybrid stepper motor

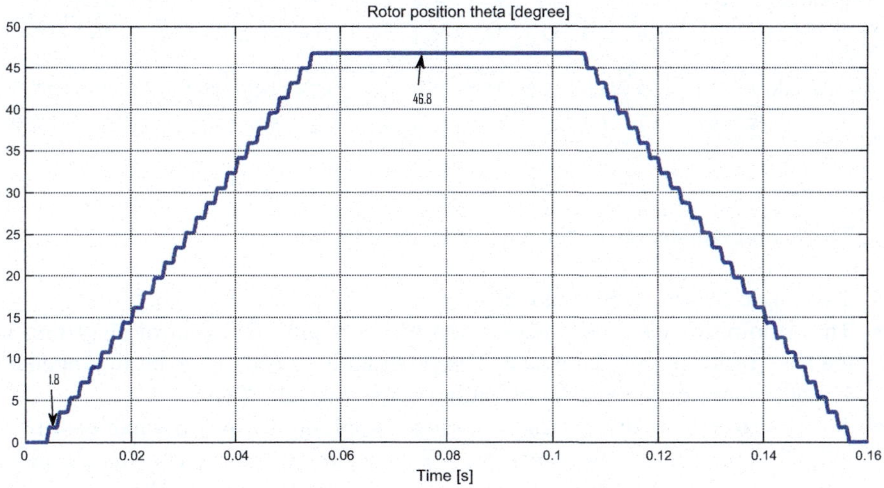

Fig. 4 Rotor angular position of stepper motor

3 Experimental Implementation of the Stepper Motor Position Control System

For the experimental implementation of the motor position control system, a hybrid stepper motor was used, the data of which are given in Table 1. Figure 5 shows the implementation setup with all components to ensure the step position control of the motor.

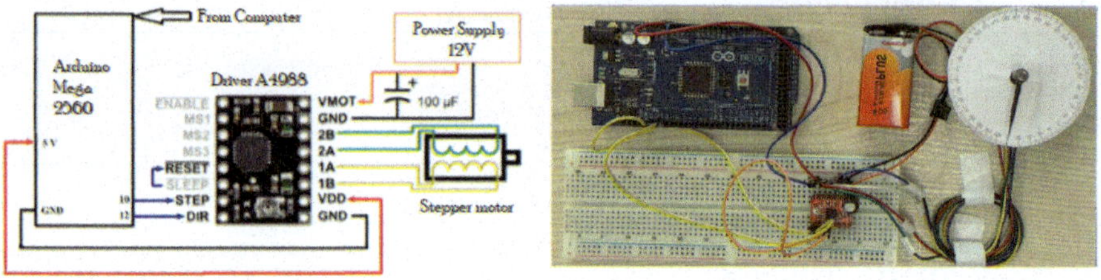

a) b)

Fig. 5 Experimental setup, (**a**) electrical diagram; test bench (**b**)

Table 2 Measured rotor angular position at steady state operation

Angular position [degree]	0	1.8	3.7	5.4	7.1	9	10.8	12.6	14.3	16.4	18.1	19.9	21.8	23.6
	25	27	28.6	30.4	32.3	34.3	36.1	37.7	39.6	41.3	43.4	45.2	46.7	**CW**
Angular position [degree]	46.7	45.1	43.3	41.5	39.8	37.7	36.2	34.2	32.3	30.6	28.9	27.1	25.3	23.2
	21.5	19.6	18.1	16.1	14.3	12.5	10.7	9.1	7.2	5.3	3.5	1.7	0	**CCW**

The main components of the experimental setup are:

- Two phase hybrid stepper motor.
- The Arduino Mega 2560 processor on which is uploaded control algorithm in the C++ programming language. The programming code is given in Appendix 1. The reference step angels are online given through PC.
- A4988 driver. This driver makes possible supplying the phase windings of the hybrid stepper motor by using the PWM technique. For each voltage pulse given to the STEP pin, the motor rotates with the step angle. The direction of rotation can also be changed by the driver.
- Other elements of the experimental setup such as the DC power supply and 100 μF capacitor.

The experimental results were obtained under the same conditions as in the case of the simulation in the Matlab/Simulink environment. From the computer, for the clockwise direction, it is given a reference step angle 46.8°, which for a step angle 1.8°, gives 26 V pulses. Angular displacement of the stepper motor was measured for each pulse separately and for the final angular position of the rotor. From the final angular position of clockwise sensing, 46.8°, the reference 0° it is given, and for the counter clockwise direction of the motor it is measured the angular position for each pulse. The results obtained from the experiment are presented in Table 2.

4 Conclusion

This paper presents a simple control system in open loop to control the position of the hybrid stepping motor. The proposed control system is robust, with low cost and very didactic for students. The control algorithm utilizes an Arduino processor Mega 2560 and PWM driver A4988 is employed to supply the hybrid stepper motor winding coils in the experimental setup. Furthermore, the ability to control the direction of rotation of the stepper motor is demonstrated. The simulation results align closely with experimental findings. Such a system is suitable to be used in those cases where precision control with high accuracy is required.

Appendix

```
const int stepPin = 10; // names pin number 10 as stepPin
const int dirPin = 12; // names pin number 12 as dirPin
int a = 180/1.8, c=0; // variable "a" is the pulse  number calculation for reference angle position
void setup() {
 pinMode(stepPin,OUTPUT); //  defines stepPin as output
 pinMode(dirPin,OUTPUT); // defines dirPin as output
 c =a; // variable "c" is the pulse number for reference angle position
}
void loop() {
 digitalWrite(dirPin,HIGH); // set dirPin in level HIGH
 for(int x = 0; x < c; x++) { // cycle
  digitalWrite(stepPin,HIGH); // set stepPin in level HIGH
  delayMicroseconds(1000); // time delay 1000 microseconds
  digitalWrite(stepPin,LOW); // set stepPin in level LOW
  delayMicroseconds(1000); // time delay 1000 microseconds
 }
 delay(2000); // time delay 2000 microseconds
 digitalWrite(dirPin,LOW); // set dirPin in level  LOW and changes the rotational direction
 for(int x = 0; x < c; x++) { // cycle
  digitalWrite(stepPin,HIGH); // set stepPin in level HIGH
  delayMicroseconds(1000); // time delay 1000 microseconds
  digitalWrite(stepPin,LOW); // set stepPin in level LOW
  delayMicroseconds(1000); // time delay 1000 microseconds
 }
 delay(1000); // time delay 1000 microseconds
 While(1){} // stop the cycle sequence
}
```

References

Hysa, G., Meşe, E.: On the behavior of five-phase induction motor drive under normal and faulty conditions. Int. J. Innov. Technol. Interdisc. Sci. **4**(3), 754–763 (2021). https://doi.org/10.15157/IJITIS.2021.4.3.754-763

Ionica, I., Modreanu, M., Morega, A., Boboc, C.: Design and modeling of a hybrid stepper motor. In: 10th International Symposium on Advanced Topics in Electrical Engineering March 23–25, Bucharest, Romania, pp. 192–195 (2017). https://doi.org/10.1109/ATEE.2017.7905081. www.ieeexplore.ieee.org/document/7905081

Iqteit, N., Yahya, K., Makahleh, F., Attar, H., Amer, A., Solyman, A., Qudaimat, A., Tamizi, K.: Simple mathematical and Simulink model of stepper motor. Energies MDPI (2022). https://doi.org/10.3390/en15176159. http://www.mdpi.com/1996-1073/15/17/6159

Pop, A., Maer, E-D.: Control technique for unipolar and bipolar step motor using Arduino and LabVIEW. In: International Ural Conference on Ural Conference on AC Electric Drives (ACED), pp. 1–5 (2021). https://doi.org/10.1109/ACED50605.2021.9462286. www.ieeexplore.ieee.org/document/9462286

Poçari, S., Londo, A.: Modelling of high precision pneumatic stepper motors for electronics and modern medicine applications. Int. J. Innov. Technol. Interdisc. Sci. **5**(1), 833–843 (2022). https://doi.org/10.15157/IJITIS.2022.5.1.833-843

Revanth Kumar, P.: Position control of a Stepper Motor using LabView. In: IEEE International Conference on Recent Trends in Electronics, Information & Communication Technology (RTEICT), pp. 1551–1554 (2018). https://doi.org/10.1109/RTEICT42901.2018.9012597. http://www.ieeexplore.ieee.org/document/9012597

Teodor-Ionut, I., Craiu, O., Popescu, L.: Analyzing a Three Hundred Teeth bi-phase hybrid stepper motor with different numbers of pole pairs. Revue Roumaine des Sciences Techniques, Série Électrotechnique et Énergétique, pp. 283–288 (2023). https://doi.org/10.59277/RRST-EE.2023.3.6. www.researchgate.net/publication/374856287

Vu, T.M., Moezzi, R., Dhoska, K.: Model predictive control with softened constraints for hybrid electric vehicle. Int. J. Innov. Technol. Interdisc. Sci. **6**(2), 1130–1149 (2023). https://doi.org/10.15157/IJITIS.2023.6.2.1130-1149

Do Artificial Intelligence Help Students to Learn More or Less? Case Study with My Students/ Teachers at SEEU

Basri Saliu

Abstract

Artificial intelligence has become topic number one in recent times, and the development of technology is no less talked in Macedonia, especially in academic circles, like universities, secondary schools, and other educational institutions. So, like for many countries and places, the development of artificial intelligence releases up new options for students of Southeast European University too. For some this is terrible, and for others it is normal and necessary. However, to find out exactly what students think about artificial intelligence and if it helps them to learn more or less, we asked students themselves. We asked young people for their opinion on how the development of technology will affect them, and also from some experts, IT teachers—how to get the most out of it. Both students and teachers happily responded and took part in this case study. Therefore, in this paper, I'll highlight and present their comments regarding ChatGPT. The focus of the paper is to measure the popularity of ChatGPT among students and their views on its use during their academic engagements. Generally, students' opinions varied widely, some found it useful for generating ideas and getting information, and some expressed concerns about the accuracy of information and potential biases in responses. However, it is worth noting that opinions can change over time and students' experiences may differ. So, in this paper we will specifically present students' perspectives on artificial intelligence in our educational settings. The most recent and specific feedback from students and some teachers will be explored in order to come up with some real time perspectives in the use of ChatGPT in our teaching at SEEU. Students' general trends will be observed and discussed with them on the use of artificial intelligence for the benefit of their own learning. As a main objective

B. Saliu (✉)
Tetovo, North Macedonia
e-mail: b.saliu@seeu.edu.mk

K. Dhoska and E. Spaho (eds.), *Bridging Horizons in Artificial Intelligence, Robotics, Cybersecurity, Smart Cities, and Digital Economy*, Sustainable Economy and Ecotechnology, https://doi.org/10.1007/978-3-031-72029-1_10

of this stands the idea if artificial intelligence can help students learn more or less.

Keywords

ChatGPT • Educational institutions • Conversation • Students of SEEU

1 Introduction to ChatGPT

ChatGPT is an advanced chatbot created using Generative Predictive Transformer (GPT) technology, developed by OpenAI. This technology allows the model to generate text in response to ideas, which allows interactive communication with the user. ChatGPT is skilled to understand and generate text in different contexts, making it approachable to different types of chats and questions. It includes general chats, help with information, creating textual content, answering technical questions, giving advice, and many others. Applications of ChatGPT are varied and include entering into websites, applications, chat platforms, educational platforms, different systems, and more.

ChatGPT includes several advantages:

- can operate 24/7 without the need for human interference, which can reduce user support expenses and the time to respond to questions.
- can handle a large number of users and conversations simultaneously.
- uses can easily access the information they need through ChatGPT without the need to browse web pages or search for information on Google.
- can be tailored to suit users' needs and preferences, making interactions stronger and simpler to the users' experience.
- can experience smart and relevant interactions through text, making it a useful tool for a variety of fields and purposes.

It took Tik-Tok nine months and Instagram two-and-a-half years to reach the number of users of the chatbot ChatGPT, created by OpenAI, a Microsoft-backed company. This is shown by the data of the Swiss bank UBS, reports Reuters. According to them, 100 million users "chatted" with the robot that writes articles, essays, jokes, and even poetry until this January, or just two months after the launch. Well this makes it the fastest growing consumer app in history.

2 How Popular is ChatGPT Among Young People in North Macedonia

In general, it is moderately popular among young people in Macedonia and more popular among students, especially students of Computers sciences. But as in many other countries, artificial intelligence is becoming more and more popular and used in the human environment. The use of artificial intelligence technologies extends to various fields such as medicine, finance, transportation, and many others. In Balkan countries, such as Macedonia, these technologies are increasingly being used to solve challenges and bring innovation to many aspects of daily life. However, in order to have a more common view of the level of use and popularity of artificial intelligence in Macedonia, it would be necessary to have a detailed analysis of the level of its use and implementation in the various sectors of the Macedonian economy and society.

However, in order to get some information about the students' opinion, if artificial intelligence helps them learn more or less, we conducted a close conversation with them and learned about their attitude. In the following part I will present their personal views on this matter.

3 Students' Attitudes Regarding Artificial Intelligence

Inspired by the latest trends in the use of artificial intelligence and with the aim of creating a clear picture of this issue at SEEU, as we mentioned above, we conducted a conversation in the form of an interview with the students of our groups. Two groups of 25 students were involved in the study and they all willingly participated in the interview.

The students were asked quite regularly about artificial intelligence and the question was straightforward, what do they think, do artificial intelligence helps them to learn more or less. In response to the question, they were free to express their opinion or by answering yes or no and elaborate it a little bit why yes or no. Indeed, most of the students answered the question based on their personal experience in the past and present. They answered the questions as follows and further elaborated their attitude.

1. Do artificial intelligence help students to learn more or less?

27 students Yes/54%

23 students No/46%

Those we polled who said No mostly say AI will make them lazier. Students with "no" answer elaborated further that Chat GPT summarizes readings for them and writes written work, whole seminars and so on. Here they further elaborated to their negative response. Their responses are summarized and put in groups based on their general reasons.

Students of group 1:

I think we will become lazier and more uncreative because everything will be handed to us on a platter, we will use only what was just handed over on a tray.

Students of group 2:

What makes us human will be lost and the soul of the person is lost and now everything is easier and therefore people will lose the empathy..

Students of group 3:

My generation we are 19 or 20 years old may not rely so much on that artificial intelligence but the young generation seem that they are getting lazier and lazier and they would submit 100% to that intelligence.

Students of group 4:

In reality I think we will become more stupid and lazier because there will be someone who will do our work for us, and everything will be through a computer, and the person will not try as hard as he should.

And the students who responded positively to the question elaborated and supported their "yes" as follows.

Students of group 1:

If you are using mobile phones and other digital devices the artificial intelligence is already built into it but people do not realize that they are already working with artificial intelligence. So, we better accept and not resist, better to educate people how to use it.

Students of group 2:

For me personally I think is sketchy but I think we need to get used to it because it's evolving through time. I think it's getting more popular among young people, with Chat GPT. I think it's really funny to use because you don't know what to expect from it.

Students of group 3:

"Each of us has a special style of writing, and if we can explain that style with a few attributes, we can convey that to the robot and he can write texts as if we had written them and the professor won't even suspect it."

4 Some Professors Shrug Their Shoulders

The professors from our university whom we asked if they knew that ChatGPT existed and perhaps without knowing it, they had already reviewed a written work in English, Albanian or Macedonian written by Chat GPT, heard about the robot for the first time from us. They were surprised, but some of them were also convinced that they knew the writing style of their students well, so they would easily recognize that they did not write the text themselves. But the students know that, so they just pass it on to the robot.

Some Professors who teach at the Faculty of Computer Sciences and Information Technologies (CST), say that it is almost impossible to detect if something is plagiarized if it is written by this artificial intelligence. And even if such a thing could be done today, it will be impossible in the future.

"It's very difficult to check because until now students had the internet at their disposal, they could take things from Wikipedia, from tutorials... And here you can search the internet to see if something has been downloaded. But ChatGPT rewrites, retells, and even if you give it the same question four times, it will create a different text for you, and that text passes the plagiarism detection tools," said most of the teachers.

However, they still believe that we can benefit from artificial intelligence much more than harm. "That's like asking me what damage a digitron could do fifty years ago when it first came into use. Of course, then mental abilities for calculations were valued, in order for the brain to work better. "We have lost some abilities with the digitron, but we no longer spend time on calculations, but on more advanced things," said the professor.

According to most of them, the tools will become more powerful and this is a harbinger of the time to come. Well, education and the whole society should adapt to the process.

Perhaps the question should be, how can we use that tool, that is, teach and direct students how to use it, to work more than they have been able to work so far. If they had to write code from scratch or a seminar from scratch, ChatGPT can give them a springboard and they can do something more complex or analyze a topic more deeply," added some other professors from CST.

They pointed out that young people must have developed critical thinking, because even this artificial intelligence is not always right and often convincingly improvises. Thus, someone who does not know will think that what he is reading is fact. They believe that professors can also benefit from this tool and with the help of artificial intelligence they could have better materials, scripts, textbooks and presentations to better convey their knowledge to students.

5 Conclusions

The use of artificial intelligence (AI) in the learning process brings several benefits and challenges that can affect the way we learn. Here are some conclusions regarding this issue:

- Personalization of Learning: IA can provide personalized content for each student based on their needs and knowledge level. This makes the learning process more efficient and suitable for each individual.

- Access to Extensive Information Resources: Through AI, students have instant access to a large amount of information and educational resources, including articles, videos, and e-books.
- 24/7 Assistance and Support: AI systems can provide assistance and answers to students' questions at any time, helping them learn even outside of traditional school hours.

Here are some conclusion regarding its challenges and Potential Risks:

- Lack of Human Interaction, excessive use of AI can reduce the social and interpersonal interaction between students and teachers, which is an important aspect of developing social and emotional skills.
- Technology Dependence: Students can become technology dependent and lose the ability to solve problems autonomously without the help of AI.
- Inequality in Access: Not all students have the same access to advanced technology, which can increase existing inequalities in education.

References

Artificial Intelligence at Google: Our Principles. Accessed 27 May 2020
Jobin, A., Ienca, M., Vayena, E.: The global landscape of AI ethics guidelines. Nat Mach (2019)
Putting Artificial Intelligence to Work. Boston Consulting Group, September 28 (2017)

Technology on Education from Albania's PISA 2022 Results

Alfons Harizaj and Olgerta Idrizi

Abstract

The integration of technology in education has been celebrated worldwide for its potential to revolutionize teaching and learning processes. Throughout the COVID-19 pandemic, technology emerged as a lifeline for education systems, facilitating remote learning and preventing a complete disruption of the academic calendar in many countries. However, the case of Albania presents a stark contrast, revealing the potential pitfalls of indiscriminate technology use. Despite the global narrative of technology's positive role in education, Albania's experience with the Program for International Student Assessment (PISA) 2022 exposes the adverse effects of hasty and uncalibrated technology integration. This abstract explores the nuanced dynamics of technology's impact on education, drawing insights from Albania's PISA 2022 results. While technology-enabled distance learning salvaged educational continuity in numerous countries, Albania's overreliance on technology appears to have backfired, resulting in diminished academic performance as reflected in the PISA scores. By analyzing the factors contributing to this discrepancy, including infrastructure limitations, pedagogical approaches, and digital literacy disparities, this study elucidates the complexities of technology implementation in the Albanian context. Through this examination, critical lessons emerge for education policymakers and stakeholders worldwide. The case of Albania underscores the imperative of a balanced and judicious approach to technology integration, one that considers the unique socioeconomic and educational landscapes

A. Harizaj (✉)
Canadian Institute of Technology, Tirana, Albania
e-mail: alfons.harizaj@cit.edu.al

O. Idrizi
The Mediterranean University of Albania, Tirana, Albania
e-mail: olgerta.idrizi@umsh.edu.al

K. Dhoska and E. Spaho (eds.), *Bridging Horizons in Artificial Intelligence, Robotics, Cybersecurity, Smart Cities, and Digital Economy*, Sustainable Economy and Ecotechnology, https://doi.org/10.1007/978-3-031-72029-1_11

of individual countries. Moreover, it highlights the necessity of comprehensive strategies that address not only the provision of technological tools but also the development of educators' digital competencies and the creation of supportive infrastructures. By synthesizing empirical evidence with theoretical frameworks, this abstract contributes to a deeper understanding of the multifaceted relationship between technology and education, offering valuable insights for guiding future educational reforms in Albania and beyond.

Keywords

PISA • Technology • Education • Reform

1 Introduction

PISA is an international assessment whose results influence the policies and educational reforms of many countries.

PISA 2022 was scheduled to be held in 2021, but was postponed by a year due to Covid-19.

The PISA 2022 results were announced in December 2023 and Albania received a low rating in all three main assessment domains. The result marks a significant decrease from the previous cycles. One of the major factors that has affected all countries has been the pandemic situation. Other factors need to be explored further to explain the results. From all of them the very sensitive one is the impact of the implementation of the new curriculum on pre-university education.

In this case, we will highlight some of the factors that can change the result with the aim of improving the achievements in the next assessment, which will be PISA 2025.

On the other hand, more and more PISA is evaluating the students' ability to learn through digital technology. This constitutes a challenge for our country that includes schools, universities, and educational institutions that promote policies in this direction.

2 Methodology

Based on the results of PISA in recent years, we will focus on the results of PISA 2022. We will identify some of the factors that have influenced the result based on the official reports of the OECD and state and educational institutions in Albania. Next, we will present which indicators will be evaluated in Pisa 2025 and how to act by schools and educational institutions to improve the results of this coming assessment.

3 Results

PISA is an international assessment of the skills and knowledge of 15-year-olds who, in most countries, are nearing the end of compulsory education. PISA assesses students' performance on "real-life" tasks that are considered important for effective participation in the company of adults and for lifelong learning. PISA is a project of the Organization for Economic Cooperation and Development (OECD). The first assessment was PISA 2000, where 30 countries participated, and the next assessment will be PISA 2025, where 90 countries will participate (ERC 2024). The published reports provide data and analysis for many indicators of achievements or failures for each country and group of countries. They are reports that are taken very seriously by governments regarding policies and reforms in education. Reports receive long-term attention from international institutions and organizations in the field of education or related to it. PISA was the most important international assessment of its kind.

PISA has moved from paper-based to computer-based assessment since 2015. In previous PISA assessments, in addition to the paper-based test in reading literacy, math and science, there were assessment options for digital skills, in mathematics and problem-solving.

Table 1 shows the results of our country in the last three cycles of 2015, 2018, and 2022 (OECD 2019a, b, c, 2023a, b, c; Maghnouj et al. 2020).

From the graph in Fig. 1, science has a continuous decrease from 2015 and reading has the lowest result among three courses.

The main domain in 2022 was mathematics and compared with 2018 the drop was 69 points (Table 2).

Table 1 PISA results 2015–2022, Albania

PISA	Reading	Mathematics	Science
2015	405	413	**427 (main domain)**
2018	**405 (main domain)**	437	417
2022	358	**368 (main domain)**	376

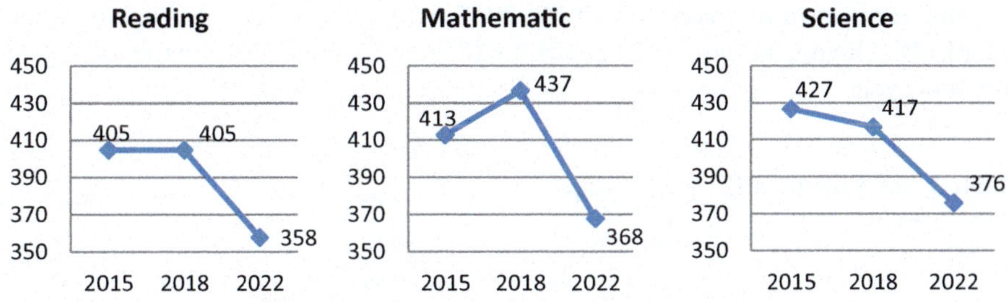

Fig. 1 PISA graph results 2015–2022, Albania

Table 2 PISA differences 2015–2022, Albania

PISA	2018	2022	Difference from 2018
Reading	405	358	−47
Mathematics	437	368	−69
Science	417	376	−41

In addition to evaluating student performance, PISA gathers comprehensive contextual data from students, parents, educators, and school administrators using context questionnaires. In Albania, the assessment in 2025 will incorporate questionnaires for students, parents, and school principals as part of this data collection process (ERC 2024).

There are evident signs that students did not approach the test and accompanying questionnaire with the required level of seriousness (Volume I) (OECD 2023b). Consequently, it is essential to exercise caution when interpreting the results.

Albania was struck by a powerful earthquake, at the end of 2019. It caused damage to infrastructure, in 12 cities.

Almost half of the students who participated in the assessment were from these countries hit by the earthquake, until April 2022, when the assessment was made, the students were learning with reduced hours and not in their schools that were damaged.

Covid-19 directly challenged the education system in Albania in 2020 and 2021. For the most part, it was transferred to the online or distance learning system.

Apart from them, other factors should be taken into consideration.

It is worth exploring if the curriculum reform that started in 2013–2014 has affected the results. This connection can be even more interesting if we compare the results of PISA Albania with PISA Kosovo (Table 3). Even there, there was a very low level of results. Kosovo and Albania unified the curricula that started to be implemented in the same period 2013–2014. Kosovo was not affected by the earthquake in Albania. Like all countries, it was affected by Covid-19. The new curriculum of Kosova was improperly implemented and schools "have remained alone in leading the implementation of the curriculum at the school level, without consolidated and continuous professional support" (Njohësit e arsimit 2023; Here comes PISA—K2.0 2022; ALBANIAN_SWD_2023_690-Albania-Report. 2024; Bota Sot News 2023; PISA Results 2024).

The innovative domain for PISA 2025 is Learning in the Digital World (Table 4). Albania has not confirmed if it will participate in this domain assessment for this cycle.

Table 3 PISA results 2022, Kosova and Albania

PISA 2022	Kosovo	Albania
Reading	342	358
Mathematics	355	368
Science	357	376

Table 4 Major and minor domains PISA 2015–2022

Cycle	Major domain	Minor domains
PISA 2018	Reading literacy	Mathematics, science, global competence
PISA 2022	Mathematics	Reading literacy, science, creative thinking
PISA 2025	Science	Mathematics, reading literacy, learning in the digital world

For PISA 2025, adaptive testing will be implemented for the assessment of science, the major domain.

Utilizing a computer-based platform enables the implementation of adaptive testing, which has the potential to enhance the precision and equity of assessments (Albania 2025). Through a multi-stage methodology, this form of testing adjusts according to each student's proficiency level. Initially, students tackle a standardized set of questions, and subsequently, based on their performance, they advance to questions of greater or lesser complexity at specified intervals.

4 Conclusions

Albania has shown before how to improve the result. In PISA 2009, she improved the result in reading by 46 points compared to PISA 2000. Reading was the main domain in both cycles. Albania was the third country with the largest increase among all participating countries.

The results of the PISA assessments in our country are not sufficiently explained related to the effect that the relevant curriculum has had.

In the past, high school mathematics education followed a structured approach, providing detailed explanations, theorems, and definitions in textbooks for each lesson. However, recent changes in the curriculum prioritize logical learning over rote memorization, aiming to foster deeper understanding and critical thinking skills.

Many teachers haven't received adequate training to effectively implement this new approach, leading to challenges in student engagement. Consequently, there's a noticeable decline in students' enthusiasm to actively participate in the learning process.

The funds allocated for increasing digital capacities have not been fully effective for students and the curriculum implemented in schools. The new digital strategy.

Covid-19, in addition to damage to educational systems, brought an experience to increase the digital capacities of schools.

5 Recommendations

The OECD has unveiled the framework for the forthcoming PISA 2025 examination, Unlike PISA 2022, which primarily assessed mathematical knowledge, PISA 2025 aims to evaluate how well nations are equipping their students in the field of science and fostering a dependable scientific understanding.

Educational institutions must implement this training profile with students for participation in PISA 2025. Based on the model of the pilot test that was held in April–May of this year, they can develop a curriculum that better integrates the scientific knowledge obtained from different courses. At the same time, they should guide and encourage how to use digital information in function of tasks and projects.

Pisa 2025 will be even more digitized, making it more challenging for the participating countries. Albania needs the use and improvement of technological capacities in teaching in order to improve results. The universities should be involved as they can help a lot in improving these capacities.

The Ministry of Education announced a plan for improving results in the upcoming PISA 2025 test []. The Albanian government has prepared a strategy and increasing digital capacities, especially for education. Educational institutions should bring these two years the support of the universities with expertise on information technology to increase the capacities of schools.

References

Albania, E.: PISA 2025 to focus on science, new assessment for foreign languages. Euronews Albania (2024). https://euronews.al/en/pisa-2025-to-focus-on-science-new-assessment-for-foreign-languages/

ALBANIAN_SWD_2023_690-Albania-Report. (n.d.). Retrieved from https://www.drejtesia.gov.al/baza-ligjore-dokumente/ 20 May 2024

Education, returning two decades later, the decline in "PISA 2022" set a record in the entire world history of the test. Bota Sot News (2023). https://www.botasot.al/en/arsimi-kthim-dy-dekada-pas-renia-ne-pisa-2022-shenoi-rekord-ne-te-gjithe-historine-boterore-te-testit/

ERC (Educational Research Centre): Who takes part in PISA? https://www.erc.ie/studies/pisa/who-takes-part-in-pisa/ (2024)

Here comes PISA—K2.0. Kosovo 2.0 (2022). https://kosovotwopointzero.com/en/here-comes-pisa/

Maghnouj, S., Fordham, E., Guthrie, C., Henderson, K., Trujillo, D.: OECD Reviews of Evaluation and Assessment in Education: Albania. OECD (2020). https://doi.org/10.1787/d267dc93-en

Njohësit e arsimit: PISA 2022—Kosova duhet të bëjë ndryshim te metodologjia e mësimdhënies. Kids of Kosova (2023). https://albanianeducation.net/njohesit-e-arsimit-pisa-2022-kosova-duhet-te-beje-ndryshim-te-metodologjia-e-mesimdhenies/

OECD.: PISA 2018 Assessment and Analytical Framework (2019a). https://doi.org/10.1787/b25efab8-en

OECD.: PISA 2018 Results (Volume II): Where All Students Can Succeed (2019b). https://doi.org/10.1787/b5fd1b8f-en

OECD.: PISA 2018 Results (Volume III): What School Life Means for Students' Lives (2019c). https://doi.org/10.1787/acd78851-en

OECD.: PISA 2022 Assessment and Analytical Framework (2023a). https://doi.org/10.1787/dfe0bf9c-en

OECD.: PISA 2022 Results (Volume I): The State of Learning and Equity in Education (2023b). https://doi.org/10.1787/53f23881-en

OECD.: PISA 2022 Results (Volume II): Learning During—and From—Disruption (2023c). https://doi.org/10.1787/a97db61c-en

PISA Results. Retrieved from https://tiranapost.al/aktualitet/rezultatet-e-pisa-manastirliu-kemi-hartuar-plan-per-permiresim-ne-tes-i530314 20 May 2024

Competition Law Enforcement in Albania: Challenges from Online Platforms

Gentjan Skara and Najada Kilic

Abstract

The rise of the digital economy has challenged the traditional enforcement of competition rules. Many countries have amended their domestic competition legislation to address the behaviour of large online platforms. Also, the EU adopted a regulation, commonly known as the Digital Market Act, to address the gatekeeper power of the largest digital companies. On the other hand, the competition bodies are restructuring their structure by establishing a unit or appointing a person to deal with large online platforms. In Albania, the situation is quite different. Addressing the challenges of technological development is a priority of the Albanian Competition Authority. However, no legislative measure has been introduced. This paper discusses the enforcement of competition rules in the Albanian digital market and the main challenges that Albanian Competition Authority faces with online platforms. The main methodology is that of doctrinal legal research, which consists of an analysis of the current competition legislative act and the national competition authority case law. After a short introduction, the second section provides an overview of the features of the digital economy and platforms. Then, the paper analyses recent legislative acts adopted in the EU and Germany, outlining the main novelties introduced (third section). The fourth section deals with the current legal framework for Albanian competition and the discusses the challenges faced by online platforms. The paper argues that the Albanian Competition Authority's role in the digital market enforcement is limited due to the absence of (i) a regulatory

G. Skara (✉) · N. Kilic
Epoka University, Tiranë, Albania
e-mail: gskara@epoka.edu.al

N. Kilic
e-mail: nkilic@epoka.edu.al

K. Dhoska and E. Spaho (eds.), *Bridging Horizons in Artificial Intelligence, Robotics, Cybersecurity, Smart Cities, and Digital Economy*, Sustainable Economy and Ecotechnology, https://doi.org/10.1007/978-3-031-72029-1_12

legal framework, (ii) a sufficient number of human resources and (iii) sufficient financial, technical and technological resources to deal with the digital market.

keywords

The digital economy • Competition law • National competition authority • Online platforms

1 Introduction

Digital services are indispensable to our daily and business activities. Once we get up in the morning, we check our email or social media to see what our friends did last night. Also, our work is dependent on technology and digital services. Computers or laptops are no longer a luxury but a necessity at work. Furthermore, digital services and technology provide real-time access to services, making our lives easier. For instance, you can access the traffic situation from an application. Or you can make a transaction online with less transaction cost and in real-time.

Businesses are changing their business models by shifting to e-commerce, app stores, online advertising, and online payment services. For instance, traditional retail corporations are embracing e-commerce, allowing customers to buy online. Banks have launched their own apps and digital wallets. Digital platforms are expanding business opportunities in many countries and have enabled the development of new products and services. Even professions where the use of technology was unthinkable are adapting themselves due to the spread of technology.

Digitalisation is profoundly transforming our economy. Technological development has changed the way of doing business and opened new avenues for innovation. As estimated by the Digital Economy Report 2019, the digital economy accounts for between 4.5 to 15.5% of the global Gross Domestic Product (UNCTAD 2024). Four out of the five largest companies in the world by market capitalisation in 2023 are tech companies ranked as follows: Apple (no 1), Microsoft (no 2), Alphabet (no 4), Amazon (5). Taking together market capitalisation, these tech companies are valued at 7,480.64 billion U.S. dollars (Statista 2024). Having said that, digitalisation has opened more business opportunities, increasing consumer choice and competitiveness of the industry.

However, the rise of the digital economy has challenged the traditional enforcement of competition rules in two aspects: regulatory and institutional framework. Many countries have amended their domestic competition legislation to address the behaviour of large online platforms. For instance, Germany has also amended its competition legislation to tighten the control of abusive conduct in digital markets (GWB-Digitalisierungsgesetz 2021). Section 19a allows the Bundeskartellamt "to intervene at an early stage in cases where competition is threatened by certain large digital companies" (Bundeskartellamt 2024). Also, in 2023, India changed the jurisdictional thresholds in merger control to capture more digital mergers (Uberoi et al. 2023). So far, the EU has enacted the most comprehensive piece of legislation to address the gatekeeper power of the largest digital companies, commonly known as the Digital Market Act (Regulation (EU) 2022a).

On the other hand, the competition bodies are restructuring their structure by establishing a unit or appointing a person (officers) to develop expertise and deal with digital markets. For example, in the UK, the Digital Markets, Competition and Consumers Bill establishes a Digital Markets Unit within the Common Market Authority (The Digital Markets 2024). Similarly, the United States and Canada established a unit to deal with digital markets in October 2019 and the end of 2021 (Kearney and Manji-Knight 2023). On the other hand, Australia or India opted for a specialist on digital platforms to conduct further work relating to digital platform markets (Klamka et al. 2023). Recently, the EU established a Chief Technology Team near the Directorate–General for Competition (DG COMP) and has opened a vacancy for Chief Technology Officer (Apply for Management Positions at the European Commission: Chief Technology Officer in DG COMP 2004). The Chief Technology Officer's responsibility is to advise the Commissioner responsible for Competition and the Director-General of DG COMP on the digital strategy, priorities, and projects.

In Western Balkan countries, including Albania, the situation is quite different. Addressing the challenges of technological development is a priority of the Albanian Competition Authority. However, no legislative measure has been introduced. This paper discusses the enforcement of competition rules in the Albanian digital market and the main challenges that the Albanian Competition Authority faces with online platforms.

The main methodology is doctrinal and comparative legal research. The doctrinal legal research analyses the EU primary and secondary competition legislation, respectively Articles 101 and 102 TFEU and Digital Market Acts, the German competition law and the Albanian legal framework. In addition, the recent EUCJ cases, Commission Decisions and Albanian Competition Authority decisions are analysed. The comparative legal research provides to what extent Albanian competition law is harmonised with the EU competition acquis, more specifically to address the behaviour of tech companies. In addition, the paper will consult the antitrust doctrinal literature such as books, scientific articles, reports, and other relevant publications.

The paper is structured by this introduction and 4 sections. The second section provides an overview of the features of the digital economy and platforms. The third section analyses recent legislative acts adopted in the EU and Germany, outlining the main novelties introduced. The fourth section deals with an overview of the Albanian competition law and discusses the role of the Albanian Competition Authority in the digital environment. In the final section, the paper concludes that Albanian Competition Authority's role in the digital market enforcement is limited due to the absence of (i) a regulatory legal framework, (ii) a sufficient number of human resources, and (iii) sufficient financial, technical and technological resources to deal with the digital market.

2 Features of the EU Digital Economy and Online Platforms

The digital economy is used to describe the shift from traditional business way towards digital technologies. Goods and services are offered through e-commerce. Digitalisation and technological developments have brought many benefits for citizens by creating entirely new categories of products and services. As society is moving towards cashless, technology allows people to buy online with just one click. On the other hand, the digital economy has increased consumer choice (Crémer et al. 2019). Citizens can review and buy high-quality products at low prices from their homes. Moreover, by googling on respective business websites, citizens have access to the product quality or can use price comparative websites or applications to find better prices.

The rise of the digital economy has led to the emergence of new possibilities and business models which compete across a wide range of products and services. Currently, at the EU level, more than 10,000 online platforms are operating in the EU's digital economy. Most of these online platforms are Small and Medium Enterprises (SME); whereas only a small number of online platforms "capture the biggest share of the overall value generated" (Proposal for Digital Markets Act 2020). Benefiting from the characteristics of the sector, these large online platforms act as gatekeepers between business users and end users by raising the following concerns about potential economic harms as a result of the concentrated structure (Prado 2020).

The main features of the large online platforms are as follows. Firstly, large online platforms lead to the dependency of many business users to grow their business model. Such dependency may lead to unfair behaviour vis-à-vis these business users. The main features of online platforms, combined with unfair practices by undertakings, that undermine contestability and fairness are as follows: "(i) extreme scale economies, which often result from nearly zero marginal costs to add business users or end users; (ii) nearly zero marginal costs to add business users or end users; (iii) very strong network effects; (iv) an ability to connect many business users with many end users through the multi-sidedness of these services; (v) a significant degree of dependence of both business users and end users; (vi) lock-in effects, a lack of multi-homing for the same purpose by end users, (vii) vertical integration, and (vii) data driven-advantages" (Regulation (EU) 2022b).

Secondly, gatekeepers have created immense power as a result of the market failure in the digital market due to the high concentration and a conglomerate ecosystem around their core platform services. Gatekeepers are often characterised by concentration as they have a natural tendency to tip towards monopolisation (Stucke and Grunes 2016). Thus, it is difficult for the new online platforms to enter into the market. As outlined by Martens:

> exclusive control over the data allows gatekeepers to control the ecosystem and generate significant value for their intermediation services. They can impose excessive entry and

access conditions, and exclusive dealing rules preventing sellers from promoting their offers outside the gatekeeper's platform (Martens 2021).

3 The Digital Market Act and Its Impact on EU Digital Economy

The rapid development of technology and digitalisation has presented many challenges for competitive law enforcement. In this context, several studies and reports have been produced to address the topics of digital competition and digital platforms. Some of these reports are as follows:

- the Commission's report on "Competition Policy for the Digital Era" (Crémer et al. 2019);
- the UK Furman Report on "Unlocking Digital Competition"18;
- the UK Competition and Markets Authority's "Market study on online platforms and digital advertising" (Competition and Markets Authority 2024);
- the German "Competition Law 4.0 report" (Bericht der Kommission Wettbewerbsrecht 2019);
- the Italian report "On big data" (Big Data: Indagine Conoscitiva Congiunta 2019);
- Franco-German report "On big data and competition law" (Competition Law and Data 2016);
- the French study "On competition and e-commerce" (Autorité de la Concurrence 2020a);
- the French report "On digital economy competition challenges" (Autorité de la Concurrence 2020b); and
- Australian Competition and Consumer Commission's (ACCC) 'Digital Platform Services Inquiry—Interim Reports' (the latest edition published in September 2023) (ACCC 2021).

While these reports found that existing competition rules provide a solid basis for protecting competition in the digital age, however, a growing concern is to regulate these online platforms. For instance, the Commission report "Competition Policy for the Digital Era", commissioned by Vestager and published in 2019, noted the importance of EU competition rules in protecting competition in a broad variety of market settings (Crémer et al. 2019) and emphasising the need to adopt new rules due to specific characteristics of online platforms, digital ecosystems and competition enforcement issues. In the Commission's view, "the challenges stemming from the rise of the Internet, the "new economy" and, today, the digital economy do require an adaptation of the way this basic framework is applied. The specificities of platforms, digital ecosystems and the data economy require adaptation and refinement of established concepts, doctrines and methodologies, and competition law enforcement itself" (Crémer et al. 2019).

In this situation, the Commission started a wide consultation on a broad range of online platform-related issues (Commission, Impact Assessment Report 2020a). Overall, public consultation feedback was in favour of strong support for a legislative intervention to tackle unfair practices in the digital market. Then, the Commission published the impact assessment, which acknowledged shortcomings of the EU's existing competition law dealing with large online platforms. According to the Commission, firstly, the EU existing competition law is insufficient to deal with these market failures "resulting from the behaviour if gatekeepers in the absence of some preconditions, such as the existence of an anticompetitive agreement in the case of Article 101 TFEU or of a dominant position in the case of Article 102 TFEU" (Commission, Impact Assessment Report 2020b). Secondly, the Commission noted that "market failures associated to tipping markets cannot be tackled on the basis of the existing competition rules, notably where market tipping is triggered primarily by the market structure" (Commission, Impact Assessment Report 2020c). Thirdly, the existing EU competition rules do not necessarily capture all the unfair business practices by large digital gatekeepers (Commission, Impact Assessment Report 2020d). Fourthly, the existing remedial competition tools stipulated by the Regulation 1/2003, respectively prohibition (fines) and commitment decisions (Arts 7 and 9); sector inquiries (Art 17) and interim measures (Art 8) can not address the gatekeeper relating to market failures (Commission, Impact Assessment Report 2020e). Finally, none of the ongoing reviews of EU competition law instruments (block exemptions Regulations for horizontal and vertical agreement, Market Definition Notice) are likely to change the situation (Commission, Impact Assessment Report 2020f).

To address these shortcomings, in December 2020, the Commission proposed a Regulation on the contestability and fairness of big tech in the digital market, known as the Digital Markets Act (DMA). On 24 March 2022, Parliament and Council negotiators agreed on the final DMA proposal (European Parliament 2022). It was signed into law on 14 September 2022 and entered into force on 1 November 2022. The DMA becomes applicable, for the most part, on 2 ay 2023 and fully applicable by March 2024 (Regulation (EU) 2022c).

The DMA general objective is "to contribute to the proper functioning of the internal market by laying down harmonised rules ensuring for all businesses, contestable and fair markets in the digital sector across the Union where gatekeepers are present, to the benefit of business users and end users" (Digital Markets Act, Recital 7 and Art 1(1)).

The DMA set out the following specific obligations: (i) address market failures to ensure contestable and competitive digital markets; (ii) gatekeepers' unfair conduct; and (iii) enhance coherence and legal certainty to preserve the internal market. Thus, the DMA aims to prevent these gatekeepers from imposing unfair conditions on businesses and end users and ensure the openness of important digital services.

The DMA targets only a small number of large undertakings that provide core platform services. As defined in Article 2(2), core platform services include: (i) online intermediation services; (ii) online search engines; (iii) online social networking services; (iv) video-sharing platform services; (v) number-independent interpersonal communications services; (vi) operating systems; (vii) web browsers; (viii) virtual assistants; (ix) cloud computing services; and (x) online advertising services. While Article 2(2) of the DMA provides an exhaustive list of what should be considered as core platform services, the core platform services definition can be expanded upon a market investigation by the Commission "for the purpose to examine whether an undertaking providing core platform services should be designated as a gatekeeper pursuant to Article 3(8), or in order to identify the core platform services to be listed in the designation decision pursuant to Article 3(9)" (Digital Markets Act, Art 17).

Once an undertaking falls under the definition of a "core platform service", then it is analysed its gatekeeper position. The DMA introduces a two-step test to determine the status of the gatekeepers either with reference to clearly circumscribed and appropriate quantitative metrics or based on a case-by-case qualitative assessment by means of a market investigation. Firstly, Article 3 formulates the conditions under which providers of core platform services should be designated as gatekeepers by introducing quantitative and qualitative criteria. According to Article 3(1), which is the base definition of a gatekeeper for the purposes of the DMA, an undertaking is assigned a gatekeeping status if it fulfils the following three cumulative criteria: (i) it has a significant impact on the internal market; (ii) it provides a core platform service that is an important gateway to the end customer; and (iii) it enjoys an entrenched and durable position in its operations or it is foreseeable to have one in the near future (Digital Markets Act, Article 3(1)). The core concepts of "significant impact on the internal market", "(important) gateway", "entrenched and durable position" or "foreseeable to have one in the near future" have not been defined in the DMA. All these concepts beg for interpretation and clarification (Fernández 2021). As the qualitative criteria describe the general characteristics of a gatekeeper, the EU legislator introduced quantitative criteria to define the gatekeeper position based on: (i) the turnover/market capitalisation/market value; (ii) the provision of a core platform service in at least three Member States; and (iii) the number of active users.

Secondly, where undertakings do not satisfy the quantitative thresholds, the Commission may designate an undertaking providing core platform services as gatekeepers following a market investigation and considering the different elements of a qualitative nature set out in the DMA. This occurs only when the undertaking providing core platform services does not satisfy qualitative criteria but meets the qualitative criteria. In this situation, the Commission shall consider several qualitative criteria such as: (i) the size of the provider; (ii) the number of business users; (iii) network effects and data-driven advantages; (iv) any scale and scope effects from which the undertaking benefits; (v) conglomerate corporate structure; (vi) business users and user lock-in; and (vii) other structural business or services characteristics (Digital Markets Act, Art 3(8)).

On 6 September 2024, the Commission, under the DMA, designated the following six gatekeepers: Alphabet, Amazon, Apple, ByteDance, Meta, and Microsoft.

In total, 22 core platform services have been listed as meeting the quantitative threshold for those particular services divided as follows:

- **social network**: Tiktok, Facebook, Instagram and LinkedIn;
- **intermediation services**: Google Maps, Google Play, Google Shopping, Amazon Marketplace, App Store and Meta Marketplace;
- **ads**: Google, Amazon and Meta;
- **browser**: Chrome and Safari;
- **operating system**: Google Android, iOS and Windows PC OS;
- **video sharing**: YouTube;
- **search**: Google Search; and
- **messaging services**: WhatsApp and Messenger (Commission, Digital Markets Act 2024).

In parallel, the Commission opened four market investigations to further assess whether Apple's messaging service iMessage, Microsoft's online search engine Bing, the web browser Edge, and Microsoft Advertising could be designated as core platform services (Commission, Digital Markets Act 2024). After closing the market investigation, the Commission found that Apple and Microsoft should not be designated as gatekeepers for the abovementioned core platform services (Commission closes market investigations on Microsoft's and Apple's services under the Digital Markets Act 2024). In addition, on 29 April 2024, the Commission designated Apple's iPadOS as a core platform service under the Digital Markets Act (Commission designates Apple's iPadOS under the Digital Markets Act 2024).

Currently, Booking, ByteDance and X have notified their potential gatekeeper status the Commission, about their services potentially meeting the threshold under the DMA. The Commission has 45 working days to decide whether to designate the companies as gatekeepers. If the Commission decides to design as gatekeeper, the latter will have 6 months to comply with the requirements in the DMA (Booking, ByteDance and X notify their potential gatekeeper status to the Commission under the Digital Markets Act 2024).

Once the gatekeepers are designed, they must comply with the obligations set out in Articles 5, 6, and 7 of the DMA. Articles 5 and 6 contain 18 obligations which are applicable to all designated gatekeepers and are distinguished into two groups. The first group contains self-executing obligations, which do not require further specification by the Commission. The second group is self-executing obligations that might require further specification by the Commission. These obligations derive from ECJ settled cases dealing with competition law or recent cases decided by the EU national competition authority.

The Commission is the sole enforcer of the DMA rules and monitoring gatekeepers' conduct and their compliance with the listed obligations (Article 26). It has the power to launch market investigations into systemic non-compliance by particular gatekeepers and open proceedings (Article 20). In the context of the opening proceedings, the Commission has the power: (i) to request information (Article 21); (ii) to carry out interviews and take statements (Article 22); (iii) to conduct inspections (Article 23); (iv) to adopt interim measures or commitment (Articles 24 and 25); (v) to impose fines up to 20% of their worldwide turnover in the preceding financial year, periodic penalty payment (Articles 30–33). In addition, the Commission may impose behavioural or structural remedies or structural separation in cases of systematic non-compliance by a designated gatekeeper (Digital Market Act, Art 18 (1)).

4 The Role of Albanian Competition Authority in the Digital Environment

Law 9121/2003 established an independent authority, named Albanian Competition Authority (ACA), to enforce competition rules. ACA is composed of the Commission and the Secretariat. The Commission is the decision-making body of the authority. It is composed of 5 members and acts as a permanent collegial body (Law 9121/2003 as amended, Art 20). Whereas the Secretariat is the Authority's investigatory body, which conducts the administrative investigations according to the Code of Administrative Procedures, the competition law and other legislation in force (Law 9121/2003 as amended, Art 20).

The Albanian competition law is considered lex generalis, which applies to all sectors of the Albanian economy without any economic activity excluded or exempted from its scope of application (UNCTAD 2019a). As stated in Art 1 of the Albanian Competition Law, the scope of the law is the protection of free and effective competition in the market. Since its establishment, on March 1, 2004, ACA's role has increasingly augmented, by intervening in many sectors of the economy (energy, telecommunications, banking, insurance, public procurement and concessions, transport, fuels, pharmaceutics, and food), in case anti-competitive practices or abuse of the dominant position have been identified and detected.

While the ACA has intervened in different sectors of the economy, the situation is quite different in the digital market (Skara 2022). So far, to the author's best knowledge and relying on a thorough search on the ACA website until 31 December 2022, no decision has been taken by ACA in relation to problems raised in the digital market. Thus, it is impossible to evaluate ACA's response to the concerns raised by the digital market. Nevertheless, the authors have identified only 2 cases, in which ACA had the opportunity to recognise the relationship between competition law and digital platforms.

The first case relates to an in-depth investigation for the period 1 January 2016 to 31 December 2018 concerning four well-known second-level banks, and it investigated whether there is a restriction of competition or not. In its analysis, ACA considered, inter alia, the online services (Automatic Teller Machine/ ATM and e-banking) offered by these banks. According to ACA, these banks have 41% of the market share in online services. In its reasoning, ACA recognised the increasing use of online platforms. However, it did not elaborate further on whether online platforms have any impact on restricting the competition (ACA Decision No 723 2020). In the follow-up decision concerning whether the banks mentioned above have fulfilled the obligation set out by Decision 723/2020, ACA pointed out an increase of online services in the banking sector, due to the digital strategy of the banking sector and COVID-19 pandemic restrictions (ACA Decision No 779 2021).

The second case relates to concentration. On April 8, 2022, ACA authorised the concentration of the Mooney Group by companies Bank 5 S.p.A. and Enel X S.r.l. (ACA Decision No 875 2022). The Mooney Group is a joint stock company established under Italian company law and offers services, *inter alia*, through digital platforms. Bank 5 S.p.A. is part of the Intesa Sanpaolo S.pA, one of the main banks operating in Italy. Intesa SanPaolo S.pA, is present in Albania through Intesa Sanpaolo Bank Albania ShA. After assessing the documents submitted, ACA decided to authorise this concentration. In these decisions, the issue between competition and digital platforms was just mentioned without providing any analysis. There are several arguments to explain this situation.

First, unlike EU or other countries which have introduced specific legislation to address the behaviour of online platforms, the existing Albanian competition law or ACA's by-laws do not address competition issues arising from digital platform. Therefore, a legislative reform should be undertaken to provide more legal certainty and confer proper enforcement tools to the competition authority.

Second, the digital market requires appropriate tools to analyse anticompetitive practices. In the world of the digital economy, driven by platforms and other similar technologies "a rethinking and reconfiguration of some core concepts of competition law in terms of whether and if so, how they should apply to particular practices of technology companies in these non-traditional market contexts?" (Akman 2019). In the case of Albania, only recently, in 2021, for the first time in the annual speech addressed before the Assembly, the Chairperson of ACA set out as a priority for 2021, the involvement in the digital sector (Latifi 2021). Since 2021, the addressing issues relating to the digital markets has been a priority. However, this is difficult in the absence of a clear regulatory legal framework or bylaw. Thus, ACA should clearly understand how the digital market operates and, most importantly, the appropriate tools to analyse anticompetitive practices.

Third, ACA should have the human resources capacity to act effectively and timely due to the dynamic and constantly changing nature of the digital markets. Many competition authorities are establishing or appointing experts in digital markets' units to address the challenges of fast-growing technological innovation. For example, a Digital Market Unit in the UK has been established within the Competition Market Authority (Jeffs and Dhondt 2021). The same follows for Australia, which within the competition authority has set up a specialist Digital Platform (Klamka and Low 2021) or Canada, which in July 2019 hired its first-ever Digital Enforcement Officer "to help the Bureau monitor the digital landscape, identify and evaluate new investigative techniques, and boost its digital intelligence gathering capabilities" (Kearney et al. 2021).

Based on its organisational structure, ACA has no directories dealing with fast-growing technological issues or a specialist responsible for the digital market. Nevertheless, ACA has two IT experts in its organisational structure. For a small market like Albania, this is a good practice. However, as online platforms are increasing, more human resources specialised in the digital environment are needed. Another challenge relates to the lack of specialist and appropriate tools for data collection from dominant platforms. As noted by the Intergovernmental Group of Experts on Competition Law and Policy, this challenge may "be more prominent in developing countries, where platforms do not have a physical presence and whose markets may not be considered as significant for such platforms" (UNCTAD 2021). In this regard, it is appropriate for ACA to design the appropriate Directory with adequate tools to address the challenges of the digital economy or appoint a specialist for the digital market.

5 Conclusions

The rise of the digital economy has challenged the traditional enforcement of competition rules. Many countries have amended their domestic competition legislation to address the behaviour of large online platforms. Also, the EU adopted a regulation, commonly known as the Digital Market Act, to address the gatekeeper power of the largest digital companies. On the other hand, the competition bodies are restructuring their structure by establishing a unit or appointing a person to deal with large online platforms.

The situation is quite different in Albanian regarding the applicability of competition rules towards online platforms. Unlike the EU and many EU Member States, which have introduced legislation to address the challenges raised by tech companies, Albania has no legislation to address the challenges raised by the growth of digital platforms and e-commerce. In this context, Albania needs to update their competition legislation to address the behaviour of online platforms. Moreover, Albania Competition Authority lacks the necessary human, financial, and technical resources to deal with these kinds of cases, which, even for the Commission or consolidated EU NCAs, has been a challenge. On the other hand, the ACA has made a significant step in recognising the impact of digital platforms on distortion of competition. In conclusion, the ACA should intensify the focus on the digital markets.

Acknowledgement This paper is an ongoing study. Parts or ideas of this study have been presented in ASCOLA Athens 2023 conference and most recently in the 16th International Conference Economies of the Balkan and Eastern European Countries (EBEEC 2024) held on 17–19 May 2024 in Vilnius, Lithuania. The title of the paper presented at EBEEC 2024 was "Competition law enforcement in Western Balkan Countries: Challenges from online giants". The authors thank reviewers for their valuable feedback.

References

ACA Decision No 723 of 17 November 2020 for the closure, with clauses and obligation, of the in-depth investigation in the bank sector for Raiffeisen Bank, National Bank of Tirana, Credins Bank and Intesa Sanpaolo Bank (2020). Official Bull, para 57–58

ACA Decision No 779 of 4 March 2021 on monitoring the decision Decision No 723 of 17 November 2020 for the closure, with clauses and obligation, of the in-depth investigation in the bank sector for Raifffeisen Bank, National Bank of Tirana, Credins Bank and Intesa SanPaolo Bank for the period 2019–2020 and giving some recommendation (2021). Official Bulletin, paras 6(12)

ACA Decision No 875 of 8 April 2022 on the authorization of the concentration realized through the acquisition of the Mooney group S.P.A. from the companies Banka 5 S.P.A. and Enel X S.R.L. (2022)

ACCC: Digital Advertising Services Inquiry—Final Report (2021). https://www.accc.gov.au/system/files/Digital%20advertising%20services%20inquiry%20-%20final%20report.pdf. Accessed 15 April 2024

According to ACA official website, the ACA has employed only two IT expert. One is working at the Directory of Market Analysis and Methodology and the other at the Directory of Legal, Integration and Judicial Affairs

Akman, P.: Editorial: an agenda for competition law and policy in the digital economy. J. Eur. Compet. Law Pract. **10**(10), 589–590 (2019)

Autorité de la Concurrence and Bundeskartellamt: Competition Law and Data (2016). https://www.bundeskartellamt.de/SharedDocs/Publikation/DE/Berichte/Big%20Data%20Papier.pdf?__blob=publicationFile&v=2. Accessed 15 April 2024

Autorità Garante della Concorrenza e del Mercato, Autorità per le Garanzie nelle Comunicazioni e Garante per la Protezione dei dai Personali: Big Data: Indagine Conoscitiva Congiunta (2019). https://www.agcm.it/dotcmsdoc/allegati-news/Big_Data_Lineeguida_Raccomandazioni_di_policy.pdf. Accessed 15 April 2024

Autorité de la Concurrence: Competition and e-commerce (2020a). https://www.autoritedelaconcurrence.fr/sites/default/files/concurrence-commerce-en-ligne-en.pdf. Accessed 15 April 2024

Autorité de la Concurrence: Contribution de l'Autorité de la concurrence au débat sur la politique de concurrence et les enjeux numériques (2020b). https://www.autoritedelaconcurrence.fr/sites/default/files/2020-02/2020.02.28_contribution_adlc_enjeux_num.pdf. Accessed 15 April 2024.

Bericht der Kommission Wettbewerbsrecht 4.0: Ein neuer Wettbewerbsrahmen für die Digitalwirtschaft (2019). https://www.bmwk.de/Redaktion/DE/Publikationen/Wirtschaft/bericht-der-kommission-wettbewerbsrecht-4-0.pdf?__blob=publicationFile&v=12. Accessed 29 April 2024

Bundeskartellamt: Amendment of the German Act against Restraints of Competition. https://www.bundeskartellamt.de/SharedDocs/Meldung/EN/Pressemitteilungen/2021/19_01_2021_GWB%20Novelle.html?nn=3591568. Accessed 11 May 2024.

Crémer, J., de Montjoye, Y.A., Schweitzer, H.: Competition Policy for the Digital Era (Directorate-general for competition, commission), p. 39 (2019)

Commission, Apply for Management Positions at the European Commission: Chief Technology Officer in DG COMP (grade AD14—Temporary Agent)—COM/2024/10446. https://commission.europa.eu/jobs-european-commission/job-opportunities/managers-european-commission/apply-management-positions-european-commission/chief-technology-officer-dg-comp-com-2024-10446_en. Accessed 10 May 2024

Commission, Booking, ByteDance and X notify their potential gatekeeper status to the Commission under the Digital Markets Act (2024). https://digital-markets-act.ec.europa.eu/booking-bytedance-and-x-notify-their-potential-gatekeeper-status-commission-under-digital-markets-2024-03-01_en. Accessed 12 May 2024.

Commission, Commission closes market investigations on Microsoft's and Apple's services under the Digital Markets Act (2024). https://digital-markets-act.ec.europa.eu/commission-closes-market-investigations-microsofts-and-apples-services-under-digital-markets-act-2024-02-13_en. Accessed 12 May 2024.

Commission, Commission designates Apple's iPadOS under the Digital Markets Act (2024). https://digital-markets-act.ec.europa.eu/commission-designates-apples-ipados-under-digital-markets-act-2024-04-29_en. Accessed 12 May 2024.

Commission, Digital Markets Act, Commission designates six gatekeepers. https://ec.europa.eu/commission/presscorner/detail/en/IP_23_4328. Accessed 12 May 2024

Commission, Impact Assessment Report, Commission Staff Working Document, 2020, SWD (2020a) 363 final

Commission, Impact Assessment Report, Commission Staff Working Document, 2020, SWD (2020b) 363 final, para. 119

Commission, Impact Assessment Report, Commission Staff Working Document, 2020, SWD (2020c) 363 final, para. 120

Commission, Impact Assessment Report, Commission Staff Working Document, 2020, SWD (2020d), 363 final, para. 121

Commission, Impact Assessment Report, Commission Staff Working Document, 2020, SWD (2020e) 363 final, para. 122

Commission Impact Assessment Report, Staff Work. Document, SWD (2020f), 363 final, para. 123

Competition and Markets Authority: Online platforms and Digital Advertising: Market study interim report. https://assets.publishing.service.gov.uk/media/5dfa0580ed915d0933009761/Interim_report.pdf. Accessed 29 April 2024

Digital Markets Act, Art 17

Digital Markets Act, Recital 7 and Art 1(1)

Digital Market Act, Art 18(1)

Digital Markets Act, Art 3(8)

Digital markets act, Article 31

The Digital Markets: Competition and Consumers Bill is pending at the House of Lords. The Digital Markets, Competition and Consumers Bill. https://publications.parliament.uk/pa/bills/cbill/58-03/0294/220294.pdf. Accessed 12 May 2024

Digital Competition Expert Panel: Unlocking Digital Competition: Report of the Digital Competition Expert Panel (2019). https://assets.publishing.service.gov.uk/government/uploads/system/uploads/attachment_data/file/785547/unlocking_digital_competition_furman_review_web.pdf. Accessed 1 April 2024

European Parliament: Deal on Digital Markets Act: EU rules to ensure Competition and more choice for users (2022) https://www.europarl.europa.eu/news/en/press-room/20220315IPR25504/deal-on-digital-markets-act-ensuring-fair-competition-and-more-choice-for-users. Accessed 1 April 2024

Fernández, C.: A new kid on the block: how will competition law get along with the DMA. J. Eur. Compet. Law Pract. **12**(4), 271 (2021)

GWB-Digitalisierungsgesetz—GWB Digitalisation Act (2021)

Jeffs, C., Dhondt, N.: Introduction: why digital markets? In: Jeffs, C., Sokol, D., Ning, S. (eds.) Digital Markets Guide. Law Business Research, vol. 1, p. 1 (2021)

Kearney, E., Manji-Knight, A.: Canada: a deep dive into forthcoming legislative reforms to the digital market. In: Jeffs, C., Sokol, D., Ning, S. (eds) Digital Market Guide (2023)

Kearney, E.K., Manji-Knight, A., Hollenberg, J.: Canada. In: Jeffs, C., Sokol, D., Ning, S. (eds.) Digital Markets Guide. Law Business Research, pp. 206–207 (2021)

Klamka, L., Low, A.: Australia. In: Jeffs, C., Sokol, D., Ning, S. (eds.) Digital Markets Guide. Law Business Research, p. 250 (2021)

Klamka, L., et al.: Australia: key reforms and their effects on digital economic competition. In: Jeffs, C., Sokol, D., Ning, S. (eds.) Digital Market Guide (2023); Uberoi, N.K., et al.: India: nurturing competition in a rapidly evolving digital space. In: Jeffs, D., Sokol, D., Ning, S. (eds.) Digital Market Guide (2023)

Latifi, J.: Fjala e Kryetares së Komisionit të Autoritetit të Konkurrencës, Znj. Latifi, në Raportimin Vjetor "Për veprimtarinë e Autoritetit të Konkurrencës për vitin 2020 dhe prioritetet për vitin 2021 në Komisionin e Ekonomisë dhe Financave (2021). http://www.caa.gov.al/publications/read/id/190. Accessed 24 June 2023; Biba, D.: Fjala e Kryetarit të Komisionit të Autoritetit të Konkurrencës në seancën plenare ku paraqiti Raportimin Vjetor Për veprimtarinë e Autoritetit të Konkurrencës për vitin 2021 në Komisionin e Ekonomisë dhe Financave (2022)
Law 9121/2003 as amended, Art 20.
Martens, B.: An economic perspective on data and platform market power. JRC Digital Economy Working Paper 2020–09 (2021). https://www.researchgate.net/publication/349179464 Accessed 1 May 2024, p. 13
Prado, T.S.: Assessing the market power of digital platforms. Quello Center Working Paper, TPRC48: The 48th Research Conference on Communication, Information and Internet Policy (2020). https://ssrn.com/abstract=3747793. Accessed 1 May 2024
Proposal for a Regulation of the European Parliament and of the Council on contestable and fair markets in the digital sector (Proposal for Digital Markets Act) COM (2020), 842 final, p. 1
Regulation (EU): 2022/1925 of the European Parliament and of the Council of 14 September 2022 on contestable and fair markets in the digital sector and amending Directives (EU) 2019/1937 and (EU) 2020/1828 (Text with EEA relevance) (2022a) OJ L 265/1
Regulation (EU) 2022/1925 of the European Parliament and of the Council of 14 September 2022 on contestable and fair markets in the digital sector and amending Directives (EU) 2019/1937 and (EU) 2020/1828 (Text with EEA relevance) (2022b) OJ L 265/1, Recital 2
Regulation (EU) 2022/1925 of the European Parliament and of the Council of 14 September 2022 on contestable and fair markets in the digital sector and amending Directives (EU) 2019/1937 and (EU) 2020/1828 (Text with EEA relevance) (2022c) OJ L 265/1, Art 54.
Skara, G.: Europeanisation of Private Enforcement of Competition Law: The Case of Albania Springer, pp. 245–254 (2022)
Statista: The 100 largest companies in the world by market capitalization in 2022. https://www.statista.com/statistics/263264/top-companies-in-the-world-by-market-capitalization/. Accessed 11 May 2024
Stucke, M., Grunes, A.: Big Data and Competition Policy. Oxford University Press, p. 336 (2016)
The report Competition Policy for the Digital Era emphasised that EU competition rules, as embedded in Articles 101 and 102 of the TFEU, continue to provide a sound and sufficiently flexible basis for protecting competition in the digital era. Crémer, J., de Montjoye, Y.A., Schweitzer, H.: Competition Policy for the Digital Era (Directorate-General for Competition, Commission), p. 39 (2019)
Uberoi, N.K., et al.: India: nurturing competition in a rapidly evolving digital space. In: Jeffs, C., Sokol, D., Ning, S. (eds.) Digital Market Guide (2023)
UNCTAD: Competition issues in the digital economy. intergovernmental group of experts on competition law and policy, 8th session, Geneva (2019a). https://unctad.org/system/files/official-document/ciclpd54_en.pdf. Accessed 4 July 2023, p. 13.
UNCTAD: Digital Economy Report (2019b). https://unctad.org/system/files/official-document/der2019_overview_en.pdf. Accessed 11 May 2024, p. 4
UNCTAD: Competition Law, Policy and Regulation in the Digital Era (2021)

A Systematic Review of Database Systems in Science and Engineering

Shefqet Meda and Hiqmet Kamberaj

Abstract

This paper attempts to fill this gap by presenting the current landscape of data management solutions/database systems used in science and engineering, providing insights into common technologies, benefits and limitations, looking for emerging trends, and identifying potential white spots that need to be addressed. It also tries to find out how these systems benefit, what the limitations are and how they are used and applied in the set of scientific and engineering fields. The present systematic review also covers a wide range of database systems, including relational databases, object-oriented databases, NoSQL databases, and graph databases, by structuring them into subdomains. The paper continues by identifying and comparing the features, the data model and structure, the query language, the scalability, and performance, the suitability for several scientific and engineering applications domains such as data integration, usage for data analytics, data visualization, and simulation. Finally, the paper concludes by providing a series of general considerations regarding the most appropriate database system or systems for the carriedout needs and type of data (such as challenges or open research questions to be considered, especially distributed and security-related concepts), future research and developments in the database systems field in science and engineering.

S. Meda (✉)
Canadian Institute of Technology, Tirana, Albania
e-mail: Shefqet.meda@cit.edu.al; meda.shefqet@ibu.edu.mk

S. Meda · H. Kamberaj
International Balkan University, Skopje, North Macedonia
e-mail: hkamberaj@ibu.edu.mk

K. Dhoska and E. Spaho (eds.), *Bridging Horizons in Artificial Intelligence, Robotics, Cybersecurity, Smart Cities, and Digital Economy*, Sustainable Economy and Ecotechnology, https://doi.org/10.1007/978-3-031-72029-1_13

Keywords

Database • Systematic • NOSQ • Visualization • Security • Relational • Performance • Query • Metrics • Strengthen • Weaknesses • Machine learning/ Deep learning

1 Introduction

This research, which presents a novel and comprehensive analysis, aims to delve into the current state of data management applications/database systems used in science and engineering. The paper not only examines these systems' benefits, limitations, and applications in various scientific and engineering fields but also provides a detailed description of the most used technologies, their strengths and weaknesses, emerging trends, and potential gaps that need to be addressed in big data analysis and processing and the application of artificial intelligence today.

As a significant practical implication of this research, we have developed an extraction procedure from database systems specifically for Machine Learning/ Deep Learning applications. This tool, which is not just a theoretical concept but also a practical guide, can significantly enhance users' understanding and use of these systems, thereby making the research more relevant and applicable in the real world.

2 Considering a Comprehensive Systematic Review, the Database Systems Range in Science and Engineering

This research, as a comprehensive and thorough systematic review, will cover a wide range of database systems in science, including relational databases, object-oriented databases, NoSQL databases (Harrison 2015), and graph databases. The paper reviews the available data models and query languages, scalability and performance for different database systems, and the generalization of the applications, such as data integration, data analysis, data visualization, and simulation. The research will also compare and contrast the database systems on different factors such as ease of use and implementation, extensibility, the support of machines in the distribution system and compatibility. We will try to point out the challenges and questions that should be considered in the research of this field, such as privacy, security and data quality issues.

A comprehensive systematic review of database systems in science and engineering research is essential as it describes the framework of knowledge on the ease of decision-making and adoption of database systems. In the end, we will convey a list of recommendations or considerations that must be considered for research and development activities in the field of database systems.

3 The Significance of Database Systems in Science and Engineering

Database systems (Hector Garcia-Molina 2008) are very important for science and engineering for several reasons: they offer a systematic and structured way to store, manage, and retrieve data. Database systems are important not only as an instrument of development and progress but also as research for development perspectives. We have listed below reasons why database systems are vital in science and engineering:

a. Data Management:
   - Organization and Storage (Avi Silberschatz 2019):
     Database Systems provide a structured, organized method of storing large amounts of data generated from scientific experiments and engineering projects. It ensures efficient storage, retrieval, and management of data.

b. Data Integration:
   Collaboration and Interdisciplinary Research: They facilitate collaboration by integrating data sets from various organizations. Databases simplify the method since the data types are different and complex to integrate in an interdisciplinary study.

c. Data Analysis and Retrieval:
   Querying and Analysis (Avi Silberschatz 2019): Database Systems allow Engineers and Researchers to inquire and analyze data using query languages; it enables complicated analyses and the ability to extract knowledge from large datasets through data mining.

d. Scalability: Handling Large Datasets-Scientific and engineering datasets continually increase in size and complexity. Database Systems are scalable, allowing individuals to handle and process substantial data.

e. Time-Series and Temporal Data: Temporal Analysis- Most scientific experiments involve time-series data. Database systems developed for temporal data may efficiently analyze and visualize changes over time.

f. Complex Relationships: Graph Databases may effectively represent and analyze the relationships in complexities such as biological networks, social activities, or engineering systems.

g. Data Security and Integrity: Integrity database systems are critical for scientific research and research for information integrity. Database systems inform researchers by using transaction and ACID (Atomicity, Consistency, Isolation, and Durability) properties.

To fulfil the above-mentioned demands, a database system is designed to support various activities, including spatial data handling for projects with geographical or spatial data, machine learning integration, APIs, and a programming language for economic research or analysis. About the above discussion, the purpose of database systems may also include other objectives such as the following (we have included applications):

h. Geographical and Spatial Research (Obe and Hsu 2015): spatial databases are equipped with specialized tools that facilitate storage and general efficient retrieval.
i. Machine Learning Integration: With applications with the integration of in-database analytics and machine learning, researchers are given the added advantage of conducting research analytics directly among other real-time decision-making processes and prediction modeling.
j. Reproducibility and Transparency: Database systems facilitate the validation of experiments for others by aiding the preservation of dataset structure and sharing.
k. Data Governance and Compliance: application in designing their fields, vast disciplines requiring strict regulatory procedures including pharmaceuticals and environmental sciences. For that reason, it can be permuted that a database is the backbone of research fields.
l. Resource Optimization: application database systems that are structured dynamically enable researchers to manage peripheral data accordingly; this reduces processing and saves on cost. Therefore, in research fields, cost-saving is one of their goals in a certain period.
m. Historical Data and Versioning: Since research is conducted over longer periods, datasets ought to be stored and always in databases; hence, the database can hold over a billion datasets. For records in history, all versions of field experiments are vital.
n. Data Visualization (Wilke 2019): Integrating visualization tools with database systems allows researchers to represent their findings visually in a meaningful way.

Overall, database systems are integral to scientific and engineering research as they provide a tool to organize, analyze, and gain insights from huge and complex datasets in a structured and efficient way.

4 Applications of DB Systems

As mentioned above, database systems have found numerous and diverse applications in industry and business, in agencies and the research field, and this is thanks to the features they have, but not limited to efficiently storing, managing, and retrieving data, keeping records, facilitates online business transactions, financial analysis, etc. Here, we have categorized and included some of the main implementations in the areas of life. We have approached each application first with the key aspects of the application and, secondly, the benefits of these applications.

(a) **Business and Finance**:
   i. Managing customer information, financial transactions, inventory, banking, etc.

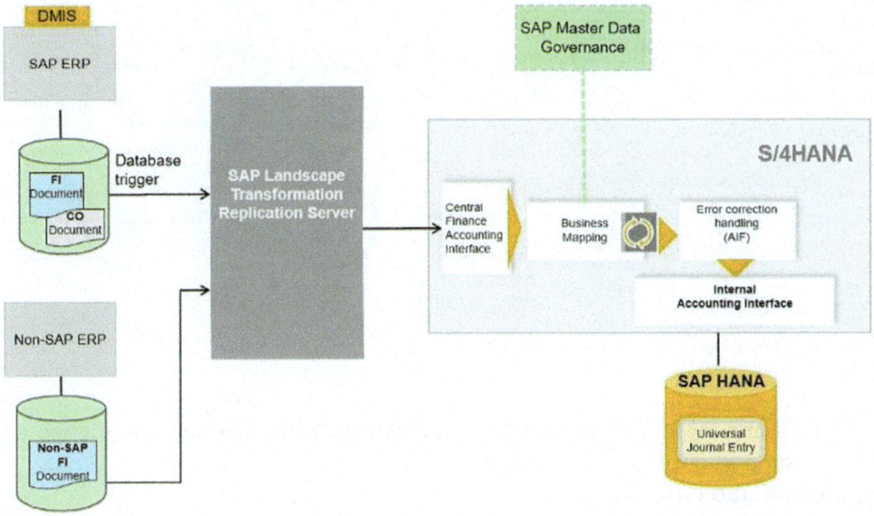

Fig. 1 SAP finance system (sap.com 2021)

ii. Allows for accurate maintenance of records, sources secure online transactions, and supports informed financial decisions (Fig. 1).

(b) **Healthcare**:

iii. Records patients' information and history, and appraisal.

iv. Source for academics' information on patients, facilitates clinical work assists in medical research discourse (Fig. 2).

(c) **Education**:

i. Source school records, course details, student information, and projects.

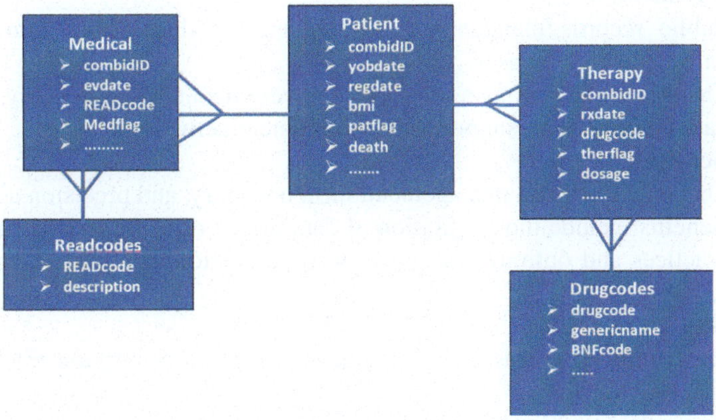

Fig. 2 An example of a longitudinal healthcare database entity relationship diagram (Reps 2014)

Fig. 3 Logistics DB application (bigdataanalyticsnews.com 2023)

ii. Source academic information on students, minimize academic jobs, and support academic research discourse.

(d) **Retail and e-commerce**:
   i. Control goods in store, source sales, and customers.
   ii. Source for marketing and assists in the easy flow of goods source for marketing and assists in the easy flow of goods

(e) **Manufacturing and logistics**:
   i. Production control, source for quality control, and supply chain management.
   ii. Efficiency in manufacturing processes, product quality, and optimizations of logistics and distribution (Fig. 3).

(f) **Telecommunications**:
   i. Anything that concerns subscriber databases, network management, and billing systems.
   ii. Benefits include managing customer subscriptions and monitoring network performance and supporting a major billing system.

(g) **Government**:
   i. Public records management, tax systems, and anything involving law enforcement databases.
   ii. The other area was also to promote government service delivery, ascertain data integrity, and support law enforcement activities.

(h) **Agriculture**:
   i. That concerns crop management, farm inventory, and precision agriculture.
   ii. Benefits include the facilitation of data-driven decision-making in farming practices and optimization of resource allocation and support agricultural research.

(i) **Energy and Utilities**:
   i. The concerns are asset management, energy consumption tracking, and grid monitoring.
   ii. Responsible maintenance of infrastructure, monitoring energy consumption, and supporting a predictive system.

(j) **Science and Research**:
   i. This involves experiment data management, a research collaboration platform, and scientific data analysis.
   ii. Additional areas of help support included sharing research data, reproducibility of experiments, and the science discovery system.

(k) **Media and Entertainment**:
   i. Media content management, digital asset storage, and user profiling to enable content recommendation.
   ii. Database system helps in the optimization of content delivery, the process of personalizing user experience, and the monetization of various forms of content.

(l) **Environmental Monitoring**:
   i. Management of climatic data and sensors, environmental sensor networks, and the databases of living creature diversity.
   ii. Supports environmental and its data collection research, environmental and conservation data, and general influence on climate.

(m) **Human Resources**:
   i. Used to manage employee records and payrolls and is essential in recruitment exercises.
   ii. Helps maintain the human resources database and supports accurate payroll processes and workforce planning.

(n) **Social Networking**:
   i. User data, connection, and content-sharing platforms.
   ii. Enhance users' social interaction, help in targeted content distribution, and foster user engagement.

(o) **Automotive Industry**:
   i. Help in vehicle tracking, managing parts inventory and control of manufacturing processes.
   ii. Optimize production, enhance quality control, and undertake easy supply chain management.

As can be easily ascertained, various applications demonstrate the benefits and advantages of database systems in various fields. They contribute to efficiency, innovative decisions, and scientific discoveries.

5 The Role of Databases in Managing and Analyzing 10 Vs of Data Produced in Science and Engineering

The role of databases in managing and analyzing big data produced in science and engineering is pivotal. Considering 10 Vs—Volume, Velocity, Variety, Variability, Veracity, Validity, Vulnerability, Volatility, Visualization, Value prescribed by Firican (2017) (Fig. 4).

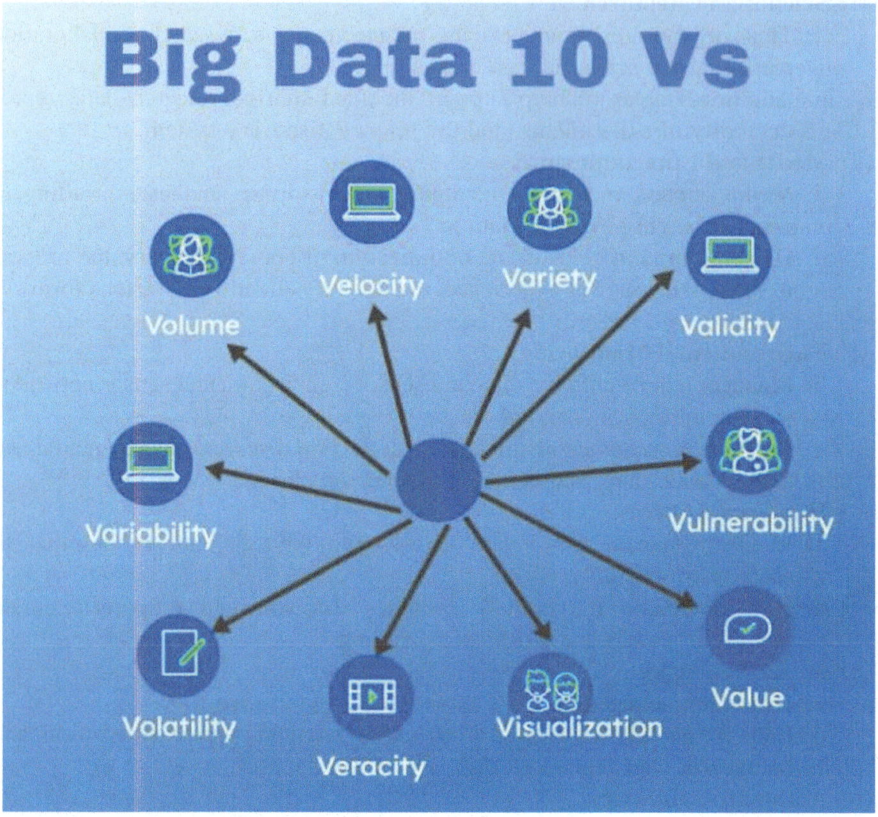

Fig. 4 Author

Let's list and briefly explain below the factors and their importance in these areas.:

a. Data Storage and Organization (Avi Silberschatz 2019): Structured Storage—ease in managing vast amounts of science-generated experimental data or engineering projects. The structured way to store uncountable experimental results, sensor reading, simulation, and other suitable data.
b. Efficient Data Retrieval: Quick Access to Information—Querying languages enable researchers and engineers to extract specific data subsets or even records from a database. As a result, it is easier to quickly access the required information for analysis.
c. Data Integration: Holistic Analysis—Databases also allow data integration from several sources. For a holistic analysis, integration is critical, especially in science and engineering, where data is often collected from varying instruments, simulations, or experiments.

d. Scalability: Handling Large Databases—Given that the amount of data in science and engineering keeps growing, databases will help scientists handle huge datasets better. In addition, more data is generated from experiments, simulations, and sensor networks.
e. Support for Complex Relationships: Graphical Representation—graphical databases will help in scenarios where data has complex relationships. This is commonly noted in biological networks or engineering systems cases.
f. Temporal Data Handling: Time-Series Analysis—Temporal databases assist in analyzing time series data. This includes conducting experiments and engineering processes that help establish how some variables change with time.
g. Data Security and Integrity: Maintain Accuracy—Some databases have transactions and ACID features that help maintain database security and integrity. It is essential in scientific research where data validity and reliability are key.
h. Machine Learning Integration: In-Database Analytics—Some databases support integration, meaning that in-database analytics can be conducted. With this feature, one can calculate important factors for decision-making, model performances, and extract the necessary knowledge or insight.
i. Spatial Data Handling: Geospatial Analysis—Projects involving geographical distribution data use spatial databases. These can be used to manipulate geographical data by storing and retrieval methods for geospatial data analysis.
j. Collaboration and Reproducibility: Shared Access—Databases allow researchers to access and share data from a central platform fostering collaboration as other scientists can reproduce the experiments.
k. Resource Optimization: Efficient Resource Usage—Databases index and organize data, optimizing the use of computational resources thereby reducing time and cost spent in processing data.
l. Historical Data and Versioning: Tracking Changes—Databases store all experiments' data, even where changes have been made, and previous versions, enabling scientists to track changes over time.
m. Data Visualization (Wilke 2019): Visual Representation—Databases integrate with visualization tools enabling researchers to develop visuals on experiments to ease interpretation and communication.
n. Regulatory Compliance: Adherence to Standards—Databases enhance data governance especially in fields with strict regulatory compliance standards, contributing to the credibility of the research findings.

As we noted above, databases are essential in managing and analyzing big data (Harrison 2015). The structured, scalable, and optimized platform has revolutionized science and engineering, allowing researchers to make new discoveries and solve complex problems.

6 Methodology

In this paper, we answered the following questions:

- What are the most used database models and systems in science and engineering?
- How do the current existing database systems tackle the challenges facing them considering the magnitude of data in science and engineering?
- Which emerging trends in database systems can be relevantly used in science and engineering?

We did it through:

- Identifying databases and journals, scientific publications, conference proceedings, and technical reports.
- Using the inclusion and exclusion criteria, facts and findings, we analyzed the literature, and studies while considering the type of research, year of publication, and relevance of the source data to the research question were used to include and exclude information.
- We extracted relevant information from the selected studies, such as the database models, systems, use cases, challenges, and solutions, and proceeded to analyze the information to identify the trends, patterns, and knowledge gaps.
- In the appendix of this paper, we have described an extraction procedure using structured format (primary facial expressions, emotions) video frames and audio frames to capture key information. This can also serve as a guide for those who want to extract information from such databases. We chose this procedure because we considered the application of databases in big data and artificial intelligence.

Therefore, we performed a useful systematic review of database systems in science and engineering and contributed to the understanding and improvement of data management in these fields.

7 The Most Common Database Models and Systems Used in Scientific and Engineering Research

In scientific and engineering research, the database models and systems used are based on the specific requirements and characteristics of the data. Some of the common database models and systems include:

a. Relational databases (Avi Silberschatz 2019): This type of database organizes data into tables where the tables are rows and columns, and the relationships between tables are specified using the keys. The use cases include structured

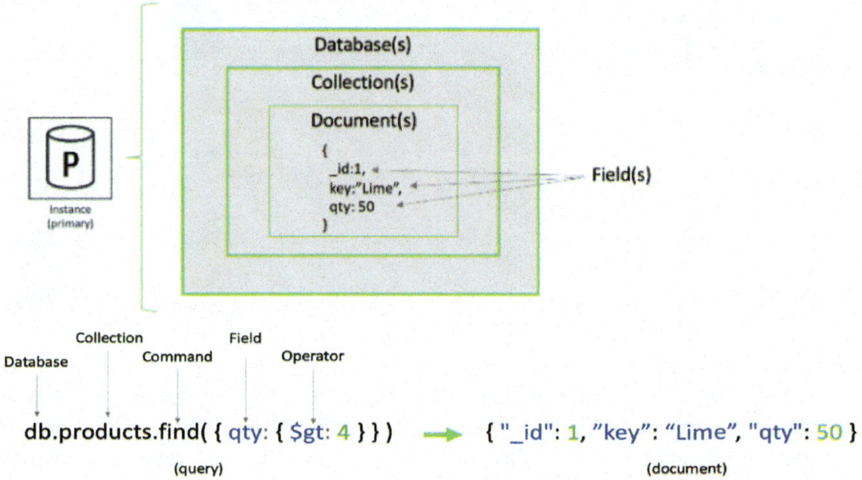

Fig. 5 A document database as a type of NoSQL database (aws.amazon.com, What Is a Document Database? 2023)

data with defined relationships, which are predominant in engineering applications and scientific experiments. The common systems include MySQL, PostgreSQL, Oracle database, and Microsoft SQL server.

b. Document-oriented databases: The type of database stores and retrieves the data in the form of documents instead of tables. In this case, the data is organized as documents such as JSON, BSON, and XML, among others where each document is a unit that is self-contained with dynamic and flexible schemas. They handle semi-structured data that is diverse and common in scientific research. The common systems include MongoDB, CouchDB, and Elasticsearch for semi-structured data (Fig. 5).
c. Graph databases (Ian Robinson 2015): These are designed to store and query data with complex and varying relationships between entities. They use nodes, edges, and properties to represent and store data to create structured data models. Use cases of Graph databases include various scientific domains in relational data models. Examples of Common systems are Neo4j, Amazon Neptune, and OrientDB (Fig. 6).
d. Time-series databases: they are designed for time-stamped rows with a unique time stamp. Commonly used in applications where monitoring is required. Use cases include Applications involving temporal data, such as monitoring or communications. Examples of Common systems are InfluxDB, Prometheus, and OpenTSDB.
e. NoSQL databases (Harrison 2015) store data in columns rather than rows, suitable for handling large amounts of data with high write throughput. Are used in big data applications to provide scale and high write rates. Common systems include Apache Cassandra, HBase, and Amazon DynamoDB. Amazon

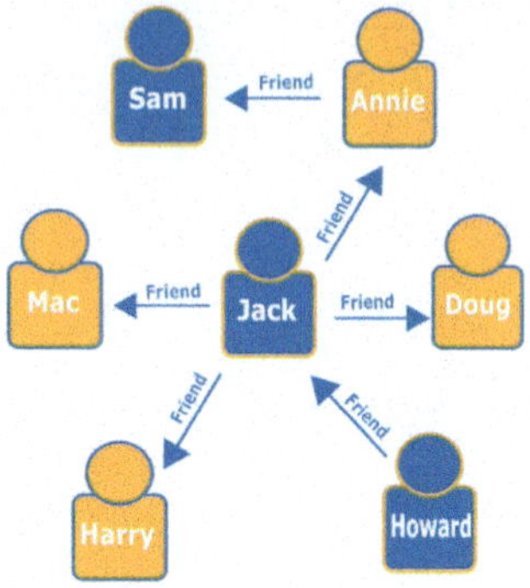

Fig. 6 An example of a social network graph (aws.amazon.com, What Is a Graph Database? 2023)

DynamoDB offers column-family storage solutions. NoSQL is a serverless fully managed NoSQL database used in applications with response bases in small microseconds. The following image shows the core feature of Amazon DynamoDB (Fig. 7).

f. In-memory databases: They store data in the computer's main memory, primarily in data and information. Use cases include In-memory Databases, Crunch-based applications, and applications that require low queries. Examples of Common systems are Redis, Memcached, and Apache.

g. Databases such as spatial databases are used in storing and querying spatial data or data that have geometry data. Use-cases include Vector Graphic systems in Engineering projects, Crossing and Routing applications. Examples of Common systems are PostGIS, Presoq, and Oracle.

h. Cloud-Based Databases: A specialized type of database system designed to deploy in cloud environments for on-demand and remote access. Example use cases include multi-centre scientific experiments, collaborative research environments, and distributed engineering tasks. Example systems: Azure SQL Database, Amazon RDS, and Google Cloud SQL.

Fig. 7 The core features of Amazon DynamoDB (aws.amazon.com, Amazon DynamoDB 2023)

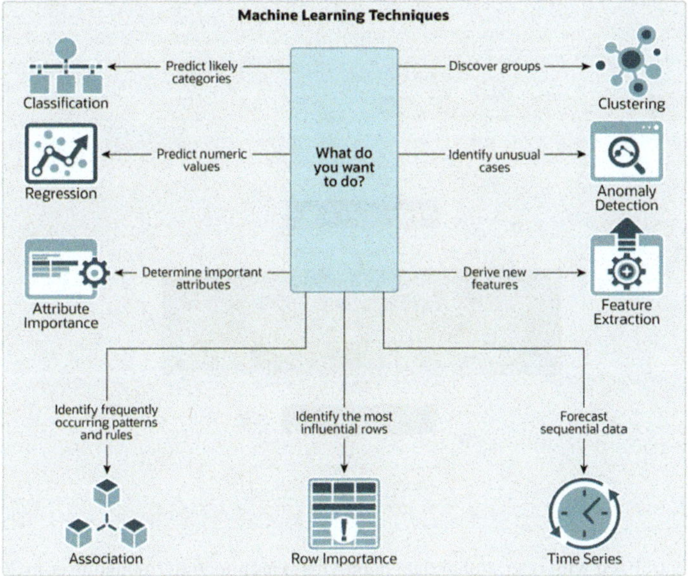

Fig. 8 How to use machine learning techniques (docs.oracle.com 2023)

i. Machine Learning Databases: Databases integrated with machine learning frameworks, allowing in-database ytics and real-time decision-making. Applications - Scientific research involving predictive modeling, data-driven decision-making, and machine learning applications. Common Systems: Google BigQuery ML, Oracle Autonomous Database with Machine Learning (Fig. 8).
j. Blockchain Databases (Avi Silberschatz 2019): Decentralized databases using blockchain technology for secure and transparent data management. Ensuring data integrity, traceability, and security in scientific and engineering transactions. Common Systems—Ethereum Hyperledger (Fig. 9).
k. Multidimensional Databases: Organizes data into multidimensional structures, suitable for analytical and business intelligence applications. Used for Data warehousing, OLAP, and scientific data analysis with multiple dimensions (Fig. 10).
l. Text Databases (Avi Silberschatz 2019): Designed for handling and analyzing textual data, including natural language processing. Use Cases—Textual analysis in scientific publications, engineering documentation, and sentiment analysis.
m. Columnar Databases: Stores data in columns rather than rows, optimized for analytics and querying specific columns. Used for data warehouses, analytics, and applications requiring fast query performance.

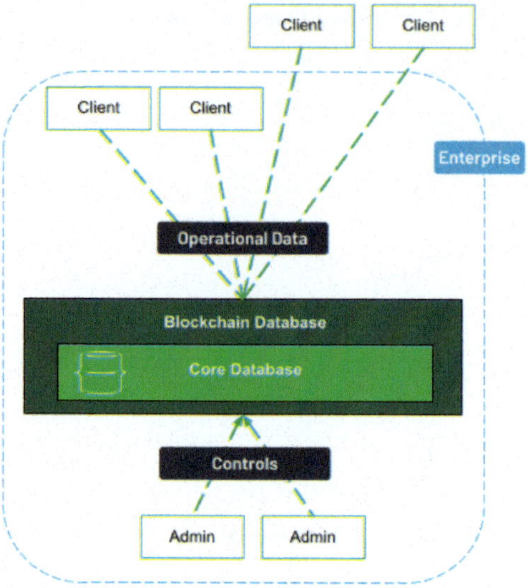

Fig. 9 A centralized with operational data deployment architecture (mongodb.com 2023)

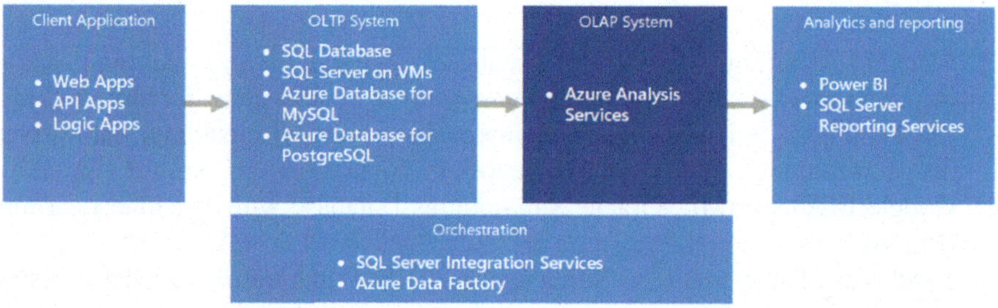

Fig. 10 Online analytical processing (OLAP)—Azure Analysis Services (microsoft.com 2023)

n. Distributed Databases (Avi Silberschatz 2019): Spreads data across multiple nodes or clusters for enhanced scalability and fault tolerance. Use Cases for large-scale scientific experiments, collaborative science projects, and engineering simulations.

At the end of this comprehensive analysis of the applied database models, we emphasize the importance of considering various factors when choosing a database system. These include the quality of the data structure, the system's complexity, its scalability, and its ability to integrate. By keeping these in mind, we can confidently select a system that best fulfils our needs and requests.

8 Identifying Trends and Gaps

So far, we have delved into the importance and role of these systems, their applications in different fields, and their various implementation models. Now, let's shift our focus to the trends and gaps in these systems. Understanding these trends, such as their applications in big data, artificial intelligence, cloud services, and blockchain, can help us stay ahead in the ever-evolving landscape of database technologies (Avi Silberschatz 2019; Harrison 2015).

A high-level overview of trends and gaps in the database technologies landscape is prescribed here:

Trends:

(a) NoSQL Databases (Harrison 2015): NoSQL databases continue to grow in popularity, especially where scalability, schema-less data models, and flexibility are required. Document-oriented databases such as MongoDB and CouchDB are adapted to work with unstructured or semi-structured data.
(b) Graph Databases (Ian Robinson 2015): With the growing trend of data connectedness, graph databases such as Neo4j are prevalent for applications with complicated relationships—social networks, fraud detection.
(c) Time-series Databases: The need to store and access time-series data quickly for IoT, monitoring, and finance led to a spate of time-series databases such as InfluxDB and Prometheus.
(d) Blockchain Databases: For secure, decentralized data management applications, blockchain-styled databases such as Ethereum and Hyperledger Fabric are garnering attention.
(e) Cloud-Based Databases: Managed cloud database services are increasingly popular due to scalability, ease of use, and cost – their effectiveness. For instance, Amazon Aurora, Google Spansr.
(f) Machine Learning Integration: Databases that integrate with ML frameworks, allowing in-database analysis and real-time decision-making, become more popular.

Gaps:

(a) Interoperability: The seamless integration between different database systems, particularly across hybrid and multi-cloud setups, continues to be a challenge.
(b) Security and Privacy: Given the increasing sophistication of data breaches, new measures such as encrypted storage, access control, and compliance features need to be constantly implemented to enhance database security and privacy.
(c) Scalability Challenges: It is critical to keep pace with the exponentially growing real-time volume of data, something made difficult by challenges scaling up or out in environments such as distributed systems.
(d) Data Integration and ETL: Efficiently integrating data from various sources and properly handling Extract, Transform, and Load operations continue to

prove difficult, further complicated by the growing number of various data formats.

(e) Data Governance: Organizations have yet to fully cover the gap regarding data quality, lineage, and adherence to data laws and regulations across highly diverse databases and systems.

(f) Resource Efficiency: Optimizing resource usage and consumption is critical, especially when operating in cloud environments, to ensure high performance and cost-effectiveness.

At the end of these findings and assessments, referring to trends and gaps, we note that the focus is on the need to implement these systems in big data, in the cloud, as well as the applications of ML/DL, Blockchain, etc. Technological developments (infrastructure in general, hardware, etc.) do not go in sync with those of upgrading database systems and software services. More needs in receiving, processing, analyzing, processing, and outputting data in big data, supervised and unsupervised, make this crucial. It remains the consideration of scientific researchers to deal with these issues mainly related to the weaknesses these systems display. We will deal further in this paper with some of the orientations where these researches should be focused.

9 Evaluating Strengths and Weaknesses

Following the issues addressed above, let us continue the analysis by evaluating the strengths and weaknesses of various database systems. This will involve examining various criteria, however, let us consider using relevant criteria such as performance, scalability, the data model they support, query execution, robustness, and ease of use. Based on these criteria, let's see and analyze the strengths and weaknesses of some of the databases mentioned so far. Let's see and analyze the strengths and weaknesses of some of the databases mentioned so far. For each category of databases, we have selected some of the most important ones for their applications and have referred to the main parameters such as performance, scalability, model support, query capabilities, robustness, and ease of use.

Performance

a. Relational Databases: PostgreSQL, MySQL (mysql.com 2024):

   Strengths: best performance with structured data, specifically in a read-intensive scenario.

   Weaknesses: Read and write-intensive distributed settings may have scalability challenges.

b. NoSQL Databases (Harrison 2015): MongoDB, Cassandra:

   Strengths: Exemplary read and write, horizontally scalable.

   Weaknesses: Acquits transactional consistency inadequately in contrast to relational databases.

c. In-Memory Databases: Redis, Memcached:
   Strengths: Fast read and write is accomplished because the data is contained in RAM.
   Weaknesses: Data is limited to the available RAM, and it is unsuitable for large datasets.

Scalability

a. Distributed Databases (Avi Silberschatz 2019): Apache Cassandra, Amazon DynamoDB:
   Strengths: The system is designed to scale horizontally, across multi-node nodes or clusters. Weaknesses: Consistency is difficult to achieve in distributed systems.
b. Cloud-Based Databases (eg. Amazon Aurora, Google Cloud Spanner)
   Strengths: Easy scaling with managed services fully automatic allocation of resources. Weaknesses: Cloud provider resource dependence and potential latency in accessing data.

Data Model Support

a. Relational Databases
   Strengths: It is best suited for structured data with a clear schema definition. Weaknesses: It is not as flexible as possible for semi- and non-structured data.
b. Document-oriented databases (eg. MongoDB, CouchDB)
   Strengths: Perfect for semi-structured or unstructured data with flexible schemas. Weaknesses: It may not be as efficient in solving complex relationship queries.

Query Capabilities (Avi Silberschatz 2019).

a. SQL-Based Databases (eg. PostgreSQL, MySQL)
   Strengths: SQL-based querying language allows you to write complex transactions and queries. Weaknesses: It is less suitable for unstructured data types, or non-stranded quotes.
b. NoSQL Databases
   Strengths: It depends on the type, but some offer rich quote performance, while others work on simplicity and scaling. Weaknesses: Not all NoSQL databases function similarly, and no standard quote languages exist.

Robustness

a. ACID-compliant databases (eg.PostgreSQL, Oracle)
   Strengths: All properties responsible for ACID transaction consistency. Weaknesses: Sometimes it harms distribution performance.

b. BASE (Basically Available, Soft state, Eventually consistent) Databases (e.g., Cassandra, DynamoDB):
   Strengths: Dealing with multiple locations, accepting damage tolerance, and high availability. Weaknesses: Strong tolerance is not as consistent as feasible when trying to be active.

Easy to Use

a. Relational databases
   Strengths: Many tools and documentation are available for mature communities. Weaknesses: Poor support and higher complexity for distributed locations.
b. NoSQL Databases
   Strengths: Better, easier-to-use for developers, locally available-source environments. Weaknesses: Low documentation and tools compared to relational database communities.

Now at the end of this analysis, we have a clearer picture of the strengths and weaknesses of these systems. As can be seen, the weaknesses mainly have to do with applications in distributed systems, but not only. Of course, there are other aspects that could be evaluated for these systems, but we focused on some of the most important ones, such as Performance, Scalability, Data Model Support, Query Capabilities, Robustness and Easy to Use.

10 The Performance Metrics Used to Evaluate Database Systems

We explained a lot about models, applications, strengths and weaknesses, characteristics, and what we should consider when choosing a suitable model that fulfils our purpose and tasks, but we have not clarified what parameters we will use to determine performance and evaluate a database system. Here, we will convey some of these parameters. We will explain what they measure and what that measured parameter means in relation to the relevant system. Performance metrics are equally important for measuring the database system's efficiency, reliability, and effectiveness. The choice of performance metrics entirely depends on the application's requirements and the organization's goals and objectives. The following are some of the important performance metrics used to measure the DB systems:

a. Throughput:
   i. It measures the number of transactions or queries a database system processes in a time period.
   ii. High throughput is used for systems that require a large number of transactions to process high throughput.
b. Latency:

   i. It measures the time takes to get a response after submitting the query.
   ii. Low latency system is used for real-time, near real-time responses because this system provides feedback in real-time to fulfill the gaming sector and financial trading systems.

c. Response Time:
   i. Response time measures the time it takes for the system to respond to a user query or transaction.
   ii. It is very important because it ensures that satisfies the user experience and is relevant to web servers, and customer-centric applications.
d. Concurrency and Scalability:
   i. It measures the number of transactions processed concurrently or helps evaluate the scalability of a DB System.
   ii. It is important for multi-user systems because it guarantees developers and users get their answer
e. Resource Utilization:
   i. Resource utilization metrics include CPU usage, memory consumption, and disk I/O. These metrics indicate how efficiently the system uses hardware resources.
   ii. It helps to determine the bottleneck and optimizes and allocates the resources.
f. Transaction Durability.
   i. Transaction durability ensures that committed transactions are permanent and survive system failures.
   ii. This metric is critical for systems that demand data consistency and reliability, including financial databases and healthcare systems.
g. Availability and Uptime:
   i. Availability measures the percentage of time the database system is operational. Uptime is the opposite of downtime.
   ii. High availability is critical for mission-critical systems to reduce disruptions and maximize uptime.
h. Fault Tolerance and Reliability:
   i. These metrics quantify the database system's ability to continue functioning in the presence of hardware failures or other disruptions.
   ii. These metrics are especially critical in applications that require data integrity, such as scientific research databases and control systems.
i. Backup and Recovery Performance:
   i. Backup and recovery performance is the speed and efficiency with which databases perform backup and recovery processes.
   ii. This metric is essential for enabling organizations to restore data quickly following a system failure.
j. Security Performance:
   i. Security performance measures the effectiveness of authentication, authorization, and encryption measures.
   ii. This metric is critical to securing sensitive data from breaches and unauthorized access.

k. Indexing Performance:
    i. Indexing performance measures the efficiency of indexing structures in enhancing query execution performance.
    ii. Indexing performance is essential in the performance of query execution, especially for systems with large datasets.
l. Compliance and Regulatory Metrics:
    i. Compliance and regulatory metrics ensure the database system adheres to the industry's regulations and standards.
    ii. Compliance with the metrics is essential in healthcare, banking, and other sectors mandated to follow regulatory standards.
m. Data Retrieval Performance:
    i. Data retrieval monitors how quickly and efficiently the system retrieves data, especially for complex queries.
    ii. It is crucial in systems that require quick responses to analytical and reporting queries.
n. Query Optimization: (Peter Bailis 2015)
    i. The measure assesses the degree to which the database system possesses mechanisms for improving the execution of query compilation.
    ii. Optimized queries enhance the whole system's performance, especially systems with complicated queries and massive data.
o. Data Compression Efficiency:
    i. The level of compression or decompression that the database system offers.
    ii. The more efficient the compression, the less the storage space that might enhance data transfer, given limited bandwidth factors.
p. Adaptability and Flexibility:
    i. How the system adjusts to changes in workloads and structure of data in the databases.
    ii. The ability is significant, especially in versatile areas where utilization patterns may change over time.

Evaluating the database systems on these performance measures is vital since it guides the organization in system selection, optimization, and resource deployment for a system that addresses the organization's data needs to achieve its objectives.

11 The Common Challenges Faced in Implementing and Using Database Systems

Various challenges are faced in implementing and utilizing database systems (Kamberaj 2021)that the organizations and the IT personnel must be aware of. These challenges may start from the design to the implementation and up to the maintenance and evaluation of the system. The common implementation and use challenges are:

(a) Data Security: Keeping data safe and secure is a significant challenge due to the ongoing threats of unauthorized access and breaches. Data security is among an organization's core responsibilities.
(b) Data Quality and Integrity: Keeping data accurate and consistent is another challenge, especially in large and frequently accessed data. The major issues in data quality are data duplication, information of certain fields, and errors in data entry.
(c) Scalability Issues: The scalability of a database system is another challenge. As the amount of data increases, it is difficult to scale database systems to satisfy increasingly heavy workloads without a decrease in performance, which would necessitate constant re-optimization.
(d) Performance Optimization: This involves optimizing the system's performance in terms of query response times and transaction throughput. Badly written queries or ineffective indexing can result in poor overall performance.
(e) Data Migration and Integration: Complex data integration and data migration are both challenges. Moving data to another system or integrating data from different sources is daunting.
(f) Complex Query Optimization (Avi Silberschatz 2019): Complex or vague queries may not execute effectively. Optimizing queries for quick response times and minimal resource usage is difficult.
(g) Compatibility Issues: Compatibility issues arise when integrating the new database with existing software, applications, or hardware.
(h) The selection of the appropriate database system also involves challenges related to Cost management transactions, data backup and recovery, concurrency control, and vendor lock-in.
(i) Data Backup and Recovery: As a matter of data safety, it is paramount to institute robust backup and recovery strategies. This involves defining how often data is backed up, the time it takes for recovery, processing speed, and whether data alterations are preserved.
(j) Concurrency Control: An aspect of concurrency, marking individual access between various users or transactions, is vital. This aids in preventing an anomaly, for instance, inconsistency of data or a deadlock.
(k) Vendor Lock-In: This is the inability to change the database vendor once you have invested in a specific vendor. Thus, it is crucial always to have a long-term view of the implications of selecting a specific vendor.
(l) Data Privacy Compliance: Adhering to data privacy regulations and compliance standards, such as GDPR or HIPAA, is challenging due to the nature of these standards. The database should comply with all applicable regulations at all times. Challenge: Nothing can be done except ensure all regulations are followed.
(m) User Training and Adoption: While the use and maintenance of the database are user-friendly, training users and administrators can be difficult. It is difficult for people to accept change and adapt to new things. Challenge: There are no opportunities other than pushing users to utilize the new system.

(n) System Downtime: Whether it occurs for maintenance or because of unexpected issues, downtime is inevitable. It is necessary to minimize it while still supporting the system regularly. Challenge: No opportunity other than attempting to make the best out of a bad situation.
(o) Data Archiving and Purging: It is not always easy to handle historical files and purging and archiving data to solve storage problems. Challenge: No other opportunity except to purge and archive only the necessary data.
(p) Monitoring and Troubleshooting: Identifying and fixing an issue quickly is essential to avoid future performance-related problems. Challenge: No opportunity other than being vigilant and seeking help when needed.
(q) Data Fragmentation: Over time, data becomes fragmented, slowing the system down. Defragmentation is required while working.

Ok, we know the challenges now, but how can we operate and address them to overcome them? What we recommend is that these challenges be addressed through:

- careful pre-planning,
- regular occurrences monitoring,
- periodic service, and
- staying aware of database technology innovation.

12 Best Practices

After we have referred to the main issues related to these systems, we consider a guide for best practices in making informed decisions. Let us list here some of the practices that we should follow when we consider the specific characteristics of our data and needs:

(a) Understand Data Characteristics
   - Data structure: structured data with well-designed relationships; relational databases: data structured with such interaction; semi-structured or unorganized data: NoSQL or document databases are more often used and may be more suitable.
   - Data Volume: We need to know how much data our application will process to determine whether we need a distributed database or a traditional single-node database.
   - Data Relationships: Graph databases might be the solution if our data is highly relational or involves complex relationships and interconnectedness.

(b) Consider Workload Patterns:
   - Read versus Write Operations:

     Databases optimized for read performance should be considered if the application has a read-heavy workload. Databases supporting high write throughput are preferable for write-intensive applications.

- Query Complexity: SQL-based databases are especially strong when tasks become more complex while NoSQL can be the preferred choice when simpler queries are involved and high-volume transactions become vital.

(c) Scalability Requirements:
- Horizontal versus Vertical Scaling: Always think about whether the application might have to scale horizontally across many nodes or vertically by adding more resources to a single node.
- Cloud versus On-Premises: Some facilities might align with a trendy cloud-based database service.

(d) Consistency and Availability Trade-offs:
- ACID vs. BASE: Understand that strong consistency (ACID) comes with the price of system availability (BASE- Basically Available, Soft state, Eventual consistency) considering our server might crash or some part of your infrastructure could fail, and so on.
- Eventual Consistency: If our application can sustain eventual consistency, NoSQL could be the right candidate because it can leverage the infamous availability and partition covers.

(e) Performance Considerations:
- In-Memory versus Disk Storage: We should also consider whether we need in-memory databases for latency overhead.
- Caching can also be an excellent ally when read performance has to be optimized, especially for specific types of data that tend to be frequently fetched. Implement caching strategies to enhance read performance, especially for frequently accessed data.

(f) Security and Compliance:
- Data Encryption: We must ensure that the database of our choice supports encryption for data at rest and in transit.
- Access Controls: Make sure only specific members can access our data to prevent high risks and extremely sensitive information from having unauthorized access.

(g) Operational Considerations:
- Ease of Maintenance: When using complex systems that might or might not even be distributed, make sure that the setup, configuration, and maintenance are as frictionless as possible.
- Monitoring and Management Tools: Choose a database system with robust monitoring and management tools to enable proactive system management.

(h) Community and Ecosystem: Community Support
- Consideration: Check the size and activity level of the community around the database system to ascertain adequate support and issue resolution.
- Ecosystem Integration: Consideration: Choose a database system with a rich, well-integrated ecosystem of tools, libraries, and frameworks.

(i) Future-Proofing:
- Scalability and Flexibility: Choose a database system that will permit evolution as data and business requirements change.

- Vendor Lock-in: Consideration: Avoid dependencies that will lead to vendor lock-in, especially in cloud-based implementations.

(j) Testing and Benchmarking:
- Benchmark Performance: Test the database system's performance thoroughly using benchmarking and under real-life conditions.

13 For Future Research and Development Activity in the Field of Database Systems in Science and Engineering

At the end of this paper, after we have addressed the most important issues that we thought were relevant regarding these systems. We also have recommended and provided guidelines above on what we should consider when we choose or plan to apply such systems. We have addressed the challenges we face when we need to process and analyze big data, cloud applications, and artificial intelligence for supervised and unsupervised data. Here, we list some activities that should be undertaken to complete this research framework.

Some of them include:

(a) Blockchain Integration into Scientific and Engineering Databases Research Focus: The potential of integrating blockchain technology into scientific and engineering databases for enhancing data provenance and traceability.
(b) Temporal and Historical Data Management: Research Focus: Develop a new model using a temporal database and suitable indexing methods to optimize how historical data is managed; it could support time-based analyses and research in science and engineering.
(c) Explainable AI in Query Optimization Research Focus: The application of AI in query optimization by developing methodologies that can reveal how queries are optimized
(d) Edge and Fog Computing Integration: Research Focus: Given that edge and fog computing integration is expected, my proposed research will narrow down to database systems that can easily integrate with them. It will especially focus on how data can be stored, processed, and synchronized between edge devices and central databases in scientific and engineering databases.
(e) Hybrid Cloud Database Architectures: Research Focus: The future database will rely on the hybrid cloud as the best available storage technique, and as such, my research will focus on developing innovative hybrid cloud structures. The structure will allow the benefits associated with on-premises storage and those of storage on the cloud. The research will also investigate the techniques of data migration, data replication, and running data servers on a hybrid cloud basis.
(f) Cognitive Database System: Research Focus: The database of the future will combine artificial and cognitive thinking, and as a result, my research focus

will be the following: I will investigate the possibility of integrating cognitive thinking to enhance intelligent data handling. The research will also investigate how cognitive artificial thinking can increase data processing and classification abilities.

(g) Privacy-Preserving Data Management: Research Focus: In a future where all data are available for public use, my research focus will be understanding how I can keep my data private. Add knowledge to receive private database management skills while you learn knowledge. Mine for the research will focus on creating a technique to permission all data as my own.

(h) Integration of Quantum Databases (Sudip Roy 2013): Research Focus—Quantum computing (Hai et al. 2024) power will be drawing near, and it will come with numerous databases to handle it. Therefore, my research focuses on the realization and the potential that quantum computing (Velamuri 2923) may be handling the future of databases. Investments will hence to understanding the principles behind quantum computing in terms of storing, retrieving, and interpretation.

(i) Energy-Efficient Database Processing: Research Focus: Data will be overwhelming in the near future, and it will be essential to have a resource to handle it. Mine will be developing a technique to optimize the database using only the necessary power. The research will be investigating software and hardware installed techniques for optimal power use.

(j) Graph Database Algorithms for Scientific Networks (Ian Robinson 2015): Research Focus: Data will no longer be handled; instead, rows will be fields and vice versa. My research will focus on learning ways of developing a graph database that will make it easier to manage it. The efficient concept of analyzing will involve graph-rich fields that will field all connections necessary.

(k) Intelligent Data Compression Techniques: This research focuses on developing intelligent data compression techniques that can adapt to the peculiarities of various types of data and workloads. The algorithms should ensure the efficient use of storage space while guaranteeing that the compressed data remains easily accessible.

(l) Automated Schema Evolution and Versioning: This area of research should focus on developing automated approaches to schema evolution and versioning in database systems. The developed techniques enable systems to accommodate changes made to metadata with maximum efficiency while ensuring backward compatibility.

(m) Natural Language Query Processing: This area of research should focus on the development of data accessibility by advancing natural language processing techniques. The research should result in systems that can comprehend users' requests in natural language.

n) Cross-Domain Data Integration: This research field should explore efficient cross-domain data integration methods. The developments should aid in facilitating collaboration and data exchange between scientists and engineers from various fields.

(o) Autonomous Database Management for IoT: This research area should focus on developing autonomous database management for the Internet of Things environments. The systems should be designed to optimize healing and adaptation to real-time data.
(p) Human-Centric Database Interaction: This research should focus on developing more user-centric interaction mechanisms with the database systems. The systems and their interfaces should be designed to be as intuitive as possible for researchers and other non-expert users.
(q) Spatial–Temporal Data Analytics: This research area should help develop the advanced capabilities of spatial–temporal data analytics. It should help develop better systems and indexing algorithms to facilitate the analysis of large-scale spatiotemporal datasets.

14 Conclusions

The aforementioned areas of future research and development show the magnitude of innovation and progress in database systems, which present trends that are in line with the changing dynamics of scientific and engineering fields. Exploring the provided directions may result in advancement that increases the efficiency, flexibility, and intelligence of database systems within the given sectors.

Appendix

Develop a Data Extraction Form to Capture Key Information

Step 1
Information is needed before developing a data extraction form. To create an effective form, we need to know:

1. **What type of data are we extracting?** (e.g., text, numbers, images, audio)
2. **What is the source of the data?** (e.g., websites, documents, surveys, databases)
3. **What are the key pieces of information we need to capture?** (e.g., names, dates, prices, categories)
4. **What format do we want the extracted data to be in?** (e.g., spreadsheet, database, text file)
5. **Are there any special considerations or challenges we need to consider?** (e.g., formatting inconsistencies, missing data, language barriers)

Step 2

Develop a Data Extraction Form for Images to Capture Key Information
Designing a data extraction form for images requires more context than just knowing it's an image. To create a helpful form, we need some additional details:

1. **What type of images are we working with?** (e.g., photos, screenshots, documents, diagrams)
2. **What specific information do we want to extract from the images?** (e.g., text, objects, locations, emotions, sentiments)
3. **Are the images structured or unstructured?** (e.g., forms with defined text fields, free-form images with no specific layout)
4. **Do we have any examples of the images we want to extract data from?**
5. **What format do we want the extracted data to be in?** (e.g., spreadsheet, database, text file, JSON)

Step 3
Information is needed to design a data extraction form specifically tailored to our needs. The form might include various elements like:

- **Text boxes**: For capturing extracted text or metadata associated with the image.
- **Multiple choice or checkbox options**: For classifying objects, locations, or other aspects of the image.
- **Drawing tools**: If we need to manually trace or highlight specific areas of interest.
- **Free text fields**: For detailed descriptions or observations about the image.

Step 4

Develop a Data Extraction Database Form Structured (Emotions) to Capture Key Information

Capturing Emotional Echoes: A Data Extraction Form for Structured Emotion Analysis
To effectively design a data extraction form for structured emotion analysis, let's delve deeper into our specific needs. Questions to be answered to tailor the form to our requirements:

1. **Data Source**:
   - **What type of media are we analyzing for emotions?** (e.g., written text, audio recordings, facial expressions, physiological signals)
   - **Is the data pre-processed or raw?** (e.g., transcripts from interviews, raw voice recordings, raw images of faces)

2. **Emotion Granularity**:
   - **What level of emotional detail are we interested in?** (e.g., primary emotions like joy, anger, sadness, fear, surprise; secondary emotions like frustration, disappointment, excitement)
   - **Do we want to capture intensity levels for each emotion?** (e.g., low, medium, high)
3. **Contextual Information**:
   - **Are we interested in any contextual factors influencing the emotions?** (e.g., speaker demographics, topic of discussion, physical setting)
   - **Do we need to track the temporal dynamics of emotions?** (e.g., how emotions change over time within a single source)
4. **Output Format**:
   - **How do we want the extracted emotional data structured?** (e.g., spreadsheet, database table, JSON format)
   - **Do we need any metadata associated with the extracted emotions?** (e.g., timestamp, source reference)
5. **Additional Points**:
   - **Are there any specific emotion analysis tools or frameworks we plan to use with this data?**
   - **Do we have any examples of the type of data we want to analyze?**

   Step 5

- **Source Information**: Identifying the data source (e.g., speaker ID, conversation topic, image file name).
- **Emotional Analysis**: Capturing the detected emotions, their intensity levels, and any additional emotion categories we require.
- **Contextual Elements**: Recording relevant situational factors like environment, demographics, or specific triggers for emotional shifts.
- **Temporal Progression**: If needed, tracking the evolution of emotions throughout the data source (e.g., emotions per sentence, per minute of audio).
- **Additional Notes**: Leave space for any further observations or interpretations of the extracted emotional data.

Step 6

Develop a Data Extraction Database from a Structured Format (Primary Facial Expressions and Emotions) to Capture Key Information

Capturing the Face of Emotion: A Structured Form for Analyzing Primary Facial Expressions

Let us examine our specific needs more deeply to design a powerful data extraction form for structured analysis of primary facial expressions.

1. **Image Source**:
   - **What type of images are we analyzing?** (e.g., static photos, video frames, live video feed)
   - **Are the faces in the images isolated or within a crowd?** (e.g., close-up portraits, group photos)
2. **Facial Expression Focus**:
   - **Are we interested in all six primary facial expressions (joy, sadness, anger, fear, surprise, disgust)?** Or do we have a specific subset in mind?
   - **Do we want to capture the intensity levels of each expression?** (e.g., low, medium, high)
3. **Facial Feature Analysis**:
   - **Do we want to analyze individual facial features independently?** (e.g., eyebrows, eyes, nose, mouth)
   - **Are we interested in any specific action units (AU) associated with these features?** (e.g., AU12—raised eyebrows for surprise, AU4—lip corner puller for sadness)
4. **Contextual Information**:
   - **Are we interested in capturing any contextual factors influencing the expressions?** (e.g., surrounding scene, body language, audio cues)
   - **Do we need to track the temporal dynamics of expressions across multiple images or video frames?**
5. **Output Format**:
   - **How do we want the extracted expression data structured?** (e.g., spreadsheet, database table, JSON format)
   - **Do we need any metadata associated with the extracted data?** (e.g., image timestamp, subject ID)
6. **Additional Points**:
   - **Are we using any specific facial expression analysis tools or frameworks with this data?**
   - **Do we have any sample images we want to analyze?** Sharing examples can refine the form design.

Step 7
Now can design a tailored form with sections like:

- **Image Information**: Identifying the image source (e.g., filename, frame number) and subject details (if applicable)
- **Primary Expression Analysis**: Marking the detected primary expression (joy, sadness, anger, fear, surprise, disgust) and its intensity level.
- **Facial Feature Breakdown**: Optionally, analyzing individual features like eyebrows, eyes, and mouth, recording any relevant action units observed.
- **Contextual Elements**: If required, note any scene details, body language cues, or audio elements influencing the expressions.

- **Temporal Tracking**: For video analysis, tracking the evolution of expressions across images or frames.
- **Additional Notes**: Leave space for any further observations or interpretations of the extracted expression data.

Step 8

Develop a Data Extraction Database from Structured Format (Primary Facial Expressions Emotions) Video Frames to Capture Key Information

Capturing the Evolving Face: A Structured Form for Analyzing Primary Facial Expressions in Video Frames
To design a powerful data extraction form for structured analysis of primary facial expressions in video frames, let's delve deeper into our specific needs.

1. **Video Source and Analysis Focus**:
   - **What type of video data are we analyzing?** (e.g., movie clips, interviews, surveillance footage, sports recordings)
   - **Do we need to analyze the entire video or specific segments?** (e.g., scene-by-scene analysis, focus on individual speakers)
2. **Frame Rate and Sampling**:
   - **What is the video frame rate?** (e.g., 24, 30, 60 fps)
   - **At what frequency do we want to extract expression data?** (e.g., every frame, every second, every Nth frame)
3. **Expression Analysis Level**:
   - **Are we interested in all six primary facial expressions (joy, sadness, anger, fear, surprise, disgust)?** Or do we have a specific subset in mind?
   - **Do we want to capture the intensity levels of each expression?** (e.g., low, medium, high)
4. **Facial Feature and Action Unit Analysis**:
   - **Do we want to analyze individual facial features independently?** (e.g., eyebrows, eyes, nose, mouth)
   - **Are we interested in any specific action units (AU) associated with these features?** (e.g., AU12—raised eyebrows for surprise, AU4—lip corner puller for sadness)
5. **Contextual Information and Temporal Tracking**:
   - **Are we interested in capturing any contextual factors influencing the expressions?** (e.g., audio cues, speaker identity, scene setting)
   - **Do we need to track the evolution of expressions across multiple frames?** (e.g., capturing emotional shifts over time)
6. **Output Format and Additional Points**:
   - **How do we want the extracted expression data structured?** (e.g., spreadsheet, database table, JSON format)

- **Do we need any metadata associated with the extracted data?** (e.g., video timestamp, frame number)
- **Are we using any specific facial expression analysis tools or frameworks with this data?**
- **Do we have any sample video clips we want to analyze?** Sharing examples can refine the form design.

Step 9
Based on this, we can design a tailored form with sections like:

- **Video Information**: Identifying the video source (e.g., filename, scene identifier) and analysis parameters (e.g., frame rate, sampling frequency).
- **Frame-by-Frame Expression Analysis**: Marking the detected primary expression for each frame (joy, sadness, anger, fear, surprise, disgust) and its intensity level.
- **Facial Feature Breakdown (Optional)**: Analyze individual features like eyebrows, eyes, and mouth per frame and record relevant action units.
- **Contextual Elements (Optional)**: Noting any scene details, audio cues, or speaker information influencing the expressions.
- **Temporal Tracking**: Charting the evolution of expressions across frames, highlighting sudden shifts or gradual changes.
- **Additional Notes**: Leave space for any further observations or interpretations of the extracted expression data.

Step 10

Develop a Data Extraction Database from Structured Format (Primary Facial Expressions and Emotions) Video Frames to Capture Key Information. Movie Clips Entire Video, 60 fps, Every Second Frequency

Capturing the Cinematic Spectrum: A Structured Form for Analyzing Facial Expressions in Movie Clips
Here's a data extraction database form tailored for analyzing primary facial expressions in movie clips with specific parameters:

Movie Information:

- **Movie Title**:
- **Scene Number**:
- **Speaker/Character**: (Optional)
- **Frame Number**: (Automatically populated based on analysis frequency)

Expression Analysis:

- **Primary Expression**: (Joy, Sadness, Anger, Fear, Surprise, Disgust)
- **Expression Intensity**: (Low, Medium, High)
- **Confidence Level**: (Percentage value indicating certainty of the detected expression)
- **Duration (Optional)**: Time spent exhibiting the expression (in seconds)

Facial Feature Details (Optional):

- **Eyebrows**: Raised, Lowered, Furrowed
- **Eyes**: Open, Closed, Squinted, Tears
- **Nose**: Wrinkled, Flared
- **Mouth**: Smile, Frown, Open, Pucker

Action Units (Optional):

- **Specific action units we're interested in** (e.g., AU12 – Raised Eyebrows for Surprise)

Contextual Elements (Optional):

- **Dialogue/Audio Cues**: Keywords or phrases related to the detected emotion
- **Scene Setting**: Location, lighting, atmosphere
- **Body Language**: Gestures, posture, movement

Additional Notes:

- Any subjective observations or interpretations of the expression
- Potential ambiguity or uncertainty in the analysis
- Interesting moments or patterns to explore further

Technical Details:

- **Video Filename:**
- **Frame Rate:** 60fps
- **Analysis Frequency:** Every second

Output Format:

- Choose our preferred format for extracted data (e.g., spreadsheet, database table, JSON)

Additional Points:

- Consider incorporating video timestamps for each frame analyzed.
- If using specific facial expression analysis tools, ensure the form aligns with their output format.
- Tailor the form further based on our specific research questions or movie genre focus.

References

Avi Silberschatz, H.F.: Database System Concepts. McGraw-Hill (2019)
aws.amazon.com.: Amazon DynamoDB (2023). Retrieved from https://aws.amazon.com/dynamodb/
aws.amazon.com.: What Is a Document Database? (2023). Retrieved from https://aws.amazon.com/nosql/document/
aws.amazon.com.: What Is a Graph Database? (2023). Retrieved from https://aws.amazon.com/nosql/graph/
bigdataanalyticsnews.com.: Reducing Costs and Waste: The Role of Big Data Analytics in Logistics Industry (2023). Retrieved from https://bigdataanalyticsnews.com/big-data-analytics-in-logistics-industry/
docs.oracle.com.: Oracle Machine Learning (2023). Retrieved from https://docs.oracle.com/en/database/oracle/machine-learning/
Firican, G.: The 10 V'S BIG DATA (2017)
Harrison, G.: Next Generation Databases: NoSQL and Big Data. Apres, Berkeley, CA (2015)
Hector Garcia-Molina, J.U.: Database Systems: The Complete Book. Pearson (2008)
Hai, R., Hung, S.H., Feld, S.: Quantum data management: from theory to opportunities. In: 40th IEEE International Conference on Data Engineering (ICDE 2024) (2024)
Ian Robinson, J.W.: Graph Databases, 2nd edn. O'Reilly Media, Inc. (2015)
Kamberaj, H.: Lecture Notes in Implementation of Database Systems. IBU (2021)
microsoft.com.: Online Analytical Processing (OLAP) (2023). Retrieved from https://learn.microsoft.com/en-us/azure/architecture/data-guide/relational-data/online-analytical-processing
mongodb.com.: Blockchain Database: A Comprehensive Guide (2023). Retrieved from https://www.mongodb.com/resources/basics/databases/blockchain-database
mysql.com.: Documentation (2024). Retrieved from https://dev.mysql.com/
Obe, R.O., Hsu, L.S.: PostGIS in action. Manning (2015)
Peter Bailis, J.M.: Readings in Database Systems, 5th edn. Red Book (2015)
Reps, J.M., Aickelin, U.: Tuning a Multiple Classifier System for Side Effect Discovery using (2014). arXiv:1409.1053v1[cs.LG]
sap.com.: SAP finance system (2021). Retrieved from https://community.sap.com/t5/enterprise-resource-planning-blogs-by-sap/central-finance-lessons-learned-sap-landscape-transformation-replication/ba-p/13493144
Sudip Roy, L.K.: Quantum databases. Creative Commons Attribution. http://creativecommons.org/licenses/by/3. https://www.cs.cornell.edu/~sudip/QuantumDB.pdf?trk=article-ssr-frontend-pulse_little-text-block (2013)
Velamuri, R.S.: Databases in the quantum computing world (2023)
Wilke, C.O.: Fundamentals of Data Visualization: A Primer on Making Informative and Compelling Figures. O'Reilly Media (2019)

The Situation of Digitalization in Enterprises: Analyzing Current Trends and Practices

Momeena Mehdi

Abstract

This paper explores the state of digitalization in enterprises by analyzing prevalent trends and practices. Employing qualitative analysis methodology and drawing upon secondary sources, the study examines the general characteristics of enterprises and key performance indicators (KPIs) such as workforce management, financial performance, market positioning, and innovation readiness. Additionally, it investigates the transformative impact of digital technologies on market structures, emphasizing the shift towards digitally-powered business models and the emergence of new entrepreneurial ecosystems. The evaluation of digitalization within enterprise systems reveals implications for operational processes and employment mechanisms, highlighting the imperative for organizations to embrace digital transformation. Furthermore, the study delves into the conditions conducive to technology development and implementation, emphasizing the pivotal role of factors such as financial resources, supportive government policies, technological infrastructure, and institutional reforms. The findings underscore the multifaceted nature of digitalization within enterprises, offering implications for strategic decision-making, policy formulation, and future research endeavors. Recommendations include fostering innovation ecosystems conducive to technological advancement and industrial growth. Overall, the research provides valuable insights into the complex dynamics of digitalization in enterprises and its broader implications for economic development and organizational competitiveness.

M. Mehdi (✉)
Canadian Institute of Technology, Tirana, Albania
e-mail: momeenamehdi@outlook.com; mehdi.momeena2021@khazar.org

Khazar University, Baku, Azerbaijan

K. Dhoska and E. Spaho (eds.), *Bridging Horizons in Artificial Intelligence, Robotics, Cybersecurity, Smart Cities, and Digital Economy*, Sustainable Economy and Ecotechnology, https://doi.org/10.1007/978-3-031-72029-1_14

Keywords

Digitalization • Development • Enterprises • Innovation • Technologies

1 Introduction

Digitalization has emerged as a critical driver of organizational success in the ever-evolving business landscape. This paper discusses the macro themes driving organizations towards a digital operating environment and the intricate dynamics of digitalization within enterprises. Shedding light on prevalent trends, transformative practices, and the broader implications for economic development and organizational strategy. By employing a qualitative analysis methodology and drawing insights from secondary sources, this study offers a comprehensive exploration of digital transformation in contemporary organizations.

The digital revolution has ushered in a new era characterized by unprecedented technological advancements and disruptive innovations. Enterprises worldwide are navigating this digital terrain, leveraging emerging technologies to enhance operational efficiency, drive innovation, and gain a competitive edge. Understanding the KPIs is essential for assessing the impact of digital technologies on organizational structures and strategic decision-making.

One of the focal points of this research is the transformative impact of digital technologies on market structures. The advent of digitalization has led to a paradigm shift, with traditional business models giving way to digitally-powered ecosystems and new entrepreneurial opportunities. This shift is not merely technological but also strategic, as organizations reconfigure their operations, offerings, and customer interactions in response to digital imperatives.

Further, the evaluation of digitalization within enterprise systems reveals profound implications for operational processes and employment mechanisms. As organizations embrace digital transformation, they must navigate challenges and opportunities arising from digital disruptions, workforce reskilling, and integrating digital tools into core business functions. This evolution underscores the need for organizations to adapt and innovate in a digitally driven environment.

This study also explores the external conditions conducive to technology development and implementation. Factors such as financial resources, supportive government policies, technological infrastructure, and institutional reforms play a pivotal role in shaping the digital landscape for enterprises. Understanding these contextual factors is essential for fostering innovation ecosystems that drive technological advancement and industrial growth.

Given the thematic discourse, this research aims to offer valuable recommendations for enterprises and policymakers. By nurturing innovation ecosystems that support technological advancement, organizations can improve their resilience, agility, and competitiveness in the digital age. This paper adds to the current discourse on digitalization by offering actionable insights for strategic decision-making, policy development, and future research efforts.

2 Analyzing the Current Situation of Digitalization in Enterprises

2.1 General Characteristics of Enterprises and Analysis of Their Activity Indicators

Enterprises are entities that engage in economic activities to generate profit. They can be companies, partnerships, sole proprietorships, or non-profit organizations, among other types. Enterprises are characterized by the creation of products or services, utilization of resources, and desire for financial gain, regardless of their size or purpose (Singh 2017). Small and medium enterprises are considered the major growth drivers for emerging economy countries by contributing to local development, innovation practices, and employment opportunities. SMEs are recognized for their entrepreneurial nature and shown to make a positive impact on boosting sales, in comparison to larger corporations (Rana and Choudhary 2019). These entities operate in various industries such as services, manufacturing, agriculture, and technology.

Wisconte et al. (2020) emphasize on company's assessment using key indicator system; key performance indicators (KPIs) constitute the efficiency profile of an organization. It allows strategic priorities to link with the current operational activities, set quantitative targets during planning, and assess the effectiveness of measures at the control phase. Understanding the general characteristics and analyzing KPIs is essential for effectively managing and leading the enterprise toward success. These indicators include factors such as the number of employees, annual revenue or sales, profitability, market share, customer satisfaction, and innovation levels (Meshram and Rawani 2019). Mainly, activity indicators provide valuable insights into the health and performance of any enterprise. By examining them, stakeholders can make informed decisions and take appropriate goal-driven actions. For example:

- Number of employees reflects the scale of the business and the effectiveness of its human resource management (Meshram and Rawani 2019).
- Annual revenue or sales indicate the company's capacity to earn income and its overall performance (Umadia and Kasztelnik 2020)
- Profitability determines financial stability—the enterprise's ability to generate profit after subtracting costs and expenses (Rehman et al. 2021)
- Market share denotes the company's standing and level of industry competition (Zeng and Zhang 2018)
- Customer satisfaction is a critical metric, it signifies customer loyalty and the quality of products or services (Kostin et al. 2021)

- Innovation levels show enterprises's preparedness to compete and adapt to changing market trends (Samirae et al. 2020)

Verevka (2018) maintains that these indicators provide a comprehensive view of the enterprise's performance. It forms a guideline for management decisions regarding growth, profitability, and sustainability. In summary, understanding and analyzing key indicators is crucial for identifying opportunities and gaps for improvement, i.e. effectively managing and leading an enterprise.

2.2 Transformation of Markets Through the Penetration of Digital Technology

Digital transformation is a new wave in management, using innovation to foster business models and productivity patterns, causing disruption in traditional market structures and the formation of new ones. In the last three decades, digital technologies have seen exponential growth in their globalized usage, inducing systematic digital disruption to transform the economy and society (World Bank 2016). Economic Commission for Latin America and the Carribean's report on the digital technologies for a new future, puts forth the stance for digitalization as the need of the hour for the universal welfare in the wake of challenges and crises.

Referring to the empirical findings of case studies and scholarly debate on ICT transforming the business models convinces us that the survival of physical markets faces a threat. Studies have shown that adopting technological infrastructure for transactions, digitalized sales, and marketing presents opportunities for businesses. The traditional wholesale markets are losing appeal for both buyers and sellers, leading to an inevitable transformation of the conventional ecosystem into a digitally-powered entrepreneurial one (Song et al. 2021).

Sadeghi et al. (2021) in their literature investigate and prove the importance of digital transformation for creating value through technological market expansion and entrepreneurship in response to the growing digitalization. The effects of digital transformation on value creation through technology require increased use of digital touchpoints for communication and consumption. Digital technology is thus a key business strategy for gaining a competitive advantage.

ECLAC (2021) examines that the dimensions of development continually evolve through the acquisition and integration of advanced digital technologies; putting businesses at the forefront of a changing ecosystem. This requires the market to adapt to digital changes and for businesses to integrate digital technology within their operating models, not just for survival but also for thriving in competitive environments.

Meanwhile, responding to environmental turbulence is recognized as an agile property. Li et al. (2021) in this regard have given the concept of digitally mindful organizations, which sense and respond to external opportunities and threats, embracing digital technologies to capacitate and facilitate market agility process.

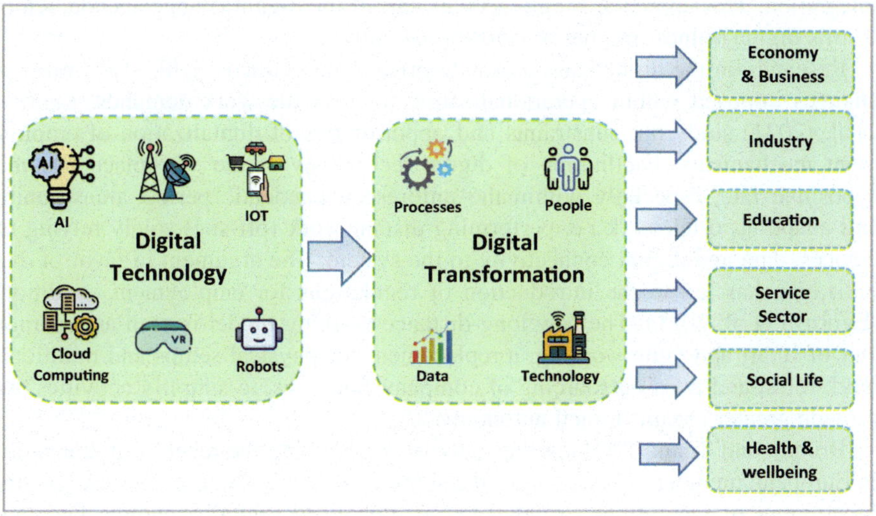

Fig. 1 Conversion of technology, digital transformation across various sectors. Source: Self-produced based on multiple literature concepts

Practical implications of the studied firms show that digital technology offers greater scope to meet business needs, leading to improved management infrastructure and enabling quicker market response systems (Li et al. 2021). The irregular shifts in consumption and production levels caused by the COVID-19 pandemic have significantly increased the use of digital technologies across various sectors, thereby amplifying its impact on markets. These market shifts impact not only businesses, but also have broader implications for society, the environment, and institutions (Fig. 1).

2.3 Evaluating Digitalization of Activities Within Enterprise Systems

KMPG measures the impact of the pandemic on enterprises' digital transformation to evaluate digital activity. According to most U.S. enterprise CEOs, digital acceleration is now worth years rather than months and there is no turning back. Digital transformation is a major determinant for competitive survival. Newer technologies and the pace of digitalization have the potential to challenge organizations' existing business models (Obukhova et al. 2020). With organization's life expectancy now falling short to fifteen years, there is a need for restructuring and revolutionizing their outdated models.

Ahmad et al. (2021) present a conceptual view of digital transformation, emphasizing that data is an integral component for digital enterprises. Digitalization relies on data to generate, absorb, and convert it into an asset through the process of data integration. This approach is embraced as part of the digital change agenda, which instills digital values into the organizational culture.

Research on the digitalization of enterprise systems sheds light on the massive changes imposed within operationalization to meet the work demands. Egorova et al. (2021) study the constraints and opportunities of digitalization of employment mechanisms. Facilitation of digital technology at the workplace has met a positive fate, especially within the unforeseen pandemic period, transitioning and adapting to telework i.e. performing distant work (off-site) solely relying on devices, data and digital connectivity to the system. The argument in favor of digitalization has led to the introduction of regulations for employment conditions (Egorova et al. 2021). The new long-distance working model is seen as a parameter of smart and agile working; a replacement for physical setups and traditional work roles, and even outsourcing of company functions, to simplify activities and provide greater organizational autonomy.

Brudny and Tutak (2021) in their article on assessing the level of digitalization in European member states, analyze the significant changes in market structure and functioning of institutions to upkeep in a globalized digital economy. Dynamic changes within enterprises are obvious due to the high degree of digital readiness; less dependency on human factors, and automation of processes and physical activities. This creates an exceptionally favorable environment for a digitalized industry.

For enterprises, the ultimate objective to employ digital technology is transforming internal processes and creating innovative solutions to generate value. As a result, organizations are eager to digitize to optimize tasks and enhance user experience.

2.4 Conditions for Technology Development and Its Implementation

Technology development and implementation in industries depend on a variety of factors. In addition to external factors, internal factors also play a significant role. Access to adequate financial resources, a skilled workforce, supportive government policies, and an open market that fosters innovation are essential factors (World Bank). Miller and Miller's (2019) research on technological integration emphasizes that effective collaboration among stakeholders such as developers, industry experts, and end users is critical for ensuring the success of technology development and implementation.

One important factor that contributes to the facilitation of technological development is the availability of information and communication technology infrastructure. This comprises reliable internet services and mobile subscriptions, which offer the vital connectivity required for businesses to operate efficiently (Masa'deh et al. 2019). Having a robust infrastructure is vital for industries like fintech, where the latest technology and supporting infrastructure are essential for startups to flourish.

Governments can play a crucial role in providing the necessary support and incentives to attract tech investments. Research published in the UNCTAD report on the G-24 discussion paper series aims at providing technical support for developing countries to bring about institutional reforms. Sanjay Lall confers the role of government policies in the wake of global competitiveness. As discussed by Lall (2004), government policies are essential in providing institutional, infrastructural, financial, and educational support to the private sector. These policies help in addressing market failures, increasing profit ratios, promoting investments, and fostering cooperation between the government and the private sector. Supportive government policies and regulations vital for enabling a conducive environment for not only technology development, safeguarding intellectual property rights, and integrating technology in various industries. Moreover, the economic and political climate of a country or region influences the diffusion of technology and its implementation.

Industrialized countries (e.g. East Asian Tigers) recognize the importance of public policies in promoting the establishment and expansion of new enterprises based on science and technology. OECD also countries allocate substantial amounts of funds to support technological infrastructure and thus foster a robust ecosystem and widespread innovation.

Case Study: PM Lee launched the Singapore Smart Nation Initiative in November 2014. A comprehensive roadmap aimed at leveraging technology to "support better living, stronger communities, and create more opportunities, for all". The SG government support included funding for research and development, regulatory environment for emerging technologies, and initiatives to promote innovation and digital literacy. Projects such as the National Digital Identity system, Smart Cities initiatives, Smart Urban Mobility and Smart Nation Sensor Platform, along with re-engineering healthcare services signify the impact of supportive government measures taken for technology adoption and implementation.

3 Methodology

This study employs a thematic analysis approach to investigate the state of digitalization in enterprises and analyze its impact on various operational aspects. The research design incorporates the analysis of existing theoretical work and secondary sources to gain comprehensive insights into the digital transformation landscape within organizations.

3.1 Data Collection

Data used here, is second-party data. Our data collection process focuses on gathering existing theoretical work, scholarly articles, industry reports, government publications through official websites, and case studies related to digitalization

Table 1 Digitalization trends in enterprises

Trends/impacts	Description
Increased automation	Adoption of AI, robotics, and automated processes to streamline operations and enhance efficiency
Cloud integration	Shift towards cloud-based solutions for data storage, collaboration, and scalability
Digital customer experience	Emphasis on enhancing customer interactions through digital channels and personalized experiences
Data analytics	Utilization of big data analytics for informed decision-making and predictive insights
Cybersecurity measures	Implementation of robust cybersecurity measures to protect digital assets and mitigate risks

Source: Self-made. Derived from data analysis by author

in enterprises. These secondary sources provide a wealth of information and insights into digital transformation trends, practices, challenges, and outcomes across different industries and organizational contexts.

3.1.1 Data Analysis

Thematic analysis is utilized to systematically analyze and interpret the data collected from theoretical work and secondary sources. The analysis involves identifying common themes, patterns, and key findings related to digitalization, organizational strategies, technology adoption, digital transformation drivers, barriers, and best practices. The thematic analysis process allowed a structured exploration of the digital transformation landscape, generating a valuable perspective of the evolving nature of digitalization within enterprises.

4 Results

In this section, we present the key findings derived from our data analysis. The results revealed the following key findings regarding digitalization in enterprises (Table 1).

Emerging Trends in Digital Transformation:

- AI and Machine Learning Integration: Continued adoption for predictive analytics.
- Blockchain Applications: Exploration for secure transactions and data integrity.
- IoT Adoption: Increased utilization for data collection and process optimization.
- Digital Ecosystems: Growth of digital partnerships and ecosystems for innovation.

5 Conclusion

In summary, this study has shed light on the intricate landscape of digitalization within enterprises, revealing prevalent trends, transformative practices, and their broader implications. The digital revolution has prompted organizations to adapt and innovate in response to evolving market dynamics, leveraging digital technologies to enhance operational efficiency and drive innovation. Organizations face the dual challenge of navigating digital disruptions while seizing opportunities for growth and competitiveness.

While our analysis offers valuable insights, it is essential to acknowledge certain limitations. Firstly, the findings may not be entirely generalizable across all industries or regions due to variations in digital maturity and contextual factors. Additionally, relying solely on secondary sources may have restricted the depth of our analysis, and there could be biases inherent in the interpretation of data.

Despite these limitations, our study provides actionable recommendations for enterprises and policymakers. Firstly, fostering innovation ecosystems is paramount stakeholders to drive technological advancement and industrial growth. Secondly, investment in digital infrastructure is crucial to facilitate technology adoption and implementation. Thirdly, continuous learning and skill development are essential to equip the workforce with digital competencies. Fourthly, adaptive regulatory frameworks should balance innovation with data privacy and cybersecurity concerns. Lastly, strategic partnerships can accelerate digitalization efforts and enhance competitiveness.

Acknowledging and navigating through these policy recommendations, enterprises can better position themselves for success in the digital age.

References

Ahmad, A., Alshurideh, M., Al Kurdi, B., Aburayya, A., & Hamadneh, S.: Digital transformation metrics: a conceptual view. **24**, 1–18 (2021). Retrieved from https://www.researchgate.net/publication/353527038

Brodny, J., Tutak M.: Assessing the level of digitalization and robotization in the enterprises of the European union member states. PLOS ONE **16**(7), e0254993. Retrieved from https://doi.org/10.1371/journal.pone.0254993

ECLAC.: Digital Technologies for a New Future (2021). Retrieved from https://www.cepal.org/sites/default/files/publication/files/46817/S2000960_en.pdf

Egorova, M., D'Aponte, M., Ponomareva, D., Minbaleev, A.: The impact of digitalization on the enterprise system: Russia and Italy and the discipline of smart working between constraints and opportunities. LeXonomica **13**(2), 151–166 (2021). Retrieved from https://doi.org/10.18690/lexonomica.13.2.151-166.2021

Jie, W.S.: Singapore's smart nation initiative—a policy and organisational perspective: 1–12. ScholarBank@NUS Repository (2017). https://doi.org/10.25818/vjdp-1gqf. https://www.smartnation.gov.sg/files/publications/smart-nation-strategy-nov2018.pdf

Kostin, K.B., Steinbiß, K., Petrinovic, O.: Determining the KPIs of the German engineering industry based on the evaluation of contemporary business models. Strateg. Manag. **26**(3), 3–36 (2021). https://doi.org/10.5937/StraMan2103003K

Lall, S.: Reinventing Industrial Strategy: The Role of Government Policy in Building Industrial Competitiveness, G-24 Discussion Papers 28, United Nations Conference on Trade and Development (2004). Retrieved from https://unctad.org/system/files/official-document/gdsmdpbg2420044_en.pdf

Li, H., Wu, Y., Cao, D., Wang, Y.: Organizational mindfulness towards digital transformation as a prerequisite of information processing capability to achieve market agility. J. Bus. Res. **122**, 700–712 (2021). Retrieved from https://doi.org/10.1016/j.jbusres.2019.10.036

Masa'deh, R., Almajali, D.A., Alrowwad, A., Obeidat, B.Y.: The Role of Knowledge Management Infrastructure in Enhancing Job Satisfaction: A Developing Country Perspective (2019). https://doi.org/10.28945/4169

Meshram, S., Rawani, A.M.: Analyzing success factors of small and medium enterprises (SMEs): a study in Indian context. In: Shanker, K., Shankar, R., Sindhwani, R. (eds.) Advances in Industrial and Production Engineering. Lecture Notes in Mechanical Engineering. Springer, Singapore (2019). https://doi.org/10.1007/978-981-13-6412-9_78

Miller, A., Miller, M.: Study of the problems of technological integration in the manufacturing industry in Russia (2019). https://doi.org/10.5937/straman1903033m

Obukhova, A., Merzlyakova, E., Ershova, I., Karakulina, K.: Introduction of digital technologies in the enterprise. In: E3S Web of Conferences (2020) Retrieved from https://doi.org/10.1051/e3sconf/202015904004

Rana, G., Choudhary, R.: Micro Small and Medium Scale Enterprises—'Hidden and Helping Hand in Economic Growth' (2019). Available at https://doi.org/10.2139/ssrn.3354268

Rehman, S.U., Elrehail, H., Nair, K., Bhatti, A., Taamneh, A.M.: MCS package and entrepreneurial competency influence on business performance: the moderating role of business strategy. Eur. J. Manag. Bus. Econ. **32**(1), 1–23 (2021). https://doi.org/10.1108/EJMBE-04-2020-0088

Sadeghi, V.J., Perez, A.G., Candelo, E., Couturier, J.: Exploring the impact of digital transformation on technology entrepreneurship and technological market expansion: the role of technology readiness, exploration and exploitation. J. Bus. Res. (2021). Retrieved from https://doi.org/10.1016/j.jbusres.2020.11.020

Samirae, Z., Alshibly, M.S., Alghizzawi, M.: Excellence in drawing up marketing mix strategies for small and medium enterprises (SMEs) and their impact on the marketing performance. Bus. Manag. Econ. Res. **63**, 30–36 (2020). https://doi.org/10.32861/bmer.63.30.36

Singh, K.D.: Enterprise system in a mid-sized Indian manufacturing organisation—a case study. Int. J. Ind. Cult. Bus. Manage. **14**(3), 257–273 (2017). https://doi.org/10.1504/IJICBM.2017.083230

Song, Y., Escobar, O., Arzubiaga, U., De Massis, A.: The digital transformation of a traditional market into an entrepreneurial ecosystem. Rev. Manag. Sci. **16**(1), 65–88. Retrieved from https://doi.org/10.1007/s11846-020-00438-5

Umadia, K.S., Kasztelnik, K.: The financial innovative business strategies of small to medium scale enterprises in developing country and influence for the global economy performance. Soc. Econ. Challenges **4**(3), 20–32 (2020). https://doi.org/10.21272/sec.4(3).20-32.2020.

Verevka, T.: Key performance indicators of high-tech enterprises. In: SHS Web of Conference, vol. 44, p. 00077 (2018). https://doi.org/10.1051/shsconf/20184400077

Wisconte, S.A., Musina, D.R., Gerasimova, M.V., Yangirov, A.V., Nasyrova, S.I.: Company efficiency assessment using key indicator system. In: Proceedings of the International Scientific Conference "Far East Con" (ISCFEC 2020) (2020). https://doi.org/10.2991/aebmr.k.200312.43

World Bank.: World Development Report 2016: Digital Dividends. Economic Commission for Latin America and the Caribbean (ECLAC) (2016). Retrieved from https://caribbean.eclac.org/publications/world-development-report-2016

Zeng, C., Zhang, P.: The construction of performance evaluation index system for strategic emerging enterprises in China. In: Proceedings of the 2018 International Conference on Economics, Business, Management and Corporate Social Responsibility (EBMCSR) (2018). https://doi.org/10.2991/ebmcsr-18.2018.48

The Impact that Professional Judgment of Materiality and Audit Risk Has on the Issue of the Auditors Opinion

M. S. C. Julian Saraci and Uarda Allushi

Abstract

This study delves into the influence of materiality and audit risk judgments on the reliability and content of auditor's reports, pivotal aspects of the auditing process that shape the nature, timing, and scope of audit procedures. Despite their significance, there is a limited quantity of research on how these judgments impact the final audit report. Using a qualitative methodology centered on questionnaires and extensive literature review, this study gathers insights from experienced auditors. The questionnaire aims to reveal factors affecting materiality and audit risk judgments, the challenges auditors encounter in making these judgments, and the perceived effects on the quality and credibility of auditor's reports. The study expects to unveil the primary drivers behind materiality and audit risk judgments, including regulatory demands, client attributes, and auditor's professional judgment. The findings are expected to offer valuable insights for auditors, standard setters, and regulators, aiding in enhancing the quality and relevance of audit reports. By comprehending the influence of materiality and audit risk judgments, auditors can refine their audit procedures' efficacy and bolster the credibility of their reports. This study also contributes to the existing literature by furnishing empirical evidence on the factors shaping materiality and audit risk judgments in the auditing field.

Keywords

Materiality judgment • Audit risk judgment • Auditor's report • ISA

M. S. C. J. Saraci (✉) · U. Allushi
University College, Canadian Institute of Technology, Tirana, Albania
e-mail: julian.saraci@cit.edu.al

U. Allushi
e-mail: uarda.allushi@cit.edu.al

K. Dhoska and E. Spaho (eds.), *Bridging Horizons in Artificial Intelligence, Robotics, Cybersecurity, Smart Cities, and Digital Economy*, Sustainable Economy and Ecotechnology, https://doi.org/10.1007/978-3-031-72029-1_15

1 Introduction

Being able to adequately interpret and apply audit risk and materiality are important skills for an auditor. These judgments are important in determining the nature, timing, and scope of audit procedures, and, in turn, affect the nature and reliability of the auditor's reports. It is important to say that auditors face several challenges when making judgments about materiality and audit risk. These assessments are difficult to make because the environment in which auditors operate is complex, dynamic, and there are differences of opinions with respect to these concepts.

Auditors depend on the idea of materiality so that they can determine significance of financial misstatements or omissions which might affect the economic decisions by users from financial statements. Materiality determination is naturally subjective and therefore calls for professional judgment in relation to its quantitative and qualitative aspects for purposes of determining thresholds that would influence the scope, nature, and timing of audits. On the other hand, audit risk- the possibility that an auditor may fail to appropriately modify their opinion on materially misstated financial statements—requires a careful assessment of any errors or frauds that might have an impact on the audit's outcome. The environment of the entity, its internal controls, and the auditor's personal risk of material misstatement are all taken into account when assessing audit risk; each of these factors significantly depends on the auditor's capability for evaluating and interpreting complicated situations.

The study aims to provide a detailed investigation into the relationship between professional judgment, materiality, and audit risk by conducting a thorough scrutiny of available literature, standards, and frameworks. It also includes several specific objectives—to determine the internal and external factors for auditors' professional judgment, to assess how judgmental variations affect the consequences of auditing, to analyze the implications of judgmental variations on auditors' opinions and stakeholders' decisions, to determine the challenges for auditors in applying judgment, and to make suggestions aimed at improving the consistency and quality of auditing.

2 Literature Review

Materiality is a basic auditing concept that is supposed to identify the significance of an omission or misstatement that may have an influence on the decisions of those who depend on financial statements. The International Standards of Auditing (ISAs), specifically ISA 320, emphasize that the determination of materiality has to consider quantitative and qualitative factors.

Auditors are required, under ISA 315, to identify and assess the risks of material misstatements in the financial statements that may have arisen from fraud or error. This would involve an essential understanding of the client's business and its operation environment, including the particular industry that it operates in. Part of this understanding is the assessment of the internal controls for the client's accounting systems to prevent, detect, and timely correct material misstatements. The auditor derives the assessment of the risk of material misstatements from the effectiveness of the internal controls. In particular, this would include the inherent risk—the risk of an error or fraud occurring—and the control risk—the risk of the misstatement going undetected by the entity's internal controls. The primary purpose of this risk assessment is to enable auditors to determine how extensive their audit procedures need to be, what they need to focus on, and when they need to be conducted.

Auditors then need to adjust their audit procedures according to the nature, timing, and extent as necessary by the risks assessed. This is done by analyzing inherent risks with regard to particular transaction classes or account balances and determining to what extent reliance can be placed on controls in planning substantive procedures. In case effective controls are expected, or in cases that substantive procedures alone are likely to be inadequate, ISA 330 requires that tests of controls be designed and performed in order to obtain sufficient appropriate evidence about operating effectiveness.

After obtaining the evidence, auditors conclude whether, in their opinion, the financial statements give a true and fair view of (or present fairly, in all material respects) the financial position, financial performance, and cash flows of the entity in accordance with the applicable financial reporting framework. In case auditors decide that they are unable to get sufficient appropriate audit evidence or if there is a material misstatement that is not corrected, then this will affect the type of opinion they will give.

Recent research, such as Smith and Higgs (2020), indicates that more experienced auditors are likely to set higher materiality thresholds because of their ability to better understand risk with respect to material misstatements and hence bring about audit efficiency. Audit risk assessment is still another area where professional judgment may highly influence the audit outcome. According to Lee and Choi (2018), auditors' perceptions of client integrity and industry risks have a direct influence on inherent and control risk assessments. Such assessments feed into the overall audit strategy and determine the nature, timing, and extent of the audit procedures. Additionally, Peterson (2021) examined how auditors change their risk assessments when they have clients in a dynamic industry, such as technology or renewable energy. The study showed that auditors frequently assess the risks as higher and adjust audit plans accordingly to reduce the perceived threats. In another related study, Martin and Steele (2019) reviewed cases in which auditors issued opinions under conditions of high uncertainty, as during financial crises. They noted that auditors' professional judgment is directed toward more conservative opinions in order to safeguard against potential backlash from incorrect assessments.

3 Methodology

This study will employ a qualitative research approach to examine the breadth and complexity of auditors; professional judgment on materiality and audit risk, as well as how these judgments impact their audit opinions. The qualitative technique was chosen in order to capture the interpretive and subjective aspects of auditor decision-making processes. Forming the survey is done electronically from the Google Forms platform and is divided into sections. The cover consisted of an introduction and consent form. The participants were exposed to the aim of the study and nature of the survey. It was also emphasized that to participate, they must have knowledge in auditing area. Each section of the survey contains mostly Likert scale questions (attitude measurement) demographics were placed first, since the questionnaire was anonymous and doesn't affect on how people fill out the survey. They were asked about their level of education, years of experience in auditing, the industry they are focused on and the size of entities they audit.

The Determining Materiality section consisted of 8 Likert scale questions. Auditors are offered a series of statements from (1 = Strongly agree to 5 = Strongly disagree), with which they can agree or disagree based on their professional judgment. Main parts consisted of Company's size of total assets or total capital, Company's Organizational Complexity, Company's Capital Structure, Company's Profitability, Company's Revenu size, Company's Industry Type, Conducting first-time audit for auditors and Type of business (profitable/non-profitable, public/private).

The Risk Assessment and Auditor's Response section consists of 16 Likert scale questions. Auditors are given different statements into which they can show their opinion (1 = Strongly agree to 5 = Strongly disagree) on how different factors can impact the auditors' responses to risk assessment procedures. Factors affecting risk judgment were measured by looking at: Company's Financial Structure, Company's Regulatory Requirements, Company's Inherent circumstances and Industry, Company's Internal Controls, Company's ability to perform audit procedures, Family-run businesses, Dominant or uncooperative management, Size of transactions, Existence, type and size of transactions with related parties, Formal Policies and hierarchical systems, Regulatory or tax audit reports in the past year, Opinion from previous auditor, Expertise level of employees, Specific items like Revenue in profitable companies, Evidence of potential or actual fraud, Frequent changes in key management personnel.

This study has collected primary and categorical data. The data is extracted from a questionnaire. The sources of data collection is Google Forms. The data is gathered from people who work on audit. The level of population is all practicing auditors. The target population of this study is experienced auditors with more than 2 years in the industry, and the sample size selected for our studies is 50 auditors.

The strategy used is purposeful sampling. Participants will be selected focusing on participants who are experienced with auditing. Another strategy used is Snowball Sampling. It was required of the participants to distribute the questionnaire to other professionals, their colleagues, or other members of their professional network who are involved in auditing.

4 Data Analysis and Results

Table 1 is the compilation of responses about various factors that govern auditor decision of materiality levels from strongly disagreeing to strongly agreeing.

The vast majority of auditors believe that "Capital structure" and "Profitability levels" are the two most important factors. Profitability Levels were closely monitored, with 76% agreeing (41%) or strongly agreeing (35%) that profitability impacts materiality assessments. Profitability affects not only the risk of material misstatements but also determines how the audit scope is decided, and more profitable companies may receive higher attention. Auditors closely watched capital structure, with 70% agreeing (39%) or strongly agreeing (31%). The composition of a firm's debt and equity can have a huge effect on the risk assessment and materiality resulting from financial instability and obligations.

Size of Total Assets/Capital showed a significant influence on materiality judgment, with 68% (33% agree, 35% strongly agree) of auditors acknowledging the importance of this variable. Given its possible impact on financial statements, this finding is perhaps more predictable than others in that a firm's capital or absolute asset size seems to play an important role in defining materiality. Type of Business, like the latter, is equally important with 68% of auditors agreeing on its importance (35% agree, 33% strongly agree) that the nature of business significantly

Table 1 Auditor's level of agreement in percentages

Variables	1-strongly disagree (%)	2-disagree (%)	3-neutral (%)	4-agree (%)	5-strongly agree (%)	Agree + strongly agree (%)
Size of total assets/capital	2	4	27	33	35	68
Organizational complexity	4	10	39	22	24	46
Capital structure	6	4	20	39	31	70
Profitability levels	2	2	20	41	35	76
Size of revenue	2	10	31	31	27	58
Industry type	2	16	14	41	6	47
Conducting first-time audit	2	6	27	39	27	66
Type of business	4	6	22	35	33	68

molds audit considerations, notably in matters of operational risks and industry-specific practices. Conducting First-Time Audit was identified by 66% of auditors (39% agree, 27% strongly agree) as a factor that heightens materiality considerations. New audit engagements require a deeper understanding of the business, often leading to a more cautious approach in determining materiality levels. Comparatively, with general agreement level Size of Revenue was an important factor for 58% of auditors (agree 31%, strongly agree 27%). The scale of a company's revenue can dictate the breadth of audit procedures due to its direct impact on financial outcomes and potential areas of significant misstatements.

The organizational complexity and industry type, while they do influence auditor judgments, are less likely to have a direct influence compared to other factors. Industry Type was noted by 47% of auditors, 41% agreeing, and 6% strongly agreeing. It seems that the industry can have an effect on the approaches in an audit because of specific risks and regulatory environments. However, it might be a less immediate concern when it comes to other financial and structural aspects. Organizational Complexity—Perceived to be less influential by 46% of auditors: agree 22%, strongly agree 24% that it influences materiality. Complex organizational structures blur a clear understanding and provide an increasing difficulty for making an assessment in terms of material misstatements; hence, detailed audits are required.

Table 2 expresses in percentages the auditors opinion on how different factors impact their risk judgment from a scale of 1 to 5 (1-Strongy disagree, 5-Strongly agree).

The highest extent of agreement was on the transaction with related parties, where 90% of auditors agreed 59% or strongly agreed 31%, an indication that auditors feel that this is a very important element. This implies that related-party transactions are thought to be extremely dangerous because of possible financial misstatements and conflicts of interest. The other important findings are the Size of Transactions or Accounts, with 84% of respondents agreeing (53%) or strongly agreeing (31%) with this factor. This has a much greater effect on financial impact, hence indicating how important they are in audit risk assessments. Very important, also, is the factor of Clients Lacking Formal Policies, with a result showing

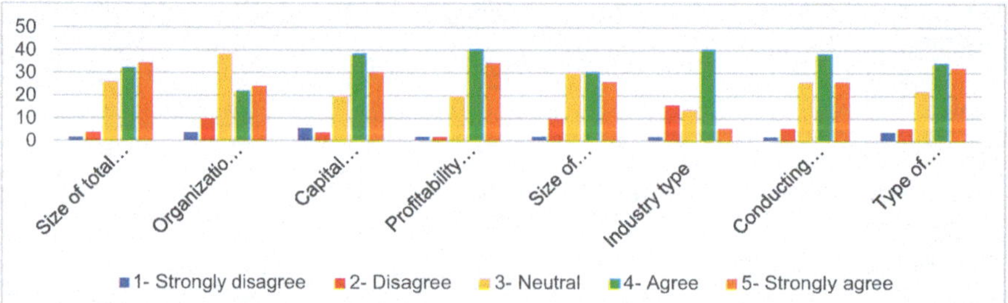

Graphic 1 Level of agreement/disagreement for factors affecting materiality judgment

Table 2 Auditor's level of agreement in percentages for risk judgment

Variables	1-strongly disagree (%)	2-disagree (%)	3-neutral (%)	4-agree (%)	5-strongly agree (%)	Agree + strongly agree (%)
Financial structure	2	2	16	39	41	80
Regulatory requirements	2	14	24	43	16	59
Inherent circumstances and Industry	2	4	27	53	14	67
Internal controls effectiveness	2	6	16	43	33	76
Inability to perform high-quality audit procedures	2	4	16	55	22	77
Lack of formal procedures in family-run businesses	2	6	10	51	31	82
Dominant or uncooperative management	2	0	18	47	33	80
Size of transactions or accounts	0	2	14	53	31	84
Transactions with related parties	2	0	8	59	31	90
Clients lacking formal policies	2	6	8	47	37	84
Regulatory or tax audit reports in the past year	2	10	14	51	22	73
Opinion from a previous auditor	6	8	20	43	22	65
Expertise levels of employees	2	2	18	53	24	77

(continued)

Table 2 (continued)

Variables	1-strongly disagree (%)	2-disagree (%)	3-neutral (%)	4-agree (%)	5-strongly agree (%)	Agree + strongly agree (%)
Special items like revenue in profitable companies	2	4	22	47	24	71
Potential or actual fraud	0	6	14	33	47	80
Frequent changes in key management	0	2	24	41	33	74

that 84% of auditors agreed (47%) or strongly agreed (37%). Unstructured policies could result in financial reporting anomalies, thus increasing audit risk. Lack of formal procedures in family-run businesses was another major concern: 82% of auditors recognized the risk by agreeing (51%) or strongly agreeing (31%). Family-run enterprises often have informal systems of governance that can pose serious risks and challenges for auditors.

The complexities of the financial structure were an issue among 80% of auditors regarding their approaches: agree 39%, strongly agree 41%. This indicates a serious concern with how the organizational financial frameworks introduce audit complexities and risks. Dominant or Uncooperative Management was looked at skeptically by 80% of auditors. This is reflected in their responses: agree 47%, strongly agree 33%. Such management behaviors can work to hinder effective auditing and increase the risk of undetected misstatements. Potential or Actual Fraud was another principal area of concern, with 80% of auditors acknowledging its importance through agreement at 33% or strong agreement at 47%. Fraud inherently calls for a broader and more intensive audit scope. Inability to Perform High-Quality Audit Procedures garnered 77% concurrence, split between agree (55%) and strongly agree (22%). This indicates concern over the auditor's ability to perform effective audits, which is important to reliable financial reporting. Expertise Levels of Employees was flagged by 77% of respondents (agree 53%, strongly agree 24%), pointing to the competency of employees as an important factor in ensuring effective internal controls and accurate financial reporting. Internal Controls Effectiveness drew 76% agreement (agree 43%, strongly agree 33%) from auditors, underlining the fundamental role of strong internal controls in managing audit risks.

Frequent changes in key management: 74% of auditors felt that this would enhance audit risk, agreeing 41% or strongly agreeing 33%, that such instability potentially affects the financial reporting and operational environment. Regulatory or Tax Audit Reports in the Past Year: 73% of respondents cited this as an audit risk increaser, with 51% agreeing and 22% strongly agreeing that such histories justify more cautious audit approaches. For items such as revenue in profitable companies: 71% of auditors agreed, 47% agree and 24% strongly agree, that high-stakes revenue areas merit particular attention because of their marked effect on business outcomes and financial reports. Inherent Circumstances and Industry Considerations—recognized by 67% of auditors as the need to focus audit approaches on sectorial risks that may be identified. Opinion from a Previous Auditor influenced the risk assessments of 65% of auditors, agreeing 43%, strongly agreeing 22%, demonstrating the influence of prior evaluations on current audit strategies.

The impact of this was considered critical by 59% of auditors: agree 43%, strongly agree 16%. Compared to immediate impact, direct financial and operational risks were considered more important than regulatory requirements.

5 Summary and Conclusions

Our study has identified a host of factors that influence the judgments by auditors regarding materiality and audit risk, which in turn affect the formation of an auditor's opinion.

Materiality is a cornerstone concept in auditing, and auditors consider several company attributes in determining the materiality threshold for an entity. Our results indicate that materiality is more likely to be influenced by a company's profitability and size of its total assets or capital among other factors. This would mean that bigger and more profitable companies are likely to receive the most stringent scrutiny due to the possibility of misstatements to influence the stakeholder's decisions. Size of Total Assets/Total Capital also deemed crucial by auditors because the absolute size of a firm's assets or capital is directly linked to the financial standing of the company, influencing both the risk assessment and the calculation of materiality thresholds. Type of Business and Capital Structure are considered important by many auditors as they directly affect the operational risks and financial stability of the company. The nature of the business can dictate specific industry-related audit considerations, while the capital structure impacts the financial risk profile of the entity. Another significant factor is conducting first-time audits. Industry type influences the audit approach due to specific risks and regulations, but it may not always directly affect materiality calculations. Organizational complexity, though impacting the clarity of financial statements, often requires detailed audits to understand the potential for material misstatements.

On risk assessment, factors such as "Evidence of potential or actual fraud", "Existence, Type, and Size of Transactions with related parties", and "Financial structure" are given the highest importance by auditors. This suggests that auditors are highly concerned about the risk of fraud, complexities in financial structuring, and the intricacies of transactions involving related parties due to their direct implications on the integrity and accuracy of financial reporting. Other significant factors influencing audit risk include "Larger transactions and accounts", "Client's Formal policies", and "Frequency of Changes in key management personnel". These factors can introduce uncertainties or vulnerabilities in the audit process, thereby raising the risk profile of the audit. Issues like "Dominant or uncooperative management" and the operation of "Family-run businesses" are identified with considerable agreement among auditors. While factors like "Regulatory requirements" and "Opinions from past auditors" show higher variability and somewhat lower mean scores, they still play a role in shaping audit risk assessments. These factors influence the auditor's approach, particularly when historical audit reports indicate previous issues or when navigating complex regulatory environments.

These findings show a direct association between the assessment of risk by the auditors and decisions on materiality, eventually influencing their opinion. The lower the risk perceived, the higher the threshold for materiality; hence, the audit procedures applied become less stringent. On the other hand, high-risk assessment lowers materiality thresholds, so the auditor becomes conservative and rigorous in the audit. This very conservative approach, however, increases the audit cost and workload by reducing the risk of issuing an inappropriate opinion. Finally, our study brings out the complex interplay of multiple factors in shaping auditors' professional judgment. In the view of these dynamics, audit practices can be refined, and the reliability of audit outcomes enhanced to instil greater trust in the financial reporting systems.

References

Creswell, J.W.: Research Design: Qualitative, Quantitative, and Mixed Methods Approaches, 3rd edn. SAGE Publications (2009)

Creswell, J. W., Clark, V.L.P.: Designing and Conducting Mixed Methods Research. SAGE Publications (2007)

Gray, D.E.: Doing Research in the Real World, 3rd edn. SAGE Publications (2014)

Grix, J.: The Foundations of Research, 2nd edn. Palgrave Macmillan (2010)

International auditing and assurance standards board (IAASB) (2019) ISA 315, identifying and assessing the risks of material misstatement. Revised. IFAC. ISBN: 978-1-60815-418-9. https://www.ifac.org

International auditing and assurance standards board (IAASB) (2020a) ISA 320, materiality in planning and performing an audit. IFAC. ISBN: 978-1-60815-420-2. https://www.ifac.org

International auditing and assurance standards board (IAASB) (2020b) ISA 330, the auditor's responses to assessed risks. IFAC. ISBN: 978-1-60815-421-9. https://www.ifac.org

Jones, T., Lee, M., Phillips, A., Thompson, G.: Discrepancies in materiality judgments among different auditing firms. Auditing: A J. Pract. Theory **38**(2), 159–179 (2019). https://doi.org/10.2308/ajpt-52503

Lee, H., Choi, Y.: Auditors perceptions of client integrity and industry risks: implications for audit risk assessment. J. Account. Public Policy **37**(4), 301–320 (2018)

Louwers, T. J., Blay, A. D., Sinason, D. H., Strawser, J. R., Thibodeau, J. C.: Auditing Assurance Services 7e (2018)

Martin, R., Steele, A.: Auditors professional judgment and opinions during financial crises: a review of cases. Auditing: A J. Pract. Theory. **38**(3), 177–196 (2019)

Nguyen, T., Fargher, N.: Auditor risk assessments and the issuance of modified audit opinions. Acc. Organ. Soc. **58**, 85–101 (2017)

Ormrod, J.E., Leedy, P.D.: Practical Research: Planning and Design, 11th edn. Pearson education (2015)

Ormrod, J. E., Leedy, P.D.: Practical Research: Planning and Design, 11th edn. Pearson Education (2015)

Peterson, A.: Dynamic industry environments and auditors' risk assessments: a study of technology and renewable energy sectors. Contemp. Account. Res. **38**(2), 210–228 (2021)

Saunders, M., Lewis, P., Thornhill, A.: Research Methods for Business Students, 7th edn. Pearson Education (2016)

Smith, J., Higgs, C.: Experience and materiality thresholds in auditing: a study on risk assessment and auditor efficiency. J. Account. Econ. **69**(3), 456–478 (2020). https://doi.org/10.1016/j.jacceco.2020.101121

Watts, R.L., Zimmerman, J.L.: The role of auditor judgment in determining materiality. J. Account. Res. **24**(1), 123–140 (1986). https://doi.org/10.1111/j.1475-679X.1986.tb00560.x

Windelband, W.: Präludien: Aufsätze und Reden zur Philosophie und ihrer Geschichte. Mohr, J.C.B (1912)

Woodwell, D.R.: Quantitative and Qualitative Research Methodologies in the Economic and Administrative Sciences. Academic Press (2014)

Sustainability of Societies in an Era of Rapidly Changing Technology

Lobna A. Okashah, Dae Yong Han, and Said Abdala Almusawi

Abstract

Every country has a national development plan that has certain elements that need to be focused on to reach certain national goals for the next 5, 10, or 30 years. Some of these elements include the global positioning of the country, human capital, health care, living environment, economy, infrastructure, etc. The rapid change in technology plays a major role in shaping the national development plans and therefore forcing societies to change their ways of living, which include changes in their dwellings (i.e. smart cities), education (i.e. online learning, use of AR and VR), jobs (i.e. remote access, AI applications as in ChatGPT, robotics, etc.), food (i.e. smart farming), health (i.e. AI in diagnosing and monitoring), infrastructure (i.e. 5G and/or 6G, alternative energy sources), and economy (i.e. digital economy, cybersecurity). This framework is based on the authors' experience in their countries, and it shows the basic requirements needed to be adopted by societies to make them responsive to those rapid changes. These requirements will mandate changes in the current systems such as education, health, and economic development systems, which will be discussed in this paper. The framework proposed in this paper will require synchronized coordination between several entities in the country, governmental and private, to achieve the defined overall objectives of the national plan. Keeping in mind that the framework presented may not be applicable in every country due to limitations in the local infrastructure such as the lack

L. A. Okashah (✉)
Independent Consultant, Kuwait City, Kuwait
e-mail: lobnaokashah@gmail.com

D. Y. Han
Vision Group Inc., Republic of South Korea, Buchon, South Korea

S. A. Almusawi
Ministry of Education, Zahra, Kuwait

K. Dhoska and E. Spaho (eds.), *Bridging Horizons in Artificial Intelligence, Robotics, Cybersecurity, Smart Cities, and Digital Economy*, Sustainable Economy and Ecotechnology, https://doi.org/10.1007/978-3-031-72029-1_16

of enough telecommunication coverage in all areas, economic limitations, etc., nonetheless, the framework can be used as a guide toward the design of an attainable national plan with the need to make it sustainable.

Keywords

Artificial intelligence • 6G • AR • Smart cities • Education • Sustainability

1 Introduction

In 2019, the world was plagued by COVID-19. The virus caused a pandemic worldwide, and it changed the rules of the games in every aspect of our lives. It forced education organizations to shift to online learning overnight. It forced businesses to perform their tasks through online meetings and to have their employees perform their daily tasks from home. It overwhelmed medical organizations, forcing the installation of rapid tracking and monitoring applications to keep track of the movement of infected populations.

The pandemic was a wake-up call to countries to start moving faster toward the digitization of their systems. The authors experienced that within the systems in their countries and were part of the emergency response teams formed by the government to provide quick and effective responses to the challenges faced due to the pandemic. The two major sectors that were not ready for the pandemic were education and health, which required rapid changes and adaptation of systems that the population was not ready to implement.

There are more challenges in the future that will face all nations, and every nation will have to deal with them according to their capabilities, resources, and readiness to face such challenges. One of the major challenges is based on the new technologies that are invading our societies at a faster rate than before plus having to deal with the new generation, which is more technology savvy and expects faster responses to their needs than the generation that preceded them. This will require serious reviews and modifications to any existing strategic plans done by the authorized entity in any government planning to stay compatible and competitive against other nations.

This paper will present what the authors experienced during the pandemic that resulted in creating a framework that can be used as a model of sustainability of societies in an era where the changes in technology are very rapid and becoming cheaper, more affordable, and more obtainable to nations than before.

2 Elements of National Development Plan

The United Nation Development Program (UNDP) has listed 17 goals named Sustainable Development Goals (SDG) that nations can use a guideline for the development of plans to eradicate poverty, ensure human rights and commit to improving the environment for the human race. Those elements are available on their website https://www.undp.org/destination-2030-accelerating-progress-sustainable-development-goals.

Many countries committed to those goals and modified them according to their areas that they needed to focus on to achieve the objectives that UNDP had for humanity.

In 2019, Kuwait published its overall plan, which can be viewed via the link https://www.newkuwait.gov.kw/plan.aspx. Kuwait identified 7 elements that the nation needed to focus on to achieve a new Kuwait by 2035 that are tied to UNDP 17 SDG. Kuwait aimed to become a leader in finance, and a cultural destination in the region. Those 7 elements defined by Kuwait were global positioning, human capital, healthcare, living environment, infrastructure, economy, and public administration.

Those pillars can be seen via the link https://s3.amazonaws.com/assets.newkuwait.gov.kw/wp-content/uploads/2016/09/KNDP_Inforgraphics_ProjectOverview_ENG.pdf.

As Kuwait was starting to kick off the efforts to work on developing action plans toward achieving its 2035 goals, COVID-19 happened, and Kuwait, like many countries, was forced to shift its priorities and focus on preventing their health and educational systems from collapsing. The crisis related to health stemmed from not having enough medical staff and facilities to deal with infected patients. It also didn't have a strong infrastructure and mobile application to monitor areas that had a high concentration of infected people similar to the system that existed in the Republic of South Korea that was developed by Korea Telecommunication (KT). This resulted in the spread of the disease and resulted in overwhelming the medical establishment to the brink of a major disaster.

The crisis related to education was created by the lack of the ability to switch to online learning quickly enough. The Ministry of Education did not have the means necessary to shift to online learning and many homes either did not have computers or had many children and not enough computers for all of them to attend online classes, submit homework, and answer online exams. The Ministry of Education struggled to create lessons to load on its YouTube Channel with poor preparation and execution, which ended up being removed because of the complaints of students and their parents.

Although the pandemic was global and outside the control of Kuwait; however, the country still would not have been able to reach its goals due to the ill-defined global positioning goals. Kuwait chose to be a financial and cultural leader in the region, when those global positioning goals were identified and implemented by neighboring countries decades ago. This shows how crucial it is for a nation to define its global positioning after performing a very comprehensive analysis of the neighboring nations' development plans, as well as understanding global objectives defined by entities such as UNDP. It is very important to a nation to create a niche for itself in the region where it physically resides and build alliances and collaborate with other nations further away from its region.

So, the question becomes what a nation should consider in determining what it needs to smartly plan for its future. The following section will describe a framework that can be used to help a nation in its planning to reach their futuristic goals.

The framework suggested by the authors adopted the terms of 4 out of the 7 elements mentioned above and added 2 new ones, which are Resources and Changes Required, bringing the total to 6 elements. The suggested framework also shows the connectivity between those terms as shown in Fig. 1.1. This framework accounts for the rapid changes in technology with a consideration of potential global crises like the pandemic.

The most important element is the first element, which is Global Positioning, and that entails deciding what areas are recognized as the major business sectors that are important for the nation and will distinguish it from its competitors whether in the region or globally. The Resources are the second element that will require a massive totality of all the resources available in the nation and that

Fig. 1.1 Framework of national development plan

includes natural resources as well as human resources. Without accurate information regarding such resources, the next element, Changes Required, will not be planned accurately. The Living Environment element will have to support the previous elements and will drive the modifications needed to the existing infrastructure that will create the atmosphere needed for the success of the previous steps. The final element will be the vehicle that will be required to finance the steps prior to it.

The following example can be used to demonstrate these elements. Suppose a country decided to become a state-of-the-art tourism destination as its main global positioning goal. Stating that would not be enough; further detail is needed to decide what kind of tourism is to be considered. Is it going to be health (medical) tourism, or historical, or sports tourism such as hiking or diving, etc.?

The second step would be to determine what resources are available and what is lacking to achieve that goal. Resources include human capital, natural resources, facilities, etc. Major steps must be taken to maintain the availability of such resources and plans to increase it to meet future demands. This will lead to consideration of the installation and modification of existing infrastructure to support the goal. The financial vehicle must be modified to meet the demands that will be imposed on it based on the goal.

To illustrate the idea, suppose the goal of a country is to become a leader in health (medical) tourism. In this case, one of the major required resources is a well-educated, trained medical staff. That will require the establishment of medical staff trained in multiple languages; a comprehensive cultural awareness of other nations, constantly updated regarding the latest research in various fields; and trained in the utilization of the latest technologies used in the medical field and beyond. This will lead to the modification of the educational curriculum of medical staff, especially medical doctors. Colleges and universities will have to include courses in cultural awareness, language courses–using translation technology is valid but will not capture the emotional connection and support needed between the patient and his treating physician–hospitality, and etiquette in treating patients, and courses in the use of technology such as AI and its use in diagnosing and treating patients. The extensive use of VR and AR will aid in understanding and performing surgeries and the use of e-game modes can ensure trainees reach expert levels before being allowed to perform any real surgery.

The living environment will have to be modified to accommodate the tourists and to create dwellings that will be beneficial for the well-being of the tourists. It will require creating housing for the medical staff closer to the hospitals to increase the efficiency of the services provided. There will be a need for farms, parks, meditation places, houses of worship, etc.

The infrastructure element covers many areas that will be modified to accommodate the required changes. As an example, changes in education in medical schools, technical colleges, and hospitals will require changes in the infrastructure of the facilities in such establishments. Many VR/AR labs will be required and equipped with top-of-the-line accessories to be used in such facilities. Strong 5G/6G network capabilities to access information related to the patients, such as test results, and communication with the treating physician in the patient's original country are just a basic requirement for a more advanced network. In a facility that will opt to use drones for the transportation or delivery of essential medicine, the facility will have to be designed to accommodate such technology.

Other infrastructure that will be required are an easy and speedy transportation system to connect the airports to the medical facilities. It will also require a well-established waste management system to handle hazardous medical waste including radioactive waste from radiation treatment, etc.

The need to structure organic smart farms is essential to meet the dietary needs of the patients. That will require allocation of land, funds, expertise, transportation systems, etc.

There will be a need to build cities that accommodate both the locals and the tourists. This will lead to building smart cities that are designed to support the goal. These cities will have their own sources of energy since it is crucial not to lose power in hospitals which will require them to be on a separate power grid from other cities and have redundance capacity to ensure the availability of power 24 hours a day, 7 days a week. The waste management system will require more specific treatment plans as mentioned above. In many areas, the use of robotics and AI-based systems will be a must to use in operating various applications. It was clear during the pandemic that robots were very helpful in providing infected patients with their food and medicine to reduce human-to-human contact. Not to mention that it is preferable to utilize robots in delicate surgeries, and in dealing with dangerous and hazardous materials and situations.

The financial element will require determining the costs of implementing all the changes required, the financial vehicles needed and what profitability will be obtained from such goals and how to invest the returns in a way to ensure the sustainability of the overall objective. The financial institutions will have to conduct speedy transactions with international entities in various countries, create more advanced technology protocols to avoid hacking, and have a very strong cybersecurity system to protect tourists from fraud and leaks of their financial data.

3 What is Needed to be Ready

From the example above, there are more demands for smarter systems, faster wireless communications, and a need to upgrade to a 6G-based network. Samsung issued a report in 2020 listing the requirements needed to support a 6G network. It stated that Terahertz (THz) bands will be needed to support 6G, and that will require solid state electronics, new materials for new types of antennas, new protocols, etc., Samsung Research (2020).

There will be a need to modify education to produce a workforce capable of developing and operating systems with new technologies. This will require introducing students to critical thinking, machine learning, AI, game theory, software development, etc., from an early age. There must be a consideration to customize the curriculum to match the capabilities of every student and reassess the effectiveness of the mass education approach as something from the past and replace it with one that challenges every student according to their individual capabilities and areas of interest Okashah et al. (2023). Applied research will be expected to be able to create new materials that can be used in new technologies. There will be more demand for satellites and high-altitude platforms instead of towers, especially in remote areas.

There will be more demand for a more sophisticated cybersecurity protocol to ensure the protection of information and assets.

A need might arise to collaborate with other nations to create a platform for unified members to avoid the limitations and control of other platforms influenced by political directions of interest groups.

The success of any plan will require the collaboration of all concerned parties in the nation, which includes the public and private sectors in addition to not-for-profit organizations.

4 Conclusion

Strategic planning is a very important task for any nation to achieve major objectives such as growth and sustainability in the global economy, hence, every country defines elements that it needs to focus on to achieve such objectives. The rapid technological changes will play a major role in achieving those goals and therefore, there must be a plan to ensure the adaptation to new technology whenever possible.

References

Okashah, L. A., Hamid, A., Kim, J., Rubin, E.: Changes required in education to prepare students for the future. In: The Sustainable University of the Future Reimagining Higher Education and Research, ch. 6, pp. 107–117. Springer Nature, Switzerland. Print ISBN: 978-3-031-20185-1 (eBook) ISBN: 978-3-031-20186-8 (2023). https://doi.org/10.1007/978-3-031-20186-8

Samsung Research: 6G The Next Hyper-Connected Experience For All (2020)

The Importance of Computer Science to Financial Firms Around the World

Asemina Meraj, Brisida Sefa, and Sidorela Pako

Abstract

The relationship between computer science and finance has become increasingly important in today's digital age, transforming the way financial institutions operate globally. This abstract explores the significance of computer science in changing the global financial services sector. It also aims to investigate the strength of the link between Financial Firms and the IT sector and to determine if it is as important as it is perceived to be. More specifically examining, how Financial Firms operating in Albania view computer science in their daily activities and what can be done to enhance IT Security. Computer science plays a critical role in creating competitiveness, advancing the evolution of financial organizations, and meeting the changing needs of clients in a rapidly growing digital world. To obtain a satisfactory answer, a literature review was conducted, and a questionnaire was sent to people working in the financial field to gain a better understanding of the real situation. Due to technological advancements, many challenges arise regarding security and adapting to innovations. The study investigates whether this is related to financial firms by examining patterns between the daily activities of financial firms and computer science, along with IT Security. Finally, the paper summarizes important results achieved from the collected data.

A. Meraj (✉) · B. Sefa · S. Pako
Canadian Institute of Technology, Tirana, Albania
e-mail: asemina.meraj@cit.edu.al

B. Sefa
e-mail: brisida.sefa@cit.edu.al

S. Pako
e-mail: sidorela.pako@cit.edu.al

K. Dhoska and E. Spaho (eds.), *Bridging Horizons in Artificial Intelligence, Robotics, Cybersecurity, Smart Cities, and Digital Economy*, Sustainable Economy and Ecotechnology, https://doi.org/10.1007/978-3-031-72029-1_17

Keywords

Financial Firms • Computer Science • IT Security • Computers

1 Introduction

The world of finance has been going deeper into the process of computerization, hence forming a strong link with IT and Computer Science as a whole. Accountants, economists, and financial specialists use different accounting programs in their daily activities which are crucial when it comes to reporting financial statements. With time, these accounting programs (such as Financa 5, Alpha, etc.) have had some improvements in order to minimize problems. This being said, having strong IT support and advancements has become somehow a necessity for all sectors, especially for the financial sector.

Computer science has not only a supporting role in the modern financial landscape, but a crucial component that fosters competitiveness and progress. Computer science is used by the financial sector to maintain strong security measures, improve client experiences, construct sophisticated financial models, and increase operational efficiency.

Researchers see IT as a strong weapon against the competition, which slowly helps the given firm achieve innovation more easily. Several researchers have conducted studies regarding the role of IT and CS on firm performance; such as (Fadhilah and Subriadi 2019) who argue on the fact that IT contributes to the growth of a company positively and indirectly at the same time. They also debate the fact that in today's environment, firms are obliged to find ways to create new opportunities to surpass the competition by utilizing IT and using it as a competitive advantage. There have been other researches that reason on the statement that a positive relationship exists between IT investment and five variables on firm performance, respectively; market competitiveness, partnerships with providers, firm growth, customer relationship, and lastly operational efficiency (Lee et al. 2016). However, in the same up-mentioned study it is brought to attention that the IT itself does not matter, it is the business that creates the right environment for innovations, and at the same time, growth comes slowly. Nevertheless, firms (in this case financial firms) have to consider carefully investment options in the field of Computer Science and IT Security.

Financial firms, whether they are national or international, value the privacy of data, and want to achieve high customer satisfaction. To do this a firm will need a strong background in IT Security and CS. This paper focuses on Albania's financial world, exploring how computer science and IT security are used and what problems Albanian financial firms face.

2 Literature Review

Because of the increase in the importance of IT not only within firms but also in other sectors, there has been an increased risk regarding data breaches, loss of information, or cybercrime.

Arcuri et al. (2017) investigate the effect that breaches of information security have on stock returns by examining reports of cyberattacks for the years 1995 to 2015 for 110 companies. They find that firms operating in the financial sector suffer more than other firms from cybercrime.

Other researchers link Computer Science and IT with an increase in firm performance which later translates into higher profit for firms.

Fadhilah and Subriadi (2019) by examining the literature trends, found that IT does have an indirect positive effect on a firm's performance. They also stated some other insightful conclusions from other papers, such as IT investments improving profitability, lower long-term costs, increase business support in the decision-making process, and lastly an increase in the ability of firms to respond very rapidly to opportunities and threats. In the same line (Wei et al. 2013) based on data collected from 252 Chinese firms, build a model to measure the interactiveness of supply chain integration and IT and their influence on firm performance. Their findings state that different types of IT investments have different effects on a firm's performance. They suggest that only improving IT infrastructure is not enough to improve firm performance. Moreover, managers have to be equipped with basic IT knowledge in order to value more and facilitate collaborative planning.

On the other hand (Subriadi et al. 2013), find that IT resources, rather than having an indirect effect on a firm's performance, have a large direct effect. They used a sample of 101 Rural Banks, from which 94 were considered valid for the study.

Others have suggested the use of machine learning and computer science to detect potential audit fraud or to even predict bankruptcy and help in making better decisions affecting future periods. Nevertheless, as can be seen, the literature review provides evidence that computer science, IT, IT security, and investments in the IT sector have an impact on the future development of a firm. Based on the literature concluding that investments in IT are very important to the growth of a firm, this study tries to peek into the reality of FFs operating in Albania in order to better understand the stance of employees of these firms toward the importance and the need for CS and IT in their daily operations.

3 Research Design

Main Hypothesis: The effective integration of computer science and it security measures means fewer challenges regarding security and adapting to innovations.

3.1 Methodology

In this research paper, secondary data are used to conduct the literature review combined with primary data collected by the authors through a questionnaire. Secondary data were collected from different resources to create the knowledge framework to help understand more about the use of Computer Science and IT Security in financial firms. The questionnaire was delivered to the participants through Google Forms (primary data).

This paper is based mostly on financial firms operating in Albania by offering financial services. The sampling was not random. The majority of the responders were professionals working in Finance/Accounting (36.4%), IT/Technology (27.3%), Operations (27.3%), and Sales (9.1%). The reason for not choosing a random sample is that the results might have not been as reliable as in the case when individuals knowing how accounting and IT are linked with one another responded, providing in this way more significant results. The number of participants in the study was 11, however, since the participants chosen were well-immersed in the world of financial firms and were able to provide a good representation of the majority of FFs the results are valid.

Most of the participants are less than 24 years old (54.5%), and the remaining percentages belong to the ages 25–34 and 55–44. In regards to education, 45.55 have obtained a master's degree, 36.4% have obtained a bachelor's degree, and 18.2% are divided equally between a high school diploma or equivalent and a doctoral degree.

4 Results and Analysis

From the study, it resulted that the respondents value computer science as "Very Important" and "Important" in the daily operations of financial firms in Albania. This indicates the crucial role of computer science in facilitating various aspects of organizational functioning, including data management, process automation, and decision-making processes. Within financial firms, computer science makes it possible to automate time-consuming and repetitive tasks. Automated procedures simplify work, and decrease the chance of human error.

On the other hand, regarding the cybersecurity threats and IT Security challenges, the majority of participants in the study reported experiencing challenges frequently (36.4%) and occasionally (36.4%). On the other hand, 27.3% reported facing IT Security challenges in daily activities very frequently. This high percentage along with no responders saying that they rarely or never face challenges highlights the critical need for IT Security measures to deal with risk and protect sensitive information effectively.

In terms of the main challenges in IT Security, 54.5% of respondents identified a lack of IT Security awareness as a primary challenge. This indicates a critical need for awareness programs to educate personnel about potential risks, and the importance of security protocols.

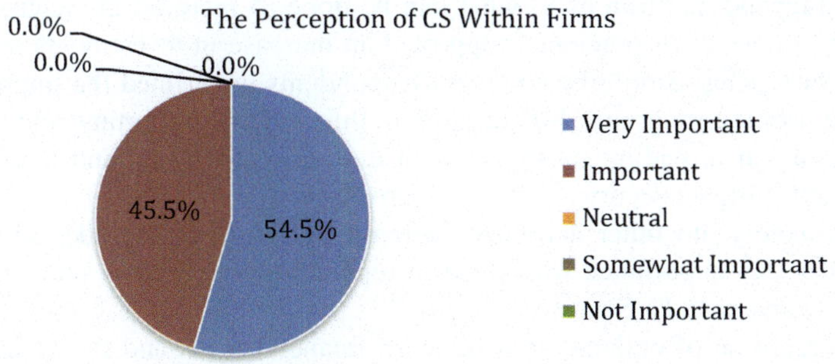

Graph 1 Presents the perception of CS within firm

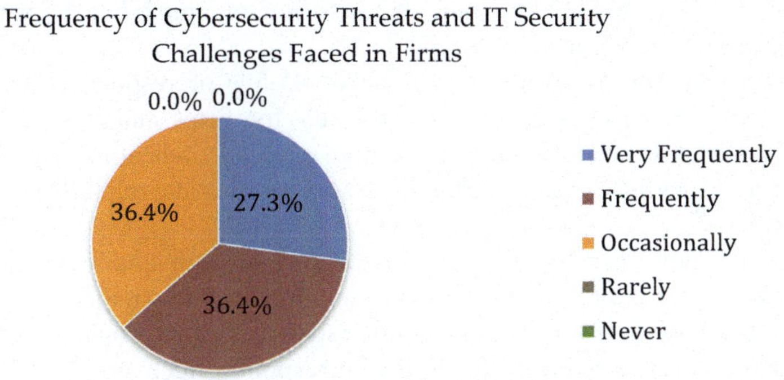

Graph 2 Shows the frequency of cybersecurity threats and IT security challenges faced in firms

Moreover, 27.3% of respondents chose cyber-attacks as a significant concern, and approximately 18.2% of respondents highlighted the complexity associated with incorporating new IT Security technologies as a challenge.

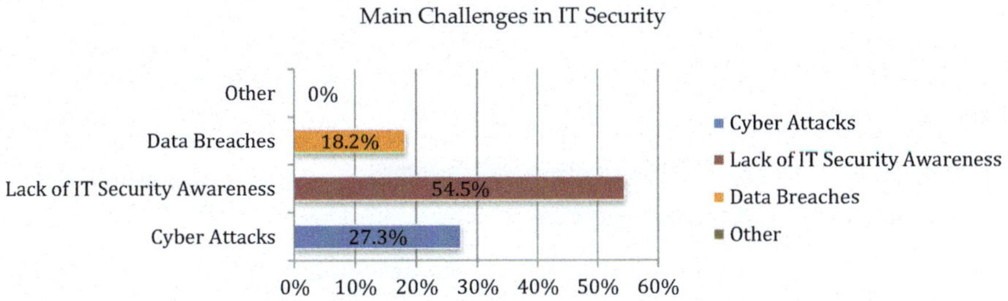

Graph 3 Presents the main challenges in IT security

Regarding the question as to what can be done to increase IT Security within their firm, 36.4% of respondents, supported an increase in investments in cybersecurity technologies. Similarly, 63.6% of respondents underlined the importance of employee training programs, highlighting in this way the important role of continuous training in providing personnel with the necessary skills and knowledge to identify and mitigate security threats effectively.

Unfortunately, no other action to increase IT Security was identified by the participants. No importance was given to regular security audits and assessments as a way to increase IT Security.

The perception of collaboration between financial firms and the IT sector was generally characterized as minimal; with 63.6% of respondents expressing this view, and 36.4% view it as a moderate collaboration. This suggests potential areas for improvement in the level of collaboration between Firms and the IT sector in Albania. Since the IT Sector is valued as an important actor in daily activities of a firm, it is crucial to bring into the attention of employers to increase the collaboration between firms and the IT Sector.

Moreover, the survey results indicate that 45.5% of responders believe that investing in IT Security measures is a priority for firms and financial firms in general, and 54.5% believe that it is of high priority, indicating a positive attitude toward mitigating cybersecurity risks and protecting the reliability of financial operations.

Employees view investment in IT Security as of high importance. Since the need of such sector is once again brought into attention, it is understood that IT is an important sector in need of investments, especially in the context of Albania.

Confidence levels related to the ability to respond effectively to cybersecurity incidents varied among participants. While a significant 9.1% and 34.6% expressed high confidence and moderate confidence, the remaining 54.5% identified areas for improvement. This indicates room for improvement for the majority of firms.

This study also noted several barriers to adopting new IT Security technologies and practices 54.6% of respondents said that the lack of skilled personnel (27.3%) and budgetary constraints (27.3%) are seen as barriers. Resistance to

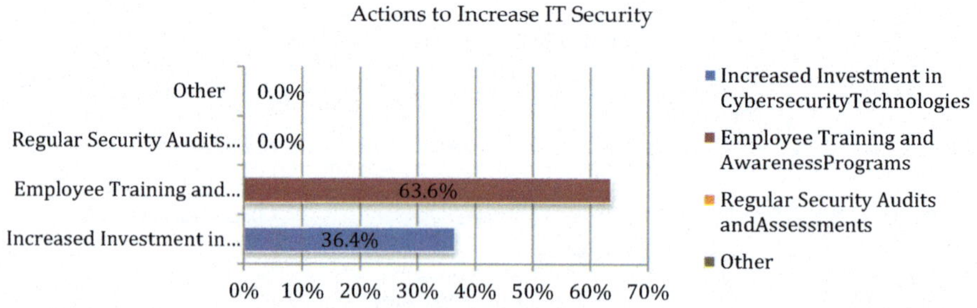

Graph 4 Shows the actions to increase IT security

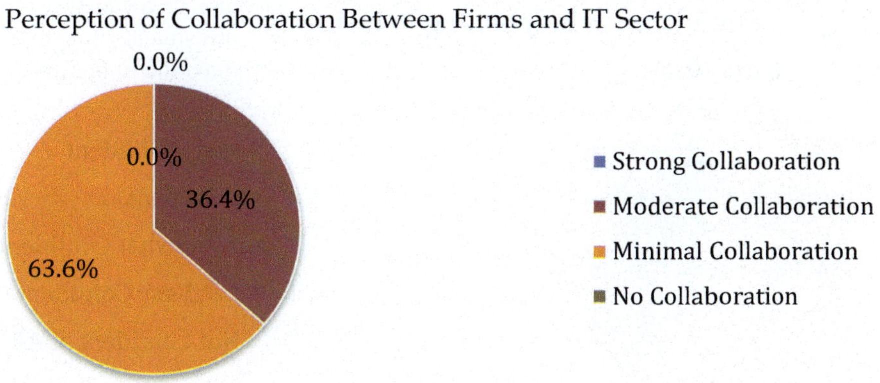

Graph 5 Presents the perception of collaboration between firms and IT sector

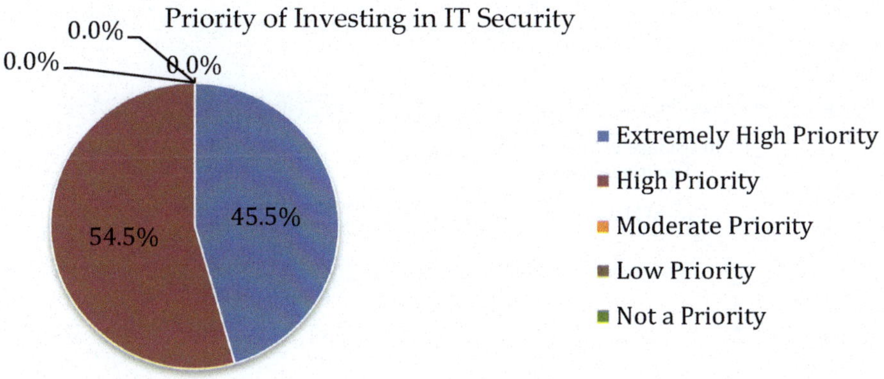

Graph 6 Shows the priority of investment in IT security

change from traditional practices was also identified as a barrier by 9.1% of respondents, emphasizing the importance of cultural shifts toward innovation. To add to this, 36.3% considered a problem in adopting and implementing new IT Security technologies and practices the complexity of integrating new technologies because of factors such as numerous regulations by regulatory entities in Albania, difficulty in training employees, and huge amounts of initial investments.

Employees should be provided with frequent security awareness training so they can identify malware, phishing scams, and other online dangers. In additions, employees should be informed on new security procedures and possible dangers.

When asked 'how often the firms educate the personnel regarding IT Security', approximately 72.7% of respondents admitted that their firms provided such education to the personnel. However, 18.2% reported a lack of such training, and 9.1% said such training happens occasionally.

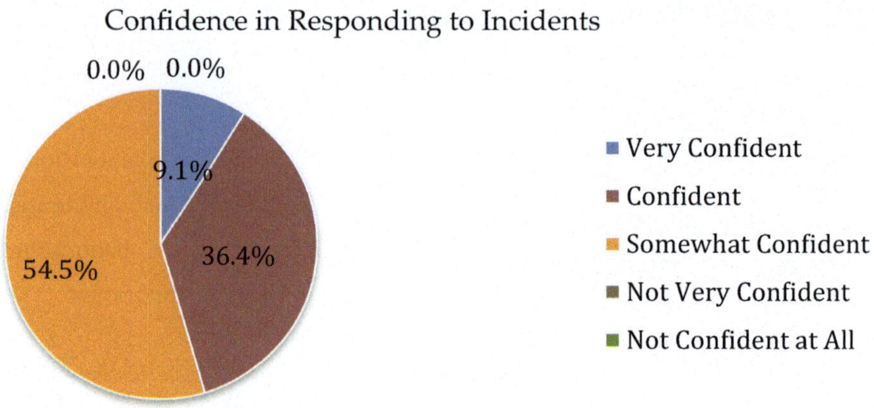

Graph 7 Presents the confidence in responding to incidents

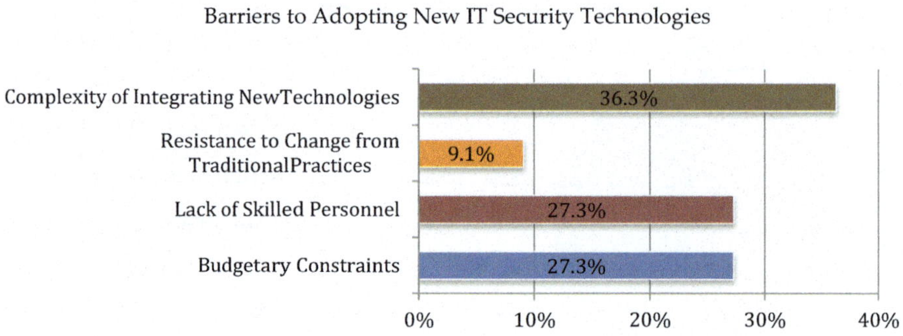

Graph 8 Shows the barriers to adopting new IT security technologies

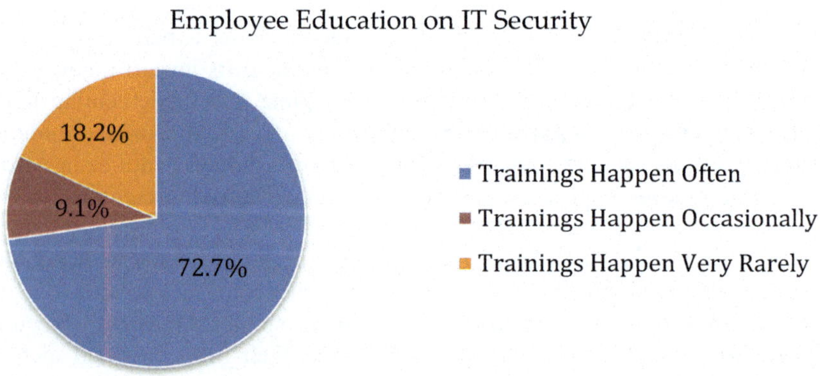

Graph 9 Shows the rate of employee education on IT security

Table 1 Summary of study findings

Aspect	Findings
Importance of Computer Science	Computer science is valued as "Very Important" and "Important" for daily operations in financial firms
IT Security Challenges	63.7% face frequent IT Security challenges; main issues include lack of awareness (54.5%) and cyber-attacks (27.3%)
Main Challenges in IT Security	Primary concerns: awareness gaps (54.5%), cyber-attacks (27.3%), and complexity of new technologies (18.2%)
Actions to Increase IT Security	36.4% support increased cybersecurity investments; 63.6% support employee training programs
Collaboration Between Firms and IT Sector	63.6% perceive minimal collaboration; 36.4% see moderate collaboration, indicating room for improvement
Priority of Investing in IT Security	45.5% prioritize IT Security; 54.5% consider it a high priority, indicating strong support for cybersecurity measures
Confidence in Responding to Incidents	9.1% highly confident, 34.6% moderately confident; 54.5% see room for improvement in incident response capabilities
Barriers to Adopting New IT Security Technologies	Major barriers include lack of skilled personnel (27.3%) and budget constraints (27.3%); resistance to change is also noted (9.1%)
Employee Education on IT Security	72.7% receive regular IT Security education; 18.2% lack such training, and 9.1% receive it infrequently

All in all, the findings from the questionnaire underline the critical importance of measures, collaboration, and investment in strengthening IT Security to protect against evolving cyber threats (Table 1).

5 Conclusions and Future Work

The study examines the critical role of computer science in daily operations and the IT security challenges within financial firms in Albania. Key challenges identified include a lack of awareness among employees regarding IT Security; cyber-attacks, and complexity in adopting new technologies. Respondents underline the importance of investments in cybersecurity technologies and employee training programs. Minimal collaboration between financial firms and the IT sector suggests potential areas for improvement. Investing in IT security measures is considered a high priority for financial firms, however, there are barriers to adopting new technologies or even improving the current situation, such as a lack of skilled personnel, budgetary constraints, and resistance to change. While most firms provide IT security-related training, there's a need for consistent training programs.

The hypothesis of the study is supported by the findings indicating the high importance placed on computer science in daily operations and IT security within financial firms leads to fewer challenges regarding security and adapting to innovations. Training of employees lowers the chances of losing important information because of a lack of knowledge. However, findings suggest that there is a long way to improvements for FFs operating in Albania. Additionally, the results suggest a need for increased investments in cybersecurity technologies, employee training programs, and improved collaboration between financial firms and the IT sector to mitigate cybersecurity risks effectively.

As a conclusion, in order for financial institutions to remain competitive in a market that is changing quickly, computer science integration is essential. Computer science makes significant and varied contributions, ranging from boosting decision-making and operational efficiency to boosting customer experience and guaranteeing strong cybersecurity. As technology develops further, financial institutions need to be creative and flexible while utilizing computer science to boost their development and advancement in the future as well as continue to train their staff regarding IT Security.

Future research should focus on improving and implementing thorough security policies on network security, data protection, incident response, and other areas of IT security. To add, it is important for employers to invest in cybersecurity technologies and IT measures in order to protect sensitive data and maintain stakeholder's trust. Also, it is important to continuously inform employees on new security procedures and possible dangers. Finally, to improve network security and promptly detect and handle attacks, firewalls and intrusion detection and prevention systems (IDS/IPS) should be installed.

References

Arcuri, M.C., Brogi, M., Gandolfi, G.: How does cyber crime affect firms? The effect of information security breaches on stock returns. In: First Italian Conference on Cybersecurity (ITASEC17), (p. 19). Venice, Italy (2017). From https://citeseerx.ist.psu.edu/document?repid=rep1&type=pdf&doi=6b8de4e18eb74cb66ac5eff6f1569cccdcfc74ff

Fadhilah, A.N., Subriadi, A.P.: The role of IT on firm performance. Procedia Comput. Sci. **161**, 258–265 (2019). https://doi.org/10.1016/j.procs.2019.11.122. From https://www.researchgate.net/publication/338348066_The_Role_of_IT_on_Firm_Performance

Fadhilah, A.N., Subriadi, A.P.: The role of IT on firm performance. In: The Fifth Information Systems International Conference 2019 (p. 8). Elsevier B.V (n.d.). From https://pdf.sciencedirectassets.com/280203/1-s2.0-S1877050919X00174/1-s2.0-S1877050919318320/main.pdf?X-Amz-Security-Token=IQoJb3JpZ2luX2VjEDkaCXVzLWVhc3QtMSJIMEYCIQCvfz%2BT9MemZQ9vwp%2BlgzToBuGPaH2HWuhofGILRnHGmwIhAP0ZA3yKaHLSbMRKRvOWvlKBWeTHQDEtKLIIeLtl

Lee, H., Choi, H., Lee, J., Min, J., Lee, H.: Impact of IT investment on firm performance based on technology IT architecture. Procedia Comput. Sci. **91**, 652–661 (2016). https://doi.org/

10.1016/j.procs.2016.07.164. From https://www.researchgate.net/publication/305953456_Impact_of_IT_Investment_on_Firm_Performance_Based_on_Technology_IT_Architecture

Subriadi, A.P., Hadiwidjojo, D., Djumahir, M.R., Sarno, R.: Information technology productivity paradox: a resource based view and information technology strategic alignment perspective for measuring information technology contribution on performance. J. Theor. Appl. Inform. Technol. **12** (2013). From https://www.jatit.org/volumes/Vol54No3/19Vol54No3.pdf

Wei, S., Liu, H., Ke, W., Wei, K.-K., Hua, Z.: the interaction effects between supply chain integration and IT capabilities on firm performance. In: Pacific Asian Conference on Information Systems (PACIS). AIS Electronic Library (AISeL) (2013). From https://core.ac.uk/download/pdf/301359674.pdf

The Role of Active Economic Enterprises and Their Impact on Employment in Albania

Blerta Kristo Nazarko

Abstract

This study aims to present the evaluations, the role, and the distinguishing features of the activity performance of active economic enterprises that operate on a prefecture and regional basis. The turnover of the annual balance according to the economic activity is related to the size and the weight that they occupy, which affects the reduction of unemployment by encouraging the economic progress of the development of the region. The investigation of the differences encountered in the progress of economic and social development (as many distinctive phenomena are found in the structure of the employment rate according to the branches of the economy) enables the drawing of attention in the design of management strategies and visions. The selection of variables is aimed at studying the effects that influence the increase in the efficiency value of active economic enterprises. This paper presents a generalization of the aspects and value of the role of active enterprises in employment according to the branches of the economic structure and the improvement of the economic well-being of each prefecture. The study enables the recognition of best practices as well as inefficiencies in the performance value by applying the DEA method. The study extends to the period 2018–2022. This study presents an alternative to continuous and comprehensive managerial studies.

Keywords

Employment • Economic structure • Annual turnover • DEA method

B. K. Nazarko (✉)
European University of Tirana, Tirana, Albania
e-mail: kristoblerta@yahoo.com

K. Dhoska and E. Spaho (eds.), *Bridging Horizons in Artificial Intelligence, Robotics, Cybersecurity, Smart Cities, and Digital Economy*, Sustainable Economy and Ecotechnology, https://doi.org/10.1007/978-3-031-72029-1_18

1 Introduction

Looking at the general picture of the progress and economic development of Albania (2018–2022), its dynamics, in the macroeconomic structure and stability, economic business enterprises (Small, Medium Enterprises-SMEs) remain important factors in the integration process itself economic and social of Albania toward the EU. The average increase in percentage of GDP for the period 2018–2021 in Albania is 2.93 (Vjetari rajonal Statistikor 2020, 2021; Vjetari Statistikor Rajonal 2022). GDP per capita in PPS (purchasing power standards) for 2022 is as much as 34% of the average of the European Union from 31% that was in 2021, although there has been an increase, it still remains low even compared to the countries of the region, according to Serbia has GDP per capita in PPS 44% for 2022, Montenegro has 50% and Macedonia has 42% of the EU average (https://ec.europa.eu/eurostat/databrowser/view/tec00114/default/table?lang=en). This proves that, despite the positive economic growth as a period, we still have a low level of GDP per capita in PPS in relation to the average of the European Union and the region. This dictates the need for more targeted reforms toward the European Union, considering every factor that has an important weight in economic development, such as the development of SMEs. Economic business enterprises are a factor of employment and well-being. In Albania, the labor force participation rate for 2022 is 62.4 and the employment rate is 55.5, the unemployment rate is 10.9 (Tregu i Punës 2023).

Unemployment is one of the most important macroeconomic problems that all countries of the world want to solve (Turgut 2021). It is important to assess the risk of unemployment. The risk of unemployment varies in different ways between sectors with a high risk that dictates the importance of economic, technological, and institutional heterogeneities at the industry level (Cassandro et al. 2020). Temporary jobs have a weak tendency for product innovation (Guarascio et al. 2019). Dobrin (2015) emphasizes that even small enterprises are involved in international business and the process of internationalization, the phenomenon of globalization is not a specific feature only for large enterprises. Many researchers analyze the processes of macroeconomic development in each country and deal with the complexity of the business field, also mentioning risk factors with their specific impacts on the wide area of business. The development of a conditioned and long-term enterprise in the market should be seen as a qualitative phenomenon related to the introduction of continuous innovations as a product, as a structural process, as well as with innovations in the organizational and managerial fields (Duda and Skalna 2019). In 2022 in the European Union, the enterprises that employed (1–249 employees) employed around 64.4% of Europeans in the non-financial business economy (https://www.statista.com/statistics/939710/eu-enterprise-employment-share-by-enterprise-size/#:~:text=By%20comparison%2C%20medium%2Dsized%20businesses,to%209%20workers)%2029.43%20percent), while in Albania this figure is 82% (https://www.statista.com/statistics/939710/eu-enterprise-employment-share-by-enterprise-size/#:~:text=By%20comparison%2C%20medium%2Dsized%20businesses,to%209%20workers)%2029.43%20percent), while the enterprises that employed over 250 employees in the European Union employed 35.4%, while in Albania this figure is 18%. The general aim of this paper is to give a general picture of the sensitivity of the influence of SMEs in the economic development of the prefecture and region. Also, this study enables the analysis of competitive abilities that appear using the DEA method, it shows the economic development trends of prefectures and regions depending on the activity and business development opportunities of SMEs, recognizes and evaluates the influences that promote development and the role of SMEs that stimulate progress. In this study, we focus on the analysis and consideration of the following hypotheses:

H1—The influence of active economic enterprises (SMEs) is distinguished by visible changes in the mitigation of unemployment, the reduction of unemployed job seekers in prefectures and regions.

H2—The influence of economic enterprises (SMEs) is clearly distinguished in the improvement and growth of the employment rate and the level of payment according to prefectures and regions.

H3—The influence of economic enterprises is also distinguished according to the structure of their composition with a noticeable impact on business development according to prefectures and regions.

H4—The influence of economic enterprises shows the relationship between the size and "weight" of the enterprises and the structure of net sales.

The novelty of this application lies in the contribution of passing on the possibilities of recognition using the DEA method to better recognize the influence of variable factors and increase the effectiveness of their optimization in the economy.

2 Methodology

In this study, the DEA method will be used to evaluate the performance of decision-making units, a non-parametric method first presented by Charnes et al. (1978), where DMUs can use many inputs and many output. The VRS (Variable return to scale) model, in addition to CRS model is used to evaluate pure technical efficiency and scale inefficiency (Banker et al. 1984; Cooper et al. 2002). The input vector can be written $\boldsymbol{x}_j = (\boldsymbol{x}_{1j}, \boldsymbol{x}_{2j}, \ldots, \boldsymbol{x}_{mj})$, which is used with the possibility to obtain the output vector $\boldsymbol{y}_j = (\boldsymbol{y}_{1j}, \boldsymbol{y}_{2j}, \ldots, \boldsymbol{y}_{rj})$, where the vectors $\boldsymbol{x}_i \neq 0$ and $\boldsymbol{y}_j \neq 0$. The set of homogeneous decision making units n DMUs operates on the production possibility set using a non-negative combination of DMUs, $P = \{(x, y) | x \geq \sum_{j=1}^{n} \boldsymbol{\lambda}_j \boldsymbol{x}_j, 0 \leq \boldsymbol{y} \leq \sum_{j=1}^{n} \boldsymbol{\lambda}_j \boldsymbol{y}_j, \boldsymbol{\lambda} \geq 0\}$, $\lambda = (\boldsymbol{\lambda}_1, \boldsymbol{\lambda}_2, \ldots, \boldsymbol{\lambda}_n)^T$ is called the intensity vector (Cooper et al. 2011). From the DEA method, both its basic models are used, the CRS model (Constant returns to scale) and the VRS model (Variable returns to scale), where the second model differs from the first by using the condition that $\sum_{j=1}^{n} \lambda_j = 1$. These models are presented mathematically as follows (Cooper et al. 2002; Zhu 2014):

CRS Model:

Min θ

$$\text{S.t.} \sum_{j=1}^{n} \lambda_j x_{ij} \leq \theta x_{io}, i = 1,2, \ldots, m;$$

$$\sum_{j=1}^{n} \lambda_j y_{rj} \geq y_{ro}, r = 1,2, \ldots, s;$$

$$\lambda_j \geq 0, j = 1,2, \ldots, n,$$

VRS Model:

Min θ

$$\text{S. t} \sum_{j=1}^{n} \lambda_j x_{ij} \leq \theta x_{io} i = 1,2, \ldots, m;$$

$$\sum_{j=1}^{n} \lambda_j y_{rj} \geq y_{ro} r = 1,2, \ldots, s;$$

$$\sum_{j=1}^{n} \lambda_j = 1$$

$$\lambda_j \geq 0 j = 1,2, \ldots, n$$

In addition to the DEA method, linear regression is also used. Among the variables selected as outputs is unemployment, starting from the assumption of DEA that inputs should reduce and outputs should raise, these outputs are called "unwanted" outputs, which turn into desired outputs (Zhu 2014) with the formula: $\overline{y}_{rj} = -y_{rj} + \eta > 0$, where η satisfies the condition, $\eta = \max\{y_{rj}\} + 1$. To evaluate the effectiveness of the impact on the efficiency value of each input and output variable, it is evaluated using the formulas gEf0(Ii) $= \frac{Ef_0(I_i)}{\sum_{I=1}^{m} Ef_0(I_i)}$ for each input and gEf0(Or) $= \frac{Ef_0(O_r)}{\sum_{r=1}^{s} Ef_0(O_r)}$ for each output (Nazarko 2024). Also, in this paper, comparative tables and graphs are used for a better understanding in the study analysis. Since the study is extended throughout the period 2018–2022, the efficiency chain evaluation is used to calculate the relative technical efficiency (Ef^{CRS}) and the efficiency of (Ef^{vRS}) called local pure technical efficiency by also calculating the scale efficiency (scale efficiency) with the aim of better recognizing the inefficiencies that have influenced the inefficiency of a DMU. Taking into account the "rule" used in DEA, with the condition that the number of DMUs is not less than three times the sum of the number of inputs and outputs taken together, the compositions of the groups connecting the inputs to the outputs are used. The overall efficiency value for each model for the period 2018–2022 is estimated with the harmonic mean $E^{H}_{fCRS} = \frac{T}{\frac{1}{E^{CRS}_{ft1}} + \frac{1}{E^{CRS}_{ft2}} + \frac{1}{E^{CRS}_{ft3}} + \frac{1}{E^{CRS}_{ft4}} + \frac{1}{E^{CRS}_{ft5}}}$ $(T = 5,$ and since the efficiency values are relative technical efficiency values).

3 Numerical Application

In this study, the set of DMUs includes 12 prefectures and three regions of Albania for the period 2018–2022. The regions are defined, the northern region (AL_{01},) which includes 5 prefectures which are AL_{011} (Dibër), AL_{012} (Durrës), AL_{013} (Kukës), AL_{014} (Lezhë), AL_{015} (Shkodër); the central region (AL_{02},) includes two prefectures which are AL_{021} (Elbasan), AL_{022} (Tiranë) and the southern region (AL_{03}), includes 5 prefectures which are AL_{031} (Berat), AL_{032} (Fier), AL_{033} (Gjirokastër), AL_{034} (Korcë), AL_{035} (Vlorë).

The variable factors selected as inputs and outputs are:

- The total number of active economic enterprises per 1000 inhabitants
- The number of active economic enterprises producing goods per 1000 inhabitants
- The number of active economic enterprises of service producers per 1000 inhabitants
- Unemployment rate in percentage
- Unemployed jobseeker (15–29 years) (%)
- Employment rate in percentage
- Average monthly salary for employed persons (thousand Lekë)

The data are taken in Vjetari Statistikor Rajonal (2022, 20232019), Vjetari rajonal Statistikor (2020, 2021); Tregu i Punës (2020, 2021, 2022, 20232019) and processed by the author.

Data related to employment according to employment status, the rate of participation in the labor force and the rate of employment by gender, long-term unemployment rate by gender are also used in the construction of statistical tables and Figures.

For the linear regression, the structure of net sales according to the prefectures related to the "weight" of the size of the active economic enterprises in relation to their total is used. Table 1 gives the average values of variable sizes for each prefecture and region.

These groupings of input–output connections are used, based on the respective objectives of the study. Grouping A (3 × 2) was determined with the aim of studying the sensitivity of the influence of small, and medium enterprises (SMES) business activity in reducing unemployment in general and among young people.

Table 1 The average data values (2018–2022)

DMU	Active economic enterprise						EMPL rate	UNEMP rate	UNEJS 15–29 years	G
	A	B	C	D	E	F				
AL_{01}	41.5	13.6	27.9	20.8	20.5	21.4	49.8	13.3	24.8	464.2
AL_{011}	26.3	10.4	15.9	1.9	2.2	1.7	46.8	11.2	23.1	503.9
AL_{012}	44.1	9.4	34.7	8.1	6.1	9.7	49.2	14.3	27.4	464.4
AL_{013}	26.7	12.2	14.5	1.2	1.6	1.0	58.3	6.0	26.4	505.2
AL_{014}	42.4	12.8	29.6	3.2	3.0	3.4	39.7	21.1	16.2	439.7
AL_{015}	51.7	22.8	28.9	6.4	7.7	5.5	55.4	11.7	26.6	440.1
AL_{02}	58.1	11.6	46.5	43.3	28.0	52.4	51.4	11.9	18.1	583.3
AL_{021}	47.4	22.3	25.1	7.9	10.2	6.5	61.7	6.6	17.2	478.2
AL_{022}	61.2	8.5	52.7	35.4	17.8	46.0	48.4	13.8	18.7	622.8
AL_{03}	67.9	36.0	31.9	35.9	51.5	26.2	58.9	10.2	21.7	467.2
AL_{031}	70.1	41.2	29.0	5.2	8.2	3.3	61.9	7.4	19.4	448.7
AL_{032}	71.6	42.3	29.3	12.6	20.0	8.1	59.3	8.8	20.1	484.3
AL_{033}	75.7	38.4	37.3	2.7	3.8	2.1	62.8	6.1	22.5	504.5
AL_{034}	65.4	38.3	27.1	8.1	12.6	5.3	61.6	8.3	23.9	452.3
AL_{035}	60.6	19.6	41.0	7.1	6.9	7.4	52.3	17.9	22.4	457.3

A. The average value of the total number of active economic enterprises per 1000 inhabitants for the period (2018–2022)
B. The average value of the number of active economic enterprises, producers of goods per 1000 inhabitants (2018–2022)
C. The average value of the number of active economic enterprises, service producers per 1000 inhabitants (2018–2022)
D. The number of economic enterprises of the prefecture and region in ratio to the total number of enterprises (in %)
E. The number of economic enterprises of the prefecture and region producing goods in ratio to the total number of producers of goods (in %)
F. The number of economic enterprises of the prefecture and region producing services in ratio to the total number of service producers (in %)
EMPL Rate-Employment rate
UNEMP Rate-Unemployment rate
UNEJS 15-29 YEARS- unemployed jobseeker (15–29 years)
G. Average monthly salary for employed persons (thousand Lekë)

In this grouping, the following are used as inputs: I_1: The total number of active economic enterprises per 1000 inhabitants, I_2: The number of active economic enterprises producing goods per 1000 inhabitants, I_3: The number of active economic enterprises of service producers per 1000 inhabitants and as outputs O_1: Unemployment rate in percentage, O_2: Unemployed jobseeker (15–29 years) (%).

For group B (3×2) are used as outputs: O_1: Employment rate in percentage, O_2: Average monthly salary for employed persons (thousand Lekë) and the inputs are those of group A. Grouping B was determined with the aim of investigating the sensitivity of the influences of business development by SMEs in the

Table 2 Summary results for both groups

DMU	Grouping model A			Grouping model B			Frequency RTS					
	E^H_{fCRS}	Frequency (t_i)		E^H_{fCRS}	Frequency (t_i)		G. (A)			G. (B)		
		SE > E_{fVRS}	SE < E_{fVRS}		SE > E_{fVRS}	SE < E_{fVRS}	I	D	C	I	D	C
AL_{01}	0.47975	4 (t_1, t_2, t_3, t_5)	1 (t_4)	0.64004	5(t_1, t_2, t_3, t_4, t_5)	–	5	–	–	3	2	–
AL_{011}	0.94775	–	2 (t_4, t_5)	0.98202	1 (t_4)	1 (t_5)	2	–	3	1	1	3
AL_{012}	0.46069	1 (t_5)	4 (t_1, t_2, t_3, t_4)	0.76092	2 (t_3, t_5)	1 (t_5)	5	–	–	2	2	1
AL_{013}	1.00000	SE = E_{fVRS}		1.00000	SE = $\boldsymbol{E_{fVRS}}$		–	–	5	–	–	5
AL_{014}	0.98344	–	1 (t_5)	0.58235	5(t_1, t_2, t_3, t_4, t_5)	–	–	1	4	5	–	–
AL_{015}	0.37436	5(t_1, t_2, t_3, t_4, t_5)	–	0.50835	5(t_1, t_2, t_3, t_4, t_5)	–	5	–	–	4	1	–
AL_{02}	0.78382	1 (t_5)	3 (t_1, t_2, t_3)	0.70567	3 (t_2, t_3, t_4)	2 (t_1, t_5)	1	3	1	–	5	–
AL_{021}	0.97978	–	1 (t_3)	0.60676	2 (t_2, t_3)	3 (t_1, t_4, t_5)	–	1	4	1	4	–
AL_{022}	0.72861	1 (t_5)	–	0.81358	–	1 (t_5)	1	–	4	–	1	4
AL_{03}	0.51003	5(t_1, t_2, t_3, t_4, t_5)	–	0.45818	5(t_1, t_2, t_3, t_4, t_5)	–	5	–	–	3	2	–
AL_{031}	0.72857	3 (t_2, t_3, t_5)	2 (t_1, t_4)	0.52910	1 (t_3)	4 (t_1, t_2, t_4, t_5)	2	3	–	–	5	–
AL_{032}	0.66098	5(t_1, t_2, t_3, t_4, t_5)	–	0.50660	4 (t_2, t_3, t_4, t_5)	1 (t_1)	4	1	–	–	5	–
AL_{033}	0.48144	2 (t_1, t_2)	3 (t_3, t_4, t_5)	0.41633	1 (t_1)	4 (t_2, t_3, t_4, t_5)	2	3	–	–	5	–
AL_{034}	0.57264	4 (t_2, t_3, t_4, t_5)	1 (t_1)	0.57286	1 (t_3)	4 (t_1, t_2, t_4, t_5)	4	1	–	–	5	5
AL_{035}	0.34607	4 (t_1, t_2, t_3, t_5)	1 (t_4)	0.45373	5(t_1, t_2, t_3, t_4, t_5)	–	5	–	–	2	3	–
Avg	0.66920	35 cases	19 cases	0.63577	40 cases	21 cases						

increase of the employment rate as well as the increase of the monthly payment by determining the performance of each DMU, inefficiencies, and inefficiency factors.

Table 2, classifies the efficiencies into four quartiles, in such a way that the first and second quartiles are below the average value of the harmonic efficiency and the third and fourth quartiles are above the average value (for group A, the average value is 0.66920 and for group B it is 0.63577) of it. For group A quartile $Q_1 = [0.34607; 0.50764)$, $Q_2 = [0.50764; 0.66920)$; $Q_3 = [0.66920; 0.8346)$ and $Q_4 = [0.8346; 1]$, while for group B the quartiles are $Q_1 = [0.45373; 0.52605)$, $Q_2 = [0.52605; 063{,}577)$; $Q_3 = [0.63577; 0.81789)$ and $Q_4 = [0.81789; 1]$. The decision-making units that have the efficiency values belonging to the first quartile Q1 are evaluated as "very weak", the second quartile as "weak", the third quartile as "moderate average" and the fourth quartile as "good and very good". In group A, there are 5 decision-making units rated very weak and 3 decision-making units rated weak, also in group B there are again 5 decision-making units rated very weak and 3 decision-making units rated weak. For group A, the decision-making units rated "weak" and "very weak" belong to the first and third regions. It is the same picture for the second grouping as well. This shows that the efficiency of business development of SMEs appears weakest in the first and third regions. Investigating the scale efficiencies, it is found that in both groups where SE > E_{fVRS} is almost twice as large as in the case where SE < E_{fVRS} (respectively for group A, 35 cases with 19 cases and for group B, 40 cases with 21 cases), this shows that the primary cause of inefficiency is the managerial inefficiency of DMUs. Looking at the returns to scale, group A is increasing (I) in 42 cases and decreasing (D) is in 13 cases. According to the DEA rule, this indicates that the efficiency of inefficient decision-making units (in the case where the returns to scale are increasing) must be related to the possibility of reducing inputs (keeping the same output values) to make the decision-making unit more efficient. Looking at the expansion coefficients of active economic enterprises per 1000 inhabitants, it can be further said that the effectiveness of influence in these units is very weak, which has not had enough impact on the reduction of unemployment in these decision-making units. For group B, the increasing returns to scale appear in 22 cases and the decreasing returns to scale in 41 cases. According to the DEA rule, it can be said that these inefficient units to become efficient must increase their outputs. A more detailed analysis can be done for each DMU separately, starting from the above table, which is also detailed in the dynamic progress of economic development according to years (t_1, t_2, t_3, t_4, t_5 of the period 2018–2022). The sensitivity of the influence of the decision-making units in the development of the business can also be judged from the correlative relationship between the structure of net sales (y) and the "weight" of the number of economic enterprises in ratio to the total number of enterprises (in %) (x) for each prefecture separately. These relationships are presented below, looking at linear regression tables for the years 2018, 2019, and 2020, where the same trend is found for the entire period.

From linear regression we get, multiple $R = 0.884984$, $R^2 = 0.783197$, adjusted $R^2 = 0.765131$, $F = 43.34989$, significance $F = 2.59724\text{E}{-}05$.

The value in the Fisher table $F_{0.05,1,12} = 4.75$ where it is found that $F = 43.34989 > F_{0.05,1,12} = 4.75$, so relationship is statistically significant.

Even for the following two years (2019, 2020) it is shown that the relationship is statistically significant.

For the year 2019, multiple $R = 0.865355$, $R^2 = 0.748839$, adjusted $R^2 = 0.727909$, $F = 35.7782$, significance $F = 6.4\text{E}{-}05$.

For the year 2020, multiple $R = 0.857174$, $R^2 = 0.734747$, adjusted $R^2 = 0.712642$, $F = 33.23979$, significance $F = 8.94\text{E}{-}05$.

The same trend is found for the next two years, which shows that throughout the period 2018–2022, the relationship between the structure of net sales (y) and the "weight" of the number of economic enterprises in ratio to the total number of enterprises (in %) (x) is statistically significant. The effectiveness of active economic enterprises can be studied by determining the "yield" coefficient $\text{Cef} = \frac{\text{The number of enterprises as a percentage of the total}}{\text{net sales as a percentage of the total}}$. This indicator evaluated in the tables of Figs. 1, 2, and 3 shows that the active economic enterprises that operate in the second region where this coefficient is 145%, then the first region with 99.8%, have a high effectiveness in the sales structure and is presented the third region is weak with a coefficient of 50% (author processing). Among the prefectures with the highest efficiency are Tirana with 177% and Durres with 161%. This phenomenon is easily ascertained from the tables given above. Looking at the level of the average net salary for an employee, it is noticed that there are obvious differences in the level of salaries. The employee in Tirana has an average salary in relation to the general average salary of 23% higher. The structure of net sales is also an assessment of the effectiveness of the turnover of the enterprise. These results according to each table with the data of the values of the variable factor sizes against the evaluation of the efficiencies according to each grouping show the role and sensitivity of their special influence on each prefecture. Looking at Fig. 4, it can be concluded that in Albania only about 50% of the employed are employees. The number of enterprises employing 10–49 workers is only 5.7% of the total number. These enterprises have a coefficient of 2.56 per 1000 inhabitants, much less than the average of the European Union countries. Also in Fig. 4, a high percentage of informal employment can be found.

Using DEA, the sensitivity of the influences of each variable factor to the efficiency values for each decision-making unit and as a whole was studied. The results of the sensitivity of the influences of each variable factor to the efficiency values are given in the Table 3.

From Table 3, it is found that the highest sensitivity of the influence to the outputs for both groups is shown by Input 2 and 3 together, which represent the inputs according to the economic structure (production of goods and production of services). The average of their influence from Table 3 is found to be over 50%, respectively 54.7% (Model A) and 55.8% (Model B). Even here, in the two prefectures of Durres and Tirana, the greatest impact of the influence of these inputs is observed. Figure 5 shows the Labour Force participation rate and employment rate (men, women) and a still high gender gap is observed, in addition to a softening

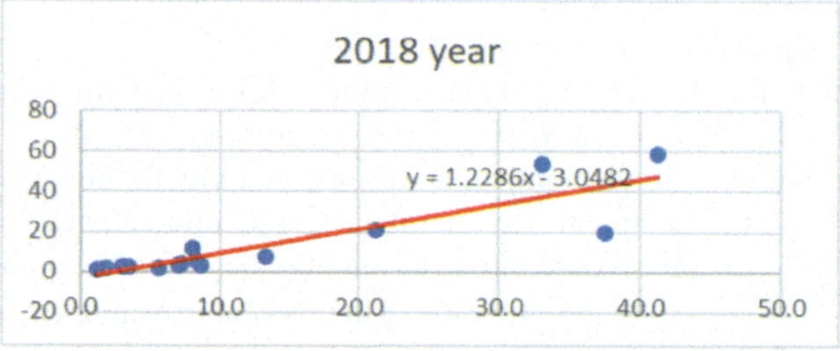

DMU	x	y
AL01	**21.2**	**21**
AL011	1.8	2
AL012	7.9	12.1
AL013	1.2	0.9
AL014	3.3	2.5
AL015	6.9	3.5
AL02	**41.3**	**59**
AL021	8.2	5.2
AL022	33.0	53.8
AL03	**37.5**	**20**
AL031	5.5	1.8
AL032	13.3	7.7
AL033	2.9	2.8
AL034	8.6	3.6
AL035	7.2	4.1

Fig. 1 Linear regression for the year 2018 of the structure of net sales (*y*) with the "weight" of the number of economic enterprises in ratio to the total number of enterprises (in %) (*x*)

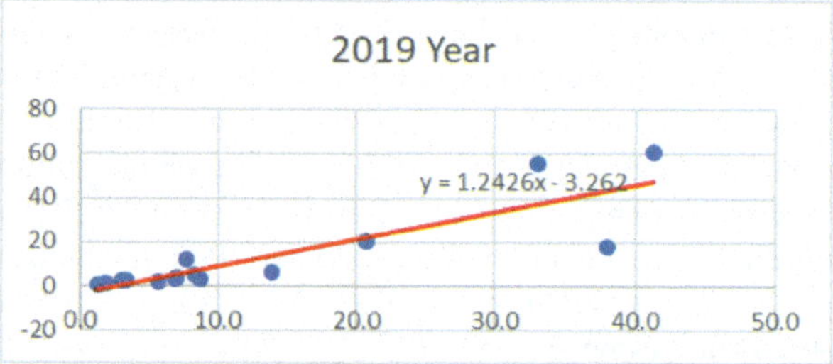

DMU	X	Y
AL01	20.7	20.4
AL011	1.8	1.5
AL012	7.7	12.2
AL013	1.3	0.8
AL014	3.2	2.7
AL015	6.8	3.2
AL02	41.3	61.1
AL021	8.3	5.1
AL022	33.0	56
AL03	38.0	18.3
AL031	5.6	1.8
AL032	13.9	6.6
AL033	2.9	2.6
AL034	8.7	3.5
AL035	7.0	3.8

Fig. 2 Linear regression for the year 2019 of the structure of net sales (*y*) with the "weight" of the number of economic enterprises in ratio to the total number of enterprises (in %) (*x*)

trend in 2022 it is 11.9%, while the gap found in the employment rate for 2022 is 11.1%.

Figure 6 shows the narrower gender gap related to unemployment.

DMU	X	Y
AL01	20.5	20.9
AL011	1.8	1.6
AL012	7.6	13.1
AL013	1.3	0.8
AL014	3.2	2.5
AL015	6.6	2.9
AL02	41.0	58.8
AL021	8.4	3.4
AL022	32.6	55.4
AL03	38.5	18.3
AL031	5.8	1.9
AL032	14.1	6.6
AL033	2.9	2.8
AL034	8.7	3
AL035	7.0	4

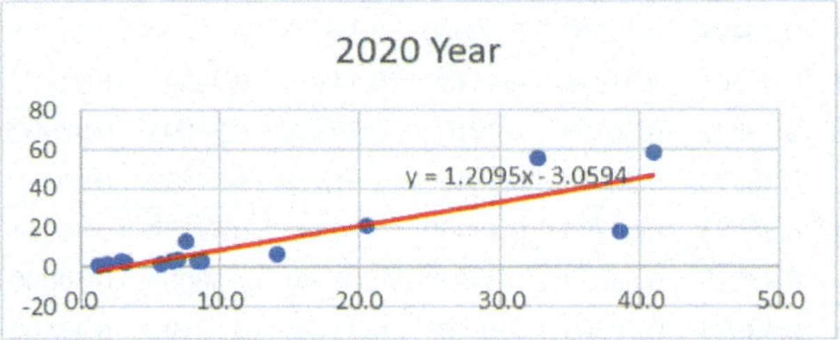

Fig. 3 Linear regression for the year 2020 of the structure of net sales (*y*) with the "weight" of the number of economic enterprises in ratio to the total number of enterprises (in %) (*x*)

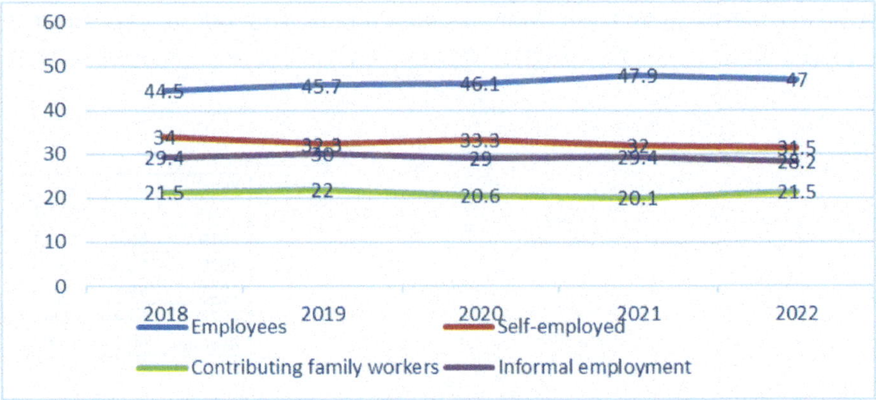

Fig. 4 Employment (15+) by status in employment, 2018–2022

Table 3 The sensitivity of the influences of each variable factor

DMU	Model A				Model B			
	Influence effectiveness				Influence effectiveness			
	I_1	$I_2 + I_3$	O_1	O_2	I_1	$I_2 + I_3$	O_1	O_2
AL_{01}	0.45831	0.54169	0.58089	0.41911	0.44250	0.55750	0.52705	0.47295
AL_{011}	0.50000	0.50000	0.49111	0.50889	0.50000	0.50000	0.48175	0.51825
AL_{012}	0.32834	0.67166	0.77762	0.22238	0.35621	0.64379	0.54819	0.45181
AL_{013}	0.50000	0.50000	0.70498	0.29502	0.50000	0.50000	0.50000	0.50000
AL_{014}	0.50000	0.50000	0.08102	0.91898	0.45122	0.54878	0.48828	0.51172
AL_{015}	0.49991	0.50009	0.63842	0.36158	0.49162	0.50838	0.53098	0.46902
AL_{02}	0.42185	0.57815	0.43974	0.56026	0.38409	0.61591	0.50622	0.49378
AL_{021}	0.50000	0.50000	0.35695	0.64305	0.49420	0.50580	0.52756	0.47244
AL_{022}	0.38656	0.61344	0.43323	0.56677	0.34640	0.65360	0.50000	0.50000
AL_{03}	0.45881	0.54119	0.40697	0.59303	0.46400	0.53600	0.52196	0.47804
AL_{031}	0.42998	0.57002	0.39368	0.60632	0.43221	0.56779	0.54440	0.45560
AL_{032}	0.42930	0.57070	0.38717	0.61283	0.43012	0.56988	0.51469	0.48531
AL_{033}	0.46680	0.53320	0.50338	0.49662	0.47588	0.52412	0.51869	0.48131
AL_{034}	0.42553	0.57447	0.52965	0.47035	0.43323	0.56677	0.54132	0.45868
AL_{035}	0.49227	0.50773	0.32090	0.67910	0.42503	0.57497	0.54335	0.45665
Average	0.45318	0.54682	0.46971	0.53029	0.44178	0.55822	0.51963	0.48037

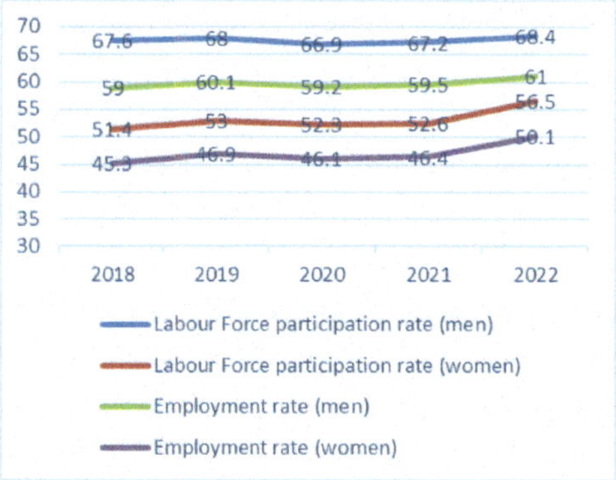

Year	a	b
2018	59.4	52.1
2019	60.4	53.4
2020	59.5	52.5
2021	59.8	52.9
2022	62.4	55.5
Average	60.3	53.28

a – Labour force participation rate
b- Employment rate

Fig. 5 Labour Force participation rate and Employment rate (men, women), 2018–2022

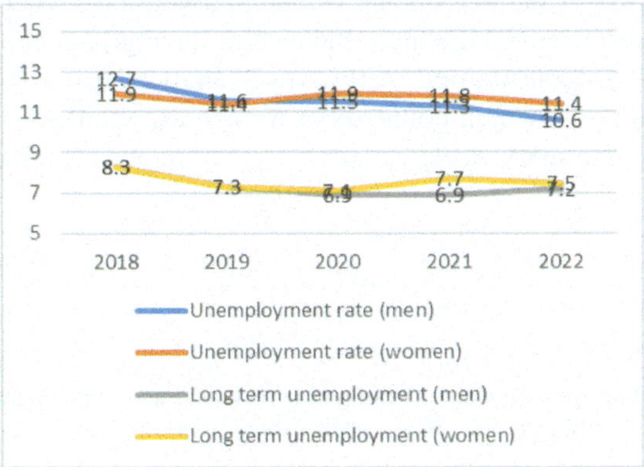

Year	c	d
2018	12.3	8.3
2019	11.5	7.3
2020	11.7	7
2021	11.5	7.2
2022	10.9	7.3
Average	11.58	7.42

c- Unemployment rate
d- Long-term unemployment

Fig. 6 Unemployment rate and long-term unemployment, 2018–2022

4 Conclusion

In this study, a general picture was presented on the sensitivity of the influences of active economic enterprises in the dynamics of economic development progress (2018–2022) according to prefectures and regions related to employment and unemployment. It was found that: the sensitivity of the influence of the economic activity of SMEs is distinguished by visible changes according to the prefectures in the mitigation of unemployment, in the reduction of unemployed job seekers, the increase in the employment rate, etc. The sensitivity of the influences of the business activity of economic enterprises (SMEs) is also distinguishable according to variable factors, especially according to the structure of their economic composition. The sensitivity of the influence of business development activity by SMEs is found to have a significant relationship between the size and the weight of SMEs to the total number of enterprises with the structure of net sales according to prefectures and economic activities (2018–2022). A relatively not very large production of their product is noticed, which requires more attention both in the absorption of investments and the application of the most advanced technologies in the business activity. This study highlights that competitive abilities should be better analyzed in order to have a more realistic and positive perception of the role that SMEs have and play in the country's economic development. The number of enterprises that employ 10 to 49 workers should be increased, so that this coefficient (the number of enterprises per 1000 inhabitants) from 2.56 approaches the average of the European Union countries with 3.5. In Albania, enterprises with foreign co-owners for the year 2022 occupy only 5.3% of the total number of enterprises, but which enables the absorption of investments, more advanced technologies and an increase in the employment rate. The role of active economic enterprises should also be seen with the increase in the level of qualification and education that enables the reduction of the percentage of unemployed job seekers aged 15–29. Active economic enterprises, SMEs are a balance in the regional economic developments of the country. Contribution of this study analysis using the DEA method gave a general panorama, which could serve to increase managerial care and the construction of more visionary strategies on a prefecture and regional basis.

References

Banker, R.D., Charnes, A., Cooper, W.W.: Some models for estimating technical and scale inefficiencies in data envelopment analysis. Manage. Sci. **30**(9), 1078–1092 (1984). https://doi.org/10.1287/mnsc.30.9.1078

Cassandro, N., Centra, M., Esposito, P., Guarascio, D.: What drives employment-unemployment transitions? Evidence from Italian task-based data, GLO Discussion Paper, No. 563, Global Labor Organization (GLO), Essen (2020). Available: https://www.econstor.eu/bitstream/10419/218860/1/GLO-DP-0563.pdf

Charnes, A., Cooper, W.W., Rhodes, E.: Measuring the efficiency of decision making units. Eur. J. Oper. Res. **2**(6), 429–444 (1978). https://doi.org/10.1016/0377-2217(78)90138-8. Available: https://www.sciencedirect.com/science/article/abs/pii/0377221778901388

Cooper, W.W., Seiford, L.M., Tone, K.: Data Envelopment Analysis, A Comprehensive Text with Models, Applications, References and Dea-solver Software. Kluwer Academic Publishers, New York, Boston, Dordrecht, London, Moscow (2002). ISBN 0-7923-8693-0

Cooper, W.W., Seiford, L.M., Zhu, J.: Handbook on data envelopment analysis. Int. Ser. Oper. Res. Manage. Sci. **164** (2011). ISBN: 978-1-4419-6150-1. https://doi.org/10.1007/978-1-4419-6151-8

Dobrin, G.I.: Types of enterprises—main risk and impact factors specific to the complex business area. J. Public Admin Finance Law 7. Available: https://www.jopafl.com/uploads/issue7/TYPES_OF_ENTERPRISES_MAIN_RISK_AND_IMPACT_FACTORS_SPECIFIC_TO_THE_COMPLEX_BUSINESS_AREA.pdf

Duda, J., Skalna, I.: Functioning and development of enterprises contemporary challenges. Published by AGH University of Science and Technology Press. ISBN 978-83-66364-36-3. Available: http://www.b2b.zarz.agh.edu.pl/J.%20Duda,%20I.%20Skalna,%20Functioning%20and%20development%20of%20enterprises.pdf

Guarascio, D., Cetrulo, A., Cirillo, V.: Weaker jobs, weaker innovation. Exploring the effects of temporary employment on new products. Appl. Econ. 1–28. https://doi.org/10.1080/00036846.2019.1619015

https://ec.europa.eu/eurostat/databrowser/view/tec00114/default/table?lang=en

https://www.instat.gov.al/al/temat/industria-tregtia-dhe-sh%C3%ABrbimet/statistikat-strukturore-t%C3%AB-nd%C3%ABrmarrjeve-ekonomike/#tab2

https://www.statista.com/statistics/939710/eu-enterprise-employment-share-by-enterprise-size/#:~:text=By%20comparison%2C%20medium%2Dsized%20businesses,to%209%20workers)%2029.43%20percent

Nazarko, B.: Performance evaluation in macroeconomics based on DEA malmquist index with a new approach for the efficiency evaluation in a two-stage process. WSEAS Trans. Inform. Sci. Appl. **21**, 169–185 (2024). https://doi.org/10.37394/23209.2024.21.17

Tregu i Punës, 2018: INSTAT (2019)

Tregu i Punës, 2019: INSTAT (2020)

Tregu i Punës, 2020: INSTAT (2021)

Tregu i Punës, 2021: INSTAT (2022)

Tregu i Punës, 2022: INSTAT (2023)

Turgut E (2021) Unemployment and factors affecting unemployment in developing countries. In: Discussions Between Economic Agents: Panel Data Analysis. Published by İksad Yayınevi, August 2021, Available: file:///C:/Users/user/Downloads/DISCUSSIONS-BETWEEN-ECONOMIC-AGENTS-Panel-Data-Analysis-sayfalar-1-312-43.pdf

Vjetari rajonal Statistikor: INSTAT (2020). ISBN 978-9928-188-89-2

Vjetari rajonal Statistikor: INSTAT (2021). ISBN 978-9928-188-94-6

Vjetari Statistikor Rajonal: INSTAT (2019)

Vjetari Statistikor Rajonal: INSTAT (2022). ISBN 978-9928-390-02-8

Vjetari Statistikor Rajonal: INSTAT (2023). ISBN 978-9928-390-01-1

Zhu, J.: Quantitative models for performance evaluation and benchmarking. Int. Ser. Oper. Res. Manage. Sci. **213** (2014). Springer. ISBN 978-3-319-06646-2. https://doi.org/10.1007/978-3-319-06647-9

The Impact of Transformational and Charismatic Leadership on Employee Job Satisfaction and Organizational Commitment

Mimoza Kasa-Jashari and Aleksandra Janeska-Iliev

Abstract

Every company cultivates its own unique organizational footprint which mainly is arising from leadership style. In this study, we will explore the impact of transformational and charismatic leadership on employee job satisfaction. Previous research has demonstrated that specific leadership styles can significantly motivate employees. In particular, transformational and charismatic leadership have been shown to foster high levels of commitment, performance beyond expectations, and overall job satisfaction. Applying the right leadership style is crucial for managers aiming to enhance their followers' satisfaction. An effective leader earns the respect and trust of their team. When employees feel satisfied with their work and their organization, they are more likely to be committed, resulting in increased individual and organizational productivity. Research indicates that transformational and charismatic leadership can boost employee satisfaction and positive workplace behavior, directly contributing to the overall well-being of the organization. This manuscript aims at countering the layout for further discussion and providing a relevant path toward research.

Keywords

Leadership • Leader • Organizational commitment • Job satisfaction • Digital transformation

M. Kasa-Jashari (✉) · A. Janeska-Iliev
Faculty of Economics-Skopje, Ss.Cyril and Methodius University in Skopje, Skopje, North Macedonia
e-mail: moza_kasa@hotmail.com

A. Janeska-Iliev
e-mail: aleksandra.janeska-iliev@eccf.ukim.edu.mk

K. Dhoska and E. Spaho (eds.), *Bridging Horizons in Artificial Intelligence, Robotics, Cybersecurity, Smart Cities, and Digital Economy*, Sustainable Economy and Ecotechnology, https://doi.org/10.1007/978-3-031-72029-1_19

1 Introduction

Every company has its own specific leadership style (Al-maaitah et al. 2021; Porfírio et al. 2021; Sadler 1970). Leadership plays a crucial role in influencing the job satisfaction and organizational commitment of employees. Numerous studies have examined the relationship between leadership styles and employee outcomes such as job satisfaction and organizational commitment (Qing et al. 2020; Shurbagi 2014; Babalola 2016). In today's constantly-changing business world, one of the most important things for shaping the work environment and impacting employee outcomes is strong and successful leadership. Some studies have focused specifically on the impact of transformational and charismatic leadership on these outcomes (Xanthakis 2020).

Leadership discussions are crucial for the development and success of any organization. These discussions typically revolve around identifying and implementing effective leadership styles, understanding the impact of leadership on employee motivation and performance, and exploring ways to cultivate leadership skills within teams. Key topics often include the comparison of different leadership styles such as transformational, charismatic, democratic, and autocratic leadership, and how these styles influence organizational culture and employee satisfaction. For instance, transformational leaders inspire and motivate employees to exceed their expectations, fostering a high level of engagement and productivity (Kovjanic et al. 2013). Another important aspect of leadership discussions is the emphasis on situational leadership, where leaders adapt their style based on the needs of their team and the specific circumstances they face (Bolden et al. 2003) This flexibility is crucial for effectively navigating challenges and driving organizational success.

Overall, leadership discussions aim to enhance the effectiveness of leaders, improve team dynamics, and ultimately contribute to the growth and sustainability of the organization. Transformational and charismatic leadership styles have attracted a lot of attention due to their significant influence on employee job satisfaction and organizational commitment. This review aims to research into the complex connection between these leadership styles and important organizational results, drawing on insights from numerous scholarly studies. Do transformational and charismatic leadership positively affect employees' job satisfaction and organizational commitment? The purpose of this research is to deepen our understanding of transformational and charismatic leadership and their impact on employee job satisfaction. Specifically, this study aims to examine how these leadership styles influence employee commitment, performance, and overall job satisfaction. Previous research has demonstrated that certain leadership styles, particularly transformational and charismatic, have significant motivational effects on employees. These styles are associated with higher levels of commitment and performance that exceed expectations, contributing to greater job satisfaction.

2 Literature Review

2.1 Leadership

Organizational behavior is a field of study that explores how individuals, groups, and structure impact organizations, aiming to utilize this knowledge to enhance organizational effectiveness (Robins et al. 2009). Although organizational behavior is extremely complex and involves many inputs and dimensions, cognitive, behavioral, and social cognitive theories can be used to develop an overall framework for an evidence-based approach (Luthans 2011). In order to increase the efficiency of employees, the role of leaders is very important in directing and managing employees to achieve the company's goals. Without a leader, a company is just a group of people working together without any direction that enables them to realize the company's goals. (Andreani and Petrik 2016). Effective leaders do not possess a defined thematic list of personal traits, but, on the contrary, provide high performance in solving issues of organizational tasks and consideration for individuals who care about interpersonal relationships (Klingborg et al. 2006).

2.2 Transformational Leadership

Leaders who are able to encourage, change and use the values, beliefs, and needs of their followers in the performance of tasks in crisis and dynamic situations with major changes, are transformational leaders (Bass and Riggio 2006; Eisenbach et al. 1999; Beugré et al. 2006; Shuklev et al. 2022). Transformational leadership is achieved when leaders and followers motivate each other to higher motivation and morale. These leaders have better relationships with their superiors. They help the organization more than transactional leaders and create more value for the organization (Korejan and Shahbaz 2016). For Burns (1978), transformational leadership occurs when people engage with others so that leaders and followers lift each other to higher levels of motivation and morale. Therefore, morality was an integral part of Burns' description of transformational leadership and, he saw democracy and participation as fundamental to the success of organizational change (Burnes et al. 2016). Transformational leadership appeals to greater motivation and increases the quality of life in people and the organization. Transformational leadership provides power-generating characteristics that generate new changes for the organization (Tucker and Russell 2004). Transformational leaders influence three areas of organizational culture: (1) the internal mindset of the people in the organization, (2) the culture among the people in the organization, and (3) the culture outside the people in the organization (Tucker and Russell 2004). Transformational leadership consists of four interrelated dimensions: charisma, inspiration, individualized attention, and intellectual stimulation. Transformational leaders try and succeed in raising colleagues, subordinates, followers, and customers to a greater awareness of issues of consequence. Transformational leaders bring change, innovation, and entrepreneurship. They describe the process of corporate transformations that recognize the need for revitalization, create a new vision, and institutionalize change (Garg and Krishnan 2003).

2.3 Transactional Leadership

Unlike transformational leadership, transactional leadership tends to be based on an exchange process whereby followers are rewarded for achieving certain goals. With transactional leadership, followers are typically given rewards in exchange for achieving certain levels of performance (Jung 2001). As authors Hater & Bass point out, transactional and transformational leadership are not unrelated models. Bass (1985) however sees them as separate dimensions, emphasizing that a leader can be both transactional and transformational. He argues that transformational leadership builds on transactional leadership, but not vice versa. Transformational leadership can be seen as a special case of transactional leadership, since both approaches are related to the achievement of a goal. The models differ in terms of the process by which the leader motivates the subordinates and according to the type of goals set (Hartog et al. 1997). Unlike transformational leadership, leaders who use the transactional approach do not want to change the future, but only to keep things the same. These leaders pay attention to the work of followers in order to find errors and deviations. This type of leadership is effective in crisis and emergency situations, as well as when projects need to be implemented in a specific way (Odumeru and Ogbonna 2013).

2.4 Charismatic Leadership

The first scientist who spoke about charismatic leadership was Max Weber. Weber defines charisma as "being set apart from ordinary people and treated as endowed with supernatural, superhuman, or at least specifically exceptional powers or qualities, considered as of divine origin, and based on them the concerned individual is treated as a leader" (Judge et al. 2006). One of the most important attributes of charismatic leaders is their ability to be highly motivational. They are motivational in that they increase the intrinsic value of employees' efforts; improve self-confidence, self-esteem, and self-efficacy beliefs of employees; increasing the intrinsic value of achieving the employees' goal; instill faith in a better future; and cause employees to be more committed (Horn et al. 2015). A charismatic leader rejects traditional attitudes in seeking the status quo and does not seek new responsibilities in the workplace. The task of a leader is to change and strive for the best while operating in the most charismatic way. He does not hesitate to break tradition or negative habits to keep up with the present and achieve future growth. Charisma is a supernatural power because it is the ability to control people's minds, changing their thoughts and beliefs out of fear of the unknown; by redirecting them in the right way (Awamleh 2022). Charismatic leaders inspire others by speaking optimistically about the future and what needs to be achieved, and instilling in their followers positive ideals that are linked to desired outcomes. Charismatic leaders inspire others by speaking optimistically about the future and what needs to be achieved, and instilling in their followers positive ideals that are linked to desired outcomes.

2.5 Leadership in Digital Era

According to Mihai and Cretu (2017), the digital transformation is no longer confined to the technical department; it has moved into the boardrooms of leaders across all industries, sectors, and regions. The current digital age is a crucial stage in the evolution of a world made possible by the widespread use of technology in the workplace. This rapid pace of the digital era is fundamentally altering the way organizations, whether private or public, operate. It requires them to develop new approaches to designing operating models. Today, efficient leaders in decision-making and innovation work in rapid learning cycles. Organizational and emotional agility are increasingly critical in the digital age, as indicated by specialist literature.

Matsunaga (2024) says that leaders have the responsibility to guide their team members through the challenges of digital transformation. Poor leadership can make the uncertainty and anxiety felt by team members worse. For instance, if a leader struggles to communicate the vision for digital transformation, team members may feel disconnected from the organization's goals, leading to frustration and resistance.

The research by Prakasa et al. (2019), transformational leadership influences the creation of an organizational culture that can adapt to environmental changes, particularly those related to digital transformation, by continuously changing employee behavior. The characteristics of respondents with an undergraduate education background also support the implementation of transformational leadership in an organizational culture that is adaptive to digital change.

2.6 Job Satisfaction

According to Locke (1969), job satisfaction is a happy emotion experienced by individuals who perceive that their work fulfills their work values. This definition emphasizes that job satisfaction is unique to each individual. Job satisfaction is a pleasurable emotional state resulting from evaluating one's work as achieving or facilitating the achievement of work values. Job satisfaction begins with work values, what employees seek or desire from their work. This varies from person to person. For example, pay discrepancies may be more unsatisfactory to one person than to someone else who cares less about pay (Locke 1969).

2.7 Organizational Commitment

Organizational commitment includes three attitudes: (1) a sense of identification with the organization's goals, (2) a sense of involvement in organizational duties, and (3) a sense of loyalty to the organization. Research evidence shows that the absence of commitment can reduce organizational effectiveness. People who are committed are less likely to quit and take other jobs (Konopaske et al. 2018).

Taking into account several previous researches regarding organizational commitment, the authors Llobet and Fito (2013), say that first commitment comes with the awareness of the costs associated with leaving the organization; secondly they define commitment as an emotional attachment to the organization and thirdly they conceptualize commitment as a moral obligation to the rules. From the point of view of multidimensionality, organizational commitment is divided into three basic dimensions: affective commitment as a desire to belong to the organization, continuance commitment based on the belief that leaving the organization would be costly, and normative commitment as a feeling of obligation to the organization.

2.8 The Relationship Between Transformational and Charismatic Leadership and Job Satisfaction

Previous studies show that job satisfaction is significantly dependent on leadership style. Flexible organizations have a participative management style with an interactive environment and a satisfied workforce. Transformational leadership style is very effective in improving job satisfaction. The results of research conducted by Asghar and Oino (2017) show that, the inspirational motivational dimension of transformational leadership suggests that transformational leaders motivate and inspire their subordinates to complete challenging tasks by sharing their vision and strategies with employees. Transformational and charismatic leadership are positively related to followers' job satisfaction. In fact, job satisfaction has been found to be higher among employees whose leaders emphasize attention, support, and care for their followers (Keskes 2014).

2.9 The Relationship Between Transformational and Charismatic Leadership and Organizational Commitment

Transformational leadership will lead to close ties between employees and the organization and create an employee-centered climate. Such a climate will allow employees to feel the friendly side of the organization and will allow them to improve employee job satisfaction (Chen et al. 2021). It is hypothesized that transformational leadership can lead to team effectiveness through the process of engaging collective motivation and encouraging team members to identify with the group, recognizing member contributions and collaborative tasks, and encouraging team members to align with strategic goals or the bigger picture. Recent studies have found that over time, transformational leadership improves collective effectiveness in teams (Nielsen et al. 2009). Transformational leaders have a great ability to influence organizational commitment by promoting values related to goal achievement, by emphasizing the relationship between employee efforts and goal achievement, and by creating a greater degree of personal commitment on the part of followers and leaders to achieving the ultimate shared vision, mission and goals of the organization (Shamir et al. 1998).

3 Results Implications

The research effort summarizes previous studies on various leadership styles and their impact on organizational outcomes. In the methods section, we detail key studies conducted by other researchers. Significant studies in this field are mentioned below, with the main discussions presented in Table 1 in the appendix.

Transformational leadership is characterized by the ability to inspire and motivate followers to prioritize the organization's goals over their own self-interests. This leadership style comprises four key components: idealized influence, inspirational motivation, intellectual stimulation, and individualized consideration (Bass 1985). Research consistently shows a positive association between transformational leadership and employee job satisfaction. Studies by Avolio et al. (2004) found that transformational leaders, who demonstrate genuine concern for their employees' well-being and foster a supportive and challenging work environment, significantly enhance job satisfaction. These leaders achieve this by creating a sense of purpose and cultivating a positive work culture that meets employees' intrinsic needs for growth and achievement.

Transformational leadership also positively impacts organizational commitment. Meyer and Allen's (1984) three-component model of organizational commitment—affective, continuance, and normative—has been used to measure this relationship. Numerous studies, including those by Rafferty and Griffin (2004), have found that transformational leadership enhances affective commitment, wherein employees develop an emotional attachment to the organization. This is because transformational leaders effectively communicate a compelling vision and foster an environment of trust and loyalty.

Charismatic leadership, often seen as a subset of transformational leadership, is defined by a leader's ability to inspire and energize followers through charm, persuasiveness, and vision. Charismatic leaders are often perceived as heroic and exceptional figures who can articulate a vision that deeply resonates with followers (Conger and Kanungo, 1998). Charismatic leadership has been linked to high levels of job satisfaction. Shamir et al. (1993) propose that charismatic leaders enhance job satisfaction by providing a clear and appealing vision, instilling a sense of mission, and fostering a strong sense of belonging among employees. The

Table 1 Leadership research perspectives

Author	Year	Title	Variables
Asghar Saima and Oino Isaiah	2017	Leadership styles and job satisfaction	Transactional and transformational leadership—independent Job satisfaction—dependent
Nadeem Bhatti et al	2012	The impact of autocratic and democratic leadership style on job satisfaction	Leadership styles—independent Job-related tension-dependent
Ronit Bogler	2001	The influence of leadership style on teacher job satisfaction	Leadership styles—independent Job satisfaction—dependent
Cedwyn Fernandes and Raed Awamleh	2011	The impact of transformational and transactional leadership styles on employee's satisfaction and performance: an empirical test in a multicultural environment	Transformational, transactional, self esteem—independent Satisfaction and self-perceived performance—dependent
Shaoyi Cheng and Qingquan Pan	2019	Charismatic leadership and its influence	Charismatic leadership—independent New enterprise performance—dependent
Akinlolu Olumide Emmanuel and Zubair Hassan	2015	The effects of transformational leadership on job satisfaction: a study on four and five star hotels in Kuala Lumpur	Transformational leadership—independent Job satisfaction—dependent
Diether Gebert, Kathrin Heinitz and Claudia Buengeler	2015	Leaders' charismatic leadership and followers' commitment—the moderating dynamics of value erosion at the societal level	Charismatic leadership—independent Followers' commitment—dependent
Imen Keskes	2014	Relationship between leadership styles and dimensions of employee organizational commitment: a critical review and discussion of future directions	Leadership styles—independent Organizational commitment—dependent
Alisa Mujkić et al.	2014	Transformational leadership and employee satisfaction	Transformational leadership and other known leadership styles—independent Employee satisfaction—dependent

Sources Adopted by the authors

emotional and motivational appeal of charismatic leaders helps fulfill employees' needs for meaning and accomplishment in their work.

Charismatic leadership also positively influences organizational commitment. Research by Shamir et al. (1993) suggests that charismatic leaders foster organizational commitment by creating a strong identification with the leader and the organizational mission. This identification leads to higher levels of affective commitment, as employees feel more connected and loyal to the organization. Additionally, the emotional bond and trust that charismatic leaders build with their followers enhance normative commitment, as employees feel a moral obligation to remain with the organization.

While both transformational and charismatic leadership styles positively impact job satisfaction and organizational commitment, they do so through slightly different mechanisms. Transformational leaders focus on developing and transforming their followers, fostering an environment that supports employee growth and organizational goals. Charismatic leaders, on the other hand, rely more on their personal attributes and the emotional connections they create with followers to drive commitment and satisfaction.

Considering the regional context, the study conducted by Mujkić et al. (2014), titled "Transformational Leadership and Employee Satisfaction," focuses on Bosnia and Herzegovina and Western Germany. The main objective was to demonstrate that transformational leadership generates higher employee satisfaction compared to transactional and charismatic leadership styles. The study employed a tailored questionnaire, with one main hypothesis centered on comparing transformational leadership to other contemporary leadership styles to determine its greater contribution to employee satisfaction. In this study, the independent variable is transformational leadership, while transactional and charismatic leadership are secondary variables. The dependent variable is job satisfaction. The results confirmed that transformational leadership indeed contributes to higher levels of employee satisfaction compared to other styles.

Another significant study included in this research is by Bogler (2001), titled "The Influence of Leadership Style on Teacher Job Satisfaction," published in Educational Administration Quarterly. In 1997, a quantitative questionnaire using Likert-type scales was administered to 930 teachers. The questionnaire assessed teachers' perceptions of their principals' leadership style and decision-making, their views on the teaching profession, and their satisfaction with various work aspects. The study proposed that teachers' perceptions of their occupation, including status, promotion opportunities, self-development, and personal growth, positively influence their job satisfaction. The analysis revealed a significant relationship between job satisfaction and teachers' perceptions of their occupation. Teachers who viewed their occupation as a profession perceived their principals as transformational leaders who were participative and exhibited less transactional leadership, leading to higher job satisfaction. The most significant finding was that occupational prestige, self-esteem, autonomy at work, and professional self-development were the strongest contributors to job satisfaction.

Several theoretical frameworks underpin the relationship between leadership styles and organizational outcomes. Social exchange theory (Blau 1964) posits that the positive behaviors exhibited by transformational and charismatic leaders create a reciprocal relationship where employees feel compelled to respond with higher levels of job satisfaction and commitment. Additionally, self-determination theory (Deci and Ryan 1985) suggests that transformational and charismatic leaders fulfill employees' intrinsic needs for competence, autonomy, and relatedness, thereby enhancing job satisfaction and commitment.

4 Conclusions

Considering a large number of researches and papers conducted in this direction, it has been concluded that transformational leadership has a positive effect on the job satisfaction of employees. At the end of this paper, we can conclude that the transformational leadership style has a strong positive effect on employees' attitudes toward their work, their work environment and ultimately affects their work performance. Transformational leaders serve as an inspiration to their followers by engaging in effective communication that fosters trust, commitment, and, consequently, greater satisfaction. Taking into account the research literature regarding the influence of leadership style on job satisfaction and organizational commitment of employees, this paper confirms the positive association between transformational and charismatic leadership styles, and their influence on job satisfaction and organizational commitment. Transformational leaders are those who try to motivate the behavior of their followers in order to achieve a common goal. Managers and supervisors play a key role in employee satisfaction levels. People who leave the organization do not do so because of the company, but because of their superiors. This proves that leadership is the key to employee satisfaction. One of the ways to become a transformational and charismatic leader is to initially identify the strengths and "play" according to them. By improving the natural qualities that a leader possesses and with which they attract people, it becomes easier to lead and influence people in the organization. Operating according to those strengths ensures that leadership does not seem forced.

In conclusion, extensive research and numerous papers have established that transformational leadership positively impacts employee job satisfaction. This paper confirms that the transformational leadership style significantly enhances employees' attitudes toward their work and work environment, ultimately boosting their performance. Transformational leaders inspire their followers through effective communication, fostering trust, commitment, and greater satisfaction.

Literature consistently supports the positive association between transformational and charismatic leadership styles and their influence on job satisfaction and organizational commitment. Transformational leaders motivate their followers to achieve common goals, highlighting the critical role managers and supervisors play in employee satisfaction. It is often said that employees leave their supervisors, not the organization, underscoring the importance of leadership in retaining satisfied employees.

To cultivate transformational and charismatic leadership, it is essential to identify and leverage one's strengths. Enhancing these natural qualities makes it easier to lead and influence others within the organization. Operating according to these strengths ensures that leadership is authentic and not forced.

Leadership Research Perspectives

See Table 1.

References

Al-maaitah, D.A., Alsoud, M., Al-maaitah, T.A.: The role of leadership styles on staffs job satisfaction in public organizations. J. Contemp. Issues Bus. Govern. **27**(1), 772–783 (2021)

Andreani, F., Petrik, A.: Employee performance as the impact of transformational leadership and job satisfaction. Jurnal Manajemen dan Kewirausahaan **18**(1) (2016). Available at: https://doi.org/10.9744/jmk.18.1.25-32. Accessed 9 June 2024

Asghar, S., Oino, I.: Leadership styles and job satisfaction. Market Forces **8**(1), 1–13 (2017). Available at: https://mpra.ub.uni-muenchen.de/91137/. Accessed 9 June 2024

Avolio, B.J., Walumbwa, F.O., Weber, T.J.: Leadership: current theories, research, and future directions. Annu. Rev. Psychol. **55**(1), 421–449 (2004)

Awamleh, F.T.: Charismatic leadership style to overcome employee resistance to the management of organizational change process. Glob. J. Manage. Bus. Res. **22** (2022)

Babalola, S.S.: The effect of leadership style, job satisfaction and employee-supervisor relationship on job performance and organizational commitment. J. Appl. Bus. Res. **32**(3), 935 (2016)

Bass, B.M.: Leadership and performance beyond expectations. Free Press, New York (1985)

Bass, B.M., Riggio, R.E.: Transformational Leadership. Psychology Press (2006)

Beugré, C.D., Acar, W., Braun, W.: Transformational leadership in organizations: an environment-induced model. Int. J. Manpow. **27**(1), 52–62 (2006)

Blau, P.M.: Exchange and power in social life. Wiley, New York (1964)

Bogler, R.: The influence of leadership style on teacher job satisfaction. Educ. Admin. Q. **37**(5), 662–683 (2001). Available at: https://doi.org/10.1177/00131610121969460. Accessed 9 June 2024

Bolden, R., Gosling, J., Marturano, A., Dennison, P.: A Review of Leadership Theory and Competency Frameworks. University of Exeter, Centre for Leadership Studies (2003)

Burnes, B., Hughes, M., By, R.T.: Reimagining organisational change leadership. Leadership 1–18 (2016). Available at: https://doi.org/10.1177/1742715016662188. Accessed 9 June 2024

Burns, J.M.: Leadership. Harper & Row, New York (1978)

Chen, C., Ding, X., Li, J.: Transformational leadership and employee job satisfaction (2021). Available at: https://doi.org/10.3390/ijerph19010233. Accessed 9 June 2024

Conger, J.A., Kanungo, R.N.: Charismatic Leadership in Organizations, 1st edn. SAGE Publications, California (1998)

Deci, E.L., Ryan, R.M.: Intrinsic motivation and self-determination in human behavior. Plenum Press, New York (1985)

Eisenbach, R., Watson, K., Pillai, R.: Transformational leadership in the context of organizational change. J. Organ. Chang. Manag. **12**(2), 80–89 (1999)

Zhen, S., Yuqiang, F., Tienan, W., Luning, L.: The impact mechanism of charismatic leadership on individual's tacit knowledge sharing. Wuhan International Conference on e-Business (2016). Available at: http://aisel.aisnet.org/whiceb2016/7

Garg, G., Krishnan, R.V.: Transformational leadership and organizational structure: the role of value-based leadership. In: Value-Based Management for Indian Organizations, pp. 82–100. Sage Publications, New Delhi (2003)

Hartog, D.N.D., Muijen, J.J.V., Koopman, P.L.: Transactional versus transformational leadership: an analysis of the MLQ. J. Occup. Organ. Psychol. **70**(1), 19–34 (1997). Available at: https://doi.org/10.1111/j.2044-8325.1997.tb00628.x. Accessed 9 June 2024

Horn, D., Mathis, C.J., Robinson, S.L., Randle, N.: Is charismatic leadership effective when workers are pressured to be good citizens? J. Psychol. **149**(8), 751–774 (2015). Available at: https://doi.org/10.1080/00223980.2014.978253. Accessed 9 June 2024

Judge, A.T., Woolf, F.E., Hurst, C., Livingston, B.: Charismatic and transformational leadership. A review and an agenda for future research. **24**(4), 203–2014 (2006). Available at: https://doi.org/10.1026/0932-4089.50.4.203. Accessed 9 June 2024

Jung, D.I.: Transformational and transactional leadership and their effects on creativity in groups. Creativity Res. J. **13**(2), 185–195 (2001). Available at: https://doi.org/10.1207/S15326934CRJ1302_6. Accessed 9 June 2024

Keskes, I.: Relationship between leadership styles and dimensions of employee organizational commitment: a critical review and discussion of future directions. Intangible Capital **10**(1), 26–51 (2014). Available at: https://doi.org/10.3926/ic.476. Accessed 9 June 2024

Klingborg, D.J., Moore, D.A., Varea, H.S.: What is leadership? J. Vet. Med. Educ. **33**(2), 280–283 (2006). Available at: https://doi.org/10.3138/jvme.33.2.280. Accessed 9 June 2024

Konopaske, R., Ivancevich, J.M., Matteson, M.T.: Organizational Behavior & Management, 11th edn. McGraw-Hill Education, Dubuque, IA (2018)

Korejan, N.M., Shahbaz, H.: An analysis of the transformational leadership theory. J. Fundam. Appl. Sci. (2016). Available at: https://doi.org/10.4314/jfas.v8i3s.192. Accessed 9 June 2024

Kovjanic, S., Schuh, S.C., Jonas, K.: Transformational leadership and performance: An experimental investigation of the mediating effects of basic needs satisfaction and work engagement. J. Occup. Organ. Psychol. **86**(4), 543–555 (2013)

Kuhnert, K.W., Lewis, P.: Transactional and transformational leadership: a constructive/developmental analysis. Acad. Manag. Rev. **12**(4), 648–657 (1987). https://doi.org/10.2307/258070

Llobet, J, Fito, AM: Contingent Workforce, Organisational Commitment and Job Satisfaction. Universitat Oberta de Catalunya, Spain (2013)

Locke, E.A.: What is job satisfaction? Organ. Behav. Human Perform. **4**(4), 309–336. American Institutes for Research arid Department of Psychology, University of Maryland (1969). Available at: https://doi.org/10.1016/0030-5073(69)90013

Luthans F.: Organizatonal Behavior, edn. 12th. McGraw-Hill, New York (2011)

Matsunaga, M.: Transformational Leadership in a Digital Era (2024). https://doi.org/10.1007/978-981-99-8409-1_3

Meyer, J.P., Allen N.J.: Testing the "side-bet theory" of organizational commitment: Some methodological considerations. J. Appl. Psychol. **69**(3), 372–378 (1984). Available at https://doi.org/10.1037/0021-9010.69.3.372

Mihai, R.L., Cretu, A.: Leadership in the digital era. Valahian J. Econ. Stud. **10**, 65–72 (2017). https://doi.org/10.2478/vjes-2019-0006

Mujkić, A., Šehić, D., Rahimić, Z., Jusić, J.: Transformational leadership and employee satisfaction. Ekonomski Vijesnik (2014)

Nielsen, K., Yarker, J., Randall, R., Munir, F.: The mediating effects of team and self-efficacy on the relationship between transformational leadership, and job satisfaction and psychological well-being in healthcare professionals: a cross-sectional questionnaire survey. Int. J Nurs. Stud. (2009). https://doi.org/10.1016/j.ijnurstu.2009.03.001

Odumeru, J., Ogbonna, I.G.: Transformational vs. transactional leadership theories: evidence in literature. Int. Rev. Manage. Bus. Res. **2**(2) (2013)

Porfírio, J.A., Carrilho, T., Felício, J.A., Jardim, J.: Leadership characteristics and digital transformation. J. Bus. Res. **124**, 610–619 (2021)

Prakasa, Y., Raharjo, K., Wiratama, I.D.: Transformational leadership and digital maturity: the mediating role of organizational culture. Adv. Econ. Bus. Manage. Res. (2019)

Qing, M., Asif, M., Hussain, A., Jameel, A.: Exploring the impact of ethical leadership on job satisfaction and organizational commitment in public sector organizations: the mediating role of psychological empowerment. RMS **14**(6), 1405–1432 (2020)

Rafferty, A.E., Griffin, M.A.: Dimensions of transformational leadership: the role of identification and commitment. Leadersh. Quart. **15**(3), 341–365 (2004)

Robins, S.P., Judge, T.A., Sanghi, S.: Organizational Behavior. edn. 13th. Prentice Hall, New Delhi (2009)

Sadler, P.J.: Leadership style, confidence in management, and job satisfaction. J. Appl. Behav. Sci. **6**(1), 3–19 (1970)

Shamir, B., House, R.J., Arthur, M.B.: The motivational effects of charismatic leadership: a sefl-concept based theory. Organ. Sci. **4**(4) (1993). Available at: https://doi.org/10.1287/orsc.4.4.577

Shamir, B., Zakay, E., Breinin, E., Popper, M.: Correlates of charismatic leader behavior in military units: subordinates' attitudes, unit characteristics, and superiors' appraisals of leader performance. Acad. Manage. J. **41**(4), 387–409 (1998). Available at https://doi.org/10.2307/257080

Shurbagi, A.M.A.: The relationship between transformational leadership style job satisfaction and the effect of organizational commitment. Int. Bus. Res. **7**(11) (2014). Available at https://doi.org/10.5539/ibr.v7n11p126

Tucker, A.B., Russell, F.R.: The influence of the transformational leader. J. Leaders. Organ. Stud. (2004). https://doi.org/10.1177/107179190401000408

Xanthakis, G.: Greek economic crisis and changes in HR management the case of greek structural design engineers. In. J. Hum. Resourc. Stud. Macrothink Inst. **10**(2), 220237–222202 (2020)

Yukl, G.: An evaluation of conceptual weaknesses in transformational and charismatic leadership theories. State Univ. N. Y. Albany **10**(2), 1–305 (1999). https://doi.org/10.1016/s1048-9843(99)00013-2

Шуклев, Б., Љубомир, Д., Јанеска, И.А.: Организациско однесување", Универзитет "Св. Кирил и Методиј" во Скопје, Економски факултет – Скопје (2022)

Fine-Tuning GPT-3.5 for Hate Speech Detection in Albanian YouTube Comments: Challenges and Analysis

Hersi Kopani and Rovena Llapushi

Abstract

Hate speech on social media appears to be inevitable in our era of ever-more connectivity. This presents a distinct challenge because it is difficult to design automated techniques to identify such speech, especially for low-resource languages. We introduce the PAR database, derived from YouTube comments associated with politics, actuality, and reality show content. Our study compares the performance of the GPT 3.5 model applied to these comments, which are written in Albanian, which is considered a low-resource language for NLP models, and predicts the performance. We assess the model's ability to detect hate speech across three topics, analyze the performance on each individual topic, and appraise the effect of translating comments from Albanian jargon to standard Albanian on the model's performance. This analysis is crucial because the model utilizes the translated English version as the target language for distinguishing such comments. Lastly, we debate whether large language models like GPT 3.5 are required for this kind of task. Is utilizing a model of this scale for transfer learning that beneficial, even if the model has no knowledge of the Albanian language? Or should we focus more of our attention on engineering and text annotation?

Keywords

Hate speech • Social media • Low-resource language • Fine-tuning LLMs

H. Kopani (✉)
Department of Computer Science, Faculty of Technology and Business, Bedër University College, Tirana, Albania
e-mail: h.kopani@beder.edu.al

R. Llapushi
Department of Mathematics "Tullio Levi-Civita", University of Padua, Padua, Italy
e-mail: Rovena.llapushi@studenti.unipd.it

K. Dhoska and E. Spaho (eds.), *Bridging Horizons in Artificial Intelligence, Robotics, Cybersecurity, Smart Cities, and Digital Economy*, Sustainable Economy and Ecotechnology, https://doi.org/10.1007/978-3-031-72029-1_20

1 Introduction

The number of active users on social media has increased rapidly. Because of that, comments posted on social media platforms like YouTube or Meta are often considered offensive comments or languages that violate the identity or origin of categories. Violating these ethical standards affects the lives of people psychologically or emotionally. In Albania, numerous instances of hate speech have been reported by the media, with several suspected cases of suicide allegedly linked to this phenomenon, as per police investigations.

Coming up with a clear definition of hate speech is a challenge, as human behaviors are complex to analyze. Also, there are many factors, such as way of expression, dialects, writing style, contextual circumstances, and some trending terms, that are present on high-audience TV programs. Moreover, labeling some comments as "hate speech," which may not necessarily be accurate, infringements upon the principles of freedom of expression. The concept of hate speech has been the subject of extensive debate and scrutiny within both political and academic spheres, yet a universally accepted definition has remained elusive (Stakic 2011), despite the United Nations Strategy and Plan of Action on Hate Speech (2022) defining it as "any kind of communication in speech, writing, or behavior that attacks or uses pejorative or discriminatory language with reference to a person or a group based on who they are, in other words, based on their religion, ethnicity, nationality, race, color, descent, gender, or other identity factor."

Within the realm of artificial intelligence (AI) models, particularly in the domain of natural language processing (NLP), the Albanian language is frequently categorized as a low-resource language. Low-resource languages can be understood as languages that are less studied, lacking in resources, less computerized, less privileged, less commonly taught, or characterized by low density, among other denominations. This classification presents numerous challenges due to the limited availability of algorithms and training data resources specific to the Albanian language. These components are crucial for the development of successful AI models capable of preventing hate speech. Currently, social media algorithms prioritize languages such as English or French for the detection of hate speech, resulting in difficulties in identifying and moderating hate comments in the Albanian language. Attempting to translate Albanian content into English for moderation purposes may introduce inaccuracies, thereby underscoring the complexities associated with addressing hate speech in low-resource languages.

This paper provides a comprehensive review of current research efforts aimed at identifying hate speech in the Albanian language. Additionally, it examines the performance of the GPT 3.5 API when applied to the PAR dataset, achieving an accuracy of 83% with a standard deviation of 0.02. Consequently, based on the results obtained, the recommendations for the Albanian language are as follows: (1) construct a high-quality dataset that can serve as a benchmark for identifying hate speech in Albanian social media; (2) involve researchers and Albanian linguistic experts to contribute to the development of a set of rules for the identification of hate speech statements; and (3) develop an efficient tokenizer, which is essential for a robust language model. Working with a large, well-constructed dataset is crucial for producing a high-quality tokenizer, which in turn enables the full utilization of the potential inherent in even large models through fine-tuning processes.

2 Related Work

Based on research results and our knowledge, there have been some publications analyzing hate speech on social media, including Instagram, Facebook, and YouTube, in the Albanian language. These studies focus on creating datasets by collecting comments from a variety of social media sites, like Instagram or YouTube, running different models, and explaining accuracy by interpreting the results. Referring to the first publicly available SHAJ dataset annotated using OffensEval schema, the accuracy of different models such as NaïveBayesa as baseline, BiLSTM and BERT shows that BERT performs better with an F1 score of 0.77. Nurçe et al. (2021) concluded that it is difficult to detect hate speech in instances of irony, sarcasm, or when there is a lack of context.

The same database SHAJ is extended and used to build a set of rules applied to BERT models using the ERASER benchmark when Kacuri (2023) explained that the focus of the research was more on explainability and less on calculation and concludes that BERT models trained with rationales performed slightly better in terms of comprehensiveness but were weaker in terms of sufficiency.

In this paper, the dataset used is called PAR, and it is only fine-tuned to address the awareness of the lack of annotated data and the deficiency in knowledge due to limited data availability for language model training. This is particularly crucial because large language models (LLMs) often lack sufficient data for effective application in low-resource languages like Albanian.

3 Dataset

3.1 Purpose

In this paper, the dataset used is called PAR, and it is only fine-tuned to address the awareness of the lack of annotated data and the deficiency in knowledge due to limited data availability for language model training. This is particularly crucial because large language models (LLMs) often lack sufficient data for effective application in low-resource languages like Albanian.

Table 1 Aggregate results of dataset based on category

Category	Hate speech T	Hate speech F
Politics	185	265
Actuality	188	339
Reality show	236	239

3.2 Data Collection

Dataset PAR[1] focuses on three main topics: politics, reality, and reality shows. The goal of selecting these three topics is to provide the variety of data associated with a large audience of TV shows. Comments are coming from YouTube platforms downloaded from the three main TV channels: TopChannel, KlanTV, and News24. According to Social Blade[2] results for 2024, two of these TV channels, in particular TopChannel and KlanTV, have the highest number of subscribers on YouTube, and the contents that they share generate high interactivity between users. Meanwhile, after our investigation of the topic that was selected for politics, News24 had the third most commented video. For politics and actuality, it uses the same topic and downloads all comments written on YouTube; however, for reality shows, due to the high number of comments, it is used only on TopChannel, one uploaded series of a dating show, and extracts all comments. After downloading these comments using the external tool Export Comment, the dataset contains 450 rows of data related to politics, 527 rows of data related to reality, and 475 rows of data related to reality shows uploaded on YouTube, which is the social media platform where the comments are extracted. Regarding ethics and privacy, even though these comments are public and readable by all, on the dataset, the authors are hidden by respecting the privacy of each YouTube user. During data collection, it became evident that numerous comments in the dataset were written in dialects, with a writing style that highlighted various grammar mistakes and dialectical nuances.

3.3 Data Annotation

Before annotating the dataset, first authors wrote the comments grammatically correct, then they followed up the process by annotation rows. Each author has labeled the row T to identify a comment as hate speech and F for the opposite. After annotating all comments from two authors, the results are compared, and discrepancies are verified with the assistance of an Albanian dictionary. Below there are the category records based on annotation results if the comments were hating speech or not (Table 1).

[1] https://github.com/HersiKopani/NLP-HateSpeechAlb.

[2] https://socialblade.com/youtube/top/country/al/mostsubscribed.

4 GPT 3.5 Architecture and Methodology

4.1 GPT-N Overview

GPT-n is a language model that operates from left to right, utilizing only the decoder portion of the transformer architecture, without the cross-attention layer. The process begins with word embeddings and positional encodings. Attention mechanisms are then applied to these initial inputs. Following this, several feed-forward layers are used, with regularization steps incorporated at the end. This entire process is encapsulated within a 12-layer transformer decoder. The model is trained using the standard method of cross-entropy loss.

4.2 Fine-Tuning

The fine-tuning loss is composed of two parts: the loss specific to the task at hand, and the loss from language modeling. During the fine-tuning phase, the model's structure remains unchanged, apart from the last linear layer. The input format varies depending on the task. Our focus is on single sentence classification. To classify individual sentences, the data is fed into the model as it was during training, and the label is predicted based on the final representation of the last input token (Fig. 1).

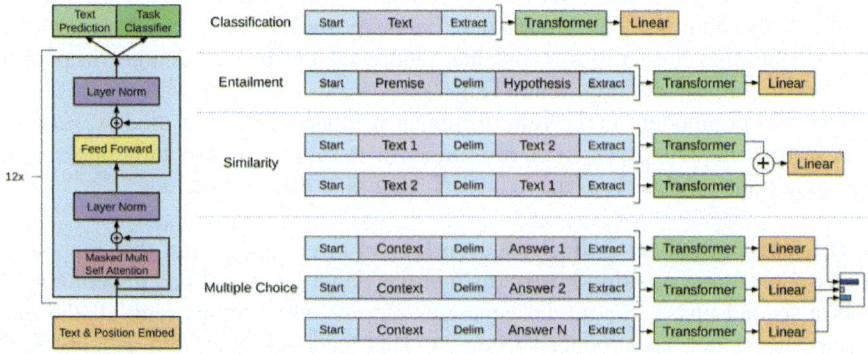

Fig. 1 The architectural structure of GPT 3.5 model (Radford et al. 2018)

4.3 The Revolution of GPT-3

The GPT-3 paper by OpenAI[3] for 2023, presents a novel concept known as in-context learning, which deviates from traditional artificial intelligence practices. In the conventional supervision paradigm, if we want to classify text as a binary problem, we first create a dataset of positive and negative examples and then train a custom model to make the binary distinction. While this model can be powerful, it may not scale to the complexity of human experience. In-context learning promises that a single, large, frozen language model can serve all these purposes. We provide the model with examples of truthful and deceptive opinions, hoping that it will learn in-context about the distinction we're trying to make. The types of n-shot learning that can be introduced to the GPT-3 model include:

- Zero-shot learning: The model can predict the answer given only the task name with no examples.
- One-shot learning: In addition to the task name and description, we provide the model with one example, and the model will be able to predict the answer.
- Few-shot learning: A few examples are introduced to the model along with the task description.

Another significant innovation is the concept of "self-supervision", a powerful mechanism for acquiring rich representations of form and meaning. In self-supervision, the model's objective is to learn co-occurrence patterns in the sequences it is trained in. The model is simply learning to assign high probability to attested sequences. These models are thought of as generators, but the generation involves sampling from the model. This is powerful because self-supervision requires minimal human effort and has facilitated the rise of another important mechanism, large-scale pretraining (Fig. 2).

The final component is the role of human feedback. The best models, referred to as Instruct models by OpenAI, are trained with more than just self-supervision. Two important aspects from the ChatGPT blog post diagram include:

- The language model is fine-tuned on human-level supervision that represents step 2 on figure, making binary distinctions about good and bad generations.
- In a second phase, the model generates outputs and humans rank all the outputs the model has produced (Step 3). This feedback goes into a lightweight reinforcement learning mechanism.

In both phases, there are important human contributions that take us beyond the self-supervision step.

[3] https://openai.com/index/chatgpt/.

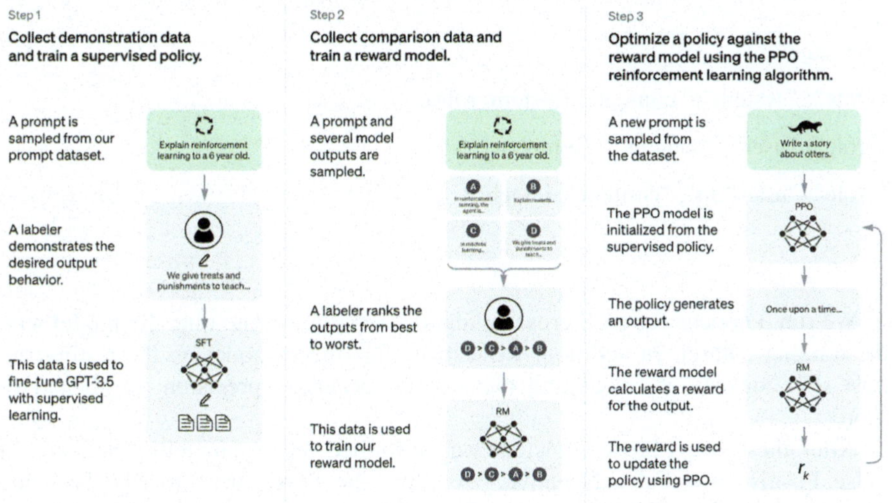

Fig. 2 Open AI's schematic representation of the GPT-n LLM (OpenAI, 2023)

4.4 Methodology

As we mentioned in the section of Dataset, the amount of data is considerably low for the task at hand. To assess the performance of the model, we decided to conduct fourfold cross validation, splitting the data set into 75% for training and 25% for testing each time, considering a trade-off between the generalization and the cost of fine-tuning such a big model. For the sake of clarity, we use the term best and worst performing model to indicate the fold for which the fine-tuned model shows higher accuracy. Each fold is created using a custom function built for this purpose benefiting from the StratifiedKFold[4] method published in 2024, ensuring that each fold has roughly the same proportions of each class. To configure GPT-3.5-turbo-0125,[5] we need to specify the following parameters in the API: System, User, and Assistant. These parameters are defined as follows:

- **System role**—"*Për një fjali të dhënë, detyra juaj është të përcaktoni nëse komenti i bërë nga një përdorues ka përmbajtje ofenduese (T) ose jo (F).*"
- **User**—the comment text
- **Assistant**—the label, True (T) or False (F)

We convert the dataset to a .json file where each row, which contains one opinion, has the following format:

[4] https://scikit-learn.org/stable/modules/generated/sklearn.model_selection.StratifiedKFold.html.

[5] https://platform.openai.com/docs/guides/fine-tuning/preparing-your-dataset.

```
{
"messages": [
{"role": "system", "content": System_role},
{"role": "user", "content": Comment},
{"role": "assistant", "content": Labels}
]
}
```

We run 3 epochs for each cross validation. Firstly, we fine-tune the model with the training dataset. In addition, we test the fine-tuned model on the testing set. Last step consists of calculating the metrics: accuracy, precision, recall and F-score.

After these steps are completed, we performed an explainability analysis of False Positive and False Negative. We show the ROC curve for the best and worst performed model. To get insights related to the understanding of the model for Albanian language we used different prompts combined with the tokenization obtained from the algorithm used by GPT-3.5-turbo-0125.[6]

5 Experiment

5.1 Modeling Politics Data

As stated, we tried, to the best of our knowledge, to rewrite the comments in a standard written language. Doing so, we could test whether a better written comment could make it understandable for the model and make the right classification. We critically analyzed the dataset to understand which of the categories contained complex comments that would make the decision to use the model more difficult. As we can imagine, the comments under the Politics topic are complicated, as the level of irony and sarcasm is at a higher level compared to the other topics. We used the row data from Politics to test our hypothesis. In the following table, we present the results of each fold, changing only the input we give to the model: first, the comment as it was written, and second, the comment corrected by us, again to the best of our knowledge. Notably, the model has a higher performance when receiving corrected versions of the comment, having as an average accuracy over the fourth folds of 87% and the low standard deviation of 0.01 shows the model's consistency across different folds in the cross-validation process. Shortly, we will report the meaning of the metrics obtained for the first fold of original comments:

[6] https://github.com/openai/openai-cookbook/blob/main/examples/How_to_count_tokens_with_tiktoken.ipynb.

Table 2 Results of GPT 3.5 on politics dataset

Metric	Original comments				Corrected comments			
	Onefold	Twofold	Threefold	Fourfold	Onefold	Twofold	Threefold	Fourfold
Accuracy	0.74	0.77	0.79	0.83	0.80	0.8	0.87	0.89
Precision	0.68	0.72	0.69	0.78	0.73	0.73	0.82	0.87
Recall	0.70	0.72	0.87	0.83	0.78	0.78	0.87	0.87
F1 Score	0.69	0.72	0.77	0.80	0.76	0.76	0.84	0.87
FP	15	13	18	11	10	16	6	6
FN	14	13	6	8	13	14	9	6
Avg Accuracy	0.78				0.87			
Std	0.03				0.01			

- Precision—the model predicts a positive class of benign cases in 68% of the cases. So, there's a 32% chance that a positive prediction is a false positive.
- Recall—the model correctly identifies 70% of all actual positive cases.
- F1-Score—this is the harmonic mean of precision and recall. An F1-Score of 0.69 suggests that the model is reasonably balanced in terms of precision and recall, but there is still room for improvement (Table 2).

Out of expectation, we see that the model performs better with the original comments in the second fold. Before moving on with the results obtained by the linguistic analysis, we will plot the ROC curve for the worst and best performing model. The AUC[7] value reported represents how much the model is capable of distinguishing between classes. It's a statistical measure that we can use to evaluate the model predictions using a probabilistic framework. Again, for the best and worst model we have 89% and 74% respectively, chance of correctly distinguishing between positive and negative class (Figs. 3 and 4).

5.1.1 Linguistic Analysis of FP and FN

To understand the reason behind this result, we made a check between FP and FN of the original comments and corrected one, to find out the differences. At this point, our investigation proceeds in two steps. Firstly, investigate further the comment for which the model gave the correct labeling to the original comment but the wrong one for the corrected comment. To fulfill this, there are a couple of actions implemented. Primary step, prompt the fine-tuned model to get the labeling. In this way, we can understand if the model has just memorized the example or if it can "reason" by itself. Then analyze the results after applying the tokenization. From this action, we can understand if the specific words are presented differently in the embedding space, depending on the way they are written. The last one asks

[7] https://www.baeldung.com/cs/ml-accuracy-vs-auc.

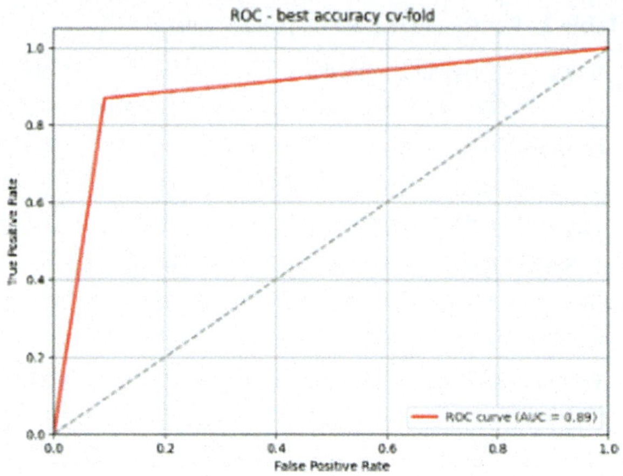

Fig. 3 ROC best accuracy cv-fold for all topics

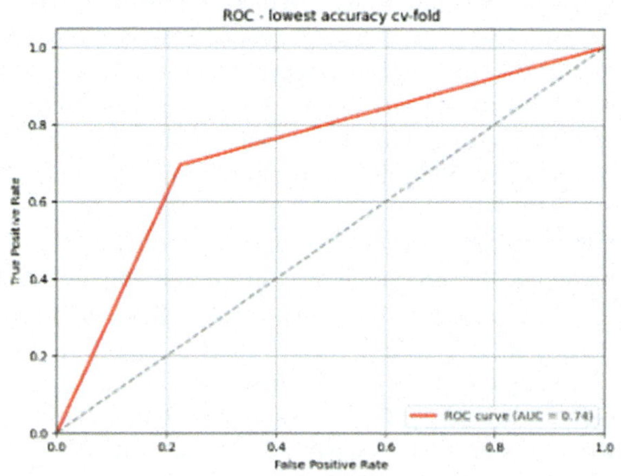

Fig. 4 ROC lowest accurate cv-fold

chatGPT to give an explanation related to the understanding the model has about the embeddings obtained. Investigate the comment for which the model gave the wrong labeling to both versions of the comment.

Secondly, investigate the comment for which the model gave the wrong labeling to both versions of the comment (Table 3). Starting with the analysis of the first comment:

The first step consists of using the prompt "Përcakto nëse fjalitë e mëposhtme kanë përmbajtje ofenduese ose shprehin urrejtje", giving as input the two sentences to the fine-tuned and gpt-3.5-turbo-0125 model, we obtain the correct labeling. Using the prompt "Bëj një analizë të fjalive dhe për secilën listo elementët që mund të kenë përmbajtje ofenduese", we obtain the main word that triggers the offensiveness "MALOKU".

The second one relies on using the encoding function for the words "MALOKU" and "maloku" we obtain a different tokenization, which could be

Table 3 The comment details and labels for analyzing step

Comment	Corrected comment	Label	Model's Label for original Comment	Model's label for corrected comment
MOJ XHULALJA NE SPAKE MALOKU SALI BERISHA KA ME SHUME SE 5 PROKURORE SPIUNE TE TIJE TE FSHEHTE PERSNALE QE SHERBIMET E DINE MOJ MALOKE DIPLOMATE E SALI BERSHURES	Moj, Xhulalja ne SPAK, maloku Sali Berisha ka me shume se 5 prokurore spiune te tij te fshehte personale qe sherbimet e dine, moj maloke diplomatike e Sali Berishes	T	T	F

Table 4 Token information of word "maloku"

Example string: "MALOKU"	Example string: "maloku"
Token_integers: [89899, 4012,52]	Token_integers: [14991, 16900]
Token bytes: [b'MAL', b'OK', b'U']	Token bytes: [b'MAL', b'OKU']

the main reason why the model didn't categorize this sentence in a correct way when receiving as input its corrected version (Table 4).

The last step focuses on using the prompt in English "What can you infer from the example below? Please pay attention to tokenization.", our aim is to gather as much as possible hints on how this model interacts with the tokens, what meaning gives, and hopefully this could be useful to give a reason why these two words are interpreted in a different word. Here is the answer: "*MAL OK U: MAL likely represents the Albanian word for mountain (mal), OK could be an abbreviation or a colloquial term, possibly meaning "okey" or indicating approval", U could be a pronoun indicating "you in English. Corrected version: mal oku: mal remains the same, meaning mountain, oku is potentially replacing OK, but it might be a form of the verb "oku" meaning "to read", or it could be a variation or slang term*". We realize that the model is mixing the knowledge from the different languages and lacks understanding of the real meaning of our word. A potential explanation for the misclassification of the corrected comment would be the interpretation of the word as "*mal*" and "*lexo*" (Table 5).

Second comment analysis:

Using the same prompt as in the first case, again we obtain correct labeling of row data, potentially, the model has memorized the examples. The tokenization is done in the same way. No need to make further analysis. We changed our approach in the analysis of this sentence. The approach is explained in the next point.

Table 5 The comment details and labels for analyzing step

Comment	Corrected Comment	Label	Model's Label for Original Comment	Model's Label for corrected comment
Love z Lubonja ❤ ❤ ❤ Intelektual, origjinal, me kurajo civile, Hila nje intelektual provincial qe I pelqen vetja 🤮	Love z.Lubonja ❤ ❤ ❤ Intelektual, origjinal, me kurajo civile! Hila nje intelektual provincial qe i pelqen vetja 🤮	F	T	T

Using the prompt "Bëj një analizë të fjalisë dhe listo të gjithë elementët që mund të kenë përmbajtje ofenduese.", in this way the model it is built, will "think" that potentially there should be an offensive element and it tries to guess, ending up in what is called hallucination (Huang et al. 2022). Here is the answer: *"Në fjalitë e dhëna, elementët që mund të kenë përmbajtje ofenduese jane: 1. ""Hila" në këtë kontekst, kjo fjalë përdoret si një shprehje me karakter ofendues për të përshkruar dikë si rëndomtë, i pafuqishëm, ose i padëshiruar. 2. "Që i pëlqen vetja" edhe kjo pjesë e fjalisë mund të përmbajë një konotacion negativ, duke implikuar arrogancë ose egoizëm ne personin që merret në pyetje"*. Unfortunately, we were not able to identify the meaning of this word in other languages. The model is clearly capturing the offensive word in a wrong way.

5.2 Modeling All Topics

Moving to the next experiment, we pass as input to the model only the corrected comment, this for two reasons. First, we saw from the first experiment that the model outperforms when having as input the corrected comments. Second, is related to the increased cost, as the tokens when using row data from all the categories are greater in number. In this experiment, we assess the performance of the model when having comments from different topics. The main question is related to the fact that the model can understand the offensive terms used for each of the topics and its ability to gain "general knowledge" for our language. Furthermore, we compare the model's performance for each category within each fold, emphasizing its performance on Politics row data. The results are reported below Table 6.

Combining all the topics together leads to an average accuracy of 3% less than when having a specific topic with a standard deviation of 0.02. This small standard deviation suggests that there is minimal fluctuation in the model's performance from one subset to another, reinforcing the reliability of the model. The best model performs on the third fold and the less performant in terms of accuracy is the fourth model. Below we report for each fold the metrics related to the topic category (Figs. 5, 6, 7 and 8).

Table 6 Metrics values of 4 folds on all categories

Metric	Corrected comments			
	Onefold	Twofold	Threefold	Fourfold
Accuracy	0.83	0.85	0.86	0.79
Precision	0.78	0.79	0.84	0.71
Recall	0.82	0.86	0.81	0.87
F1 score	0.80	0.82	0.83	0.78
FP	35	34	23	55
FN	28	22	29	20
Average accuracy	0.83			
Std of accuracy	*0.02*			

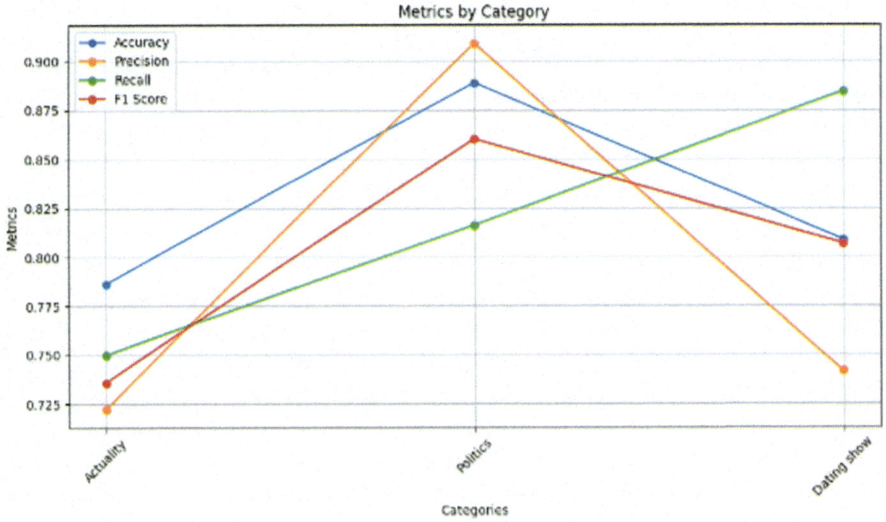

Fig. 5 Metrics of onefold of all topics

Overall, we can say that the model performs quite in the same way for the comments related to the Actuality, showing a tendency of low performance related to Dating comments that is related to the nature of the comments that according to us have a tendency of bullying. On the other hand, it shows a good performance for the topics related to Politics, reaching on the second fold an accuracy of 92%. This could be related to the fact that we first fine-tuned the model only on Politics data, and looking at the diagram of how the model works, somehow it is benefiting from the reinforcement learning scheme. Looking at the ROC curve for the best and worst model, we can say that the latter model has a higher degree of class separability, making it the more accurate model by 5% (Figs. 9 and 10).

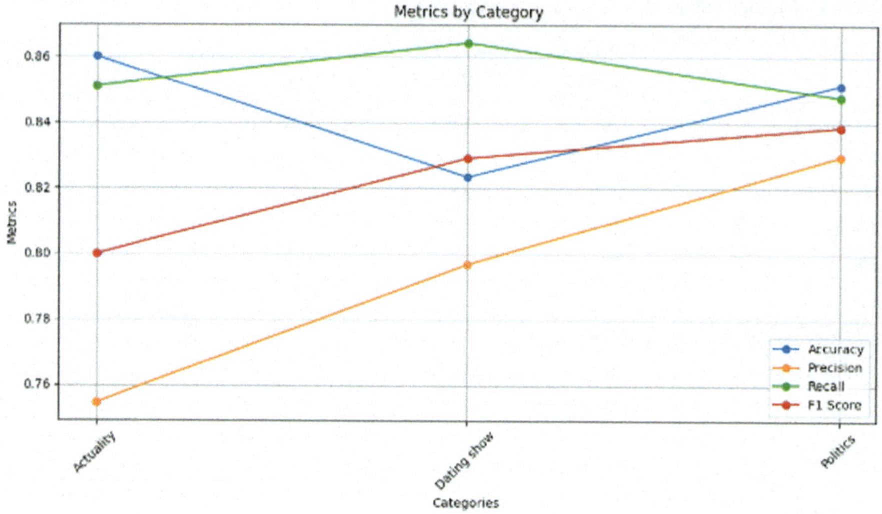

Fig. 6 Metrics of twofold of all categories

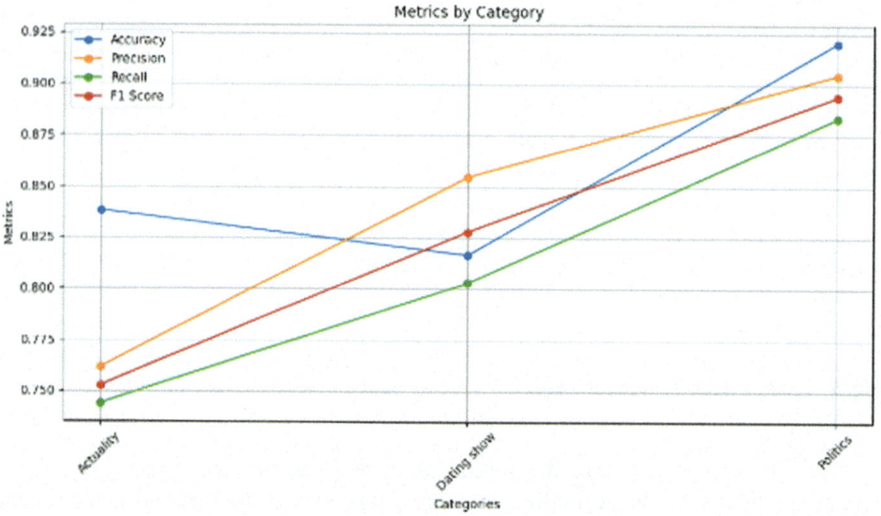

Fig. 7 Metrics of threefold of all categories

5.2.1 Linguistic Analysis of FP and FN

We analyze the FP and FN in a slightly different way in this experiment. The plots show the number of each class compared to the comment's category (Figs. 11 and 12).

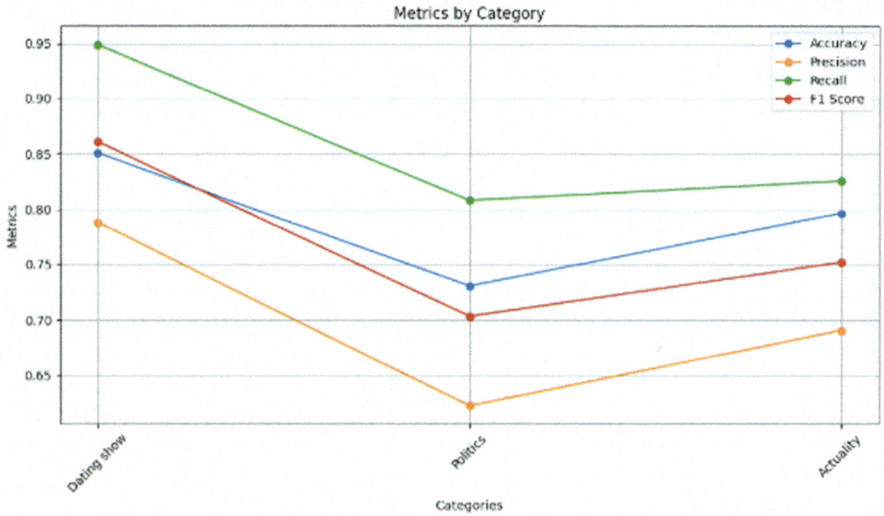

Fig. 8 Metrics of fourfold of all categories

Fig. 9 ROC best accuracy cv-fold for politics topic

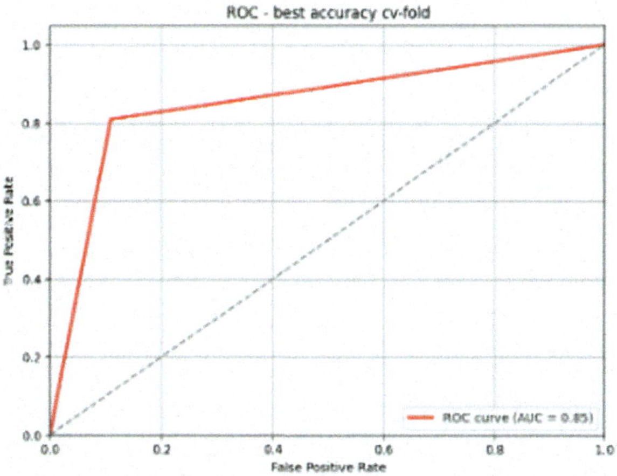

The model has a lower number of FP/FN for the Dating show category and a higher number of FP/FN for the comments related to Politics, as expected from our analysis. Overall, the model tends to predict a positive comment incorrectly, resulting in a higher number of predictions as Hate Speech when they should be labeled as False. Depending on the use case in which we use the model to classify such comments, one can prefer to have more FP than FN or vice versa. If we consider the case of filtering out the offensive comments from social networks, then missing a hate speech is more costly, then we might tolerate more FP to ensure hate speech is not overlooked. As we are dealing only with the corrected comments, we will use a different approach when analyzing the comments under

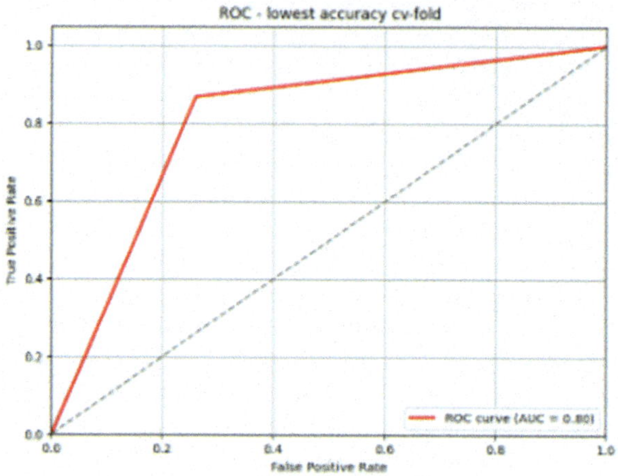

Fig. 10 ROC lowest accuracy cv-fold

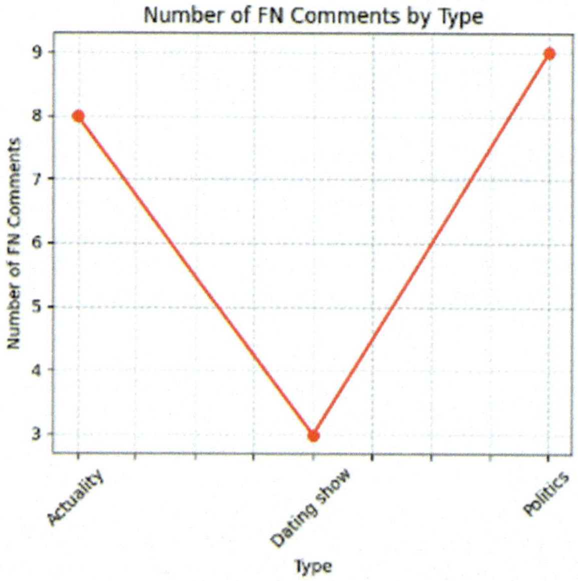

Fig. 11 Number of FP comments by topics

these categories. Below is the list of some of the comments labeled in a wrong way. Note that we selected the ones that are easy to make a distinction between HS and not. In addition, we consider the comments of the split for which the model has a low performance in terms of accuracy (Tables 7 and 8).

This investigation follows these steps:

Using the prompt "Përcakto nëse fjalitë e mëposhtme kanë përmbajtje ofenduese ose shprehin urrejtje", passing the list of the sentences reported above to the fine-tuned and gpt-3.5-turbo-0125 model, we obtained a new label for our sentences.

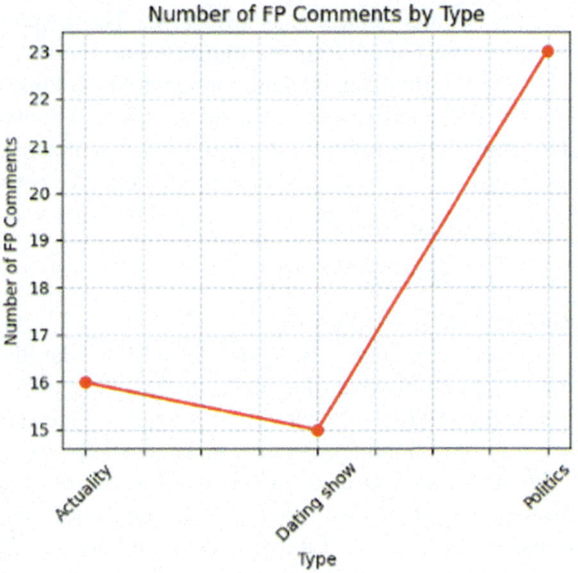

Fig. 12 Number of FN comments by category

Table 7 False positives comment sample

False Positives
Avokati i doktorit tha qe doktori eshte nje hero dhe sulmohet kot nga konkurentet! Hajde shtet, hajde! Ky duhet te ishte ne burg qe para 5 vjetesh
Çfare fitimi more avokat, e mori ferra kete vend... 😢
Taku eshte gjigand! 😂

Table 8 False negative comments sample

False negatives
Shume e neveritshme kjo Enisa
O Hila po ve dhe mendjen ndonjehere ne perdorim, jo gjithmone vetem llapen
Puuuu sa ta shpif ti o Grida me kete te folur. Fol o grua mire, thjeshte dhe qarte

In summary, the second model tends to mark all the sentences as non-offensive, while the first model correctly classified the sentences of the FP class. Still no improvement on the FP class.

Using the prompt "Bëj një analizë të fjalive dhe për secilën listo elementët që mund të kenë përmbajtje ofenduese", we want to understand what makes the model "to think" in a specific way. The reason given for the second sentence caught our attention: "Ofendimi drejtuar ndaj personit, duke përdorur fjalë vulgare".

Using the prompt "Analizo në detaje fjalinë 2 dhe listo elementët ofendues, gjithmonë nëse ka të tillë", we can get the specific word that leads the model to identify this sentence as offensive. The response is as follows "Elementët ofenduese në fjalinë 2 janë: - Fjala vulgare "ferra".

Using the function available on OpenAI cookbook we tried to analyze how this word is projected into the embedding space. The word "ferra" is a root word that does not contain suffixes. In an optimal tokenization algorithm for our language, this word should be treated as a unique one. The word is tokenized into:

Example string: "ferra"

Token_integers: [69, 14210]

Token bytes: [b'f', b'erra']

Using the prompt "Pse fjala "ferra" konsiderohet vulgare?", we aimed to get the insight from which language the meaning of this word is taken. Considering the response "Fjala "ferra" konsiderohet vulgare për shkak të përmbajtjes së saj seksuale dhe shprehjes negative që ka në shqip", we use another prompt in English language "Can you please explain what the meaning of this word is?" to finally get to the desired goal "In Albanian, the word "ferra" is a derogatory term that translates to "bitch" in English". We found out that the word "Perra" in Spanish has the same meaning and is being tokenized in the same way as the word "Ferra". In a multilingualism framework, this is a well-known problem for low-resource languages. As these languages benefit when the model is trained in multi-corpora, according to Chang et al. (2023), it can also lead to what is known as the curse of multi-lingually. As dataset sizes increase, adding multilingual data begins to hurt performance for both low-resource and high-resource languages, likely due to limited model capacity. That is exactly what is happening here. As the words have the same token bytes, they share the same position in the embedding, making them have the same meaning, which is right for one language and totally wrong for the other language.

6 Conclusion

Our study's experiments show that significant improvements can be obtained when the GPT-3.5 API is fine-tuned for the purpose of detecting hate speech on Albanian social media. In particular, the accuracy rises from 78% (with a standard deviation of 0.03) to 87% (with a standard deviation of 0.01) when comments are updated to align with the standard Albanian language. The model performs better when focusing just on comments related to the Politics category rather than when it includes comments from three distinct topic areas (83% of accuracy with std of 0.02). Since the model must comprehend and adjust to the subtleties of each issue, it becomes more difficult to detect hate speech effectively as it is exposed to a greater range of language uses, contexts, and patterns. The study reveals that the fine-tuning of the GPT-3.5 API for detecting hate speech in Albanian social media can lead to mixed interpretations of tokens, causing potential misclassifications. It also found that the model's construction might predispose it to anticipate offensive elements, leading to hallucinations or incorrect identifications. Lastly, the research highlighted the challenges of multilingualism, where words with similar token bytes in different languages can confuse the model, a phenomenon known as the curse of multilingualism.

References

Chang, T.A., Arnett, C., Tu, Z., Benjamin, B.K.: When Is Multilinguality a Curse? Language Modeling for 250 High- and Low-Resource Languages (2023). Published by Cornell University. https://doi.org/10.48550/arXiv.2311.09205. Available: https://arxiv.org/pdf/2311.09205

Huang, L., Yu, W., Ma, W., Zhong, W., Feng, Z., et al.: A Survey on Hallucination in Large Language Models: Principles, Taxonomy, Challenges, and Open Questions, pp. 6–12 (2022). Published by Cornell University. Available: https://arxiv.org/abs/2201.12345

Kaçuri, M.: Explainability of hate speech classification for Albanian language using rule-based systems and neural networks. Published by Vienna University of Technology (TU). https://doi.org/10.34726/hss.2023.105780. Available: https://repositum.tuwien.at/handle/20.500.12708/188586?mode=full

Nurce, E., Keci, J., Derczynski, L.: Detecting abusive Albanian. 10th May2022, Published by Cornell University. Available: https://arxiv.org/ftp/arxiv/papers/2107/2107.13592.pdf

Radford, A., Karthik, N., Tim, S., Ilya, S.: Improving Language Understanding by Generative Pre-Training, p. 4 (2018). Published by OpenAI, Available: https://cdn.openai.com/research-covers/language-unsupervised/language_understanding_paper.pdf

Stakic, I.: Homophobia and hate speech in Serbian public discourse: how nationalist myths and stereotypes influence prejudices against the LGBT minority. J. Balkan Near Eastern Stud. Equal Right Trust (ERT), **7**, 44–462 (2021). Published by The Equal Right Trust (ERT). Available: www.equalrightstrust.org/ertdocumentbank/ERR7_isidora.pdf

United Nations: What is hate speech?. September 2022. Published by United Nations. Available: https://www.un.org/en/hate-speech/understanding-hate-speech/what-is-hate-speech

Digital Financial Technologies and Banking: The Role of Blockchain in Improving Banking Services

Ragib Mammadli

Abstract

Since its introduction in 2009, blockchain technology has become a transformative force in the banking industry, creating unprecedented opportunities for banks. Using cryptographic techniques, blockchain technology provides banks a high level of transaction security, increases transparency and accuracy of transactions, provides clear control over all transactions and improves risk management, improves operational efficiency, detects fraud and ensures compliance with regulatory requirements. Blockchain technology has also helped reduce costs and increase efficiency in the banking sector by optimizing processes and eliminating intermediaries for cross-border transactions. Moreover, blockchain technology enables the tokenization of assets and increases the liquidity of previously illiquid assets. The use of blockchain technology in banking has led to the emergence of a number of problems as well as many advantages. One of the main problems of blockchain technology is scalability. The number of banking transactions is constantly increasing, which leads to a decrease in productivity and processing speed, delays and congestion. Other major concerns include privacy and security of data and transactions, regulatory compliance, cost of implementation, user experience issues, and interoperability issues among different blockchain systems. These issues pose significant barriers to banks seeking to harness the full potential of blockchain technology. Banks that embrace technological innovation and successfully address emerging challenges can gain an advantage over rival banks and position themselves as leaders in the digital banking sector. However, as of 2024, the adoption of blockchain technology in the banking sector has been slower than expected due

R. Mammadli (✉)
International Magistrate and Doctorate Center, Azerbaijan State University of Economics (UNEC), Baku, Azerbaijan
e-mail: mammadli.raqib.mammadali.2023@unec.edu.az

K. Dhoska and E. Spaho (eds.), *Bridging Horizons in Artificial Intelligence, Robotics, Cybersecurity, Smart Cities, and Digital Economy*, Sustainable Economy and Ecotechnology, https://doi.org/10.1007/978-3-031-72029-1_21

to various market downturns and general skepticism toward cryptocurrencies. This article will highlight the benefits of blockchain for the banking sector and discuss the disadvantages that prevent this technology from becoming an ideal option for the banking world.

Keywords

Banking • Financial technology • Digitalization • Blockchain

1 Introduction

Blockchain is a continuous and consistent chain of technological blocks that contain any information that is structured according to certain rules. The connection between blocks is ensured not only by numbering, but also by the fact that each block contains its own hash code and the hash code of the previous block. Typically, copies of blockchains are stored on many different computers that are independent of each other (Broome 2019).

The term first appeared in the Bitcoin system as the name of a cryptocurrency-related database, so the name blockchain is often mistakenly equated with transactions in various cryptocurrencies.

Currently, blockchain technologies are used in fields such as finance, healthcare, and education for the purpose of increasing management and efficiency (Fig. 1).

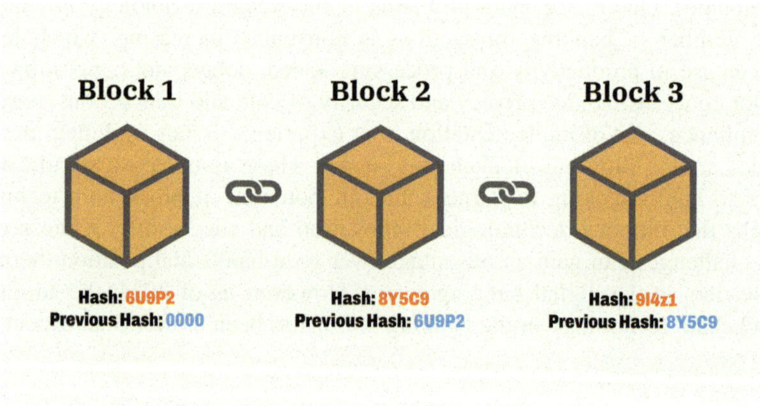

Fig. 1 Blockchain structure (Susanto 2021a)

2 Blockchain and the Banking Sector

Today, it is impossible to imagine anything without technology. In the twenty-first century, banks have transformed from traditional buildings into financial institutions based on modern technological innovations.

Along with innovation, banks face modern challenges such as security, customer satisfaction and high transaction costs. In order to eliminate problems as much as possible and attract new customers, increase the range of services and adapt to the competitive environment, banks prioritize the application of modern technologies. One of these technologies is blockchain.

There are four types of blockchain technologies in the banking sector (Guo and Liang 2019):

Public blockchain. This blockchain has no system access restrictions. Any user can submit and review transactions by logging into the system.

Private blockchain. User access to the private blockchain is limited, and only bank employees invited by the administrator can access the system.

Hybrid blockchain. Although blockchain technologies operate on a decentralized management system, this type of blockchain has both centralized and decentralized management features.

Consortium blockchain. In this type of blockchain, several banks merge and create a single blockchain platform for themselves. Participants manage the system together and are responsible for the accuracy of all transactions.

3 Advantages of Using Blockchain in Banking Services

In general, the application of blockchain in banking services has the following advantages:

Improved Security. Using cryptographic techniques, blockchain secures bank transactions and protects them from manipulation and fraud. Each transaction is specifically encrypted and connected to the previous transaction; thus a chain is created. Changing this chain is almost impossible without consensus on the network. Additionally, once a transaction is made, it is instantly recorded on the blockchain and transaction details cannot be changed or deleted later. The special security techniques used reduce the possibility of unauthorized access to information by fraudsters and make it extremely difficult to commit fraud (Kate 2023).

Transparency. Through blockchain technologies, every bank transaction is recorded in a common database accessible to all users. This transparency helps build trust among all users, as they can independently monitor transaction details.

Reduced Settlement Times. Traditional banking system requires many intermediaries and the payment process often takes a long time. By eliminating intermediaries and delays, blockchain allows transactions to be completed in near real-time (Préfontaine and Zioufi 2022).

Global Access. With the help of blockchain, banks can easily conduct transactions between countries. There is no need to exchange currencies and special intermediaries. This feature helps banks enter new markets and serve people from underserved areas.

Improved Compatibility. Blockchain technologies facilitate the process of regulatory and legal compliance through smart documents (Pinna et al. 2017). Through these technologies, banks can manage and monitor the compliance of documents such as transaction documents, tax obligations, and licenses. This reduces the risk of inconsistency and invalidity and avoids penalties.

Risk management. Through blockchain technologies, auditors can independently verify all bank transactions and identify discrepancies in advance. Some programs can even identify inconsistencies by themselves without human help. It helps banks to identify and mitigate different types of risks including credit risk, market risk, operational risk and compliance risk (Guo and Liang 2019).

Product innovation. By using blockchain technologies, banks can identify new trends and unmet needs in the market and stimulate the creation of new and innovative financial products such as DeFi platforms and tokenized assets. Blockchain enables the tokenization of illiquid assets such as real estate and commodities and increases their liquidity. Banks that effectively use blockchain technologies can gain a competitive advantage in the market by offering innovative services (Ji et al. 2022).

Customer satisfaction. Blockchain creates special opportunities for banks in faster processing of bank transactions, enhanced security, and transparency. This increases customer satisfaction and loyalty, creates strong relationships between banks and customers, and helps predict and prevent customer churn.

Cost reduction. Blockchain enables banks to reduce transaction costs by optimizing transaction processes and eliminating intermediaries. In addition, automation through smart documents leads to further reductions in operational costs. As a result, costs are significantly saved, and bank resources are used more efficiently (Farev 2019).

Opportunities to work together. Through blockchain technologies, several banks can create a single banking network. With the created infrastructure, banks can work together to expand their range of services and solve problems together.

Increased Efficiency. Blockchain technologies automate many processes, significantly reducing manual work in banks, and leading to high efficiency.

Improved Data Management. Through blockchain technologies, all financial information generated in banks is collected in a single source and data consistency is ensured. This reduces the need for reconciliation between multiple systems and ultimately simplifies data management (Broome 2019).

Increasing financial inclusion. Many people around the world do not have access to banks and banking services, depending on where they live and their circumstances. Blockchain technologies enable those people to access money and financial services by offering digital wallets and decentralized banking solutions (Ramchandra et al. 2022).

4 Disadvantages of Applying Blockchain Technologies to Banking Services

Blockchain is significantly impacting the banking industry, offering unparalleled benefits in terms of improved security, risk management and operational efficiency. Despite all the advantages that blockchain brings to the banking industry, there are several disadvantages of applying these technologies to banking services and bank management. The following can be attributed to them:

Privacy Concerns. All information related to banking operations, customers and management is stored in the blockchain system. Using cryptographic techniques, blockchain secures them and protects from manipulation and fraud (Susanto 2021a). Although blockchain provides transparency and a high level of security, there is no technology that can guarantee 100% security. In the event of any fraud, confidential information about banks and customers can be stolen, exposed and misused. Additionally, Smart documents may contain technical errors that can be used by hackers to disrupt the network or steal customers' funds (Chowdhury et al. 2021b).

Scalability. One of the main problems of blockchain technology in banking services is scalability. The number of bank transactions is constantly increasing day by day, and more space is required to store the data related to those transactions in the system. The amount of data collected and stored in the network can cause freezes and delays in operations.

High energy consumption. The blockchain database, where all transaction data is stored, is continuously active. In addition, the use of some blockchain technologies in banking services requires high energy consumption (Sims and Jayasuriya 2021). These can lead to high transaction costs.

Environmental Impact. In addition to high transaction costs, the energy consumed by some blockchain technologies can lead to carbon emissions and environmental pollution. This raises concerns about the sustainability of blockchain in banking services.

Application complexity. Application of blockchain in banking services requires special technical expertise. Banks may face challenges when implementing new blockchain technologies into existing traditional banking systems. And this can result in delays and stoppages in banking operations (Broby 2022).

Specialized staff vacancy. Effective implementation of Blockchain technologies requires specialists with special skills, along with technical expertise. Sometimes, existing employees can hinder the successful implementation and use of blockchain initiatives due to lack of relevant knowledge, lack of training, or resistance to adopting new banking technologies.

Implementation cost. Blockchain technologies bring unparalleled advantages to banks. However, the application of these technologies requires high investment. Sometimes, due to the high costs of infrastructure and training, banks delay or completely refuse to introduce new technologies (Meghani 2020).

Technological difference. Banks or financial institutions can use different blockchain technologies. Sometimes different blockchain technologies operate based on different protocols and systems. These differences may limit cross-country transactions and cooperation between banks implementing different blockchain technologies.

Regulatory Uncertainty. Blockchain technologies facilitate the regulatory and legal compliance process regarding banking management through smart documents. However, rules and standards related to the regulation of blockchain technologies have not yet been fully developed. These uncertainties sometimes create compliance and liability issues for banks and can hinder the widespread adoption of blockchain technologies in banking services (Thomsett 2020).

Lack of human control. Blockchain technologies automate many processes, and significantly reduce manual work in banks. But relying only on automated data reduces the role of human control in banks. Banks may face unwanted penalties and liabilities if any errors or fraud in the system are not detected by technology and there is no human control.

Customer experience issues. Through blockchain technology, banks can offer a wider range of services to customers. But for customers who are used to traditional banking systems, it can be difficult to adapt to blockchain technology. Especially relatively older users face navigation problems within the program and may refuse technological innovations.

5 Conclusion

Blockchain technologies have revolutionized the banking sector digitally by providing opportunities such as high security, risk management and cost reduction. By using blockchain technologies, banks can increase operational efficiency, automate and improve many processes, and provide new and better services to customers.

However, the application of blockchain technologies to banking services has led to problems such as privacy and quality issues, security risks, increased costs, and regulatory compliance.

In order to get the most out of the opportunities offered by the blockchain, banks should follow the following recommendations:

- Ensure the correctness of all bank data collected in the blockchain system;
- Recruit the necessary specialists and train existing employees in relevant fields;
- Must ensure a high level of security.

In today's digitized environment, banks must keep up with emerging technologies and consider strengthening their brands with new and powerful technological solutions to stay competitive and meet the growing demands of customers.

References

Broby, D.: Banks, digital banking initiatives and the financial safety net: theory and analytical framework. J. Econ. Sci. Res. **04**(01) (2022)

Broome, L.: Banking on blockchain. J. Law Technol. **21**(1) (2019)

Chowdhury, M.U., Khairunnahar, A., Khan, M.: Blockchain application in banking system. J. Softw. Eng. Appl. **14**(7) (2021b). https://doi.org/10.4236/jsea.2021.147018

Farev, I.: Digital banking transformation: application of artificial intelligence and big data analytics for leveraging customer experience in the Indonesia banking sector. In: 8th International Congress on Advanced Applied Informatics (IIAI-AAI) (2019)

Guo, Y., Liang C.: Blockchain application and outlook in the banking industry. Financ. Innov. **2** (2019)

Ji, F., Tia A.: The Effect of Blockchain on Business Intelligence Efficiency of Banks. Emerald Publishing Limited (2022). ISSN: 0368–492X

Kate, B.: Unveiling the power of blockchain: a comprehensive review of its role in the banking sector. Int. J. Intell. Comput. Inf. Sci., IJICIS, **24**(1), 29–41 (2023)

Meghani, K.: Use of artificial intelligence and blockchain in banking sector. Indian J. Appl. Res. **10**(8) (2020). Print ISSN No. 2249-555X. https://doi.org/10.36106/ijar

Pinna, A., Cocco, L., Marchesi, M.: Banking on blockchain: costs savings thanks to the blockchain technology. Future Internet (2017). https://doi.org/10.3390/fi9030025

Préfontaine, E., Zioufi, N.: New technologies and their impact on consumer behaviour: an investigation in the banking sector. Int. J. Bank. (2022)

Ramchandra, M., Kumar, K., Sarkar, A., Agarwa, K.: Assessment of the impact of blockchain technology in the banking industry. Mater. Today Proc. **56**(4), 2221–2226 (2022). https://doi.org/10.1016/j.matpr.2021.11.554

Sims, A., Jayasuriya, D.: Fantastic beasts: blockchain based banking. J. Risk Financ. Manag. **14**(4), 1–43. MDPI, Basel. ISSN 1911–8074. https://doi.org/10.3390/jrfm14040170. https://hdl.handle.net/10419/239586

Susanto, H.: Encrypting of digital banking transaction records: A blockchain cryptography security approach. Int. J. Comput. Appl. **174**(27), 0975–8887 (2021a)

Wewege, T.: Disruptions and digital banking trends. J. Appl. Financ. Bank. **10**(6), 15–56. ISSN: 1792–6580 (2020)

Challenging the Cybersecurity Threats Through Education Programs

Reis Mulita

Abstract

The rise of technology has brought about new security challenges that affect various institutions, including educational institutions at different levels. As cyber threats become more evident to society, it is essential to equip students with the necessary knowledge and skills to protect themselves and their assets online. This study examines cybersecurity knowledge and awareness among ICT and social science students in Tirana, Albania. The research involves comparing the two groups of students regarding their understanding of the significance of cybersecurity in today's world. The literature review and data collected will support the qualitative and quantitative findings, which will testify to the idea presented in this study. This data collection is based on a survey conducted among students from both disciplines, focusing on the Tirana and Durres districts. The work-in-process results indicate that ICT students better understand cybersecurity and its implications than social science students. The study also reveals a need for increased awareness and education about cybersecurity among social and ICT Students. The paper concludes by highlighting the importance of cybersecurity education for all students, regardless of their discipline and level of education, as cybersecurity threats are becoming increasingly common and complex. Introducing cybersecurity education in curricula programs at early stages and establishing partnerships between educational institutions and industry to provide real-world exposure is suggested as a solution.

Keywords

Cybersecurity • Education • Students • Information technologies • Social sciences

R. Mulita (✉)
Canadian Institute of Technology, Street "Xhanfize Keko". Bo 12, Tirana, Albania
e-mail: reis.mulita@cit.edu.al

K. Dhoska and E. Spaho (eds.), *Bridging Horizons in Artificial Intelligence, Robotics, Cybersecurity, Smart Cities, and Digital Economy*, Sustainable Economy and Ecotechnology, https://doi.org/10.1007/978-3-031-72029-1_22

1 Introduction

We all live in a time of technological advancements where Information and Communication Technologies drive societal activity and behaviors. In this situation, we all benefit from the advancements in ICT through the services and facilities they offer to society. But, at the same time, we suffer from its negative consequences and external costs. One of these very significant consequences is cyber threats and attacks. As we live in the digital age, access to the Internet continues to expand, along with its use for a wider variety of purposes. There are estimates that close to 4.7 billion people are active Internet users—close to 60% of the world's population (Datareportal 2024). However, many users have limited knowledge or awareness of the risks of being online and have never been involved in educational or training programs on Cybersecurity (Aiken 2016). Although the shortage of trained professionals is widely recognized, it should also be made clear that one (if not the primary) reason for this shortage is that there is not a sufficient number of students entering dedicated educational programs. In the USA, according to Federal Statistics, it has been found that the last 5–6 years reported that internet crime complaints have considerably increased, showing 515,612 complaints in 2020, while 162,091 complaints were reported previously (FBI Report 2021). In 2021, phishing and allied cyberattacks, business emails, and malware were the primary sources of data breaches. According to recent statistics, in 2022, ransomware cyberattacks surpassed phishing attacks due to the primary reason for data compromises (WSAV Report 2022).

Europe is also facing the same situation in regard to cybersecurity attacks through different modes. EU cyberthreats 2030 presents the challenges the EU faces in the period 2030 (ENISA 2023). Through this presentation, I will bring evidence of the vital role of education and skills through university and school programs to equip students with skills and knowledge capacities to face society's challenges regarding threats and attacks. Cybersecurity education in schools is not just about creating a generation of cybersecurity experts but also about shaping resilient digital citizens. By imparting essential knowledge, skills, and values, schools contribute to the holistic development of students who can navigate the digital world securely, ethically, and responsibly.

With the global diffusion of the Internet, social media, and mobile smartphones, increasing responsibilities end up with the individual user. Many individuals online are sufficiently informed to protect themselves and help protect others (Shillair 2015). Still, many others do not know how these attacks occur or how to defend themselves. For example, a survey of US households found that three-quarters (75%) of respondents could recognize a strong password. Still, less than a quarter (13%) knew what a virtual private network (VPN) does, and even fewer (10%) could recognize an example of multi-factor authentication (Olmstead and Smith 2017). The shortage of qualified cybersecurity professionals requires that institutions of learning work to retain their cybersecurity specialists and also design programs that can be implemented to increase the number of students willing to become cybersecurity professionals and leaders in the future (Finch 2020).

Educating people about cybersecurity can achieve more secure and safer digital communication for all (Amzeal News 2023a, b).

1.1 EU Policies and Cybersecurity

The EU aims to source products and services designed to meet the EU's specific needs and values while avoiding being subject to the influence of the outside world. In the digital world, such needs may encompass hardware, software, or algorithms manufactured as products and services, which should comply with the E.U. values and thus preserve a fair digital ecosystem while respecting privacy and digital rights (ENISA 2019). In short, digital strategic autonomy means the capacity for the E.U. to remain autonomous in specific areas of society where digital technologies are used (Timmers 2019). The new challenges brought about by the digitalization of the E.U. communitarian society raise questions about its capacity to retain ownership and control of European personal data (ENISA 2017). Furthermore, the COVID-19 pandemic highlighted the importance of cybersecurity and the need for the E.U. to continue to invest in education in the digital sector (EPSC 2019). Within this context, ENISA's report sets priorities for the critical research and innovation directions in cybersecurity. According to the ENISA report (ENISA 2023) was identified the E.U. key research directions follow Data security, Trustworthy software platforms, Cyber threat management and response, trustworthy hardware platforms, cryptography, user-centric security practices and tools, Digital communication security.

Also, according to official documents (ENISA 2023), Europe must ensure that its businesses and governments have access to a sufficient number of skilled people who can design, deploy, operate, and audit critical infrastructure services and that it retains these skilled people in the context of a global shortage of cybersecurity professionals.

Regarding students and cybersecurity, there is enough evidence that although the shortage of trained professionals is widely recognized, it should also be made clear that one (if not the primary) reason for this shortage is that there is not a sufficient number of students entering dedicated educational programs. Europe should thus look at additional ways to enroll students in cybersecurity curricula. Several alternative approaches could be adopted, including extracurricular programs driven by reputable diversity institutions in partnership with governmental units. Albanian society is also under the pressure of cyberattacks, which have been very present for the last two years. In these situations, the Albanian Agency for Cybersecurity (AKCESK 2022) has a strategy for cybersecurity protection. There have also been attempts to establish partnerships with universities for cyber training of university students. I bring to attention as a case in this regard the training on Cyber Drills held by the Albanian Agency of Cybersecurity trainers for CITs and other students from Tirana at the CIT premises in April 2024.

The main objective of this study is:

- to analyze the current cybersecurity threats by identifying and categorizing the most prevalent and emerging threats facing individuals and organizations today;
- to investigate the level of knowledge and skills on cybersecurity across different disciplines in universities;
- to analyze the negative costs of cyberattacks and to bring evidence of the benefits of cybersecurity awareness, education, and professional training.

The hypotheses raised in this presentation are as follows:

- There is a significant difference in the level of education and skills in cybersecurity between C.E. and Non-computer engineering students;
- Non-computer engineering students are more vulnerable to cyber threats and attacks than C.E. students;
- Education on cybersecurity curricular and non-curricular programs will make students and society more resilient and safer to cyberattacks and cyber threats. Students and Cybersecurity;
- C.E. students and professionals should contribute to other non-CE students as trainers on cybersecurity to equip other students with the necessary skills for the cybersecurity era of ICT.

2 Literature Review

The ITU defines cybersecurity as "the collection of tools, policies, security concepts, security safeguards, guidelines, risk management approaches, actions, training, best practices, assurances and technologies that can be used to protect the cyber environment and organization and users' assets" (ITU 2020). Cyber security is used, in most literature, as an all-inclusive term. Definitions of this word differ, e.g., the Merriam Webster (2024) dictionary defines it as "measures taken to protect a device or computer network from unauthorized access or attack".

Other authors such as Dan Craigen, Nadia Diakun-Thibault, and Randy Purse (2014), based on a literature review, found that the term is used broadly and its definitions are highly variable, context-bound, often subjective, and, at times, uninformative.

Experiences in the field show that cybersecurity threats are deliberate actions to compromise the integrity, confidentiality, or availability of information systems and data. As digital transformation accelerates, these threats become more sophisticated, necessitating advanced defensive strategies. This presentation will provide evidence of common types of cybersecurity threats, their impacts, and references to relevant literature. Cybersecurity safeguards digital data on networks, computers, and devices from unauthorized access, attacks, and destruction. While I.T. security protects both physical and digital data, cyber security protects the digital data on our networks, computers, and devices from unauthorized access, attack, and destruction.

The crime triangle [N. Dhanjani, B. Rios, and B. Hardin] is sometimes used to define cybercrime, which states that for cybercrime to occur, three variables must exist: a victim, a motive, and an opportunity. The victim is the person who will be attacked, the motive motivates the criminal to perform the crime, and the opportunity is when the crime will be committed.

Evidence shows that research provides empirical support that national-level efforts to improve cybersecurity education, awareness raising, and training have made a positive difference (GCSCC 2021).

2.1 Types of Cyber Security Threats

One of the most present modes of cyber threats is malware. Malware presents different modes of malicious software, such as viruses and worms, trojans, ransomware, spyware, and adware. These programs can disrupt operations, steal sensitive information, or hold data hostage (Malik et al. 2023). According to Jakobsson, phishing attacks involve tricking individuals into providing sensitive information by masquerading as trustworthy entities via email, messaging, or fake websites (Jakobsson and Myers 2006). As Mirkovic and Reiher distributed, denial of service (DDoS) attacks flood a network or service with overwhelming traffic, rendering it unavailable to legitimate users. These attacks often leverage botnets to amplify their impact (Mirkovic and Reiher 2004). Another mode of cyberattack is SQL injection, which attacks target databases by inserting malicious SQL code into a query. Halfond suggests that this can allow attackers to manipulate the database. On and ase contents or gain unauthorized access (Halfond et al. 2006). Conversely, Bilge is worried about zero-day exploits that exploit target vulnerabilities in software that are unknown to the vendor. These attacks can be highly damaging due to the lack of available patches or defenses (Bilge and Dumitras 2012).

2.2 Impacts of the Cyberthreats

Economic Costs. Cybersecurity breaches can result in significant financial losses due to theft, fraud, remediation costs, and business disruption (Anderson et al. 2013). According to the Cybersecurity Magazine (2023), it has been 6 years since a report from Cybersecurity Ventures predicted ransomware damages would cost the world $5 billion (USD) in 2017, up from $325 million in 2015—a 15X increase in just 2 years. The damages for 2018 were predicted to reach $8 billion; for 2019, the figure was $11.5 billion; in 2021, it was $20 billion—which is 57X more than in 2015.

Reputational damage is also significant since organizations that suffer breaches often face reputational damage, leading to loss of customer trust and market value, according to the Ponemon Institute (2019). Beyond these negative costs, cyber-attacks can disrupt critical operations, particularly in sectors like healthcare, finance, and infrastructure, where downtime can have severe consequences (Smith and Brooks 2013).

2.3 Cybersecurity Education Landscape

Educators have discovered that high school and college students have become increasingly interested in studying cybersecurity. Thus, in addition to introductory courses, they have designed seminars to expand the students' knowledge of cybersecurity (Triplett 2023). Universities and colleges worldwide have developed specialized programs in cybersecurity at undergraduate, graduate, and doctoral levels. These programs typically cover cryptography, network security, ethical hacking, and incident response (Whitman and Mattord 2018). Professional certifications provide practical and theoretical knowledge necessary for cybersecurity professionals (Harris 2019). Certain certifications like Certified Information Systems Security Professional (CISSP), Certified Ethical Hacker (CEH), and CompTIA Security+ are highly valued in the industry, as well as the Online Courses and Bootcamps ranging from beginner to advanced levels. Additionally, intensive boot camps provide accelerated training programs designed to equip individuals with the necessary skills quickly (Anderson 2020). It is important to bring in evidence as to why cybersecurity is essential in schools. Innovative approaches in cybersecurity education, such as Hands-on Learning and Cyber Ranges, are of high importance as they provide simulated environments where students can practice defending against cyber-attacks in real-time. Also, integrating cybersecurity education with other disciplines, such as law, business, and public policy, helps students understand the broader context of cyber threats and defenses (Craigen et al. 2014). Gamification and cybersecurity competitions (e.g., Capture the Flag) engage students and enhance learning through interactive, competitive experiences (Bishop and Frincke 2006). As technology becomes ingrained in every aspect of our lives, equipping the younger generation with the knowledge and skills to navigate the digital landscape securely is paramount. Literature also defines some defensive strategies to face cyberattacks, such as regular Updates and Patch Management. Keeping systems and software up to date helps protect against known vulnerabilities (Arora et al. 2010). Employee training and awareness, as well as educating employees on cybersecurity best practices, are also important, as well as reducing the risk of human error leading to security breaches (Furnell and Clarke 2012). Multi-factor authentication (MFA) is also another strategy for defense by cybercasters since implementing MFA adds an extra layer of security and makes it harder for attackers to gain unauthorized access (Das et al. 2014).

3 Methodology

Is there a difference in cyber security awareness between students doing a computing-related degree and those doing a different kind of degree? Surveys and interviews are designed and conducted to find the level of knowledge and skills about students' cybersecurity. The aim is to evaluate the basic cybersecurity knowledge and habits focused on engineering students and social science students. The questionnaire is composed of sections, where questions in all sections expect yes/no, and other responses are accompanied by tick-boxes. The questionnaire was designed to be completed online. The survey emphasizes the importance of cybersecurity in our modern digital era, the risks associated with cybercrime, and the significance of implementing cybersecurity best practices. In order to assess the cybersecurity knowledge of both computer science and non-computer science students, MSc students at the Canadian Institute of Technology in Tirana, specializing in Computer Engineering with a focus on Networking and Cybersecurity as well as Big Data, conducted a research project. As part of the study, the students surveyed both ICT profile students and students from other disciplines to gather primary data on the topic. Each student researcher was tasked with engaging with different categories of students. The information collected from the study included responses from both computer science (C.S.) and non-computer science (NCS) students. The data the student teams collected over the November 2023–May 2024 period was combined to collect data on cybersecurity issues among students of Non-Computer Science and Computer Science, giving an answer to the following research question.

- Is there a difference in the level of education, knowledge, and skills students are equipped with in Computer Engineering and Social Science?
- Does a computer science-specific education improve cyber resilience, while a lack of skills makes other groups more vulnerable to cyber threats and attacks?

A mixed methodological approach is used in this study to collect data and analyze the situation. Surveys are used to collect primary empirical data about students' level of education and skills on the cybersecurity level of knowledge and skills for students steering in different disciplines. This presentation addresses many questions, but it selects only those that are directly linked to the issue we are discussing.

Q.1. Have you ever dealt with a malware infection on your computer?

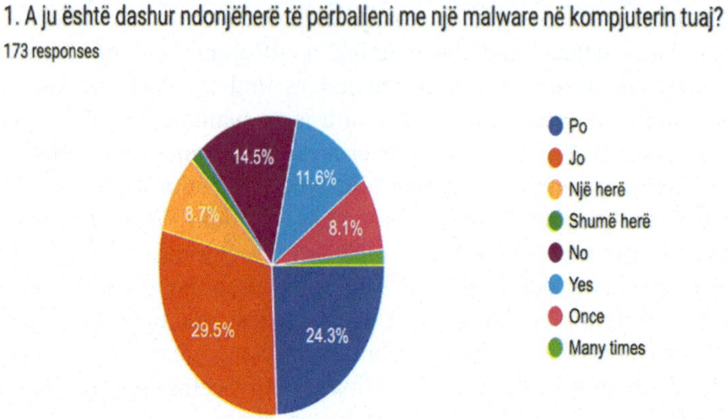

It was found that most of the students had dealt with a malware infection in their computers.

Q.2. Have you ever attended a security awareness training?

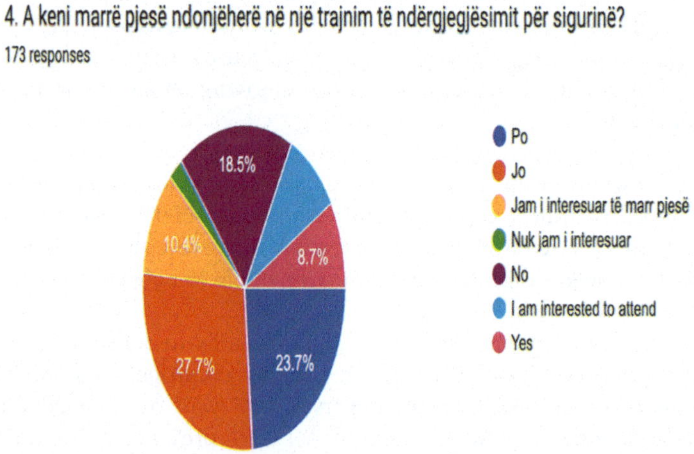

Most students haven't attended any security awareness training, even though there is a great interest in participating in these trainings.

Q.3. Would you like to acquire knowledge, education, and skills to combat Cybersecurity? Threats?

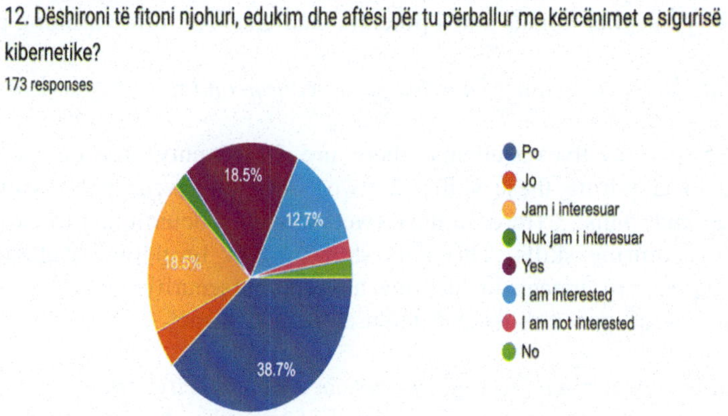

The dominant number of students want to acquire. Knowledge, education, and skills to combat cybersecurity threats.

Q.4. Cybersecurity Training: the level of required training needs. Please indicate your desired level of expertise in cybersecurity threat skills.

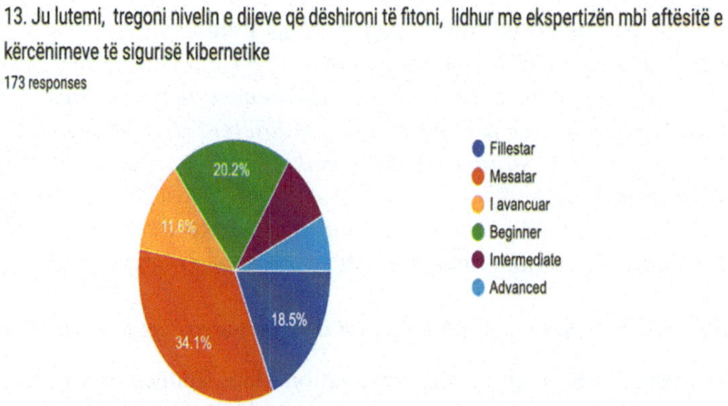

In regard to the level of required training needs, it is found that most of students want to attend an intermediate level of knowledge and skills on cybersecurity threats and skills, followed by those which need a beginner level.

Interviews. To have a clear picture and analyze the situation in regard to the level of knowledge and skills of students of different disciplines, a qualitative approach is used by applying interviews, as presented below.

Q.1. Have you ever attended a security awareness training?
Out of 25 (twenty-five) of the interviewed students, 2 (two) attended an online security awareness training. 23 (twenty-three) out of 25 (twenty-five) interviewed

students have not attended such training. Meanwhile 20 (twenty) of them expressed interest in attending one. This suggests that nearly all of the interviewees are not informed about cybersecurity through formal and nonformal training, indicating a vulnerability to cyber threats and a potential audience for future training programs.

Q.2. Would you like to acquire knowledge, education, and skills to face cybersecurity threats?
Out of 25 (twenty-five) students, there are 20 (twenty) responders who are interested in acquiring these skills. 2 (two) of interviewees possess elementary knowledge and skills. 3 (three) interviewed were not determined to express their interest in acquiring skills. This data shows a low level of engagement but a high willingness to engage in learning about cybersecurity, which is critical for improving overall cyber skills and defense against threats.

Q.3. What is your desired level of expertise in cybersecurity skills?
All of interviewees aim for a beginner level of expertise. No one aim for an Intermediate level. No one aim for an Advanced level. Data reflects a significant interest in developing elementary and foundational cybersecurity knowledge and skills.

4 Discussion

The research questions posed in this paper are as follows:

- *Is there a difference in cyber security awareness between students doing a computing-related degree and those doing a different kind of degree?*
- *Do cyber security education and skills make C.E. students more resilient than non-CE students?*
- *Does lack of cybersecurity knowledge and skills make students of non-computer education more vulnerable than students of the C.E. group?*

Findings from literature review surveys and interviews are as follows:

It is clear that a computer science education does indeed improve awareness of cyber security issues and precautions to be taken. The graphs demonstrate a clear positive impact of a computing science education in this respect. Most of the Computer Engineering students understood the cyber-attacks and their threats. Computer Engineering education improves awareness, and it does so for both genders. Students with a technology or science background are less vulnerable to phishing attacks. Upcoming generations of Albania who do not study Computer engineering Science will have insufficient awareness of privacy and security issues and will, as a consequence, be more vulnerable to cyber-attacks.

5 Conclusion

Our study revealed that cybersecurity is a topic of interest across different groups of students coming from different study backgrounds. It also testifies to the hypotheses raised in this presentation, as follows:

There is a significant difference in the level of education and skills in cybersecurity between Computer Engineering and Non-Computer Engineering students. Non-computer Engineering students are more vulnerable to cyberthreats an attack than computer engineering students. The study also reveals a need for increased awareness and education about Cybersecurity among both social and ICT Students.

Introducing cybersecurity education in curricula programs at early stages and establishing partnerships between educational institutions and industry to provide real-world exposure is suggested as a solution.

Education on Cybersecurity curricular and non-curricular programs will make students and society more resilient and safer to cyberattacks and cyber threats. Students and Cybersecurity. Computer engineering students and professionals of the area, should contribute to other non-computer engineering students as trainers on Cybersecurity, to equip other students with necessary skills for cybersecurity era of ICT.

The work results indicate that ICT students better understand cybersecurity and its implications than social science students.

Introducing cybersecurity education in curricula programs at early stages and establishing partnerships between educational institutions and industry to provide real-world exposure is suggested as a solution;

The paper concludes by highlighting the importance of cybersecurity education for all students, regardless of their discipline and level of education, as cybersecurity threats are becoming increasingly common and complex.

Fundamental cyber security life skills should be taught at primary school and delivered to all pupils, ensuring that awareness is raised in all genders equally.

References

Aiken, M.: The cyber effects. The cyber effect was selected as one of the best science picks of the week in 2016 by Nature [13]. http://www.maryaiken.com/cyber-effect (2016)

Amzeal Newes.: Cybersecurity Training Course to Protect Organizations from Data Breaches. https://amzeal.com/doc/acentec-inc-announces-cybersecurity-training-amzeal-10515780 (2023a)

Amzeal Newes.: How to Report Spam Text Messages. https://maxdalton.how/tutorial/how-to-report-spam-text-messages (2023b)

Anderson, R., Barton, C., Bohme, R., Clayton, R., van Eeten, M.J., Levi, M.: Measuring the Cost of Cybercrime. Springer (2013)

Anderson, C.: The Cybersecurity Bootcamp Manual: A Guide for Aspiring Cybersecurity Professionals. Self-published (2020)

Arora, A., Krishnan, R., Nandkumar, A., Telang, R., Yang, Y.: Impact of Patch Release on Software Vulnerability Discovery: An Empirical Analysis. Information Systems Research (2010)

Bilge, L., Dumitras, T.: Before we knew it: an empirical study of zero-day attacks in the real world. In: Proceedings of the ACM Conference on Computer and Communications Security (2012)
Bishop, M., Frincke, D.A.: Who owns your computer? [Digital rights management]. IEEE Secur. Priv. **4**(2), 61–63 (2006). https://doi.org/10.1109/MSP.2006.5
Craigen, D., Diakun, N.-T., Purse, R.: Defining Cybersecurity. TIM. Technology Innovation Management Review. https://www.timreview.ca/article/835 (2014)
Cybersecurity Magazine: Ransomware Market Report 2023. https://cybersecurityventures.com/global-ransomware-damage-costs-predicted-to-reach-250-billion-usd-by-2031 (2023)
Das, A., Bonneau, J., Caesar, M., Borisov, N., Wang, X.: The tangled web of password reuse. In: Network and Distributed System Security Symposium (2014)
Datareportal: Digital Around the World. https://datareportal.com/global-digital-overview (2024)
ENISA: Stocktaking of Information Security Training Needs in Critical Sectors. https://www.enisa.europa.eu/publications/stock-taking-of-information-security-training-needs-in-critical-sectors (2017)
ENISA: Consultation Paper, 2019. EU ICT Industrial Policy: Breaking the Cycle of Failure. https://www.enisa.europa.eu/publications/enisa-position-papers-and-opinions/eu-ict-industry-consultation-paper (2019)
ENISA: (2023)
EPSC: Rethinking Strategic Autonomy in the Digital Age. Strategic Notes. No 30, 2019. https://ec.europa.eu/epsc/sites/epsc/files/epsc_strategic_note_issue30_strategic_autonomy.pdf (2019)
Federal Bureau of Investigation. Internet Crime Report 2020. https://www.ic3.gov/Media/PDF/AnnualReport/2020_IC3rpeort.pdf (2021). Accessed 7 June 2023
Finch, A.: Cybersecurity workforce development in minority, low income, and native American reservation communities. Int. J. Smart Educ. Urban Soc. **11**(4). American Public University, USA (2020) https://orcid.org/0000-0002-0078-1973
Furnell, S., Clarke, N.: Power to the people? The evolving recognition of human aspects of security. Comput. Secur. (2012)
Global Cyber Security Capacity Centre: Cybersecurity Capacity Maturity Model for Nations. Oxford University and Oxford Martin School, U.K. (2021)
Halfond, W.G., Viegas, J., Orso, A.: A classification of SQL-injection attacks and countermeasures. In: Proceedings of the IEEE International Symposium on Secure Software Engineering (2006)
Harris, S.: CISSP All-in-One Exam Guide. McGraw-Hill Education (2019)
ITU: ITU Publications 2020. Global Cybersecurity Index. https://www.itu.int/dms_pub/itu-d/opb/str/D-STR-GCI.01-2021-PDF-E.pdf (2020)
Jakobsson, M., Myers, S.: Phishing and Countermeasures: Understanding the Increasing Problem of Electronic Identity Theft. Wiley (2006)
Malik, M.I., Ibrahim, A., Hannay, P., Sikos, L.F.: Developing resilient cyber-physical systems: a review of state-of-the-art malware detection approaches, gaps, and future directions. Computers **2023**(12), 79 (2023). https://doi.org/10.3390/computers12040079
Merriam-Webster Dictionary. https://www.merriam-webster.com/dictionary/cybersecurity (2024). Accessed Jun 2024
Mirkovic, J., Reiher, P.: A Taxonomy of DDoS Attack and DDoS Defense Mechanisms. ACM SIGCOMM Computer Communication Review (2004)
Olmstead and Smith: What Does the Public Know About Cybersecurity? Pew Research Center, The USA (2017)
Ponemon Institute: Cost of a Data Breach Report. IBM Security (2019)
Shillair, R.: Online Safety Begins with You and Me: Convincing Internet Users to Protect Themselves. Elsevier (2015)
Smith and Brooks: Security Science: The Theory and Practice of Security. Elsevier (2013). https://www.researchgate.net/publication/235688059_Security_Science_The_Theory_and_Practice_of_Security
Timmers: Policy in Focus—Strategic Autonomy and Cybersecurity, 2019 https://eucyberdirect.eu/wp-content/uploads/2019/05/paul-timmers-strategic-autonomy-may-2019-eucyberdirect.pdf (2019)

Triplet, W.: Addressing cybersecurity challenges in education. Int. J. STEM Educ. Sustain. **3**(1), 47–67 (2023). https://doi.org/10.53889/ijses.v3i1.132

Whitman, M.E., Mattord, H.J.: Principles of Information Security. Cengage Learning (2018)

WSAV.: Data Breaches, Annual Report. https://www.wsav.com/wp-content/uploads/sites/75/2022/01/20220124_ITRC-2021-Data-Breach-Report.pdf (2022). Accessed 7 June 2023

Improving Test Design for High-Stake Exams by Use of Equating Methods

Alfons Harizaj, Olgerta Idrizi, Miranda Harizaj, and Vilma Tomço

Abstract

The equating methods are used in this paper to identify which are some problems of State Matura exams or other high-stake exams related to results and test design. The test equating methods are statistical procedures used to adjust scores on assessments so that they can be compared across different forms of a test or across different testing years. When the results have discrepancies, it should be attributed to actual undergraduate performance variances and not to the level of test difficulty. Equating results through different years is a process that aims to ensure fairness and comparability of scores. This is particularly important in educational assessments where variations in test difficulty can occur. The State Matura exams are taken by students at the end of their 12th grade, and successful completion of these exams is a requirement for obtaining a high school diploma in Albania. The results are also used for ranking candidates that apply for admission in universities. Here's how equating methods could be used to compare the results in different years and improve test design in Matura or other high-stake exams.

A. Harizaj (✉)
Canadian Institute of Technology, Tirana, Albania
e-mail: alfons.harizaj@cit.edu.al

O. Idrizi
Polytechnic University of Tirana, Tirana, Albania
e-mail: olgerta.idrizi@umsh.edu.al

M. Harizaj
The Mediterranean University of Albania, Tirana, Albania
e-mail: miranda.harizaj@fie.edu.al

V. Tomço
University of Tirana, Tirana, Albania
e-mail: vilma.tomco@unitir.edu.al

K. Dhoska and E. Spaho (eds.), *Bridging Horizons in Artificial Intelligence, Robotics, Cybersecurity, Smart Cities, and Digital Economy*, Sustainable Economy and Ecotechnology, https://doi.org/10.1007/978-3-031-72029-1_23

Keywords

High-stake exams • Equating • Scores distribution

1 Introduction

The exams are important for education assessment, as even parents and a broader community believe on them (Homepage UNESCO 2024). Users of educational assessments assume that students' scores can be validly compared and assume score comparability even if those scores come from measurements taken at different times, in different places, or using variations in assessment content and procedures. Traditionally, it has been assumed that as long as test forms are constructed from the same curriculum benchmarks and based on the same specifications, they can be compared regardless of the year or occasion they are administered (Dorans 2017; DeMars 2002). Ideally, users could be assured that students with the same score possessed the same level of proficiency with respect to the domain of knowledge and skills a test was intended to measure (Ryan and Brockmann 2009).

When two test forms undergo a meticulous equating process, educators can confidently interpret performance on one test form as possessing an equivalent substantive meaning when compared to the equated score of the other test form. This ensures that the educational insights derived from each form are reliably comparable, fostering a more accurate understanding of students' proficiency levels across different versions of the test (Linden 2022). The current literature on test equating generally defines it as the process necessary to obtain score comparability between different test forms (Fitrah 2023).

This paper intends to present the importance of score equating for test developers and users of test results, as they play a critical role in making sensitive decisions such as admission in universities, certification, graduation, scholarships, and requirements. It is valuable to implement the equating process in all standardized high-stakes tests to check quality or improve test design (Sappl 2023; Zouhfong 2022). The equating produces very useful information for experts of tests development process to produce equitable tests, more professional scoring guides, and more qualitative items for item bank development. Based on these indicators are presented and analyzed results from recent Matura exams in Albania.

2 Method and Instruments

2.1 Method

To compare and equate the test results between two years linear equating method is used. The presentation how it works is given below. Linear equating assumes that the only differences between the two tests are mean and variability, measured by standard deviation or "SD". Linear equating is a linear transformation as in:

$$Y = k(X - l) + m \tag{1}$$

where Y and X—are the scores of the tests to be equated and k, l and m—known values from characteristics data of test results for both forms Y and X. The transform x scores in test form X in equivalent y scores in test form Y is applied by functions:

$$y_x = \frac{\sigma_y}{\sigma_x} \cdot x - \frac{\sigma_y}{\sigma_x} \cdot \mu_x + \mu_y \tag{2}$$

$$Y = k(X - l) + m \Rightarrow y_x = \frac{\sigma_y}{\sigma_x}(x - \mu_x) + \mu_y \tag{3}$$

$$k = \frac{\sigma_y}{\sigma_x} \tag{4}$$

$$l = \mu_x \tag{5}$$

$$m = \mu_y \tag{6}$$

where:

x	are candidates scores on test form X;
$l = \mu_x$	are the mean scores of test form X;
σ_x	is the standard deviation scores of test form X;
$m = \mu_y$	is the mean of scores in test form Y;
σ_y	is the standard deviation scores in test form Y.

Equating results by linear or other methods are useful instruments to compare or equalize the level of achievements of students (González and Wiberg 2017; Setiawan 2019; Arıkan 2020). The results of Matura exams from previous years are analyzed to interpret aspects of equity of tests results, quality of tests or standards of test design. The results that come from non-equity tests might have negative impact in the decisions of ranking candidates (Ayanwale 2023; Kolen and Brennan 2014). This is the case when the results of Matura exams are used for ranking candidates for admission in universities. The results are equated by using linear equating method (Wallin 2019; Pommerich 2016) with data that are explored from the Educational Services Center (ESC) or Ministry of Education and Sports (MoES).

2.2 Participants

The students who took the high school graduation exam in the respective academic year were considered. All students undergo the exams, ensuring the application of equating method across the entire population without requiring sampling. The data and results for all graduating students through Matura exams were obtained from the official website of the Educational Services Centre, Albania.

2.3 Data Collection and Analysis

The data are taken from official web of the Educational Services Centre (ESC). The ESC, established in 2015 under the responsibility of the Ministry of Education and Sports, Albania (MoES) and preceded by the National Examination Agency (2007–2015), is the institution responsible for national and international assessments and examinations in Albania. The ESC is fully responsible for the design, administration, quality assurance and analysis of the State Matura Examination. The data of State Matura exams are taken from period 2014 to 2023 (Homepage QSHA: Educational Services Center (Qendra e Shërbimeve Arsimore) 2014–2024).

This is considered in two periods 2014–2017 and 2019–2023 as the test forms has been the same format within each period. Starting from 2019 year and on the tests are composed based on the new format and content. The framework of state Matura exams includes: the tests are based on same content for each year; the tests are composed with the same number of multiple choice and open-ended questions; the maximum of scores is 60 points; a common marking scheme is used for scoring of student answers. These characteristics respect requirements needed for test equating.

Comparing scores across the two tests form is more convenient, when the tests are composed with the same content standards (Wainer et al. 1994; Dorans 2008; Li et al. 2010). Additionally, the tests would need to use the same blueprint that emphasized the same content areas as well as the same performance-level descriptions. Users of educational tests often seek to compare scores even if the scores were obtained at different times, in different places, or using variations in assessment content and procedures. The equated forms can be used interchangeably (Kim and Walker 2021).

The State Matura exam includes three compulsory examinations and two additional exams that students choose from a list of subjects that is approved by MoES each year. The compulsory examinations are: Albanian Language and Literature (ALL), Mathematics, and one Foreign Language from English, French, German, Italian, Spanish. The list of subjects for two additional exams includes subjects as Biology, Sociology, Philosophy, Chemistry, History, Physics, Economy, etc. This context makes it necessary using additional measurements, such as equating, to make scores across test forms comparable and equivalent (Homepage Ministry of Education and Sport of Albania-State Matura 2024).

The data are analyzed and equated for Math exams of State Matura as it is taken from all of students. This ensures the application of equating method across the entire population without requiring sampling (Homepage QSHA: State Matura Report (2014–2021)). From the list of courses of additional exams are taken the courses; Biology, Physics, Chemistry and Economy. The results of them are analyzed just to make visible the differences from equating process and the need to equalize them by using equating methods.

3 Results

3.1 Differences Results of Math Test During 2014–2017

In the Table 1 are given mean and the standard deviation of test results of Math subject for period 2014–2017 (Homepage QSHA: State Matura Report (2014–2021)).

Consider a student who achieved a result of 50 points in the math test 2016. By using the transformation Eq. (3) for rescaling the scores of Form X to the scale of Form Y the equating scores in 2017 are:

$$y_{2017} = \frac{12.68}{10.52} * 50 - \frac{12.68}{10.52} * 19.66 + 23.57 = 60.14 \qquad (7)$$

By using the same transformation equation, the result of 50 points in math for a certain year are equated in points of coming years as illustrated in Table 2.

The differences in the math test are large and variable. The difference of equating from 2016 to 2017 is too high, 10.14 scores. Referring two consecutive years it is considered that the student populations have had the same statistical characteristics. It was expected that there would be no sensitive difference or that equated scores would be very closed. This result is an indicator that difficulty level of test 2016 is higher than the difficulty level of test 2017. In addition, if a student was graduated in 2016 and applied for university in 2017 the graduate is discriminated against candidates graduated in 2017. This can bring to missed opportunity for them when the decisions are depending from these results.

Table 1 Mean and standard deviation for Mathematics 2014–2017

Years	Mathematics	
	Mean	St. deviation
2014	17.93	12.15
2015	22.14	12.22
2016	19.66	10.52
2017	23.57	12.68

Table 2 Equated 50 points of Math test 2014–2017

Year	Row scores	Equated year	Equated scores	Difference
2014	50	2015	54.39	4.39
	50	2016	47.43	−2.57
	50	2017	57.04	7.04
2015	50	2016	43.64	−6.36
	50	2017	52.48	2.48
2016	50	2017	60.14	10.14

3.2 Equating Results of Math Test During 2019–2023

Data of test results for years 2019 and 2021, are equated. For the other years is given an interpretation based on cumulative scores. Table 3 shows the mean and standard deviation of test results for some subjects for years 2019 and 2021. The distribution of results for students who completed the math test 2019 has a mean of 27.37 points and standard deviation of 12.05 points and the distribution of results for students who completed the math test 2021 has a mean of 25.89 points and a standard deviation of 10.26 points.

Python for data analyses library is used to calculate and reproduce the cumulative distributions of results of math test for period 2019–2023 presented in Fig. 1 (Homepage QSHA: State Matura Report (2014–2021); QSHA-State Matura Report: Homepage QSHA: State Matura Report 2023).

By cumulative distributions graph of math test results is able to compare the results in different years. Comparing the outcomes of two consecutive years is particularly important. It is expected for similar results in consecutive years.

By comparing the graphics of cumulative scores in Fig. 1, the conclusion is that the tests have not been equitable. It is an indicator that may have been too serious problems in respecting of standards of tests design (Issayeva 2021). If we consider that ranking between candidates for admission to universities is sensitive toward decimals this can be considered very sensitive difference (Altıntaş and Kutlu 2020; Altıntaş and Wallın 2021).

Table 3 Mean and standard deviation State Matura tests 2019 and 2021

Subjects	Year 2019		Year 2021	
	Mean—μ_x	St. Deviation—σ_x	Mean—μ_y	St. deviation—σ_y
Math	27.37	12.05	25.89	10.26
ALL	26.78	9.47	33.45	9.97
Biology	36.2	14	41.82	12.98
Physics	39.41	13.22	40.06	12.26
Chemistry	44.86	12.08	39.41	12.38
Economy	37.94	11.1	43.38	11.16

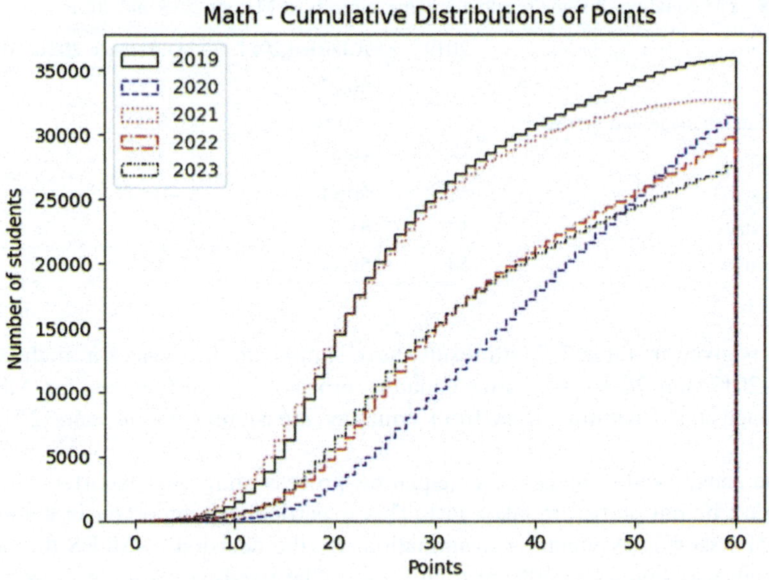

Fig. 1 Cumulative math scores 2019–2023

Table 4 Mathematics grades 2019, 2021

Year	Mathematics grades						
	4 (%)	5 (%)	6 (%)	7 (%)	8 (%)	9 (%)	10 (%)
2019	5.57	29.06	26.18	15.11	10.47	7.80	5.81
2021	1.84	26.37	30.86	19.20	9.77	6.25	5.71

The percentages of grades for the subject of Mathematics in years 2019 and 2021 are presented in Table 4.

The percentage of students that failed in 2019 was 5.57% comparing to 1.84% of students failed to in 2021. The first comment would be that level of difficulty for 2019 math test has been higher than the difficulty level of 2021 math test. By this situation the results cannot link with levels of achievements according the curriculum (Maghnouj et al. 2020).

3.3 Equating Results 2019–2023, Other Courses

The Table 5 gives equated of some results for some other subjects of State Matura exams by using the transformation Eq. (3) and data from Table 5.

The raw scores 55 on Biology 2019 have equity scores 59.25 on Biology 2021, or 4.25 points more. The raw score 56 on math 2019 have equity scores 50.27 on math 2021, or 5.73 points lower. In the same are adjusted scores for some other

Table 5 The equating of some results for some subjects of Matura 2019 and 2021

Courses	2019	Adjusted 2021	Difference 2021–2019
Math	56	50.27	−5.73
Albanian language and literature	52	60.00	8.00
Biology	55	59.25	4.25
Physics	51	50.81	−0.19
Chemistry	57	51.85	−5.15
Economy	54	59.53	5.53

subjects given in Table 5. Math results have significant differences in both periods 2014–2017 and 2019–2021 after equating process. The differences for Albanian Language and Literature tests from equating are significant in years 2019 and 2021.

The sensitive differences after equating process show that pass/fail cut-score shouldn't be put on before taken test. This is particularly important in the context of high stakes assessments or examination. By the decision of MoES the pas/fail cut-score was said to be 20% of total scores. The fixed pass/fail cut-score neither allows to produce comparisons over time nor to link results to expected levels of achievement found in the curriculum (Maghnouj et al. 2020).

4 Discussion

The findings of this paper suggest that even when tests are based on the same content and constructed with aim to be parallel forms, they can still be different in difficulty and score distribution. Both the National Basic Education Examination (Grade 9) and the State Matura Examination (Grade 12) can use comparison and equating results also for improvement of test design. They would benefit from continuous improvement to increase their reliability in assessing students' learning.

Assessment and scaling techniques should consider for the future using of digital infrastructure for taken exams from students and data analysis of their results. By doing so there are better possibilities to use IT applications for scaling and equating the results.

5 Conclusions

The results presented in this paper aim to gather evidence on the practice of equating. It is hoped that the findings will encourage the ESC or other exam boards to consider equating test forms as part of their test development and score reporting processes. The information in this paper aims to arise the interest of experts of test developers, educators, and public about the equating methods of test forms and benefits of its use.

The use of equating methods can help to improve the comparability of ranking students by ensuring fairness, reliability, and validity of the test scores. This can be particularly important in situations where ranking students based on test scores is used for important decisions such as university admissions, recruitment, or certification processes. Ignoring equating can have negative consequences for the major decisions made about students and can affect the quality of information to the public and policymakers.

Equating can improve the quality of information offered by the ESC to the public. To ensure validity and reliability, test experts must keep the same standards each year. Standardized testing needs stable measures that do not discriminate/disadvantage one group or some students.

By adjusting the scores to a common scale, it is possible to compare the scores of students who received different versions of the test. Equating methods ensure that students are not unfairly disadvantaged or advantaged based on the specific version of the test they receive. Equating methods can improve the reliability of test scores by accounting for any differences in difficulty between the different tests' versions. Equating guarantees that cores are a more accurate reflection of the students' abilities.

Equating methods can enhance the validity of test scores by allowing students to be compared across different test versions. This can help to ensure that the scores are a valid measure of the content being tested; regardless of which version of the test was used. Equating methods can improve test security by allowing test developers to create multiple versions of a test that are statistically equivalent.

State Matura exams produce a lot of data that has great impact and interest to the public. Educators, researchers, experts, and the public need more access to data. The official reports should have more significant indicators of data analysis. By doing other analyses or reports, they bring to the attention of the public and experts many important issues. The Western Balkan countries as well as other countries are all included in the Matura State Exams process. The data and process infrastructure of each country are of great interest to other countries and in some ways, they have the opportunity to increase collaboration between them.

References

Altıntaş, Ö., Kutlu, Ö.: Investigating differential item functioning of Ankara university examination for foreign students by recursive partitioning analysis in the Rasch model. Int. J. Assess. Tools Educ. **6**(4), 602–616 (2020). https://doi.org/10.21449/ijate.554212

Altıntaş, Ö., Wallın, G.: Equality of admission tests using kernel equating under the non-equivalent groups with covariates design. Int. J. Assess. Tools Educ. **8**(4), 729–743 (2021). https://doi.org/10.21449/ijate.976660

Arıkan, Ç.A: The impact of covariate variables on kernel equating under the non-equivalent groups. J. Meas. Eval. Educ. Psychol. 362–373 (2020). https://doi.org/10.21031/epod.706835

Ayanwale, M.: Test score equating of multiple-choice mathematics items: techniques from characteristic curve of modern psychometric theory. Discov. Educ. (2023). https://doi.org/10.1007/s44217-023-00052-z

DeMars, C.E.: JMU Scholarly Commons. https://commons.lib.jmu.edu/cgi/viewcontent.cgi?article=1036&context=gradpsych (2002)
Dorans, N.J.: The Practice of Comparing Scores on Different Tests. ETS Research & Development, Princeton (2008). https://www.ets.org/Media/Research/pdf/RD_Connections6.pdf
Dorans, N.J.: Contributions to Score Linking Theory and Practice. Advancing Human Assessment: The Methodological, Psychological and Policy Contributions of ETS, pp. 79–132. Springer International Publishing (2017)
Fitrah, M.: A: equating principles in assessment: a literature review in the context of education and assessment. IJECA (Int. J. Educ. Curric. Appl.) 199–209 (2023). https://doi.org/10.31764/ijeca.v6i3.19381
González, J., Wiberg, M.: Applying Test Equating Methods Using R. Springer. https://link.springer.com/book/10.1007/978-3-319-51824-4 (2017). ISBN: 978-3-319-51822-0
Homepage Ministry of Education and Sport of Albania-State Matura. https://arsimi.gov.al/matura-shteterore/. Accessed 5 April 2024
Homepage QSHA: Educational Services Center (Qendra e Shërbimeve Arsimore). https://qsha.gov.al/. Accessed 10 April 2024
Homepage QSHA: State Matura Report (2014–2021). https://qsha.gov.al/raporte-dhe-statistika/. Accessed 15 April 2024
Homepage UNESCO: https://www.unesco.org/en/articles/exams-and-assessments-covid-19-crisis-fairness-centre. Accessed 24 May 2024
Issayeva, L.: Test Score Equating and Linking. Assessment Systems Corporation, Stillwater, Minnesota, USA. https://assess.com/test-equating/ (2021)
Kim, S., Walker, M.E.: Comparisons Among Approaches to Link Tests Using Random Samples Selected Under Suboptimal Conditions. ETS Research Report Series, pp. 1–20 (2021). https://doi.org/10.1002/ets2.12328
Kolen, M.J., Brennan, R.L.: Test Equating, Scaling, and Linking Methods and Practices. Springer. https://link.springer.com/book/10.1007/978-1-4939-0317-7#overview (2014)
Li, Y., Marion, S., Perie, M., Gong, B.: An approach for evaluating the technical quality of interim assessments. Peabody J. Educ. **85**(2), 163–185 (2010). https://doi.org/10.1080/01619561003685304
Linden, W.J.: What is actually equated in "test equating"? A didactic note. J. Educ. Behav. Stat. 353–362 (2022). https://doi.org/10.3102/10769986211072308
Maghnouj, S. et al. (2020) OECD Reviews of Evaluation and Assessment in Education: Albania. OECD Publishing Paris. https://doi.org/10.1787/d267dc93-en
Pommerich, M.: The fairness of comparing test scores across different tests or modes of administration. In: Educational Assessment and Measurement. Taylor and Francis Group, London (2016). https://doi.org/10.4324/9781315774527
QSHA-State Matura Report: Homepage QSHA: State Matura Report 2023. https://qsha.gov.al/matura-shteterore-2023/. Accessed 15 April 2024
Ryan, J., Brockmann, F.: A Practitioner's Introduction to Equating with Primers on Classical Test Theory and Item Response Theory. Council of Chief State School Officers, Washington (2009). https://doi.org/10.3102/10769986211072308
Sappl, S.: Equating Tests Across Grades in Large-Scale Assessment, A Study of Two Alternate Methods in Conjunction With Methodological Innovations. The University of Western Australia. https://research-repository.uwa.edu.au/en/publications/equating-tests-across-grades-in-large-scale-assessments-a-study-o (2023)
Setiawan, R.: A comparison of score equating conducted using Haebara and stocking lord method for polytomous. Eur. J. Educ. Res. **1**, 071–1079 (2019). https://doi.org/10.12973/eu-jer.8.4.1071

Wainer, H., Wang, X-B., Thissen, D.: How well can we compare scores on test forms that are constructed by examinees' choice? J. Educ. Meas. 183–199 (1994). https://doi.org/10.1111/j.1745-3984.1994.tb00442.x

Wallin, G.: Extensions of the Kernel Method of Test Score Equating. DiVA. https://www.diva-portal.org/smash/record.jsf?pid=diva2%3A1378833&dswid=-4521 (2019)

Zouhfong, C.A.: Test score equating and item anchoring for high stakes examination. Vietnam J. Educ. Sci. (2022). https://doi.org/10.15625/2615-8957/22210401

On Relations Between Digital Development Level and ISO 27001 Standard Index, and a Comparative Analysis of Cyber Security Indicators Between Western Balkan Countries

Enriko Ceko

Abstract

In light of the increased global interest in digitalization, ISO standards, cyber security, digitalization indicators, etc., my goal in writing this essay was to demonstrate the links between the digital development level and the ISO 27001 index of the Western Balkan nations. The research methodology yielded the main findings of this study. Specifically, data were collected regarding the level of digitalization development index in the Western Balkans, as measured by the DDL, as well as data regarding the extent to which the ISO 27001 standard was applied in these countries. Additionally, data regarding ISO 27001 certificates were processed, an index of ISO 27001 certificates was created, and a regressive analysis was managed to examine the relationships between the digitalization development index and the ISO 27001 certificates index, while addressing cyber security issues in addition to looking at the route toward EU membership. The study's secondary findings focused on the comparative analysis of digitalization and cyber security indicators among Western Balkan nations, evaluating them to demonstrate each nation's relative advantage in using these indicators. This is due to the fact that having strong cyber defenses—derived from the comparison of Cyber Security Index and Digital Development Level indicators—is essential to obtaining a competitive edge in industries including information technology, artificial intelligence, digitalization, e-commerce, and e-government. The main recommendation is that Western Balkan countries should adopt ISO standards, such as ISO 27000, especially ISO 27001, and work simultaneously to improve the use of CSI and DDI indicators and investments in the field of cyber security in order to better respond to cyberattacks and cyberthreats.

E. Ceko (✉)
Business Administration and Information Technology Department, Faculty of Economy, Canadian Institute of Technology, Tirana, Albania
e-mail: enriko.ceko@cit.edu.al

K. Dhoska and E. Spaho (eds.), *Bridging Horizons in Artificial Intelligence, Robotics, Cybersecurity, Smart Cities, and Digital Economy*, Sustainable Economy and Ecotechnology, https://doi.org/10.1007/978-3-031-72029-1_24

Keywords

Digital development level • Digitalization • ISO 27001 index • Western Balkans • Information technology management • Competitive advantage

1 Introduction

Digitalization is the process of individuals, businesses, and organizations using digital tools to improve their work speed, efficiency, and value. Digitalization is particularly relevant for businesses, as it can help save time and create new growth opportunities. With internet users worldwide expanding to 5.3 billion in 2024, the digital world is now the largest it has ever been. And as we accommodate the technological practices and tools that make up daily life, businesses are at the forefront of keeping pace with the latest developments in the digital world. While digitalization has been applied in numerous organizations since the technological advances of the 1990s and 2000s, its potential uses and benefits still haven't been fully exhausted.

To measure and compare the issues related to digitalization there are developed several indexes. One of them is Digital Development Level (DDL), calculated according to the E-Government Development Index (EGDI) and Networked Readiness Index (NRI). The DDL is the average percentage the country received from the maximum value of both indexes (NCSI 2023).

These days, privacy protection, cyber security, and IT security are essential for businesses and organizations. In the current market, businesses want to give their clients trust and show that they are dedicated to protecting the security of the data they handle. To do this, a firm might gain a competitive edge by obtaining certification of an ISO standard or security regulation, which attests to the proper management of security needs in information processing operations. The ISO 27000 series of standards is one of them. This group of information security standards lays out the conditions and recommendations for putting in place an Information Security Management System (ISMS) to handle information security inside a company. The primary standard in this group is ISO 27001, which serves as the series benchmark.

Cyber security is the process of defending networks, sensitive data, and electronic devices from hostile assaults. It is often referred to as electronic information security or information technology security.

There is an increasing number of data breaches every year, and the global cyber threat is still evolving at a rapid rate. Global investment in cyber security solutions is inevitably rising as the danger posed by cyberspace is expected to continue growing in scope.

Controls are unquestionably necessary to ensure the security of devices, communication networks, and information assets given the increasing number of security events and assaults on information and IT systems that enterprises are facing.

The idea of cyber security was born out of this requirement. In the current market, businesses want to give their clients trust and show that they are dedicated to protecting the security of the data they handle. The private sector is the primary implementer of ISO standards, but public administration bodies are also involved in cyber security matters. Therefore, for a nation to be safe from cyberattacks, there needs to be a strong correlation between government cyber security initiatives and the implementation of ISO 27001 standards.

2 Literature Review

2.1 The Western Balkans as a Part of the Balkan Region

The Balkans, which roughly correlate to the Balkan Peninsula, are a region in southeast Europe that has been defined in a variety of ways throughout history (Gray and Sloan 2014). The Balkan Mountains, which encircle all of Bulgaria, gave rise to the region's name. The Black Sea borders the northeast of the Balkan Peninsula, the Adriatic Sea borders the northwest, the Ionian Sea borders the southwest, the Aegean Sea borders the south, and the Straits forms the east.

There are many definitions for the peninsula's northern boundary (Vezenkov 2017). Musala, located in Bulgaria's Rila mountain range at 2,925 m (9,596 feet), is the highest peak in the Balkans.

German geographer August Zeune, who thought the Balkan Mountains were the main mountain range in Southeast Europe, stretching from the Adriatic Sea to the Black Sea, invented the idea of the Balkan Peninsula in 1808 (Todorova 1997). In the nineteenth century, the regions of Europe that were then provinces of the Ottoman Empire were referred to as Rumelia or the Balkan Peninsula.

Its definition was more geopolitical than geographical, and it was furthered by the early twentieth-century establishment of the Kingdom of Yugoslavia. Since the technical definition of a peninsula differs from the concept of the Balkan Peninsula's natural limits, contemporary geographers reject the notion of a Balkan Peninsula, while historical researchers typically discuss the Balkans as a region. Due to several conflicting definitions, the boundaries of the Balkans are up for debate. Regarding the region's constituent parts, there is no unanimous consensus.

According to most definitions, the phrase includes all of the following: North Macedonia, Albania, Kosovo, Montenegro, Bulgaria, Greece, Bosnia and Herzegovina, and a sizable portion of Croatia and Serbia. The phrase is occasionally used to refer to Romania and the southern regions of Slovenia.

Italy is usually omitted, even though according to certain definitions it has a minor portion of its land on the Peninsula.

The economies of Albania, Bosnia and Herzegovina, Kosovo, Montenegro, the North Macedonian Republic, and Serbia are together referred to as the Western Balkans, which comprise the Balkan area, all looking toward EU integration,

Table 1 Western Balkan countries general data

	Albania	Bosnia and Herzegovina	Kosovo	Montenegro	North Macedonia	Serbia
Population	2,862,427	3,502,550	1,795,666	622,182	2,077,132	6,963,764
Area km^2	28,749	51,197	10,908	13,812	25,713	77,474
Density 100/km^2	100	69	159	45	81	91
Water area (%)	0.047	0.0002	0.01	0.0261	0.0109	0.0013
GDP nominal bln	$15.418	$20.106	$8.402	$5.424	$12.672	$55.437
GDP PPP, bln	$38.305	$47.590	$20.912	$11.940	$32.638	$122.740
GDP/head nominal	$5,373	$5,742	$4,649	$8,704	$6,096	$7,992
GDP/head	$13,327	$13,583	$11,664	$19,172	$15,715	$17,552
Gini Index	29.0	33.0	29.0	36.7	31.9	35.6
HDI	0.791	0.769	0.739	0.816	0.759	0.799
IHDI	0.705	0.658	N/A	0.746	0.660	0.710

Data of this table are drawn from the World Bank Western Balkans report (2023)

specifically discussing with the EU standards, cyber security, and ISO standards issues included (Table 1).

2.2 Digitalization

Digitalization is the process of individuals, businesses, and organizations using digital tools to improve their work speed, efficiency, and value. Digitalization is particularly relevant for businesses, as it can help save time and create new growth opportunities.

With internet users worldwide expanding to 5.3 billion in 2024, the digital world is now the largest it has ever been. And as we accommodate the technological practices and tools that make up daily life, businesses are at the forefront of keeping pace with the latest developments in the digital world.

While digitalization has been applied in numerous organizations since the technological advances of the 1990s and 2000s, its potential uses and benefits still haven't been fully exhausted.

Digitalization uses digital technologies to change a business's internal processes and provide new revenue- and value-producing opportunities. It involves digitizing existing analog information into digital formats and certain processes that companies can perform better using the latest technology and tools.

Digitalization doesn't just mean the absence of paper and the use of computers to log data—a tiny part of the digitalization process. In its full scope, digitalization encompasses a whole network of processes that use digital data to streamline tasks like invoicing, accounting, and inventory management with technology.

Utilizing digitalization is a must for any business operating in today's competitive environment, especially to save time and avoid unnecessary expenditures. Digitalizing can provide many benefits, whether you're switching from paper bills to e-invoices or overhauling your entire sales operation for online optimization.

Digitalization. Digitalization is the process of moving existing processes into digital technologies. Digitalization can control costs, increase efficiency, and create more avenues for revenue generation. An example of digitalization is implementing digital order fulfillment operations using technological tools and software.

Digital transformation involves more comprehensive changes than digitization and digitalization. It implies the integration of digital technology into all areas of a business. Digital transformation alters business models, operational processes, and business cultures to take advantage of digitalization. An example of digital transformation is cloud-based tools to help employees work remotely. There are several ways businesses use digitalization.

- Digitalization of products and services
- Digitalization of internal processes
- Digitalization of customer interactions
- Digitalization of the supply chain

2.3 Digital Development Level

The Digital Development Level is an indicator which helps comparing countries/economies on the issue of development of digitalization, and is calculated according to the E-Government Development Index (EGDI) and Networked Readiness Index (NRI). The DDL is the average percentage the country received from the maximum value of both indexes (NCSI 2023) (Table 2).

Table 2 How digital development level is measured

Source NCSI (2023)

2.4 Trends on Digital Development

Currently the awareness of the environmental impacts of digitalization increased. Technology is a major enabler of the green transition, acting as an accelerator in the circular economy, among other things, but it also increasingly contributes to greenhouse gas emissions. Internet traffic is constantly growing, and a significant portion of network usage is still powered by fossil energy. It's important, professionals, whether they contribute, are aware of the sustainability challenges associated with digital services and take this into account in their work.

Main trends for the near future are related to

1. Sustainability
2. Artificial intelligence usage in digital development
3. Data and analytics practicing in the eye of change
4. Improve security posture (Hion 2023).

2.5 ISO 27001 and Cyber Security

These days, privacy protection, cyber security, and IT security are essential for businesses and organizations. In the current market, businesses want to give their clients trust and show that they are dedicated to protecting the security of the data they handle. To do this, a firm might gain a competitive edge by obtaining certification of an ISO standard or security regulation, which attests to the proper management of security needs in information processing operations. Controls are unquestionably necessary to ensure the security of devices, communication networks, and information assets given the increasing number of security events and assaults on information and IT systems that enterprises are facing. The idea of cyber security was born out of this requirement. The goal of these assaults is to get access to access, modify, or destroy sensitive information belonging to organizations.

Adopting strong cyber security measures is a difficult task as hackers are always coming up with new ways to launch attacks thanks to the abundance of tools and technology available. Nonetheless, there is a method for putting data and information security measures into place that partly simplifies and normalizes the process of putting these IT security measures into place. These are the information security and cyber security-related ISO standards and rules. The International Standards Organization creates and disseminates ISO standards (ISO). These days, these standards are a vital component of firms' compliance programs, giving them reputation and recognition across borders. Organizations can gain a competitive edge over their rivals by implementing ISO standards because these certified standards are regularly reviewed and audited to ensure compliance, which significantly enhances the organization's reputation with stakeholders like shareholders and clients. ISO standards are categorized into families and given sequential numbers that correspond to the areas they cover. This allows standards related to comparable problems to be grouped. The procedures, rules, guidelines, skill-building, and other aspects of the field they address—security, continuity, quality, etc.—are the focus of these standards and regulations.

The ISO 27000 series of standards is one of them. This group of information security standards lays out the conditions and recommendations for putting in place an Information Security Management System (ISMS) to handle information security inside a company. The primary standard in this group is ISO 27001, which serves as the series benchmark. The criteria for creating, implementing, maintaining, and continuously improving an ISMS are outlined in this standard. Managing cyber risks might appear challenging or even unachievable given the surge in cybercrime and the ongoing emergence of new threats. Organizations may become more risk-aware and proactively detect and fix vulnerabilities with the support of ISO/IEC 27001, which advocates the information security holistic approach—vetting of people, policy, technology, etc. An operational excellence, cyber-resilience, and risk management tool is an information security management system that is put into place by this standard.

Cyber Security as a practice of defense

The process of protecting networks, computers, servers, mobile devices, electronic systems, and data against hostile intrusions is known as cyber security. It is often referred to as electronic information security or information technology security. The word may be categorized into a few basic categories and is used in a range of situations, including business and mobile computing.

- Safety of networks.
- Program security.
- Data protection.
- Operative safety.
- Continuity of operations and disaster recovery.
- End-user training.

The scale of the cyber threat

There is an increasing number of data breaches every year, and the global cyber threat is still evolving at a rapid rate. According to a risk-based security analysis, in only the first nine months of 2019, data breaches exposed an astounding 7.9 billion records. The amount of records revealed over the same period in 2018 is less than half (112%) of this statistic.

Most breaches occurred in the medical, retail, and public sectors, and the majority of the instances were caused by malevolent criminals. Because they gather financial and medical data, some of these industries are particularly attractive to cybercriminals; nonetheless, any company that uses a network might become the subject of consumer data breaches, corporate espionage, or customer assaults.

Global investment in cyber security solutions is inevitably rising as the danger posed by cyberspace is expected to continue growing in scope. According to Gartner, investment in cyber security will top $260 billion worldwide by 2026 and reach $188.3 billion in 2023. In response to the growing cyber danger, governments everywhere have released guidelines meant to assist businesses in putting into place efficient cyber security procedures.

The National Institute of Standards and Technology (NIST) in the United States has developed a framework for cyber security. To prevent malicious code from spreading and facilitate early discovery, the architecture suggests ongoing, real-time monitoring of all electronic resources.

Cyber security indicators
Cyber security indicators are managed in three main groups:

- General cyber security indicators
- Baseline cyber security indicators
- Incident and crisis management indicators

General cyber security indicators are listed below:

1. Cyber security policy development
2. Cyber threat analysis and information
3. Education and professional development
4. Contribution to global cyber security

Baseline cyber security indicators are listed below:

5. Protection of digital services
6. Protection of essential services
7. E-identification and trust services
8. Protection of personal data

Incident and crisis management indicators are listed below:

9. Cyber incidents response
10. Cyber crisis management
11. Fight against cybercrime
12. Military cyber operations

3 Research Framework, the Purpose of the Case Study

The context for this research has been Western Balkan nations' ISO 27001 applications and digital development challenges.

RQ1: Is there a connection between the ISO 27001 certifications index and the digital development level?

Based on this, two theories have been developed:

1. Ho: The ISO 27001 Standard Certifications Index and the Cyber Security Index are unrelated.
2. H1: The ISO 27001 Standard Certifications Index and the Cyber Security Index are related.

> ... taking into account that there has been little research on the relationships between the ISO 27001 Standard Certifications Index and the Digital Development Level, as well as the fact that theoretical approaches to these relationships have been developed but that there are no quantitative, statistical, or algebraic arguments about these relationships.

4 Methodology

In particular, prior empirical research has not explained how the Digital Development Level, ISO 27001 Standard Certifications Index, and quality management influence and connect, despite the importance of these factors being acknowledged in the business and entrepreneurship ecosystem. The few serious theoretical studies that have demonstrated a strong correlation between the DDL and ISO 27001 Standard Certifications Index do not use numerical, statistical, or algebraic studies. As a result, a hypothesis that is backed by research and analysis is required. One in-depth case study technique should be used in an exploratory manner to gain a thorough grasp of phenomena and enable a deeper examination of theoretical structures.

The ISO 27001 standard certifications index and the Digital Development Level (Y) were analyzed using regression. The ISO 27001 Standard Certifications Index was created by the author of this article by dividing the total number of firms in each Western Balkan nation by the number of ISO 27001 Standard Certificates issued. The DDL was derived from the NSCI Report 2022.

4.1 Data Collection

- Data for the digital development level has been gathered from the NCSI Report 2022 (NCSI 2023), an annual ranking of countries by their capacity for, and success in cyber security.
- Data for ISO 27001 standard certificates has been gathered from the ISO 2022 survey (ISO Survey 2022).
- Data for several businesses in Western Balkan countries drawn from national institutes of statistics of Western Balkan countries, and HitHorizon.

4.2 Data Analysis

A regressive analysis (inferential statistics) between the Digital Development Level and ISO 27001 standard certificates Index for 5 Western Balkan Countries (Albania, Bosnia and Herzegovina, Montenegro, North Republic of Macedonia, and Serbia) was performed (Fig. 1; Tables 3, 4, 5).

A descriptive statistic about Digital Development Level has been undertaken to analyze and compare Western Balkan countries on the issue. Results in the Tables 6, 7, 8, 9, and 10.

a. **Implications for Theory and Practice**
   - Regarding the theory, a new avenue for investigation into the relationship between the ISO 27001 Standard Certificates Index and the Digital Development Level has been made possible by the research's conclusive findings. This research views these two indices as instruments for enhancing a business's competitive advantage as well as that of a nation.

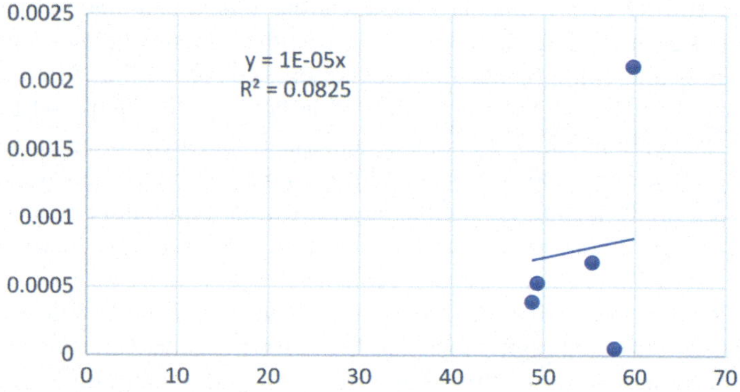

Fig. 1 Correlation between ISDO 27001 Standard Certificates Index (X) and Cyber Security Index (Y)

Table 3 Digital development level (NCSI Report 2023)

Country	DDL	ISO 27001 Index
Serbia	59.81	0.002121
Montenegro	57.79	5.04E-05
North Macedonia	55.36	0.000682
Bosnia and Herzegovina	49.31	0.000531
Albania	48.74	0.000391

Table 4 Number of ISO 27001 Standards Certificates, number of businesses per WB country, and ISO 27001 Standard Certificates Index

No	Country	Number certificates 2022	Variation in number	Variation %	No businesses	ISO 27001 Standard Certificates Index
1	Albania	49	11	29	125,222	0.0003913050
2	Bosnia and Herzegovina	52	5	11	98,000	0.0005306122
3	N. R. Macedonia	48	3	7	70,424	0.0006815858
4	Montenegro	2	0	0	39,682	0.0000504007
5	Serbia	435	8	2	205,139	0.0021205134

b. **Limitations and Further Research**
Numerous data on the ISO 27001 Standard Certificates for 2022 and the Digital Development Level have been used in this research, employing a straightforward regression approach to examine the relationship between the two indexes. The statistical analysis results lead to the conclusion of a strong correlation between the two indices, providing a clear and accessible understanding of the relationship, emphasizing the strength of the correlation as a key takeaway from the analysis, however, it is essential to acknowledge a potential limitation regarding the sample size and the temporal horizon of the data. To enhance the future study's validity and applicability, consideration could be given to expanding the sample size by incorporating data from multiple years instead of solely relying on the information from 2022. This would provide a more comprehensive understanding of the trends and relationships over time. To confirm whether these relationships hold for other eras, more investigation is required.

5 Conclusions and Recommendations

- A favorable attitude toward digitalization, cyber security concerns and ISO standards is necessary to achieve a competitive advantage.
- Taking a broader view, this study expands on the general knowledge of digitalization, cyber security and ISO 27001 standard certification, as well as the relationships between them. It suggests that future research should concentrate on developing and validating the suggested framework and look into the issue in more contexts and settings.
- A very first time ever regressive study confirmed the strong theoretical relationship between the Digital Development Level and the ISO 27001 Standard Certification Index for Western Balkan Countries.
- The primary recommendation is that, in response to the path of EU integration, businesses should apply ISO standards more broadly and the ISO 27000 family of standards in particular to become more safe, secure, and guaranteed against

Table 5 Regression between Digital Development Index (Y) and ISO 27001 Standard Certificates Index (X) table

Summary output	
Regression statistics	
Multiple R	0.752822
R square	0.56674
Adjusted R square	0.31674
Standard error	40.02297
Observations	5

ANOVA

	df	SS	MS	F	Significance F
Regression	1	8381.36	8381.36	5.232339	0.106207
Residual	4	6407.353	1601.838		
Total	5	14,788.71			

	Coefficients	Standard error	t Stat ara>	*P*-value	Lower 95%	Upper 95%	Lower 95.0%	Upper 95.0%
Intercept	0	#N/A	#N/A	#N/A	#N/A	#N/A	#N/A	#N/A
ISO 27001 Index	39,402.79	17,225.78	2.287431	0.084095	−8423.66	87,229.23	−8423.66	87,229.23

With these results, we have verified that there is a connection between the Digital Development Level and the ISO 27001 Standard Certificates Index (Hypothesis 1). $R^2 = 0.56674 > 0.50$ (50%)

Table 6 Descriptive statistics for WB countries

Digital development level	
Mean	51.21342
Standard error	1.474779
Median	51.46
Mode	#N/A
Standard deviation	18.36086
Sample variance	337.121
Kurtosis	−0.99792
Skewness	−0.06873
Range	71.63
Minimum	11.3
Maximum	82.93
Sum	7938.08
Count	155
Largest (1)	82.93
Smallest (1)	11.3
Confidence level (95.0%)	2.913409

Table 7 Western Balkans digital development level rank

Rank	Country	DDI
1	Largest	82.93000
56	Serbia	59.81000
62	Montenegro	57.79000
69	North Macedonia	55.36000
	Median	51.46000
	Mean	51.21342
81	Bosnia and Herzegovina	49.31000
84	Albania	48.74000
176	Smallest	11.00000

cyber-attacks, and filling and closing digitalization gap. This will help them remain dependable to their clients and support, improve, and protect business activities, processes, and procedures, giving the Western Balkan economies a competitive advantage.

Table 8 WB ISO 27001 certificates issued and how the ISO 27001 index has been calculated

No	Country	Number certificates 2022	Variation in number	Variation %	No businesses	ISO 27001 Standard Certificates Index
1	Albania	49	11	29	125,222	0.0003913050
2	Bosnia and Herzegovina	52	5	11	98,000	0.0005306122
3	N. R. Macedonia	48	3	7	70,424	0.0006815858
4	Montenegro	2	0	0	39,682	0.0000504007
5	Serbia	435	8	2	205,139	0.0021205134

Table 9 A comparative analysis of WB6 for NCSI, DDI, and ISO 27001 certificates index

Country	DDI	Country	ISO Index
Serbia	59.81	Serbia	0.002121
Montenegro	57.79	NRM	0.000682
NRM	55.36	B	0.000531
B	49.31	Albania	0.000391
Albania	48.74	Montenegro	5.04E-05

Table 10 Cyber security indicators prepared by the author of this article

1. Cyber security policy development

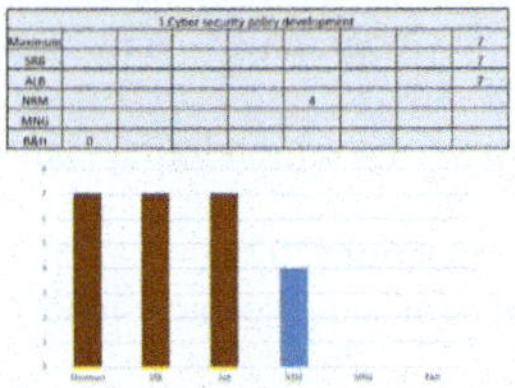

2. Cyber threat analysis and information

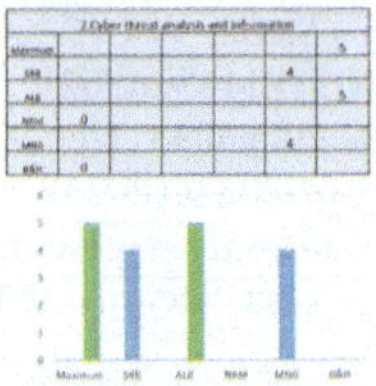

3. Education and professional development

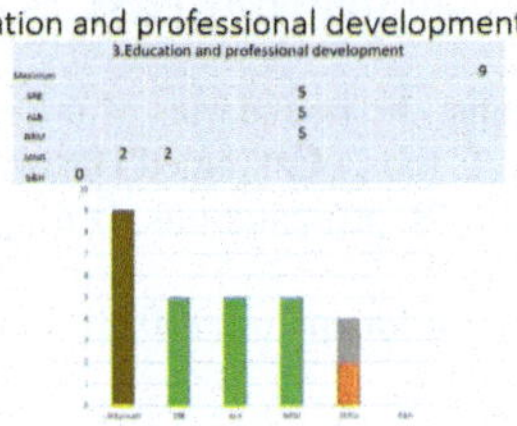

4. Contribution to global cyber security

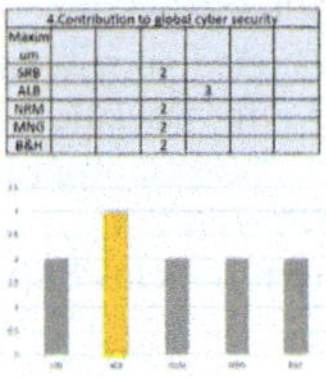

5. Protection of digital services

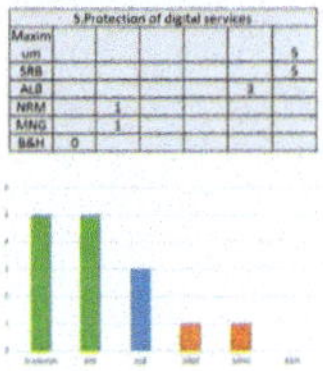

6. Protection of essential services

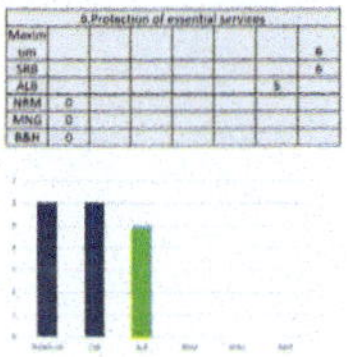

7. E-identification and trust services

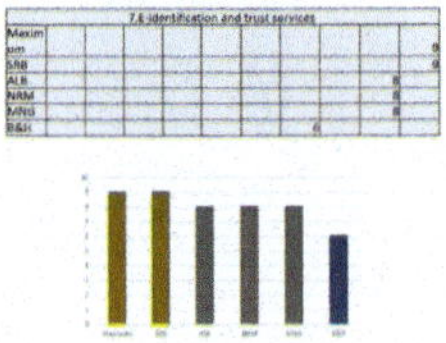

8. Protection of personal data

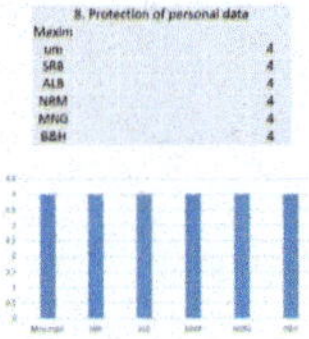

9. Cyber incidents response

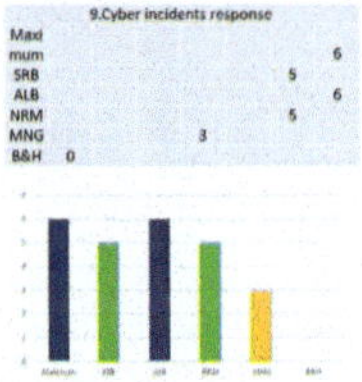

10. Cyber crisis management

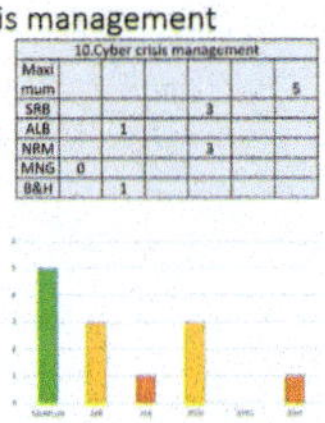

11. Fight against cybercrime

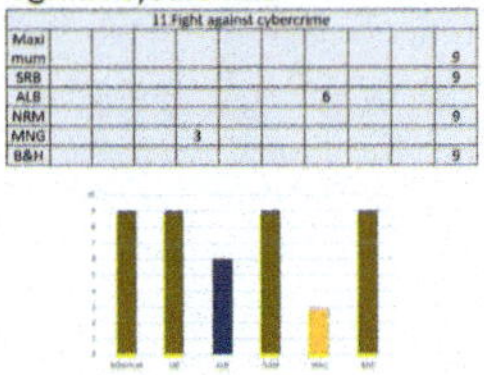

12. Military cyber operations

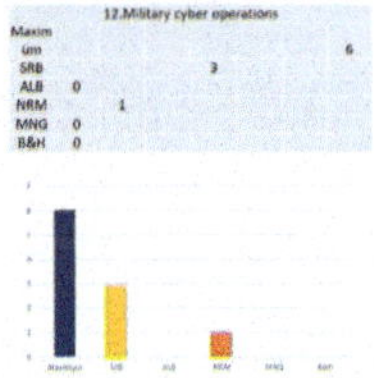

References

Gray, C.S., Sloan, G.: Geopolitics, Geography and Strategy. Routledge (2014). ISBN 9781135265021

Hion: Digital Development Trends of 2024. https://www.hiondigital.com/en/blog/digital-development-trends-of-2024/ (2023)

ISO: ISO survey. https://www.iso.org/the-iso-survey.html (2022)

NCSI: National Cyber Security Index 2022. https://ncsi.ega.ee/ncsi-index/ (2023). Retrieved 16 Oct 2023

Todorova, M.N.: Imagining the Balkans, p. 27. Oxford University Press, Inc., New York (1997). ISBN 9780195087512

Vezenkov, A.: Entangled geographies of the Balkans: the boundaries of the region and the limits of the discipline. In: Daskalov, R.D., Marinov, T. (eds.) Entangled Histories of the Balkans—Volume Four: Concepts, Approaches, and (Self-) Representations. Brill, pp. 115–256 (2017). ISBN 978-90-04-33782-4

World Bank: Western Balkans Spring report. www.worldbank.org (2023). Retrieved 16 Oct 2023

Computer Architecture Optimization Techniques for AI Workloads

Shefqet Meda and Ervin Domazet

Abstract

The recent progress in Machine Learning (Géron in Hands-on machine learning with Scikit-Learn, Keras, and TensorFlow. O'Reilly Media, 2022) and particularly Deep Learning (Goodfellow et al. in Deep learning. Cambridge, Massachusetts, MIT, London, 2016) models exposed the limitations of traditional computer architectures. Modern algorithms demonstrate highly increased computational demands and data requirements that most existing architectures cannot handle efficiently. These demands result in training speed, inference latency, and power consumption bottlenecks, which is why advanced methods of computer architecture optimization are required to enable the development of ML/DL-dedicated efficient hardware platforms (I. o. E. a. E. Engineers in 25th IEEE international symposium on high performance computer architecture, p. 734, 2019). The optimization of computer architecture for applications of ML/DL becomes critical, due to the tremendous demand for efficient execution of complex computations by Neural Networks (Goodfellow et al. in Deep learning. Cambridge, Massachusetts, MIT, London, 2016). This paper reviewed the numerous approaches and methods utilized to optimize computer architecture for ML/DL workloads. The following sections contain substantial discussion concerning the hardware-level optimizations, enhancements of traditional software frameworks and their unique versions, and innovative explorations of architectures. In particular, we considered hardware including specialized accelerators, which can improve the performance and efficiency of a computation

S. Meda (✉)
Canadian Institute of Technology, Tirana, Albania
e-mail: Shefqet.meda@cit.edu.al; meda.shefqet@ibu.edu.mk

S. Meda · E. Domazet
International Balkan University, Skopje, North Macedonia
e-mail: Ervin.domazet@ibu.edu.mk

K. Dhoska and E. Spaho (eds.), *Bridging Horizons in Artificial Intelligence, Robotics, Cybersecurity, Smart Cities, and Digital Economy*, Sustainable Economy and Ecotechnology, https://doi.org/10.1007/978-3-031-72029-1_25

system using various techniques, specifically describing accelerators like CPUs (multicore) (Hennessy and Patterson in Computer architecture—a quantitative approach. Morgan Kaufman, 2017), GPUs (Wen-Mei in GPU computing gems, Emerald Edition, 2015), and TPUs (Jouppi et al. in In-datacenter performance analysis of a tensor processing unit, 2017), parallelism in multicore architectures, data movement in hardware systems, especially techniques such as caching and sparsity, compression, and quantization, other special techniques and configurations, such as using specialized data formats, and measurement sparsity. Moreover, this paper provided a comprehensive analysis of current trends in Software Frameworks, Data Movement optimization strategies (Olson et al. in Modeling data movement performance on heterogeneous architectures. Waltham, MA, USA, 2021), Sparsity, Quantization, and Compression methods, using ML for Architecture exploration, and, DVFS (Hennessy and Patterson in Computer architecture—a quantitative approach. Morgan Kaufman, 2017), which provides strategies for maximizing hardware utilization and power consumption. The objective of implementing these optimization techniques is to largely minimize the current gap between the computational needs of ML/DL algorithms and the current hardware's capability. This will lead to significant improvements in training times, enable real-time inference for various applications, and ultimately unlock the full potential of cutting-edge machine learning algorithms.

Keywords

Computer architecture optimization • Machine learning • Deep learning • Parallelism • Sparsity • Data movement optimization • Quantization • Compression • Software framework optimization • DVFS • TPU • CPU • GPU • TensorFlow • Pytorch

1 Introduction

Machine Learning/Deep Learning are among the rapidly growing and most transformative fields of science. Image recognition, natural language processing, health care, autonomous vehicles, and several other fields of applications have seen a massive progress in recent years. As the scale and complexity of ML/DL applications continue to grow, there is an increasing demand for efficient and scalable computing infrastructure to support these workloads. However, these advancements are often bottlenecked by the limitations of traditional computer architectures. These architectures, designed for general-purpose computing, struggle to efficiently handle the massive computational demands of modern ML/DL algorithms. The inefficiency manifests in several ways:

- **Training takes too long**. Entire teams are spending days or weeks training models to find better hyperparameters and experiments in fine print.

- **High Inference Latency** (Hennessy 2017). Deploying these models on resource-constrained devices, like mobile phones and embedded systems, is often impractical due to the high computational requirements for real-time inference.
- **Significant Power Consumption.** Large-scale training of complex models incurs substantial energy costs, raising concerns about sustainability and cost-effectiveness. By optimizing the underlying hardware and software infrastructure, we consider the following objectives to be achieved:
  - **Accelerate and optimize training speeds**. This involves reducing the time required to train complex ML/DL models, which speeds up the development process and allows for more iterative experimentation.
  - **We will also aim to develop real-time inference capabilities**. This will allow us to deploy ML/DL models to inferences close to the edge devices through faster, less efficient hardware, thereby enabling a greater range of AI applications on-devices.
  - **Finally, we will optimize the power efficiency of ML/DL models**. This last objective seeks to reduce power consumption needed for training and inference, which will make it more feasible and cost-effective to deploy ML/DL models.

The optimization of computer architecture encompasses a wide range of topics, except above mentioned, including hardware design, software frameworks, and system-level optimizations, all of which are critical for achieving optimal performance in ML/DL workloads. Specialized hardware accelerators (Sze et al. 2017) such as CPUs (multicore), GPU, TPU, and custom ASICs have shown more efficiency on ML/DL computation. These accelerators are specifically designed to accelerate the matrix and tensor operations that are prevalent in ML/DL applications and further neural network computations, enabling faster model training and inference. They have made it feasible for model training to be faster.

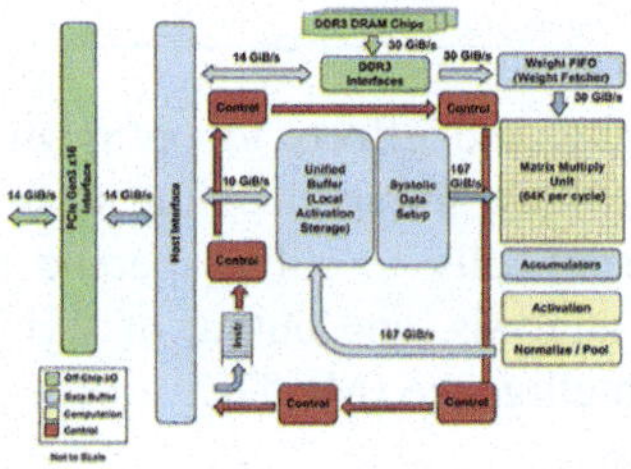

TensorFlow processing unit architecture. *Source* Google (2017)

In addition to hardware optimizations, software frameworks and runtime systems also contribute significantly to optimizing computer architecture for ML/DL applications. Frameworks such as TensorFlow (2024), PyTorch (Migacz 2024), and Apache MXNet (2024) offer higher-level abstractions and API to build, train, and deploy neural network models (Sze et al. 2017), enabling faster model training and inference.

Moreover, machine learning techniques for architecture exploration have shown promise in optimizing computer architecture for ML/DL applications. Specifically, the techniques using supervised learning, reinforcement learning, and Bayesian optimization (Lee and Pourmohamad 2021) have proved very effective in exploring the design space and discovering the optimal configuration to optimize performance and power efficiency. In conclusion, advanced optimization of computer architecture for ML/DL is a multidisciplinary field that mixes expertise in hardware, software, and machine learning.

Hardware optimization: Accelerators—CPU, GPU, TPU

Hardware optimization is a technique used to enhance the performance and efficiency of computing systems (Clark et al. 2021). It involves increasing performance of accelerators like GPUs, TPUs and FPGA, considering designing to handle heavy workloads more efficiently. Traditional ones have their limitations include: CPUs not being efficient in support for upsized vectorized instructions such as AVX, along with efficient memory access to handle large datasets and model parameters. Limitations in parallelism is slowing down performance (Hennessy 2017), in some of the highly parallel tasks like Matrix multiplications. Most of them are not fully optimized for some ML/DL operations and the high-power consumption among others.

In the below Pseudocode 1, an example of how a TPU might function to achieve matrix multiplication is presented.

```
// Pseudocode just for example
void matrixMultiply(float* X, float* Y, float* Z) {
  // Load matrices X and Y into on-chip memory
  MXE_Load(X, Y);

  // Perform matrix multiplication using MXE hardware
  MXE_Multiply();

  // Store result matrix Z in on-chip memory
  MXE_Store(Z);
}
```

There are many strategies and efforts to optimize hardware. After we have carefully studied them, we have grouped of the most optimal ones that can be considered applied not specifically to solve specific problems and tasks but to cope with the processing, analysis, and output of significant amounts of data for applications of Artificial Intelligence (ML/DL):

- **Parallelism** (Hennessy 2017). This involves utilizing the parallel processing capabilities of the hardware architecture. It enables multiple processing units within the core, CPU to compute multiple independent instructions; other techniques include SIMD and distributed computing.
- **Vectorization** (Wen-Mei 2015). It considers vector processing units such as SSE (Streaming SIMD Extensions) and AVX in CPUs and GPUs to perform operations on more than one data simultaneously. It sends one operation to the core simultaneously sent to multiple data elements.

- **Cache Optimization** (Spinczyk and Borghorst 2019). Maximizing cache utilization to reduce memory access latency and improve data locality. Techniques include cache blocking, prefetching, and data alignment.
- **Instruction-Level Optimizations**. Optimizing code to make the best use of instruction pipelines, branch prediction, and instruction scheduling.
- **Memory Optimization** (Kim 2022). This reduces memory consumption and accesses, to reduce stalls and improve overall performance.
- **Data Layout Optimization**. Organizing data structures and memory layouts to optimize for cache coherence, spatial locality, and access patterns.
- **Algorithmic Optimization**. Choosing or designing algorithms that are well-suited for the underlying hardware architecture, minimizing unnecessary computations and memory accesses.
- **Power Efficiency Build**. It is especially critical to mobile devices and some limited energy environments, particularly when predicting systems in real-time.
- **Profiling and Benchmarking**. This process allows optimizing critical codes by observing or finding the bottlenecks of performance.
- **Hardware-Specific Optimization**. Specific features such as CPU, GPU, or TPU architectures, specialized instruction sets, tensor cores, or memory hierarchies.

Furthermore, the below flowchart explains hardware specialization. While traditional CPUs serve as effective general-purpose chips, TPUs or AI accelerators have dedicated characteristics based on the specific requirements for DL.

- **Compiler Optimizations**: Making use of compiler optimizations to automatically transform code for better performance, including loop unrolling, function in lining, and auto-vectorization.
- **FPGA and ASIC Optimization**: For specialized hardware like FPGAs and ASICs, optimization involves hardware design at a lower level, focusing on logic utilization, routing efficiency, and clock frequency (Wijtvliet et al. 2019).

We can recommend another approach like algorithms specifically for FPGA execution, considering the parallel and pipelined nature of FPGAs. An example is conclusions of a group of researchers for accelerating the learning process of CRBM (Restricted Boltzmann Machine) with FPGAs and OpenCL, and conducting an extensive scalability study for different model sizes and system configurations (Cadenelli et al. 2020).

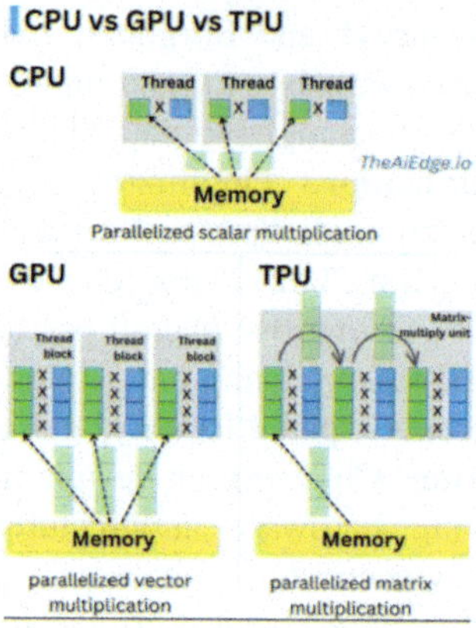

CPU, GPU, or TPU architectures. *Source* eitc.org (2023)

- **Quantization and Pruning**: Additional optimizations like for neural network accelerators such as TPUs include Quantization (Jacob et al. 2018) and pruning accelerate computation by reducing the precision of weights and activations of weights and removing redundant connections of the neural network. Give developers exercises to apply these enhancements to hardware accelerators across all computing platforms.

2 Parallelism—Multicore Architectures

Utilizing parallelism is a fundamental consideration of computer architecture optimization for ML/DL applications, especially when it comes to multicore architectures. Enabling computers to utilize parallelism is critical for expediting the compute-intensive nature of ML/DL workloads and realizing the desired performance and efficiency levels (Clark et al. 2021). There are several restrictions connected with traditional computers when it comes to parallelism: they were not designed to capitalize on parallelism in the context of ML/DL workloads by parallelizing matrix operations, convolutions, and other operations conducted by neural networks across various CPU cores.

Such multicore architectures (Hennessy 2017), enable the following benefits, over traditional one's, for ML/DL on CPUs:

- faster training and inference,
- added scalability,
- parallel processing power.

Multicore CPUs put more cores on the same chip, essentially allowing them to run more instructions at once. This leads to better performance for ML/DL tasks that can be subdivided into separate computations.

- **Core specialization**: Modern cores are rich in dedicated units, such as vector processing units. These types of units can run the same operation on multiple data elements simultaneously. This is very beneficial for ML/DL workloads that involve a lot of vector or matrix computations.
- **Multiple cores that can run in parallel**: Multiple cores on a CPU can operate at the same time. When subtasks of certain instructions can be run at the same time, multicore CPUs will show an excellent parallel processing speedup. For instance, ML/DL heavily relies on linear algebra and matrix multiplication, which can be trivially partitioned per sub-step. To implement CPU, all the processors split the workload per step and run multiple cores to speed up the whole calculation.
- **Thread-level parallelism**: Modern CPUs support additional execution threads along cores, so-called hyper-threading or simultaneous multi-threading further enhances parallelism because it keeps the cores busy even while waiting for data or cheesing on another thread. Below is an example of the application of quantization according to Round Robin in the multicore architecture and a part of the corresponding code. If the outputs for two different algorithms are compared, as we did, the superiority of quantization and multicore in execution is clearly observed.
- **Software Optimization for Multicore Exploitation**: Be able to save the workload of multiple cores properly, the framework has to be appropriately designed. Software frameworks like TensorFlow or PyTorch use python Threading to subdivide the load of data on the CPUs. They chop small slivers of data, and the computation is about a model into smaller clones of the data and add them to separate cores (sample above).

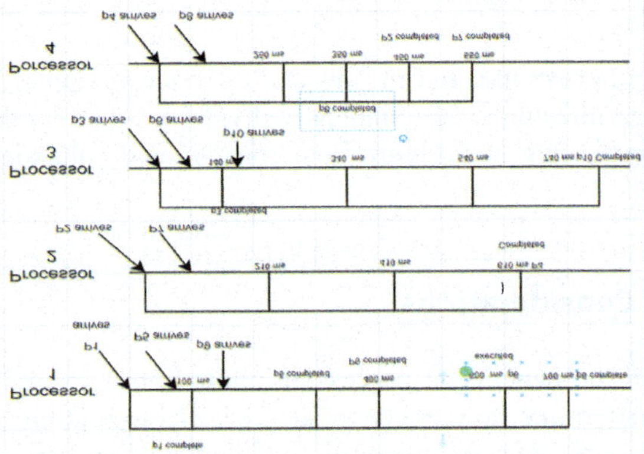

```
int main() {
  int n, quantum;
  cout << "Enter The Number of Processes: ";
  cin >> n;
  cout << "Enter The Time Quantum: ";
  cin >> quantum;

  Process processes[n];
  cout << "Enter process details:\n";
  for (int i = 0; i < n; i++) {
    cout << "Process " << i + 1 << ":\n";
    processes[i].id = i + 1;
    cout << "  Arrival Time: ";
    cin >> processes[i].arrivalTime;
    cout << "  Burst Time: ";
    cin >> processes[i].burstTime;
  }

  calculateTimes(processes, n, quantum);
  calculateTurnaroundTime(processes, n);
  calculateWaitingTime(processes, n);

  cout << "\nRound Robin Scheduling Results:\n";
  printTable(processes, n);

  return 0;
}
```

Multicore architecture running threads and code

Multicore architecture running threads and code

- **Libraries such as OpenMP or Intel Threading Building Blocks** (Stotzer et al. 2017): Provide tools for programmers to write code that takes advantage of multiple cores.
- **Data and Model Parallelism**: Both data and model parallelism techniques are enabled by multi-core architectures. Splits the training data into smaller batches and distributes them across multiple cores. Each core processes its assigned data independently, accelerating computations, especially for large datasets.
- Model parallelism divides the machine learning or deep learning model across cores. For extremely large and difficult models which would not fit on a single core, this is necessary. However, it must be done with caution, since the various cores must collaborate to ensure the model changes are coherent.
- Simple code in Python:

Pseudocode 2: For data parallel training with model parallelism on two GPUs

```
# Pseudocode for data parallel training with model parallelism on two GPUs
def train_step(data_batch):
# Split data batch across GPUs
data_batch_gpu1, data_batch_gpu2 = split_data_batch(data_batch)
# Forward pass on GPU 1 (first half of the model)
with tf.device("/gpu:0"):
activations1 = model_part1(data_batch_gpu1)
# Send activations from GPU 1 to GPU 2
activations1 = send_data(activations1, "/gpu:1")
# Forward pass on GPU 2 (second half of the model)
with tf.device("/gpu:1"):
output = model_part2(activations1)
# Backpropagation and gradient accumulation (simplified)
return output
# Training loop
for epoch in range(num_epochs):
for data_batch in training_data:
output = train_step(data_batch)
# ... update model weights
```

- Software Optimization Frameworks like TensorFlow and PyTorch are optimized in deployed on multicore architectures. Automatically divide data and models across cores for parallel processing. cores must communicate and collaborate on model changes.

2.1 Challenges and Considerations

While it is easy to understand the benefits of processing multiple tasks simultaneously, a variety of factors determine how effective such parallel execution can be. While cores can simultaneously perform in parallel, communication between them may be the bottleneck if they have to share data and any updates. Techniques such as gradient accumulation and efficient communication protocols are critical. Additionally, optimized memory access patterns are crucial to success.

Cores must be allowed to quickly and efficiently access whichever data they require from memory, and if this is not allowed or limited by memory bandwidth, the numbers of parallel operations cannot be optimized for faster performance. Finally, not all parts of an ML/DL application can be perfectly parallelized. Amdahl's law determines the maximum speedup due to parallelism. Memory bandwidth influences cannot be avoided; rather than intensified, these drawbacks are magnified.

Conclusion: so, we can conclude that, conventional CPU architectures are vital in achieving the parallelism in ML/DL applications. When combined with several cores and specialty hardware and software representations, they lead to quicker train and recast times. Optimal algorithms and communication strategies will be key to ensure the best potential from these machines as core counts proceed to expand.

We suggest that the use of high-level languages, or otherwise modern programming languages and open source platforms would be another opportunity to enable researchers and developers to provide optimal solutions for ML/DL applications and increasing performance.

3 Data Movement Optimization—Data Prefetching, Memory Hierarchy, Caching

Traditional computer architectures can have limitations in data movement that can impact the performance and efficiency of Machine Learning (ML) and Deep Learning (DL) workloads. Here are listed some aspects of how these limitations affect ML/DL specifically:

Data Prefetching. In traditional architectures, software prefetching is used, where the processor asks for the data from memory explicitly before it is needed. This can introduce latency if the request isn't anticipated correctly or if there's concurrency for memory bandwidth. In ML/DL, algorithms often exhibit predictable access patterns, repeatedly accessing specific data elements. Inefficient prefetching can lead to situations where the needed data isn't readily available when the processing unit requires it, causing delays.

Memory Hierarchy Bottlenecks (Kim 2022). Traditional architectures utilize a hierarchy of memories with varying speeds and capacities (e.g., registers, cache, main memory, disk). Accessing data from slower levels in the hierarchy takes longer. ML/DL algorithms often work with large datasets. If the frequently accessed data resides in slower memory, retrieving it for each operation can significantly impact performance.

Another limitation is—Storage I/O Bottlenecks. Traditional architectures use mechanical hard disk drives or even slower to use this data for storing load from memory, and the storage I/O bottleneck can occur in ML/DL systems. The problem with storage I/O bottlenecks is that systems require huge data and input/outing this big data makes the system slow. There are slow read/write speeds and high latency in storage devices that can prevent ML/DL storage from efficient work, especially when loading data and preprocessing it. Therefore, *these bring us to the following limitations that bottleneck ML/DL workloads*

Increased Latency and Reduced Throughput
Delays caused by inefficient prefetching, accessing slower memory levels, and cache misses all contribute to longer execution times for training and inference. Limited data movement capabilities can restrict the amount of data processed per unit time, hindering overall performance.

What we can do to optimize and avoid these limitations in performing with ML/ DL apps includes:

Data Prefetching:

- Software Prefetching: by predicting and prefetching data into the cache before being accessed by the computation to reduce latency, the computation can be overlapped with data movement.
- Hardware Prefetching: Exploit hardware mechanisms of modern processors to automatically prefetch data into cache according to access patterns observed during runtime.
- Spatial and Temporal Prefetching: Prefetch data blocks that are spatially or temporally related to the currently accessed data, such as prefetching neighboring elements in arrays or prefetching data expected to be accessed in future iterations.

Memory Hierarchy Optimization:

- Cache Awareness (Spinczyk and Borghorst 2019): design algorithms and data structures that can maximize cache usage and take advantage of spatial and temporal locality.
- Cache blocking: Partition data into blocks that fit into the cache, enabling reuse of cache lines and reducing the frequency of cache misses.
- Cache Replacement Policies: Use cache replacement policies (e.g., LRU, LFU, Optimal) that prioritize keeping frequently accessed data in the cache to minimize cache thrashing.
- Cache Compression: compress the stored data in the caches to increase the capacity of the caches and lessen the impact of cache pollution.

Caching (Spinczyk and Borghorst 2019):

- Data Reuse: identify and leverage the opportunities to reuse intermediate results and computation to minimize data movement across memory levels.
- Result Caching: considering that computations in DNN are iterative, the results can be stored in memory to prevent repetitive computations and lower latency.

– Model Caching: the trained models and parameters can be cached in memory to reduce the load on the main memory and enhance the performance. The example illustrated below is an application of Fuzzy Logic in cache optimization by applying dynamic membership. As can be seen from the result, we have an execution time almost twice as short.

RESULT AND PERFORMANCE EVALUATION

Process ID	Arrival Time (ATi)	Burst Time (BTi)	Static Priority (PTi)	Dynamic Priority (DPi)	Deadline (D)
P1	0	5	3	0.915	35
P2	0	12	5	0.576	10
P3	0	3	1	0.869	4
P4	0	6	4	0.938	18
P5	0	10	6	0.83	23
P6	0	6	2	0.738	50
P7	0	8	9	0.9	25

Comparison Table

Algorithm	Average Waiting Time	Average Turnaround Time	Average Response Time
Priority Algorithm	25.71	32.85	25.71
IFCS	18.28	25.4	18.28
AFCS	18.12	25.4	18.01

Fuzzy Logic in cache optimization with dynamic membership

Memory Access Patterns

– Sequential Access Optimization: Arrange data structures and access patterns to optimize for sequential memory access, minimizing the overhead of random memory accesses.
  - Data Layout Optimization: Organize data in memory to improve cache locality and alignment with the underlying hardware architecture, such as using row-major or column-major layouts for matrices depending on access patterns.
  - Strided Access Optimization: Minimize the stride (distance between consecutive elements) in memory access patterns to improve cache utilization and reduce memory access latency.

Data Compression and Encoding:

– Lossless Compression: Compress data before storing it in memory or transferring it between levels of the memory hierarchy to reduce storage requirements and bandwidth usage.

– Lossy Compression: Apply the lossy compression technique to sacrifice precision for better memory bandwidth utilization through varied sparsity methods for ML/DL such as quantization and low-precision data representation.

Advantages of Data Movement Optimization:

– Decrease Training and Inference Time: By minimizing wasted time waiting for data movement, these techniques can significantly accelerate the overall training and inference processes in ML/DL applications.

– Improved Resource Utilization: Efficient data movement allows for better utilization of processing power by reducing idle time due to data access delays.

4 Sparsity (Adolf et al. 2017): Some Special Techniques, Like Using Specialized Data Formats and Hardware Units

- Sparsity is essential in the computer architecture of Machine Learning and Deep Learning applications because it refers the presence of a large number of zeros in the data being processed. Some of the most relevant limitations from several issues that the traditional computer has when performing ML/DL apps are as follows:
- Sparsity Handling: Generally, sparsity is not directly addressed in traditional computer architecture design because most algorithms and hardware units are designed to work with the dense data format efficiently without specific optimizations for sparse data.
- Data Formats: Classic computer architectures support standard numerical data format, like float-point or integer, and do not directly support sparse data representation.

Hardware units: in classic computer architecture, including CPUs and GPUs, are optimized for dense computations; they are not equipped with specialized units optimally designed for sparse data. Sparse matrix–matrix multiplication and other sparse operations are performed on these general-purpose hardware units, providing suboptimal performance due to the sparse data. Specialized techniques yield significant performance improvements because of the sparsity of the data. The following are some of the aspects of computer optimization for ML/DL applications that use sparsity.

Sparse data formats: Compressed sparse formats. These arise through the use of compressed formats such as Compressed Sparse Row (CSR) or Compressed Sparse Column (CSC) to represent sparse matrices efficiently. They only store the non-zero elements, with their indices, yielding reduced storage requirements and better cache locality.

Sparse tensor representation: The data stored in these structures will be sparse, so the storage pattern must account for this. Design specialized data structures to optimize the storage and access patterns because sparse data structures such as graphs and embeddings are seen frequently in ML/DL models.

Sparse matrix operations: Sparse matrix multiplication. It can be performed by developing optimized algorithms and units for sparse matrix multiplication, as this is fundamental to ML/DL applications. Memory access reduction and computational complexity can be achieved by explicitly identifying how to achieve modular multiplication based on sparsity. We can extend hardware accelerators to support the computation of sparse tensors as the representations and operations involved in many ML/DL applications are seen as tensors.

Sparsity-aware hardware acceleration: Sparse tensor cores. Design specialized hardware units capable of efficiently handling sparse tensor operations, exploiting sparsity to reduce computational and memory requirements. These units can perform computations only on non-zero elements, improving energy efficiency and performance for sparse ML/DL models. Sparsity-Aware Memory Systems in developing memory architectures and controllers optimized for handling sparse data access patterns, with mechanisms for efficient storage and retrieval of non-contiguous data elements in sparse tensors.

Quantization and Pruning: Quantize models or model activations for sparse representation and reduce the precision for zero and near-zero values while maintaining the accuracy of non-zero elements. Sparse quantization can considerably reduce memory bandwidth and reduce memory storage overheads required to support large-scale ML/DL models. ***What we have to consider***: Remove redundant or insignificant weights from neural network models to obtain sparse weight matrices. Sparse weight matrices can be further sparsified using special storage formats and hardware units designed specifically for sparse data representation. Moreover, sparse weight matrices will reduce the memory overhead and the computational complexity of their processing during inference.

Dynamic Sparsity: Develop algorithms and hardware mechanisms that can detect sparsity and use the acquired knowledge durably at runtime. Dynamic sparsity algorithms and methods may detect variation in sparsity and adjust the computational approach based on changes in input data, sparsity models, and other factors. Integrating specialized data formats, hardware units and using sparsity-aware algorithms in computer systems can significantly accelerate ML/DL applications, improve their energy efficiency, and enhance scalability.

5 Quantization and Compression (Jacob et al. 2018)

The increasing size and complexity of machine learning-based algorithms and models create two main challenges: it may not be possible to deploy an ML/DL model on the resource-constrained devices, and training such big models may be slow.

The number of bits in which the data are represented in the model is reduced using quantization. In a traditional setting, high precision formats like 32-bit floating point numbers are used to represent the weights and activations of a model. Techniques considered in quantization include being able to represent these values in lower precision formats such as 8-bit integers with minimal effects on the model's accuracy. Compression techniques are considered to work in tandem with quantization as quantization is focused on reduction of the redundant bits within an encoding while compression thrives from a reduction in information in the representation. The limitations of the traditional computers that affect quantization and compression in ML/DL applications are: Compression and quantization are unfamiliar concepts in classic computer architecture. In classic computer architecture, compression algorithms aim at storage and transmission of the data. The data can be files, as in Huffman compression, or progression, as seen in LZ compression techniques. Sounds and images are quantized to reduce the size before being transmitted since this approach saves bits. Data quantization and compression are techniques used in many computer ways in storage, transmission, and retrieval. These two techniques have not been adequately studied since accuracy in representation is less of concern. The critical issue is reduced bit storage while assured of minimal to no information or precision loss. Identifying this optimal precision or compression level is a challenge if the problem area or the algorithms do not dictate the amount of information retained in a unit.

Techniques in ML/DL apps

- **Quantization and Compression Algorithms** to be used to reduce the precision of model parameters and activations thereby decreasing memory usage and computational complexity
- while preserving model accuracy techniques like weight pruning, quantization, and low-rank factorization are commonly employed to compress neural network models for deployment on resource-constrained devices or for efficient distributed training.

Benefits

- Reduced memory footprint. Lower precision data representations require less storage space, enabling the training of larger models on limited hardware resources
- Faster training and inference. Reduced precision leads to faster computations during the training and inference stages
- Lower power consumption. Lower power consumption due to smaller data footprints and potentially less complex computations makes these techniques attractive for mobile and embedded devices

Challenges

- Finding the Right Balance: Determining the optimal level of quantization or compression requires careful consideration, as too much reduction in precision can significantly impact model accuracy.
- Framework and Hardware Support: Not all frameworks and hardware platforms offer the same level of support for various quantization and compression techniques.
- Potential Need for Algorithm Adjustments: Depending on the technique used, algorithms within the neural network might need adjustments to maintain accuracy with lower precision data.

6 Software Framework Optimization

Software framework optimization is a critical aspect of computer architecture optimization for machine learning (ML) and deep learning (DL) applications, facilitating efficient execution of ML/DL workloads on diverse hardware platforms while maximizing performance and scalability. ML/DL frameworks provide high-level abstractions and APIs for developing, training, and deploying neural network models, abstracting away hardware-specific details and enabling portability across different architectures. (Abadi et al. 2016). This kind of optimization serves to support the efficient execution of machine learning and deep learning workloads on any hardware platform and to maximize ML/DL task performance and out-of-the-box scalability. ML/DL frameworks are high-level abstractions and APIs that let users create, train, and deploy neural network models without worrying about underlying hardware-specific details. However, before listing what might be optimized to enhance the performance of traditional computers, let's outline what such computers imply prevents such optimizations:

- Software frameworks. In classic computer architecture, software frameworks are developed for any general-purpose computing work that may apply to many different applications. They include an operating system, a programming language and libraries.
- Limited Hardware acceleration (Grel et al. 2020). Hardware acceleration to leverage optimizations for explicit instruction sets and memory-specific optimizations is mostly not used by this framework.
- Programmer Responsibility. More of the burden of optimization code is on the program, using techniques like manual vectorization or optimization of cache sometimes. Lower the frameworks offer much lower to no leverage of hardware-specific optimizations for specific instruction set and memory hierarchy.
- Lower-Level Control. The frameworks provide greater control over memory and instruction schedule but require more effort from the programmer to optimize.

Now let's see what can be optimized under given constraints

- Optimizing ML/DL frameworks to leverage hardware-specific features and optimizations, such as specialized instructions, memory hierarchy, and parallelism capabilities. Frameworks like TensorFlow, PyTorch, Keras, and Apache MXNet incorporate backend optimizations to exploit hardware acceleration, enabling efficient execution of neural network computations on CPUs, GPUs, and specialized accelerators like Tensor Processing Units (TPUs). These optimizations include auto tuning techniques, kernel fusion, and memory layout optimizations tailored to specific hardware architectures, ensuring optimal performance across a wide range of hardware platforms. They offer High-Level Abstractions, Automatic Optimization, and Focus on Parallelism.

- Focus on minimization of overhead and improvement of efficiency in data processing and model inference pipelines. This is possible due to for example model quantization, tensor slicing, or even asynchronous execution which allows for efficient memory usage and pipelining of calculations. In both cases, latency is minimized and throughput is maximized for ML/DL workflows.
- Distributed training and inference are also considered to require minimal overhead with, for example, parameter servers or communication primitives, which allow distributed computing to be scalable and fault-tolerant. This extends to runtime systems and libraries supporting ML/DL applications and offering low-level abstractions and optimizations. Libraries such as cuDNN, or math Kernel Library MKL-DNN known for optimized implementations of common neural network operations. They utilize hardware-specific optimizations and parallelism for GPUs, and CPUs that can be used during ML and DL computational tasks to improve the efficiency and performance of neural networks.

7 Dynamic Voltage and Frequency Scaling (DVFS)

DVFS is a power management technique that adjusts the voltage and frequency of a processor dynamically based on workload requirements to optimize energy efficiency while maintaining performance.

Let's see limitation of Dynamic Voltage and Frequency Scaling (DVFS) in traditional computer architecture before optimization.

- DVFS is commonly implemented at the operating system level or through hardware features provided by the CPU. The operating system dynamically adjusts the CPU voltage and frequency based on workload characteristics, system power constraints, and thermal considerations. Needs to reduce power consumption during periods of low workload while maintaining adequate performance during periods of high workload.
- DVFS in classic computer architecture is used in a variety of applications, including general-purpose computing tasks, web browsing, office productivity, and multimedia playback. We have to be focused on improving energy efficiency and prolonging battery life in mobile devices, laptops, and servers for better performance.

Considerations for Optimization in case of ML/DL Applications:

- In architectures optimized for ML/DL applications, DVFS may be implemented at multiple levels of the hardware and software stack, including specialized hardware accelerators (e.g., GPUs, TPUs) and ML/DL frameworks. DVFS techniques are used to dynamically adjust the voltage and frequency of processing units (e.g., CPU cores, GPU cores) based on the computational demands of ML/DL workloads.

- Optimization Goals: DVFS in ML/DL-optimized architectures aims to maximize the performance and efficiency of neural network computations by dynamically adjusting the voltage and frequency of hardware accelerators and processing units. The goal is to balance computational throughput with energy consumption, ensuring optimal performance for ML/DL workloads while minimizing power consumption.
- Applications: DVFS in ML/DL-optimized architectures is used to accelerate model training and inference tasks in neural network frameworks such as TensorFlow, PyTorch, and MXNet. Optimization efforts focus on improving the efficiency of ML/DL workloads on specialized hardware accelerators (e.g., GPUs, TPUs) and maximizing the throughput of neural network computations.
- Some frameworks might integrate with DVFS controls to dynamically adjust processing power based on the specific workload characteristics of the ML/DL task.

8 Utilizing Machine Learning for Design

The ever-growing complexity of Machine Learning (ML) and Deep Learning (DL) workloads demands continual innovation in computer architecture design. Traditionally, this exploration process has relied on manual design and simulation, which can be time-consuming and computationally expensive. Advanced computer architecture optimization leverages machine learning (ML) for design space exploration, accelerating the discovery of efficient hardware for ML/DL tasks. The design space for ML/DL architectures is vast, encompassing numerous parameters like memory organization, processing unit configurations, and interconnect structures. Manually evaluating all potential configurations is impractical. Simulating the performance of different architectures for various workloads is computationally expensive, hindering the exploration process. ML techniques offer a powerful approach to address these challenges. By employing techniques like reinforcement learning or evolutionary algorithms, we can automate the exploration of the architecture design space. These algorithms can:

- Efficiently Search the Design Space: ML models can be trained on historical data of architecture performance and workload characteristics. This allows them to identify promising design configurations and prioritize them for further evaluation through simulation or hardware prototyping.
- Learn from Experience: As the exploration progresses, the ML model can learn from the performance of evaluated architectures. This allows it to refine its search strategy and focus on more promising design directions over time.
- Benefits of ML-driven Exploration: Faster Design Exploration: ML can significantly accelerate the discovery of efficient architectures compared to traditional methods. This allows for quicker adaptation to the evolving demands of ML/DL algorithms.

- Improved Design Quality: By exploring a wider range of configurations, ML can potentially identify more efficient architectures compared to what might be discovered through manual exploration.
- Data-Driven Design Decisions: The exploration process becomes data-driven, with decisions based on performance evaluations and learned relationships between architecture parameters and workload characteristics.

However, utilizing Machine Learning for Design Space Exploration also Presents Challenges:

- Data Quality: The effectiveness of the exploration process heavily relies on the quality and quantity of data used to train the machine learning models.
- Interpretability: Understanding the rationale behind the proposed architecture by the machine learning model can be challenging. This limits the ability to refine the exploration process or incorporate domain knowledge from human architects.

9 Cases Where These Optimization Techniques Have Been Successfully Applied

- Google's BERT Model Optimization

  BERT model—popular state-of-the-art natural language processing model. The goal was to decrease memory size and make inference faster. Model size decreased by 60%, inference made 1.5× faster. Techniques used: sparsity handling to work with sparse tensor representations more effectively. quantization and compression techniques to reduce precision and memory usage.
- NVIDIA's GPU Optimization for Deep Learning (Grel et al. 2020)

  Techniques employed: Hardware optimization: Accelerators—GPUs, Software Framework optimization: Using CUDA—Compute Unified Device Architecture.

  NVIDIA improved its GPU architecture and CUDA software framework to be used effectively in deep learning. It resulted in reduction in training and inference time for deep learning models, more efficient use of the GPU, hence savings. It used accelerated matrix multiplication to exploit the matrix operations that power deep learning.
- Qualcomm's Snapdragon AI Engine used Hardware Optimization: AI accelerators (Hexagon DSP, Adreno GPU). Enabled on-device AI capabilities for tasks such as image recognition, natural language processing, and object detection. Applied Leveraged specialized AI accelerators (Hexagon DSP, Adreno GPU) for efficient execution of neural network computations.
- Tesla's Autopilot Hardware 3. Used Hardware Optimization - Custom hardware (Tesla FSD Chip). It enabled real-time processing of sensor data and AI-based decision-making for autonomous driving tasks.

10 Conclusion and Future Considerations

In conclusion, computer architecture optimization for machine learning and deep learning applications is essential for realizing the full potential of ML/DL workloads by enabling the execution of computation and neural network computations. As discussed in this paper, there are multiple ways that we optimize computer architecture to meet the needs of ML/DL workloads. These include hardware-level optimizations, software framework enhancements, and other novel architectures for exploration. Hardware optimizations, including the use of GPUs, TPUs, and custom ASICs, and other accelerators, have been widely employed. Researchers have also developed multicore architectures to take advantage of parallelism and have also optimized data movement. Software frameworks have also grown and improved dramatically, enabling the scaling and execution of ML and DL models. Furthermore, the application of machine learning to explore architecture is speeding up the design process, allowing researchers to explore the design space much more quickly and, with the help of few compute resources, identify optimal hardware configurations. With the use of supervised learning, reinforcement learning, and Bayesian optimization (Lee and Pourmohamad 2021), learning-based methods enable researchers to effectively probe the variety of architectural parameters and results, leading to the creation of high-performance, energy-efficient computing systems suited to ML/DL workloads. As machine learning and deep learning applications thrive, and novel technologies and computational strategies are introduced into sophisticated and varied hardware devices, the field of computer architecture optimization for ML/DL has ample room for growth.

To summarize, computer architecture optimization is critical for the successful rise of machine learning and deep learning applications, allowing an efficient manner to execute ML workloads and DL across a broad array of hardware devices and use settings. By extending existing limits of knowledge in computer architecture optimization (Hennessy 2017), researchers and engineers play an essential role in addressing the key computational constraints of the future of ML and DL applications and promoting AI's influence in advancement across different fields.

Here, some promising pathways for future exploration have been highlighted (Gao 2019):

- Co-optimizing software and hardware for even more effective optimizations.
- Exploring emerging technologies, such as neuromorphic computing, for hardware acceleration potential.
- Developing domain-specific hardware architectures designed for families of ML/DL algorithms' requirements.
- Improving Machine Learning for Design Space Exploration to generate more precise, interpretable outcomes.
- High-level languages, or otherwise modern programming languages (Hennessy and Patterson 2019).

- Security and privacy considerations: Architectures that incorporate hardware-based security measures to protect against the data privacy and model integrity threats facing modern models.
- Domain-specific architectures: Customizing hardware architectures to the needs of the specialized application domain such as image processing, NLP, recommender systems.

By pursuing these and related directions for further advancements in computer architecture optimization, we can enable the next stage of progress in ML/DL models and maximize their impact across the numerous fields.

References

Abadi, M., Barham, P., Chen, J., Chen, Z., Davis, A.: TensorFlow: a system for large-scale machine learning. In: 12th USENIX Symposium on Operating Systems Design and Implementation, pp. 265–283 (2016)
Adolf, R., Whatmough, P., Wei, G.Y., Brooks, D., Reagen, B.: Deep learning for computer architects. In: Synthesis Lectures on Computer Architecture. Springer Nature Switzerland (2017)
apache.org: Apache MXNet: A Flexible and Efficient Library for Deep Learning. [Online]. https://mxnet.apache.org/versions/1.9.1/ (2024)
Cadenelli, N., Prats, D.B., Polo, J., Garcia, J.L.B., Perez, D.C., Jakšić, Z.: A Highly Parameterizable Framework for Conditional Restricted Boltzmann Machine Based Workloads Accelerated with FPGAs and OpenCL. Elsevier, pp. 201–211 (2020)
Clark, T., Bagheri, B., Scott, L.R.: Scientific Parallel Computing. Princeton University (2021)
eitc.org: "cpu-vs-gpu-vs-tpu," [Online]. http://www.eitc.org/research-opportunities/photos1/cpu-vs-gpu-vs-tpu_012023a/image_view_fullscreen (2023)
Gao, Q.Q., et al.: An overview of machine learning in computer architecture. J. Comput. Sci. Technol. 709–731 (2019)
Géron, A.: Hands-On Machine Learning with Scikit-Learn, Keras, and TensorFlow, 3rd edn. O'Reilly Media (2022)
Goodfellow, I., Bengio, Y., Courville, A.: Deep Learning, Cambridge, Massachusetts. MIT, London (2016)
Google: An In-Depth Look at Google's First Tensor Processing Unit (TPU). [Online]. https://cloud.google.com/blog/u/1/products/ai-machine-learning/an-in-depth-look-at-googles-first-tensor-processing-unit-tpu (2017)
Grel, T., Huang, M., Nguyen, V.: Optimizing the deep learning recommendation model on NVIDIA GPUs. [Online]. https://developer.nvidia.com/blog/optimizing-dlrm-on-nvidia-gpus/ (2020)
Hennessy, J.L., Patterson, D.A.: Computer Architecture—A Quantitative Approach. Morgan Kaufman (2017)
Hennessy, J.L., Patterson, D.A.: A new golden for computer architecture. Commun. ACM 48–60 (2019)
I. o. E. a. E. Engineers: 25th IEEE International Symposium on High Performance Computer Architecture, p. 734 (2019)
Jacob, B., Kligys, S., Chen, B., Zhu, M., Tang, M., Howard, A., Adam, H., Kalenichenko, D.: Quantization and training of neural networks for efficient integer-arithmetic-only inference. In: Proceedings of the IEEE Conference on Computer Vision and Pattern Recognition, pp. 2704–2713 (2018)
Jouppi, N.P., et al.: In-Datacenter Performance Analysis of a Tensor Processing Unit. [Online]. https://arxiv.org/pdf/1704.04760 (2017)

Kim, B., Kim, T.T.H., Kim, J.Y.: Processing-in-Memory for AI: From Circuits to Systems. Springer Nature (2022)

Lee, H.K.H., Pourmohamad, T.: Bayesian Optimization with Application to Computer Experiments. Springer (2021)

Migacz, S.: Performance Tuning Guide. [Online]. https://pytorch.org/tutorials/recipes/recipes/tuning_guide.html (2024)

Olson, L.N., Gropp, W.D., Lockhart, S., Bienz, A.: Modeling data movement performance on heterogeneous architectures. In: IEEE High Performance Extreme Computing Conference (HPEC), Waltham, MA, USA (2021)

Spinczyk, O., Borghorst, H.: CyPhOS—a component-based cache-aware multi-core operating system. In: Architecture of Computing Systems—ARCS 2019 (2019)

Stotzer, E., Terboven, C., Van Der Pas, R.: Using OpenMP-the Next Step: Affinity, Accelerators, Tasking, and SIMD (Scientific and Engineering Computation). MIT Press (2017)

Sze, V., Chen, Y.-H., Yang, T.-J., Emer, J.S.: Efficient processing of deep neural networks: a tutorial and survey. In: Proceedings of the IEEE, pp. 2295–2329 (2017)

TensorFlow: Theoretical and Advanced Machine Learning with TensorFlow. [Online]. https://www.tensorflow.org/resources/learn-ml/theoretical-and-advanced-machine-learning (2024)

Wen-Mei, W.H.: GPU Computing Gems, Emerald edn (2015)

Wijtvliet, K., et al.: Accelerating machine learning workloads with OpenCL on FPGAs. In: IEEE International Conference on Cluster Computing (2019)

Enhancing Passenger Transportation Systems with Artificial Intelligence

Tarik Omerćehajić, Almir Osmanović, Mirza Bečirović, and Salko Ćosić

Abstract

Artificial intelligence systems, both textual and spoken, are gradually integrating into our daily lives. Today, we find them as help systems on websites, as interfaces for booking plane and train tickets, learning about companies on social media, executing online transactions, or making reservations. None of this would be possible without neural networks and the development of Natural Language Processing (NLP), which can not only interpret complex human phrases but also respond in the same format. Human nature favors communication by speech, which is why the future of speech applications looks promising. It is not unexpected that virtual assistants like Siri, Alexa, and Google Assistant are found in practically every home. This article will explore the fundamentals of voice processing and creation, and showcase the use of a virtual station for public transportation, highlighting the advantages, disadvantages, and potential of implementing such an application in the real world.

Keywords

Speech • Speech processing • Automation • Mechatronics • Virtual station

T. Omerćehajić (✉) · A. Osmanović · M. Bečirović · S. Ćosić
Faculty of Mechanical Engineering, University of Tuzla, Tuzla, Bosnia and Herzegovina
e-mail: tarik.omercehajic@gmail.com

A. Osmanović
e-mail: almir.osmanovic@untz.ba

S. Ćosić
e-mail: salko.cosic@untz.ba

K. Dhoska and E. Spaho (eds.), *Bridging Horizons in Artificial Intelligence, Robotics, Cybersecurity, Smart Cities, and Digital Economy*, Sustainable Economy and Ecotechnology, https://doi.org/10.1007/978-3-031-72029-1_26

1 Introduction

Currently, there are around 6,000 languages worldwide. These languages developed naturally from the human need for communication and have evolved over time as populations migrated. Language encompasses much more than just words. Given this context, the question arises: why do humans use language to communicate? People use verbal communication to share knowledge and simplify their lives. Conversation helps us understand each other and form deep social relationships. The most common purpose of verbal communication is to exchange information quickly.

Initially, communication between humans and machines (computers) was completely mechanical, using keyboards, mice, and other similar devices. This required a specific set of skills, adaptability, and learning time. Consequently, researchers began exploring the potential for speech-based communication between machines and humans, leading to the creation of the first voice user interfaces (VUIs). This form of communication gained popularity with the introduction of the first virtual assistants, such as Google Assistant, Siri, and Alexa (see Fig. 1). The key advantage of VUIs is that they allow us to communicate without using our hands or looking at the device itself, making communication safer, faster, and less distracting (Rachel 2018).

Speech user interfaces are far more difficult to design than graphical user interfaces because the user lacks a visual representation of what is in front of them and how to interact with it. Speech user interfaces typically contain a distinct verbal command used to launch the application/device, similar to start buttons (to activate Amazon's voice assistant Alexa, say the device's name "Alexa" to start listening to verbal orders). Speech user interfaces are still mostly used at home or locally, and voice assistants are not yet widely available in public places, restaurants, cafes, etc. One reason natural language processing has not yet reached that level is the

Fig. 1 Commercial virtual assistants

diversity of human dialects and potential variations, which make creating a universal solution extremely challenging. Furthermore, one of the major difficulties for the future will be the security of speech user interfaces and how to properly protect users from external influences (Jurafsky and Martin 2021).

2 Speech Synthesis into/from Textual Format

Automatic speech recognition is one form of voice recognition, but additional forms of speech recognition systems include:

- Speaker-dependent system: Before using speech recognition, you must train it by reading a sequence of words and phrases.
- Speaker-independent system: Speech recognition software recognizes the majority of users' speech without training.
- Discrete voice recognition requires the user to pause between each sound in order for speech recognition to recognize each word individually.
- Continuous speech recognition: Speech recognition is capable of understanding regular speech speeds.
- Natural language: Speech recognition can not only recognize speech but also answer questions and other inquiries.

3 Automatic Speech Recognition

Automatic voice Recognition (ASR) describes the process of converting spoken words into textual representation using computers and software-based approaches for identifying and processing human voice (see Fig. 2). Some of the most well-known speech recognition systems include dictation systems, voice assistants (Google Assistant, Siri, Alexa, Bixby, and so on), Google Translate, and the YouTube translation generator. Automatic speech recognition systems are designed and developed using a variety of components, including presentation and preprocessing, speech categorization, various feature extraction approaches, classification, databases, and overall system performance.

Acoustic Analysis (Speech Recognition)—In this step, the input sound waveform is transformed into a discrete form or feature vector. A voice signal is frequently transient rather than steady, and a range of around 20–25 ms is considered, after which it acts as a stationary signal. This brief capture of the discrete signal is known as a "frame." The acoustic model extracts specific elements from the frame. During feature extraction (see Fig. 3), all unnecessary information is removed, leaving only useful data. A phoneme can be described as the fundamental component of an audio system. Linguistic experts are responsible for breaking down words into phonemes. Each language contains approximately 50–60 phonemes. The acoustic model represents the link between the audio stream and the phonemes.

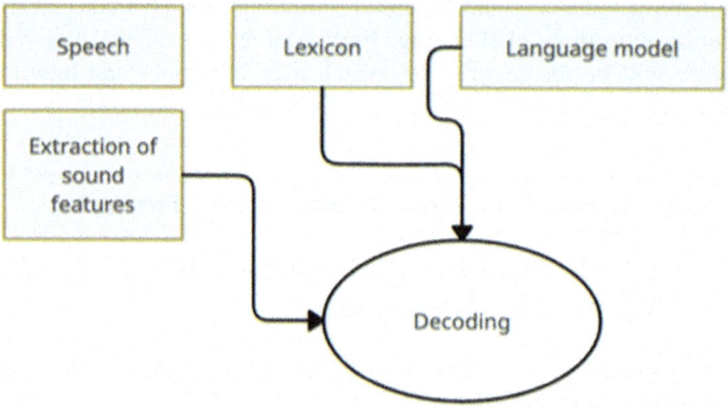

Fig. 2 Structure of the automatic speech recognition system (Swiss public transport API 2017)

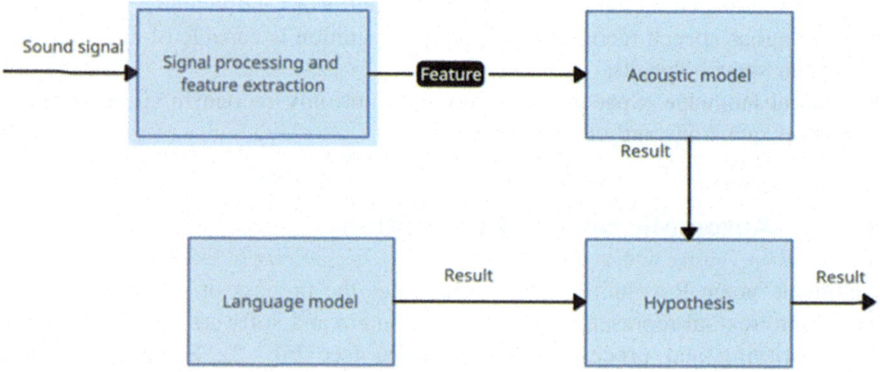

Fig. 3 Feature extraction (Woodford 2023)

Natural Language Processing (NLP), or speech processing, is a core component of artificial intelligence systems. The interface between AI systems and humans should be more human-like, and NLP allows us to extract linguistic elements from audio recordings. Human voice synthesis and analysis can be accomplished by training neural networks, specifically deep neural networks (DNN), on enormous amounts of data. The primary challenge that NLP addresses is communication between humans and machines. Computers can only handle values of 0 and 1. NLP is a computer science area that deals with the analysis, modeling, and understanding of human language. It has become an integral part of our daily life. Voice assistants, such as Google Assistant, Apple Siri, Amazon Alexa, IBM Watson, and others, enable us to verbally control various household devices, quickly create reminders and timers, check the weather forecast, and get the most recent news. Of course, Google, one of the most popular internet search engines, uses NLP to comprehend what someone is looking for and deliver more relevant search results.

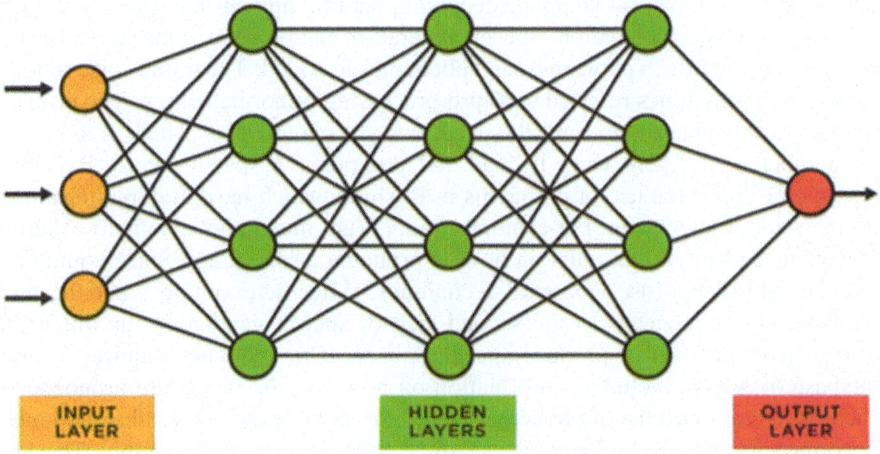

Fig. 4 An example of a neural network with an input layer, hidden layers and an output layer

In recent years, NLP has also been used in online translators, where it helps us understand the sentence, not only from a technical but also emotional standpoint, in order to produce a more accurate translation. One of the most difficult aspects of NLP is dealing with uncertainty and the ambiguity of some phrases, such as the occurrence of synonyms. Machine learning algorithms are used to analyze voice data in the same way that they are used for photos, text, and structured data. Supervised machine learning techniques, such as classification and regression, are commonly utilized for a variety of NLP tasks. Any machine learning strategy for NLP, whether supervised or unsupervised, can be broken down into three standard steps: feature extraction from text, feature representations for model learning, and model evaluation and improvement. Today's most common machine learning algorithms in NLP are Naive Bayes, Support Vector Machine (SVM), Hidden Markov Model, Conditional Random Fields (CRF), and others. In recent years, there has been an enormous increase in the usage of neural networks to handle complicated, unstructured data challenges. Language is naturally complex and disorganized. As a result, models with superior representation and learning skills are required to comprehend and solve language tasks. Deep learning methods have enabled major breakthroughs in NLP tasks, resulting in very high performance (Fig. 4).

4 Speech Generation

Text-to-speech (TTS) systems have numerous applications. They initially served to read to people with visual impairments, with the system reading the book and translating it into speech. Similar but newer methods are now used not only for reading books, but also to teach visually impaired people how to use computers and navigate user interfaces. TTS systems face a variety of issues, including semiotic text classification, natural language decoding, natural, human-like speech creation, and intelligent speech creation. Speech generation can be divided into three stages: text to words, words to phonemes, and phonemes to sound. The initial stage of text-to-speech is sometimes referred to as pre-processing or normalization. The primary purpose of this phase is to limit the occurrence of unclear text, which is to reduce the number of different ways to read and pronounce a specific word. Excellent example of this is the use of paronyms in Bosnian, which have multiple meanings but the same written form. For example, "gore" (up; mountains). Humans examine phrases to find important units on how to pronounce something. Since computers lack this skill, they use statistical techniques. After determining what the user wants to say, it proceeds to the second step of speech generation, in which the speech generator has to produce speech sounds. The computer requires a large database of words, including information on how specific words are pronounced. Each word requires a list of phonemes that make up the sound. After that, text must be formed using speech. Three ways can be utilized to accomplish this. The first includes a human recording each word and pronunciation independently, and the computer reproducing each of these audio files one at a time. The second option has the computer generate noises of various frequencies depending on phonemes. The third technique is to emulate the process of human speech. This is the most complex solution, but the end result is most similar to human speech and sounds very natural (Tan 2023).

4.1 Application of Automatic Speech Recognition

Today, automatic speech recognition is used in a variety of settings, possibly without our knowledge. Automatic speech recognition is commonly associated with voice assistants, however it is used in a far broader context. A good example of automatic voice recognition is the YouTube app (see Fig. 5), which uses technology to generate subtitles for videos that users have not given. You may enable this function by selecting the Subtitle/CC option in the settings, which causes the YouTube machine vision algorithm to process the video clip in real time and generate subtitles on your screen. Automatic speech recognition is one of the technologies that facilitates digital communication for people with partial and complete hearing impairments, as well as real-life communication, because these capabilities are always available to them due to the advent of portable devices.

5 Components of the VoiceBus Model for Practical Application—Bus Station

To create the VoiceBus model that can be used in public transportation vehicles, the following components are necessary:

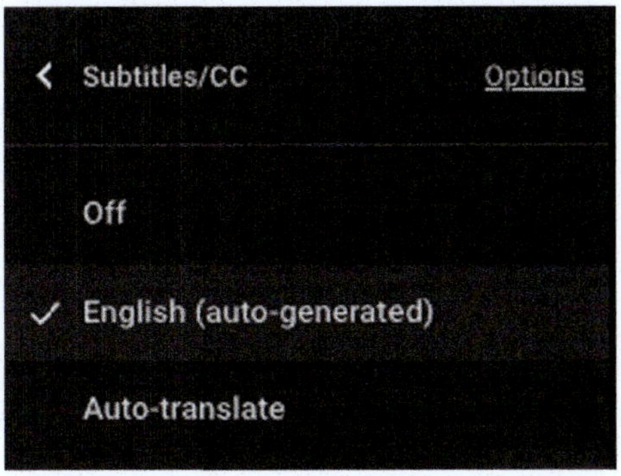

Fig. 5 Youtube auto-translate feature

- Dialogflow (speech registration and processing)
- Traffic API (public transport timetable)
- Backend services (connecting intents and API services to find the optimal timetable)
- User application—an application for visual and textual display of requested information to the user.

Dialogflow is an automatic voice recognition platform which allows the development and integration of conversational interfaces into mobile and online applications, as well as other devices. It can process text and speech inputs and respond to users. Dialogflow's key feature is that it allows for the building of speech recognition using a visual user interface, and with pre-built agents, training the speech program takes only minutes. The basic building block of a Dialogflow program is an intent (see Fig. 6). An intent categorizes a conversation initiated by the user. When the user speaks a certain command, Dialogflow performs a syntactic analysis of that sentence and searches for the corresponding intent to match it. In other words, the user's command is classified, and its purpose is determined. A basic intent consists of: a training phase, actions, parameters, and responses.

In Fig. 7, is presented an example of intent for the VoiceBus application. An intent is defined for querying the timetable for a specific bus line. If a user commands the application: "When is the next bus from Tuzla to Sarajevo," Dialogflow will map it to the intent for querying the timetable. Additionally, an intent can be defined so that useful information for the system is extracted from its content, such as information about the type of transport, departure, and destination locations. Each parameter within the intent corresponds to a type of property based on which the data obtained from human speech will be extracted. In addition to standard system properties such as location, time, and colors, other properties can be defined, such as the type of vehicle or transport for which the query is sent. By extracting the true meaning of the spoken sentence, it enables maintaining the conversational flow between the user and the computer. It is crucial to accurately predict what the user requested to provide the correct response. In the basic configuration, Dialogflow agents generate a response based on a predefined static

Fig. 6 Intent

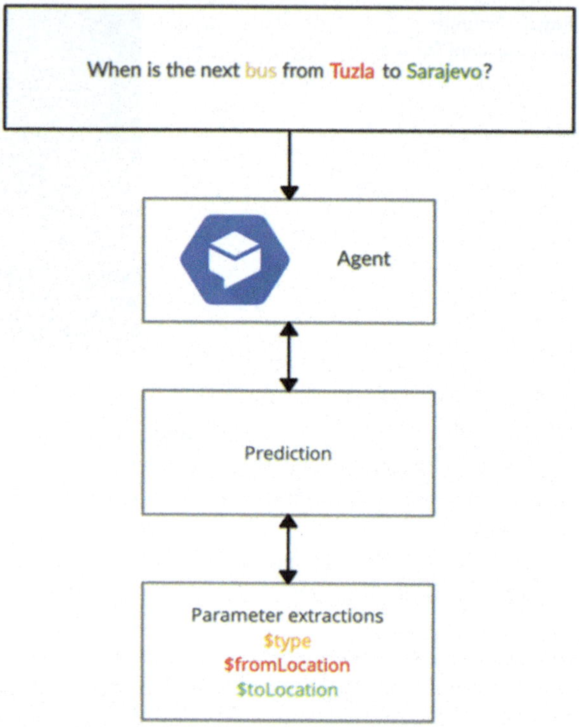

response. For the development of more complex solutions that use external data and communicate with external services, integration services are required to enable this. In the Dialogflow application, Fulfillment or integration service is used for this purpose, allowing the definition of a custom API (Application Programming Interface) to which data extracted from the intent is sent. In this case, a special external application has been created to receive data such as the type of vehicle, departure, and destination.

The model was developed using an open-source application that contained public transportation data for Switzerland. The Traffic API provides information about public transportation (bus, tram, and train) in Swiss cities and local stations. The data collected from the API is in JSON format, which makes data processing easier in the application. The backend service, or application that processes data, is built in JavaScript and runs on the Node.js server platform. It is made up of numerous components. The system's first component is Dialogflow-to-backend service communication. After successfully processing the voice query, Dialogflow sends a POST request to the server, including the received data. These data are routed to the next component, which sends them to the TRAFFIC API and waits for a response. The received response is processed and forwarded to the user's application. The user application and the backend service communicate using web

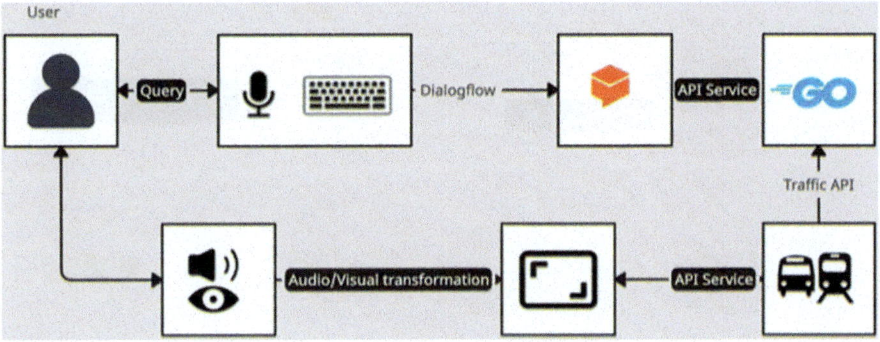

Fig. 7 Flow chart

sockets, which allows for real-time communication. When all data in the background service has been processed, a signal is given to the user interface, which processes the received data and presents it to the user in an audio-visual format. The application visually displays all received information to the end user and gives a spoken response using a text-to-speech (TTS) application. The application creates Single Page Applications (SPAs) using the Vue.js platform. The application connects to the backend application's socket service and waits to receive data to display to the user. The user receives data in three formats:

1. Textual format—the starting and destination points are displayed on the screen, along with the departure time and platform/station information if it's train transportation. The duration of the journey and the estimated arrival time are also provided.
2. On the right side of the application, users can visually see the journey route via a map with directions.
3. All the mentioned textual information is returned to the user in spoken format through the speaker, in the language the user used to query the application.

6 Designing a Model for Practical Application—Bus Station

The project's creation and implementation were based on a dataset that can be used to simulate and track the position of public transportation. During the research, it discovered a website for public transportation in Switzerland that includes information on bus, train, tram, trolleybus, and boat travel. To use the practical model, the user must first create a voice inquiry, enter text in the supplied field, or hit the help button. The help button informs the user in audio format how to operate the model and what settings are available. Touching the microphone icon causes the device to start listening and awaits the spoken command. To get the best explanation, the user should do the action in English with the following voice content:

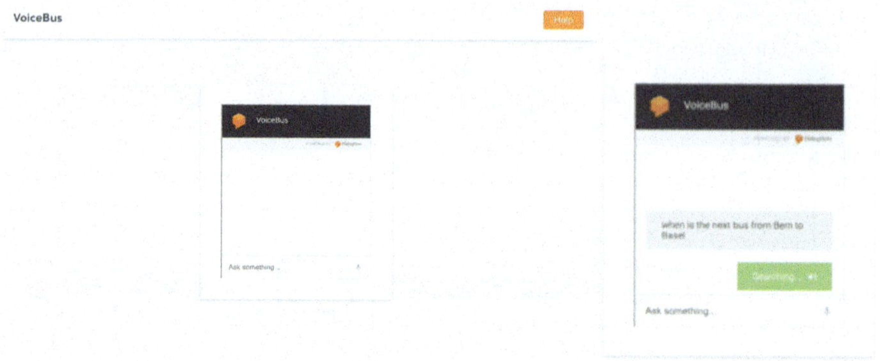

Fig. 8 Entering the requested information

"When is the next bus from Basel to Bern?" When the user has finished speaking, the textual input is processed by the Dialogflow application (see Fig. 8). Semantic analysis of spoken text transformed to textual form is carried out based on the aim and context desired.

Within the Dialogflow interface, it is defined in the intent that the user is required to enter the departure and destination locations, while the type of transportation is optional, and if not selected, the next available line is provided. If the user did not provide one of the requested parameters in their query, they will be prompted to speak and enter the missing parameter, which will also be displayed on the screen. Once all the requested parameters are available and clearly defined, they are forwarded by sending a call with the defined data to the backend service that was created. The backend service receives the data, stores it in local variables, and checks again to ensure that all necessary data is available. If everything is fine, the departure and destination locations, as well as the type of transport, are sent to the public transport API. The server response is awaited, and the received data about the next bus line is stored in local variables. This processing takes place in a fraction of a second, and the user is presented with the results after a briefly displayed message indicating that their query is being processed. The query results are presented to the user in an audio/visual format, as shown in Fig. 9. On the screen of the station, the route that the user will take during the journey is displayed. In the dialogue on the left side, a summary of their query with basic information such as the departure and destination locations, the duration of the journey, and the estimated arrival time is provided. In the case of train traffic, the platform from which the departure is taking place is also displayed to the user. Based on the data available from the Traffic API, in the future, the user may be presented with a detailed travel plan showing all the stations the transport vehicle will stop at on the way to the desired location, whether there are any alternative travel methods, whether transfers are necessary, and the use of multiple transportation routes, and so on.

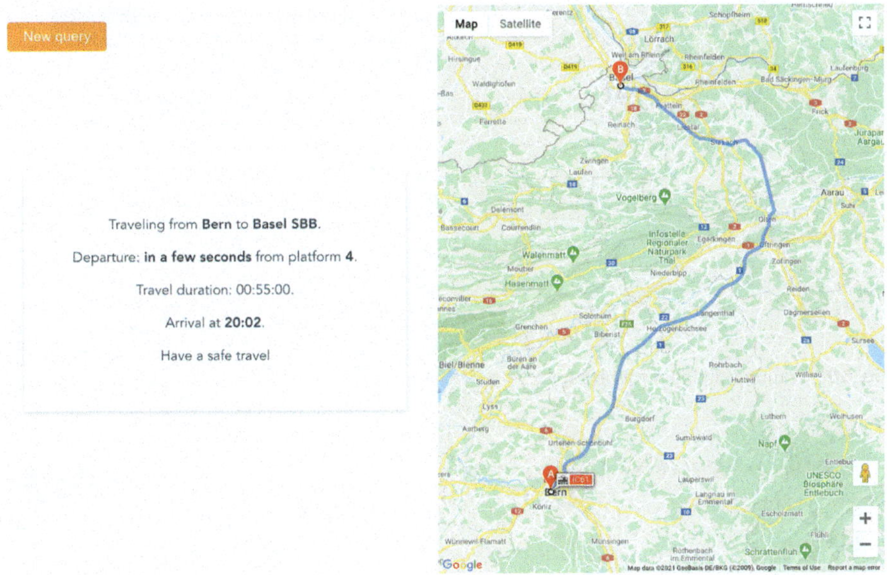

Fig. 9 End user information

6.1 Model Analysis

During the development of the virtual assistant for public transportation, several shortcomings were observed, as well as possibilities for future improvements and enhancements. From a technical perspective, the application is fully functional and meets the basic criteria, i.e., it enables audio/text input of data, and the response to the user's request is delivered in audio/visual form. The application, although appearing simple, contains many components that must orchestrate in perfect order for it to function correctly. In the development of this project, Dialogflow was used as the platform for processing human speech, extracting its characteristics, and forwarding the results of speech queries to the data processing application. However, Dialogflow also has its limitations. Specifically, during the development of this project, one of the limitations of using this application was observed. Namely, Dialogflow is capable of detecting and processing the language used by the user, but this information cannot be used as an external parameter to be utilized in the response provided to the user. Consequently, to develop the VoiceBus application for real-world use, it is necessary to use a custom solution for processing human speech, which requires more of both computational and human resources, as well as a certain amount of time needed for training and testing the neural network for speech processing. Speaking about network training, Dialogflow provides an outstanding visual method for determining user queries, validating them, and retraining based on defined corrections as demonstrated in Fig. 10. By establishing

Fig. 10 Training dialog with prompts

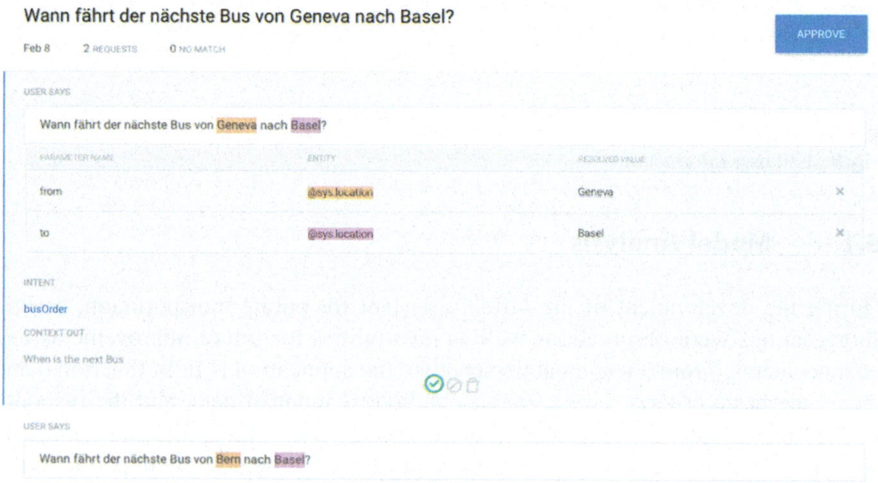

Fig. 11 Case—the program did not recognize the spoken type of vehicle

a training dialogue, all inquiries submitted by users can be checked for errors or unknown terminology, as well as the results of sound analysis.

Figure 11 shows an example of a program that failed to correctly recognize the type of vehicle specified by the user. The departure and destination locations were successfully identified and transformed into variables for transmission to the data processing service. This user activity is easily fixed by clicking on the word and selecting the appropriate intent. Using the same analogy, manual correction would be carried out, and the agent would be retrained to avoid these errors in the future.

7 Conclusion

In an era where mobile phones have become common and, as a result, a source of distraction due to the need for visual communication to utilize them, voice-controlled virtual assistants are precisely the solution that is desired and should make daily life easier. The benefits of a virtual public transportation station are evident. For example, the program does not require any physical input to activate, making it easier to use when the user's hands are occupied. The ability to communicate in several languages, both between humans and computers, is suitable for usage in global cities and tourist destinations. Reaching users in their own language makes communication easier for them. To further improve the model, it is important to create a custom automatic voice recognition system based on deep learning and machine vision. This solution should not only conduct the user's query, but also identify the language in which the question was made. Furthermore, improvements in expanding the number of available searches are required to provide consumers with a broader range of services and travel options.

References

Alavala, C.R.: Fuzzy Logic and Neural Networks, Basic Concepts and Applications. New Age International, New Delhi (2005)

Dialogflow. https://cloud.google.com/dialogflow?hl=en (2017)

Jurafsky, D., Martin, J.H.: Speech and Language Processing, 3rd edn. Prentice Hall (2021)

Rachel, B.: Voicebot and Chatbot Design: Flexible Conversational Interfaces with Amazon Alexa, Google Home, and Facebook Messenger. Packt Publishing, Birmingham, UK (2018)

Swiss public transport API: http://transport.opendata.ch/ (2017)

Tan, X.: Neural Text-to-Speech Synthesis. Springer, London, UK (2023)

Taylor, P.: Text-to-Speech Synthesis. Cambridge University Press, UK (2009)

Woodford, C.: Speech Synthesizers. https://www.explainthatstuff.com/how-speech-synthesis-works.html (2023)

Study and Analyses of Radiofrequency Radiation Measurements Level from Different Base Stations

Abdulsalam Alkholidi, Shefqet Meda, Vasil Mitezi, and Anxhela Baraj

Abstract

Since the beginning of the radio era, there have been and still are concerns about the radio frequency radiation (RFR) emitted by these radio systems and their impact on human health and the environment, as well as an invisible pollution. But nowadays these fears of RFR have doubled due to the development of new generations of wireless networks, such as the fourth and fifth generation, and there will be a sixth generation in a few years from now. The main goal of this study is to confirm that the lowest RFR exposure to human health respects the regulation standards issued by the ITU-R section and other relevant authorities. Furthermore, an additional aim of this paper is to do several radio radiation measurements in various areas. So that these results are available to readers. The measurements will be focused on public squares, kindergartens, schools, hospitals, and other facilities. The most important recommendation reached by this study is to conduct comprehensive measurements of the total power density of radiation emitted by all radio stations at their different frequencies.

Keywords

Exposure time • Transmitted power • Frequency • Distance • Measurements • RF propagation • FCC • ICNIRP

A. Alkholidi (✉) · S. Meda · A. Baraj
Canadian Institute of Technology (CIT), Kompleksi Xhura, Rruga Xhanfize Keko, 1000 Tirana, Albania
e-mail: abdulsalam.alkholidi@cit.edu.al

A. Alkholidi
Faculty of Engineering, Sana'a University, Sana'a, Yemen

V. Mitezi
Electronic and Postal Communications Authority, Tirana, Albania

K. Dhoska and E. Spaho (eds.), *Bridging Horizons in Artificial Intelligence, Robotics, Cybersecurity, Smart Cities, and Digital Economy*, Sustainable Economy and Ecotechnology, https://doi.org/10.1007/978-3-031-72029-1_27

1 Introduction

The entire range of electromagnetic radiation arranged according to wavelength or frequency is known as the electromagnetic spectrum. Radio Frequency Radiation (RFR) refers to the electromagnetic radiation that is emitted by various sources and propagates through space in the form of electromagnetic waves with frequencies ranging from about 3 kHz to 300 GHz. This range includes frequencies used for communication, such as radio waves, microwaves, and some types of radar, the electromagnetic spectrum is presented in Fig. 1.

The range of all electromagnetic radiation types is known as the electromagnetic spectrum (EMS). This spectrum includes many types of energy waves that are categorized according to their frequencies and wavelengths. It consists of:

- Radio waves are used in broadcasting and communication.
- Microwaves: Employed in microwave, engineering, satellite communications, and cooking.
- Infrared Radiation (IR): Its function is as heat and is used in thermal imaging.
- Visible Light: the narrow region of the spectrum that is observable to the naked eye.
- Ultraviolet Light: Significant in astronomy and sterilizing, beyond visible light.
- X-Rays: There are two X rays, soft and hard, utilized for security screening and medical imaging.
- Gamma Rays: Released by radioactive materials and applied to the therapy of cancer.

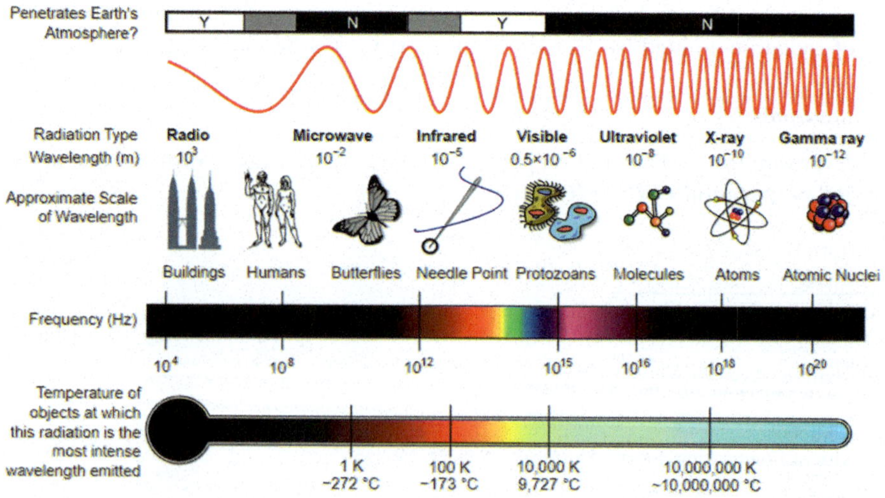

Fig. 1 The electromagnetic spectrum (https://upload.wikimedia.org/wikipedia/commons/c/cf/EM_Spectrum_Properties_edit.svg)

Each frequency band in the EMS has special qualities and uses, and they are all very important to communication, science, military, and medicine.

Emitted by radioactive materials, gamma rays are employed in the treatment of cancer.

RFR is produced by a wide variety of devices and technologies, including:

(1) Wireless communication: Cell phones, Wi-Fi routers, Bluetooth devices, and other wireless communication devices emit RFR to transmit data wirelessly.
(2) Broadcasting: Television and radio broadcasting stations emit RFR to transmit audio and video signals.
(3) Microwave ovens: Microwave ovens use RFR to heat food by causing water molecules within the food to vibrate rapidly.
(4) Radar systems: Radar systems emit RFR pulses to detect the presence, speed, and location of objects, such as aircraft, ships, and weather formations.
(5) Medical applications: Certain medical devices, such as MRI machines and certain types of diathermy equipment, utilize RFR for diagnostic and therapeutic purposes.

RFR is a form of non-ionizing radiation, which means it does not have enough energy to ionize atoms or molecules and therefore does not have the ability to directly damage DNA or cause cells to become cancerous. However, exposure to high levels of RFR can lead to tissue heating and other biological effects, depending on factors such as frequency, intensity, and duration of exposure.

2 Literature Review

Within the EMS, RF radiation has wavelengths that vary from one millimeter to one hundred kilometers. It is extensively utilized in mobile phones, television, radio broadcasting, radar, etc., among other communication devices. Its biological implications, however, have generated a great deal of discussion and investigation (Baan et al. 2011).

The health impacts of RF radiation have been extensively studied. The International Agency for Research on Cancer (IARC) classified RF radiation as a possible human carcinogen (Group 2B) in 2011, based on evidence suggesting a potential link to glioma and acoustic neuroma (IARC) (IARC Monographs on the Evaluation of Carcinogenic Risks to Humans, vol 102. Retrieved from IARC website). A notable study by the National Toxicology Program (NTP) found some evidence of carcinogenicity in rats exposed to RF radiation, though the results were not conclusive for humans (https://ntp.niehs.nih.gov/whatwestudy/topics/cellphones; ICNIRP 2020; https://www.niehs.nih.gov/newsreleases/high-exposure-to-radio-frequency-radiation-associated-with-cancer-in-male-rats). Contrastingly, large-scale epidemiological studies, such as the Interphone study, have not found consistent evidence of a significant increase in cancer risk among mobile phone users (IARC 2011).

Radiofrequency radiation's effects on health have been well researched. Based on data pointing to a probable connection between RF radiation and glioma and auditory neuroma, the IARC designated RF radiation as a possible human carcinogen (Group 2B) in 2011. In rats exposed to radiofrequency radiation, a prominent study conducted by the National Toxicology Program (NTP) revealed significant evidence of carcinogenicity; nevertheless, the results were inconclusive for humans (NTP 2018). In contrast, numerous large-scale epidemiological investigations, like the Interphone study (Interphone Study Group 2011), have not consistently shown evidence of a significant increase in cancer risk among mobile phone users (IEEE 2020; Benedict and Meyer 2017).

Guidelines are established by regulatory authorities, such as the World Health Organization (WHO) and the Federal Communications Commission (FCC), to prevent radiofrequency exposure and safeguard public health. These recommendations are based on what is currently known about the biological consequences of radiofrequency radiation. To avoid overheating tissues, for example, the FCC sets exposure level limitations (FCC 2021). The WHO notes that more study is necessary even if the available data does not definitively link radiofrequency radiation to significant health hazards (WHO 2020; FCC 2021).

Understanding the possible long-term effects of low-level radiofrequency radiation and how RF fields interact with biological systems at the cellular level are the main topics of recent research. Technological developments, including the spread of 5G networks, are drawing attention once again to the potential health risks associated with higher frequency radiofrequency radiation (Baan et al. 2018). The scientific community is still in favor of conducting thorough, peer-reviewed research to identify any possible health hazards related to radiofrequency radiation (Baan et al. 2018).

Significant improvements in wireless communication are anticipated with the introduction of 5G technology, which will provide faster speeds and more connectivity. But with more 5G base stations being installed, worries about possible health effects from increasing radio frequency (RF) radiation exposure are growing.

Particularly, research on 5G base stations is still being conducted. 5G networks can function at a variety of frequencies, including millimeter-wave frequencies (24 GHz and upwards). The majority of current research and guidelines center on frequencies from earlier wireless technology generations, namely 2G, 3G, and 4G.

Studies on lower frequency RF radiation are extensively used to infer the biological consequences of RF radiation from 5G base stations. According to these investigations, exposure to high radiofrequency radiation levels may result in thermal consequences, such as tissue heating due to RF energy absorption. Guidelines for limiting exposure to avoid such heating effects are provided by the International Commission on Non-Ionizing Radiation Protection (ICNIRP) (2020). But there is still more to learn about and studies to be done on the non-thermal effects of RF radiation, which may include possible long-term health implications (Benedict and Meyer 2021).

There isn't much epidemiological research looking at how 5G base station RF radiation affects human health. Although studies like the Interphone study point to the need for continued monitoring, research on current RF exposure levels, such as those from 2 and 4G networks, has not clearly connected RF radiation to major health consequences (Interphone Study Group 2011). The World Health Organization (WHO) states that further research is required to address potential long-term consequences of new technologies, even if current evidence does not show significant health hazards from radiofrequency radiation at levels below established recommendations (WHO 2020).

Only thermal effects, or tissue heating, are taken into consideration as potentially detrimental in the existing exposure levels that have been suggested by the International Commission on Non-Ionizing Radiation Protection (ICNIRP) and the Federal Communications Commission (FCC). On the other hand, there is growing proof that electromagnetic radiation exposure has non-thermal consequences on human populations and biological systems (Ishai et al. 2024).

3 Measurements and Discussion

3.1 Measurements of Radiation Levels in the Region of Korče and Erseke

After studying the latest research related to RFR, in this section we will present and study the measurement results of RFR and analyze them. This study was compiled on the basis of the measurements that the Administration Sector of the KK and the Regional Monitoring Offices of the SRF carried out in the framework of monitoring the frequency band used with Individual Authorization in order to verify the implementation of the standards approved by the International Commission for Defense from Non-Ionizing Radiation (ICNIRP), according to Authorization No. 63/28 Prot. date 05.11.2020. The purpose is to monitor the frequency spectrum, mainly emissions from mobile phone cells, especially near buildings of highly sensitive public institutions, such as schools and hospitals, for the perceived effect they may have on public health in their communities, as well as monitoring the level of non-ionizing radiation and receiving and processing data collected from field measurements.

Type of measurements: non-ionizing radiation level monitoring. The study summarizes as follows:

1. Monitoring of the frequency band in the city of Korçë carried out near the Korçë Regional Hospital, the general high school "Themistokli Gërmenji" and the 9-year school "Sotir Gura" to verify the level of the electric, magnetic field, and power density (EMF).
2. Monitoring of the frequency band in the city of Erseka carried out near the city's Health Polyclinic, the 9-year school "Papa Kristo Negovani" and the high school "Petro Nini Luarasi" to verify the level of the electric, magnetic, and power density.

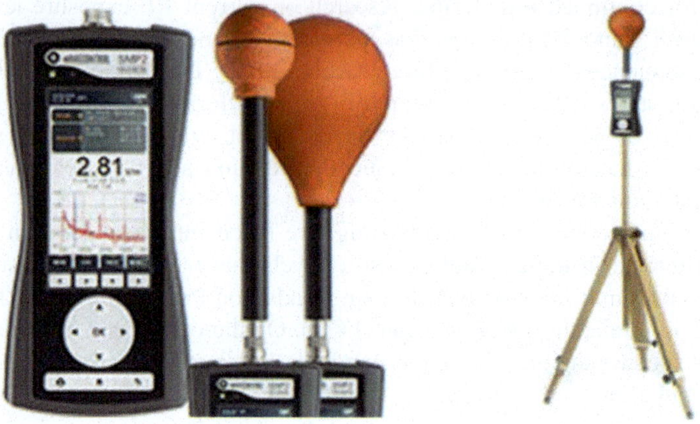

Fig. 2 Equipment used for monitoring non-ionizing radiation

The equipment used in monitoring (Fig. 2).

1. SMP2 field strength meter
2. Antenna WPF8: 100 kHz–8 GHz

3.2 Measurements Performed Near the Regional Hospital in Korça

During the monitoring of the spectrum in the premises of the Regional Hospital in Korça, measurements were made of the strength of the electric field, E (V/m). The geographical coordinates of Korça Hospital are: N: 40° 36′ 39.34″ and E: 20° 47′ 08.55″. The view from the premises of Korça Hospital is presented in Fig. 3, and the technical specification of the equipment's use is explored in Table 1.

The average values of the level of non-ionizing radiation obtained from field data are demonstrated in Table 2 and Fig. 4.

3.3 Measurements Carried Out Near the "Themistokli Gërmenji" Industrial High School in Korça

During the monitoring of the spectrum in the premises of the general high school "Themistokli Gërmenji" in Korçë, measurements of the strength of the electric field (E) (V/m) were carried out.

Fig. 3 View from the premises of Korça Hospital

Table 1 The equipment technical specification

Software version	1.0.3.1
SMP2 serial	15SN0095
SMP2 firmware	0.209
Probe serial	15WP040481
Probe frequencies	100 kHz–8 GHz
Average interval	6 min/unit
Average type	Sliding
Units	V/m

Table 2 Average values of the level of non-ionizing radiation obtained from field data

Location	Realization time	Electric field E (V/m) max	Electric field E (V/m) mes	Distanca nga BTS	Dislevel in relation to BTS
Korca Hospital	10/11/2020 03:26 PM	2.832294	0.89559	350 m	25 m

The geographical coordinates of the general secondary school “Themistokli Gërmenji” in Korça:

N: 40° 36′ 59.80″
E: 20° 46′ 38.22″

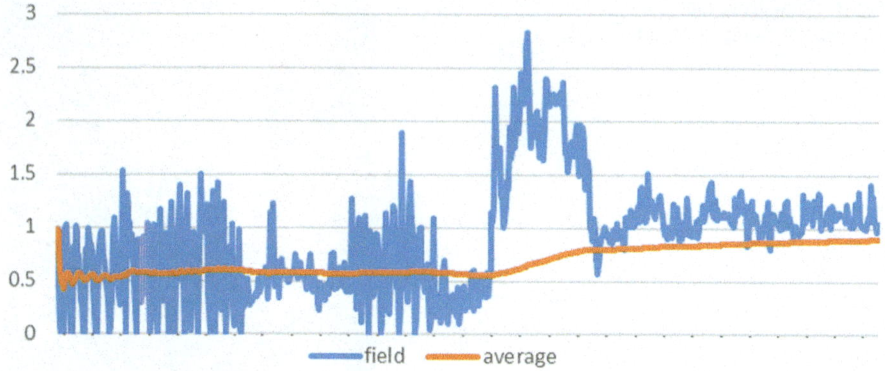

Fig. 4 Graph of the electric field strength (E in V/m) over the time interval, measured at "Themistokli Gërmenji"

The maximum value of the level of non-ionizing radiation obtained from field data is demonstrated in Table 3.

Figure 5 shows the graph of electric field strength E (V/m) in time interval.

Table 3 The maximum value of non-ionizing radiation levels obtained from field data at the 'Th. Gërmenji' School in Korçë

Location	Realization time	Electric field E (V/m) max	Electric field E (V/m) mes	Distanca nga BTS	Dislevel in relation to BTS
School "Th. Gërmenji" Korçë	10/11/2020 03:36 AM	1.763358	0.689823	280 m	25 m

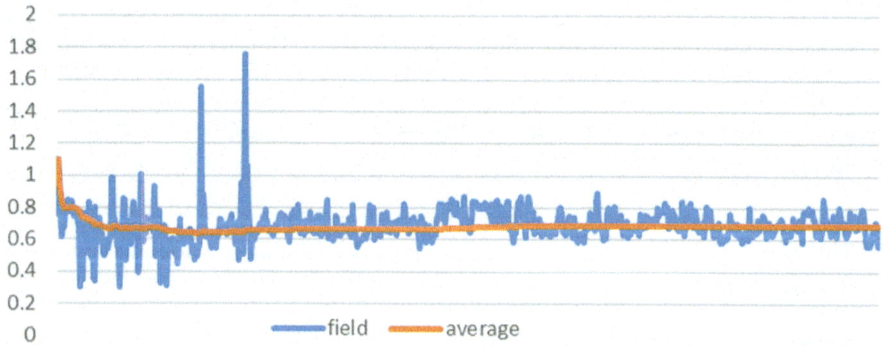

Fig. 5 Graph of the electric field strength (E in V/m) over the time interval, measured at "Near the Polyclinic of the City of Erseke"

Fig. 6 View from the premises of the Polyclinic in Erseke

Table 4 The maximum value of the level of non-ionizing radiation obtained from field data at "Erseke polyclinic"

Location	Realization time	Electric field E (V/m) max	Electric field E (V/m) mes	Distanca nga BTS	Dislevel in relation to BTS
Erseke polyclinic	10/11/2020 01:39 PM	1.053364	0.203838	230 m	18 m

3.4 Measurements Performed Near the Polyclinic of the City of Erseke

The geographical coordinates of the Polyclinic in Erseke (see Fig. 6):

N: 40°20′14.05″
E: 20°40′52.26″

The maximum value of the level of non-ionizing radiation obtained from field data is illustrated in Table 4, and Fig. 7 introduces the graph of the electric field strength E (V/m) in the time interval.

Most significant recommendations of the study

1. Claim of related authority in taking all required safety measures of RFR to protect the environment and prevent human health.
2. Continue periodic monitoring of radio radiation measurements and publishing all the obtained results online will ensure transparency and reassure the public.
3. Cooperate between academic establishments and the Albanian Radio Regulatory Authority.

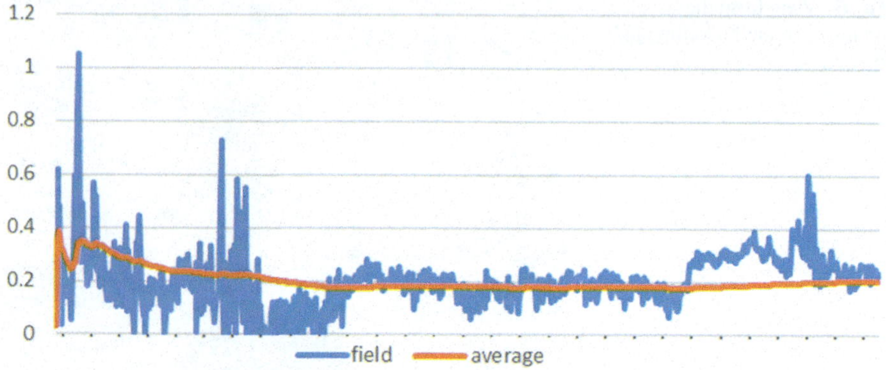

Fig. 7 Graph of the electric field strength E (V/m) in the time interval

4 Conclusion

In this paper, according to the results mentioned above, it is found that the levels are constantly much below those defined by the "ICNIRP Guidelines for Limiting Exposure to Time-Varying Electric, Magnetic, and Electromagnetic Fields (Up to 300 GHz)". The results of these measurements recorded so far show that the radiation levels emitted by mobile antennas do not exceed the safety threshold determined by the International Commission on Non-Ionizing Radiation Protection (ICNIRP). Finally, the measurements confirm the safety of the specific RF source/system according to ICNIRP guidelines and emphasize the importance of continued compliance and research to ensure public health safety in the face of evolving RF technologies.

References

Ajibare, A.T., Oladejo, S.O., Ekwe, S.O., Akinyemi, L.A., Ramotsoela, D.: Radiofrequency electromagnetic radiation exposure assessment, analysis, computation, and minimization technique in 5G networks: a perspective on QoS trade-offs. SAIEE Afr. Res. J. **114**(4), 114–127 (2023). https://doi.org/10.23919/SAIEE.2023.10319380

Baan, R., Grosse, Y., Lauby-Secretan, B., et al.: Carcinogenicity of radiofrequency electromagnetic fields. Lancet Oncol. **19**(1), 14–15 (2018)

Baan, R., Grosse, Y., Lauby-Secretan, B., El Ghissassi, F., Bouvard, V., Benbrahim-Tallaa, L., Guha, N., Islami, F., Galichet, L., Straif, K., WHO International Agency for Research on Cancer Monograph Working Group.: Carcinogenicity of radiofrequency electromagnetic fields. Lancet Oncol. **12**(7), 624–626 (2011). https://doi.org/10.1016/s1470-2045(11)70147-4. Erratum in: Lancet Oncol. **16**(8), e379 (2015). https://doi.org/10.1016/S1470-2045(15)00119-9. PMID: 21845765

Benedict, R.H., Meyer, J.A.: Radiofrequency radiation: effects on biological systems and health. J. Environ. Health Res. **23**(1), 45–56 (2021).

Benedict, R.H., Meyer, J.A.: Radiofrequency radiation: effects on biological systems and health. J. Environ. Health Res. **22**(3), 235–247 (2017)

Federal Communications Commission (FCC): Radiofrequency Safety (2021)

Field Strength Meter for EMF exposure, SMP3 Field strength meter, Wavecontrol.com: https://www.wavecontrol.com/rfsafety/images/data-sheets/en/SMP2_Datasheet_EN.pdf. Accessed on 1 July 2024

https://upload.wikimedia.org/wikipedia/commons/c/cf/EM_Spectrum_Properties_edit.svg

https://ntp.niehs.nih.gov/whatwestudy/topics/cellphones

https://www.niehs.nih.gov/newsreleases/high-exposure-to-radio-frequency-radiation-associated-with-cancer-in-male-rats

IARC Monographs on the Evaluation of Carcinogenic Risks to Humans, vol 102. Retrieved from IARC website

ICNIRP: ICNIRP note: critical evaluation of two radiofrequency electromagnetic field animal carcinogenicity studies published in 2018. Health Phys. **118**(5), 525–532 (2020). https://doi.org/10.1097/HP.0000000000001137

IEEE: IEEE Standard for Safety Levels with Respect to Human Exposure to Radio Frequency Electromagnetic Fields (2020)

IARC – International Agency for Research on Cancer, World Health Organization (WHO). "IARC Monographs on the Evaluation of Carcinogenic Risks to Humans: vol. 102. Retrieved from IARC website". https://monographs.iarc.who.int/

International Commission on Non-Ionizing Radiation Protection (ICNIRP): ICNIRP Guidelines for Radiofrequency Radiation (2020)

Interphone Study Group: Brain tumour risk in relation to mobile telephone use: results of the Interphone study. Int. J. Epidemiol. **40**(3), 670–680 (2011)

Ishai, P.B., Davis, D., Taylor, H., Birnbaum, L.: Problems in evaluating the health impacts of radio frequency radiation. Environ. Res. **243**, 115038 (2024). https://doi.org/10.1016/j.envres.2022.115038; (https://www.sciencedirect.com/science/article/pii/S0013935122023659). ISSN 0013-9351

NTP: National Institutes of Health (NIH), News Releases, Thursday, November 1, 2018, High exposure to radio frequency radiation associated with cancer in male rats. https://www.nih.gov/news-events/news-releases/highexposure-radio-frequency-radiation-associated-cancer-male-rats. (2018)

WaveControls SMP2 Electromagnetic Field Meter Specification Datasheet: https://www.raecorents.com/amfile/file/download/file/508/product/836/. Accessed on 10 June 2024

World Health Organization (WHO): Electromagnetic Fields and Public Health: Mobile Phones (2020)

How to Preserve Students' Engagement Through Google Classroom Activities? A Case Study with Legal English Students at SEE University

Rufat Osmani

Abstract

The use of technology in education lately has changed the traditional way of teaching and learning standards in the way that it has opened new chances for connection, collaboration, and participation. Although the availability of many learning management systems for schools is wide, Google Classroom has emerged as one of the most compelling tools for distance and blended learning settings. Despite many benefits, Google Classroom's most delicate challenge is how to keep students engaged over time. Preserving students' focus and motivation during Google Classroom activities is seen as a perpetual concern for teachers in a digital area where interferences are numerous and concentration time is short. On that account, there is a need for exploring and implementing adequate strategies that can retain and increase students' engagement surrounded in this digital learning environment. The focus of this study is to search for different strategies that are implemented by teachers to promote and preserve student engagement in Google Classroom activities. With a focus on contributing to an extensive overview of best practices in this field, the latest research findings, applicable knowledge from experienced teachers, and contemporary approaches will be examined.

Keywords

Google classroom • Technology • Activities • Engagement

R. Osmani (✉)
South East European University, Tetovo, North Macedonia
e-mail: r.osmani@seeu.edu.mk

K. Dhoska and E. Spaho (eds.), *Bridging Horizons in Artificial Intelligence, Robotics, Cybersecurity, Smart Cities, and Digital Economy*, Sustainable Economy and Ecotechnology, https://doi.org/10.1007/978-3-031-72029-1_28

1 Introduction

The use of technology in education lately has changed the traditional way of teaching and learning standards in the way that it has opened new chances for connection, collaboration, and participation. Although the availability of many learning management systems for schools is wide, Google Classroom has emerged as one of the most compelling tools for distance and blended learning settings. Being favourite option among teachers globally is due to its easily understood layout, simple combination with Google's workspace functionality utensils and adaptability.

Despite many benefits, Google Classroom's most delicate challenge is how to keep students engaged over time. Preserving students' focus and motivation during Google Classroom activities is seen as a perpetual concern for teachers in a digital area where interferences are numerous and concentration time is short. On that account, there is a need for exploring and implementing adequate strategies that can retain and increase students' engagement surrounded in this digital learning environment.

The focus of this study is to search for different strategies that are implemented by teachers to promote and preserve student engagement in Google Classroom activities. With a focus on contributing to an extensive overview of best practices in this field, the latest research findings, applicable knowledge from experienced teachers, and contemporary approaches will be examined.

In order to promote an advancing digital classroom community and provide a better learning experience, we need to include multimedia features and game-like elements to foster peer engagement and provide personalized feedback. It is crucial for teachers to find ways to get the most out of student participation in Google Classroom activities as we undergo the constantly changing setting of education in the digital era. By including efficient strategies and achieving the full potential of this changing platform, teachers may increase students' learning experiences.

2 Review of Literature

In the digital age, instructors suffer from the issue of maintaining student interest in online learning settings. Google Classroom, a popular learning management system, allows you to design dynamic and engaging activities that can improve student engagement. This literature review addresses approaches and best practices for maintaining student engagement through Google Classroom activities, focusing on scholarly research and practical observations.

2.1 Creating Interactive Assignments

Teachers are looking for innovative ways to involve students in the learning process as a result of the integration of technology in the classroom. Teachers can develop interactive assignments that encourage greater learning and active engagement using the platform offered by Google Classroom. This paper explores methods for creating interactive assignments with Google Classroom, backed up by both academic studies and real-world applications.

To create effective interactive assignments through Google Classroom, educators can employ the following strategies:

1. Set clear learning objectives: Clearly communicate the learning objectives and expectations for each assignment to guide students effectively and promote understanding (Clark and Mayer 2016).
2. Incorporate multimedia resources: Integrate multimedia elements such as videos, images, and interactive simulations to enhance engagement and facilitate learning (Mayer 2009).
3. Promote collaboration: Design assignments that encourage collaboration and peer interaction, fostering critical thinking and problem-solving skills (Johnson and Johnson 2009).
4. Provide timely feedback: Offer timely and meaningful feedback on student work to support learning and growth, utilizing Google Classroom's feedback features effectively (Hattie and Timperley 2007).
5. Differentiate instruction: Tailor assignments to accommodate diverse learning needs and preferences, offering choice and flexibility to empower students (Tomlinson 2001).

Teachers should provide tasks that promote interaction to sustain students' interest and participation. Making use of Google Classroom's capabilities, like discussion boards, enables peer-to-peer learning and cooperative problem-solving. Multimedia components like movies, pictures, or interactive tests can improve students' overall involvement in the learning process.

2.2 Facilitating Real-Time Communication

Whenever there is online interaction between teachers and students exchanging information simultaneously is of great importance to have effective collaboration and interaction. Coupled with this, Google Classroom with its different components and integrations helps students to communicate in real time, increases engagement and ensures inclusivity for all students. This study presents methods of taking advantage of Google Classroom to promote synchronous participation supported by different studies and examples drawn from real life situations.

With the intention of promoting real-time communication with students, teachers can use the following techniques through Google Classroom:

1. Provide clear communication criterion: Focused and productive communication requires agreeing upon clear expectations and directives for communication within Google Classroom participation criterion and a deadline for response (Baker et al. 2010).
2. Promote active participation: In live and chat sessions support active participation from students through asking questions, aiding discussions and obtaining opinions. This boosts students' involvement and also improves the quality of communication (Harasim 2012).
3. Create opportunities for collaboration: Activities that make students work together and projects that require straightway communication between students, in particular, group discussions, exchange of views and peer review activities promote teamwork and peer learning through Google Classroom (Means et al. 2010).

When learning takes place virtually, preserving the participation of students demands effective communication. In this case, Google Classroom provides real-time communication possibilities that include the section for comments and the Stream function for announcements and updates. This requires the active involvement of teachers in these conversations with the intention of encouraging a feeling of belonging and allowing them to react quickly to any issues or questions.

2.3 Delivering Feedback on Time

In online settings where direct interactions are insufficient, instant feedback is of vital importance to the development and learning of students. Google Classroom's other key point, which includes giving feedback on time, is offered to teachers as one of the several features. This study reveals ways of using Google Classroom to deliver on-time and effective feedback based on the scholarly work and experience of the teachers who participated in this study.

Students get involved when they get timely feedback. One of the tools that Google Classroom possesses is the Assignment tool where teachers give each student personalized feedback about their work. When students are evaluated for their work, they comprehend their progress, and constructive comments on their work also foster a sense of development and constant upgrading. Timely and critical remarks preserve students' interest in online learning environments. Hattie and Timperley (2007) emphasize the importance of feedback in fostering development and learning. Teachers can apply comments and explanation items in Google Classroom to provide students with instant feedback on their work, therefore enhancing participation and increasing continuing development.

When giving feedback to students in Google Classroom, the methods below can be used:

1. Be specific and effective. Provide specific and constructive feedback, focusing on students' performance with reference to learning objectives. Point out the

strengths of your students and the points that need improvement, then advise on how to further develop them in the future (Nicol and Macfarlane-Dick 2006).
2. Encourage reflection. The given feedback should encourage reflection among the learners and help them come up with ways to improve. Feedback needs to be considered as an instrument for learning and development; therefore, encourage attention to its development (Hattie and Timperley 2007).
3. Use a variety of feedback modes: With the aim of indulging different student preferences and requirements, apply a variety of feedback modes, including written comments, audio feedback, and video feedback. Different feedback forms can be used to identify what makes an impression on students (Sadler 2010).

2.4 Using Google Meet for Virtual Interaction

The sense of connection in online learning can be maintained through face-to-face communication. Google Meet allows teachers to have a live chat which when combined with Google Classroom creates a more personalized learning experience. Teachers need to ensure that they frequently conduct online chats, Q&As or team activities with their students in order to enhance student engagement.

1. Establish guidelines for communication. Determine the expectations and norms of participation, as well as the etiquette regarding interaction, respectfulness, and communication protocols (Baker et al. 2010).
2. Encourage active participation. Encourage students' thoughts during Google Meet sessions by giving them an opportunity to share their ideas and ask for insights from their peers. Students can also engage through attendance response system, quizzes, and activities that promote interaction (Harasim 2012).
3. Utilize small groups effectively. Promote student–student interaction in sub-groups that are used for small group discussions during instruction groups or projects. Ensure clear guidance and instructions about what is expected from each subgroup activity so that students will express themselves (Means et al. 2010).

2.5 Promoting Self-paced Learning

With the aim of allowing various learning styles and values, it is required to apply differentiated teaching. Through Google Classroom, students can have access to assigned tasks, resources, and exams whenever it suits them. This versatility enables students to deal with their education and learn autonomously.

Autonomous learning enables students to work through the course materials following their personal timetable, thus allowing more flexibility and a personalized learning experience. A personalized learning experience is supported by Google

Classroom through its features and instruments; therefore, this allows tailored education. This investigation examines techniques for supporting autonomous learning through Google Classroom and according to scholarly research and concrete implementation.

If given the possibility of autonomous learning and independence, students' interest and participation in Google Classroom can be increased. By the same token, Nicol and Macfarlane-Dick (2006) state that in order to empower students and encourage deep learning, teachers need to support formative assessment and self-regulated learning. For that reason, applying components like differentiated assignments, changing rate alternatives and self-assessment instruments may enable students to deal with their learning.

Given these points, by applying the below strategies, Google Classroom can be used to effectively strengthen autonomous learning:

1. Provide clear expectations. Autonomous learning expectations and guidelines that include aims, deadlines, and criteria for evaluation should be explicitly delivered. Consequently, students will know their responsibilities and adhere to the plan in their learning (Guskey and Bailey 2001).
2. Embrace flexibility. To cater to the learning needs and preferences of students, consider offering a range of options, in assignments, topics, and pacing. By empowering students to make decisions about their learning journey, you can foster engagement and motivation (Tomlinson 2001).
3. Ensure support. As students navigate their learning paths, it's crucial to provide them with guidance and resources. Supply materials, feedback, and opportunities for reflection to monitor their progress and address any challenges they may encounter (Hattie and Timperley 2007).

3 Methodology

This study was carried out at South East European University, Tetovo, Republic of North Macedonia, in the academic year 2023–2024. In order to gather data about strategies for preserving student engagement through Google Classroom activities, two surveys were administered, one for students and one for teachers. The survey for students consisted of 10 questions whereas the survey for teachers consisted of 5 questions. Teachers were requested to put forward their experiences and strategies on how to maintain student engagement through different activities and over Google Classroom.

4 Results

Fifty students from Legal English 1 and Legal English 2 participated in this study. The first question was, "How often do they use Google Classroom to complete tasks and assignments?" The outcome of this question revealed that 62% of participants said they used Google Classroom every day, 24% said they used it three to five times a week, 10% said they used it once or twice a week, and 4% said they used it less frequently than that. Thereupon, the results show that students make extensive use of Google Classroom, with most using it every day to finish assignments.

As far as the second question is concerned, "Which types of Google Classroom activities did they find most appealing and pleasant?" They were provided with some options, such as discussions, multimedia assignments, quizzes, and any other activities they considered important. In relation to this question, 42% of participants found discussions to be the most engaging activity in Google Classroom, 30% said that it was multimedia assignments for them, 18% voted for quizzes as the most engaging activity, and 10% mentioned other activities such as collaborative projects, interactive presentations, and virtual field trips.

With regard to the third question, "What features or elements within Google Classroom do you believe contribute most to keeping you engaged in learning?" Here as well, models provided covered interactive content, timely feedback, collaborative tools, and any instrument that they found practical. The data obtained indicated that 38% of respondents identified interactive content like videos, simulations, and interactive presentations as being the most engaging feature. Further on, 26% underlined timely feedback from teachers about assignments and assessments, while 20% pointed out collaborative tools such as Google Docs, which are essential for group projects and discussions. Lastly, well-organized course materials were considered significant by 16%, as were clear instructions for assignments and activities. The fourth question was, "What are your preferred methods of receiving feedback for your homework assigned through Google Classroom?" The data collected showed that written comments were preferred by a majority (58%). In response to this question, audio feedback was chosen by another group (32%) where teachers provided spoken feedback recorded as audio files, while just a few students (10%) would prefer different types of feedback. These findings were unexpected due to the fact that no teacher in the interview ever gave audio or video feedback to their students.

The fifth question asked students, "In what ways do you feel that Google Classroom activities support your learning and understanding of course material?" The options given to students contained reinforcing concepts, facilitating collaboration, providing practice opportunities, or any other contribution. A significant number of students, or 73% of participants, reported that Google Classroom activities help enhance concepts learned in class. Quizzes, discussions, and interactive assignments were also acknowledged as effective tools for supporting understanding. By including a variety of activities that support different learning styles and preferences, teachers can create active and appealing learning environments that promote deep understanding and proficiency of the material covered according to the course syllabus. One of the conclusions drawn from this question is to explore how different kinds of activities among the students have concrete consequences for student learning outcomes and involvement.

One of the conclusions drawn from this question is to explore how different kinds of activities among the students have concrete consequences for student learning outcomes and involvement.

Question six asked, "Do you feel that you have opportunities to express your thoughts and opinions during Google Classroom activities and to provide examples of how these opportunities were facilitated?" From the data collected, 72% agreed that they had occasions when they could air their views in this context, whereas 28% stated otherwise. This may be a result of students who are part-time students and who, in this case, are not obliged to attend classes as regular students.

The seventh question asked students, "What challenges do you face in staying engaged during Google Classroom activities?" The given examples included distractions, difficulty understanding instructions, and any technical issues. The responses gathered showed that 42% of participants identified interruptions as a significant problem that hindered them from staying focused throughout Google Classroom activities. These include external factors such as noise, housework duties at home, and personal gadgets; 32% mentioned technological issues like slow internet connections, device compatibility problems with particular platforms, or platform malfunctions as drawbacks to engagement; 18% referred to difficulties in comprehending instructions given within Google Classroom activities, which caused misunderstandings and withdrawal from participation; and finally, 8% indicated dissatisfaction about some Google Classroom activities being non-interactive or lacking any fun elements that would help motivate them into participating more actively, thereby paralyzing motivation for continued participation.

The eighth question asked students, "How do you manage your time and pace yourself when completing your assignments or activities in Google Classroom?" The results from this question indicate that students apply different arrangements of working time strategies to move themselves further. These include identifying priorities, creating schedules, dividing large tasks into smaller, more manageable blocks, applying efficient time usage, establishing time frames, using reminders and warnings, and taking breaks. These strategies were applied by the majority of students who conducted the survey for the purpose of optimizing and maintaining a balanced workload.

The ninth question asked students, "Do you feel that you have a sense of autonomy and control over your learning within Google Classroom?" The obtained data showed that 64% of participants indicated that they feel they have a sense of autonomy and control over their learning within Google Classroom, and the reason for this is stated as the flexibility offered by Google Classroom, allowing them to access resources, complete assignments, and engage with course materials at their own pace and convenience. Contrastingly, 36% of participants reported feeling that they lack a sense of autonomy and control over their learning because some participants expressed frustration with rigid assignment structures and deadlines, feeling constrained in their ability to manage their time and pace themselves effectively.

The tenth question asked students, "How do you communicate with your peers during Google Classroom activities, and how does this communication affect your involvement in the learning process?" The obtained findings indicate that peer interaction within Google Classroom activities positively affects students' engagement with the learning process. When students are given the opportunity to discuss, work together, and assess their peers work in virtual meetings, they feel more motivated, see an increasing understanding, and feel a sense of belonging. Moreover, the majority of the students who conducted the survey stated that they engage in discussions, ask questions, and share their perceptions with their peers.

Understanding the standpoints and methodologies of instructors is decisive for fostering powerful coaching strategies and growing student participation in various instructional environments. In this paper findings from an extensive survey given to a various institution of educators which included two high school instructors and three university instructors can be found out. Teachers' survey consisted of five questions which required them to state their opinions about the strategies that they use to improve student engagement in Google Classroom, a digital learning environment.

For the purpose of better comprehending the opinions and experiences of the teachers who participated in this survey and to reveal how they encourage students' involvement in Google Classroom, an analysis of the survey answers is done.

The first question sought to know how teachers structure their Google Classroom activities to capture the interest of students and, at the same time, maintain their engagement throughout the lesson. Respondents answered that they use different strategies, such as:

1. Blending real-world plots and multimedia for the effective purpose of relating theory to practice and meeting the diverse learning styles among students.
2. Applying a problem-based classroom model to promote involvement, encourage discussion, problem-solving, and teamwork during online meetings, and finally creating small groups in order to have peer interaction.
3. Producing tasks that encourage student autonomy and exploration by offering a variety of assignments and incorporating developmental assessment to frame the learning process and preserve students' attention.
4. Implementing audio-visual content and collaborative tools, for example, videos and simulations, in addition to opportunities for peer collaboration, to strengthen student curiosity and understanding.
5. Developing a sense of belonging using icebreakers, audio-visual presentations, engagement tools, and practical activities during classes all promote motivation and active involvement.

The second question asked teachers, "What types of multimedia or interactive content do you incorporate into your Google Classroom activities to enhance student involvement?" The obtained data revealed the following:

1. Combining different multimedia content like videos, animations, podcasts, and infographics to satisfy different learning preferences and interests, while integrating interactive elements such as quizzes, polls, and virtual simulations to promote active involvement and deeper understanding.
2. Using other forms of multimedia, such as already-recorded classes, which, upon observation, showed very good student learning; highly interactive presentations; online discussion forums; and interactive means such as Google Docs, to support interactive learning experiences.
3. Using interactive digital books, case studies, and a multimedia laboratory with educational games, simulations, and virtual reality experiences, can help them practice what they have learned and think critically.
4. The incorporation of different types of media content—educational videos, TED talks, interactive presentations, and virtual field trips—to enhance student engagement and motivation, further enhanced through multimedia projects where students create their own videos or podcasts to show understanding.
5. Integrating multimedia resources like educational videos, interactive tutorials, and virtual experiments, along with interactive whiteboard tools for live demonstrations and problem-solving sessions, and online simulations and games to reinforce key concepts and engage students in active learning experiences.

The third question asked teachers, "How do you exploit features like polls, quizzes, or discussion boards within Google Classroom to promote student participation and interaction?" The findings reveal the following:

1. Incorporating quizzes and surveys to evaluate understanding and supporting active involvement that was incorporated into lectures to get instant feedback. Also, discussion forums encourage active participation and team learning, as well as meaningful discussions over topics of the course.
2. Course tools include active assessment and online discussions using polls, quizzes, and discussion boards. Such tools provide a sense of community and intellectual exchange, assessing prior knowledge, reviewing content, and providing reflective dialogue among students.
3. Exploiting polls, quizzes, and discussion boards to stimulate participation and interaction, gathering feedback, checking understanding, and fostering collaborative problem-solving. These instruments actively engage students with course material and each other.
4. Using polls, quizzes, and discussion boards as interactive assessment tools throughout lessons to monitor students' progress and encourage participation. Discussion boards provide group learning experiences and build a friendly online community.

5. Incorporate into the delivery other poll activities, quizzes, and discussion boards to encourage involvement and cooperation in testing understanding and further encouraging student-directed discussions and ideas. These features contribute to active and engaging online learning.

The fourth question asked teachers, "Do you have any specific strategy or technique that you use to encourage less active students to participate more actively in Google Classroom activities?", Following are every teacher's answer:

1. Creating a supportive atmosphere where every student's contribution is valued and using activities that involve little risk like anonymous polls, and offering individual feedback to every student to build confidence.
2. Offering multiple ways of participation: written responses, audio recordings, or small group discussion, icebreaker activities, and team-building activities that help with belonging.
3. Implementing differentiated instruction strategies that respond to students' individual needs and forms of engagement and establishing a positive classroom climate, which creates a safe opportunity for students to experiment and share their ideas.
4. Provide clear instructions, give lessons in certain segments, and provide opportunities for peer collaboration to build confidence where contributions are applauded.
5. Individual needs are going to be attended to by incorporating customized strategies. There will be a number of choices in assignments, flexible deadlines, and the use of visual aids to enhance understanding, together with community building and mutual respect among the classmates.

The fifth question asked teachers, "How do you provide timely feedback and reinforcement to students within the Google Classroom setting to keep them motivated and engaged?" Teachers responded as shown below:

1. Applying written comments, audio feedback, and criteria for assessing the assignments to offer constructive feedback within an acceptable time limit, along with praise and recognition for exemplary work, fostering a positive learning environment.
2. Using the Google Classroom feedback features for individual feedback, being constructive, and, in the meantime, not holding back positive comments to acknowledge improvement and to encourage students.
3. Providing timely feedback that provides a ground for action through written comments and rubric-based grading, focusing on areas for improvement while drawing attention to strengths, and giving regular praise and recognition of students' efforts.

4. Focus especially on specific, constructive, and supportive feedback through various means: written comments, grading rubrics, and video and audio recordings, along with public recognition of students' efforts and accomplishments to preserve motivation and involvement.
5. Provide open and helpful feedback that will provoke action on the part of the students through written comments, explanations, remarks, and grading rubrics, recognizing and celebrating students' improvement and accomplishments with personalized notes, virtual badges, or public recognition to build a supportive learning environment.

5 Conclusion

Preserving student involvement in online classes is a challenging task and requires careful preparation and making effective use of the available resources. One of the platforms that provides teachers with most of the needed capabilities to create useful and interesting learning opportunities is Google Classroom. Teachers can create a positive and productive learning environment by applying group tasks, supporting real-time communication, giving feedback at the right time, using virtual interaction through Google Meet, and encouraging learning in which students' progress through the material at their own pace. The data collected from this study point out that, with the purpose of increasing student engagement, it is crucial to deliver cooperative and interactive exercises in Google Classroom.

On the other hand, the findings of the study stress that immediate feedback and learning support for students within Google Classroom are paramount. The study also highlights intensive interaction, personalized feedback, and the availability of learning resources to keep up student engagement. The support of autonomous learning, or learning directed by the learner, helps preserve student interest in Google Classroom. Varied assignments, opportunities to adjust time according to the learners' preferences, and permission for students to take care of their learning have a great effect.

The conclusions identified the constraints and opportunities for development for keeping student involvement in place while using Google Classroom activities. The constraints would include barriers to involvement, problems with the accessibility or usability of technology, and features that bring about distraction or disengagement among learners.

It could also contain a list of best practices and effective tactics for keeping students engaged with Google Classroom, creative technology tools, effective instructional design strategies, and teaching techniques that guide student engagement and motivation.

Finally, to conclude, we can call for more investigation and study within the fields of online learning and student involvement. We may equally point to further research with comparative, longitudinal, or exploratory investigations of new pedagogical approaches and technologies.

Overall, the findings of this study on how to sustain student interest in activities within Google Classroom would add to the knowledge base of best practices and barriers, as well as opportunities that may be acted upon to improve the online learning environment. Findings could presumably set guidelines for policymakers, administrators, and educators for maximizing online learning and promoting student achievement.

A more in-depth analysis should be carried out, with a larger number of students and more time at disposal so that the results are more reliable. For the reason that Legal English students were involved in this study, we could investigate how this method applies to all ESP and Basic English Skills students.

References

Baker, J.D., Passmore, D.L., Smith, W.R.: The impact of establishing a set of explicit, manageable communication norms in virtual teams. Group Decis. Negot. **19**(4), 371–394 (2010)

Clark, R.C., Mayer, R.E.: E-Learning and the Science of Instruction: Proven Guidelines for Consumers and Designers of Multimedia Learning, 4th edn. Wiley (2016)

Guskey, T.R., Bailey, J.M.: Developing Grading and Reporting Systems for Student Learning. Corwin Press, Thousand Oaks, CA (2001)

Harasim, L.: Learning Theory and Online Technologies. Routledge (2012)

Hattie, J., Timperley, H.: The power of feedback. Rev. Educ. Res. **77**(1), 81–112 (2007)

Johnson, D.W., Johnson, R.T.: An educational psychology success story: social interdependence theory and cooperative learning. Educ. Res. **38**(5), 365–379 (2009)

Mayer, R.E.: Multimedia Learning, 2nd edn. Cambridge University Press (2009)

Means, B., Toyama, Y., Murphy, R., Bakia, M., Jones, K.: Evaluation of Evidence-Based Practices in Online Learning: A Meta-analysis and Review of Online Learning Studies. US Department of Education. finalreport.pdf (ed.gov) (2010)

Nicol, D., Macfarlane-Dick, D.: Formative assessment and self-regulated learning: a model and seven principles of good feedback practice. Stud. High. Educ. **31**(2), 199–218 (2006)

Sadler, D.R.: Beyond feedback: developing student capability in complex appraisal. Assess. Eval. High. Educ. **35**(5), 535–550 (2010)

Tomlinson, C.A.: How to differentiate instruction in mixed-ability classrooms. ASCD. Classrooms-2nd-Edition-By-Carol-Ann-Tomlinson.pdf (santillana.com.co) (2001)

Intelligent Digital Framework for Art Investment Amid Geopolitical Conflicts: A Sustainable Economic Approach

Ornela Ramašauskaitė

Abstract

This study addresses the challenges of art investment in geopolitical conflict zones by proposing a framework for a predictive model. Utilizing meta-analysis, Q-sort methodology, and Hierarchical Cluster Analysis (HCP), the research identifies key factors influencing art valuation in unstable regions. The framework integrates political stability indicators, economic factors, cultural significance metrics, and market trends, employing machine learning algorithms to forecast price movements and assess investment risks. Results demonstrate the significant impact of geopolitical factors on art valuation and provide insights for model development. The study concludes that this framework offers a foundation for a robust tool to navigate art investment in conflict zones, fostering a more sustainable approach. Future trends impacting effectiveness include advancements in AI, the growing importance of ESG factors, and increasing art market digitization. Potential barriers to adoption include data availability in conflict zones, resistance from traditional market players, and ethical concerns regarding cultural heritage commodification. These can be mitigated through collaborative data-sharing initiatives, stakeholder education, and ethical guideline development. The framework's successful implementation could revolutionize art investment strategies in volatile regions, contributing to cultural heritage preservation and sustainable economic development.

Keywords

Art investment • Geopolitical conflicts • Risk assessment • Cultural heritage

O. Ramašauskaitė (✉)
Mykolas Romeris University, Vilnius, Lithuania
e-mail: ornelos@gmail.com; orramasauskaite@stud.mruni.eu

Women Researchers Council UNEC, Baku, Azerbaijan

K. Dhoska and E. Spaho (eds.), *Bridging Horizons in Artificial Intelligence, Robotics, Cybersecurity, Smart Cities, and Digital Economy*, Sustainable Economy and Ecotechnology, https://doi.org/10.1007/978-3-031-72029-1_29

1 Introduction

The global art market has evolved significantly from what Frey and Pommerehne (1989) initially characterized as a complex interplay between muses and markets. Today, it is a sophisticated investment vehicle attracting collectors, investors, and institutions worldwide. However, as Mandel (2009) demonstrates, art serves as an investment and a conspicuous consumption good, adding layers of complexity to its valuation in volatile regions. The fundamental question Chanel (1995) posed regarding art market predictability becomes particularly pertinent when examining investments in geopolitically unstable regions.

In the context of rapid technological transformation, as outlined by Schwab (2017) and Vidas-Bubanja (2017), traditional approaches to art investment often fail to capture the nuanced interplay between geopolitical factors and market dynamics. This limitation becomes particularly evident in conflict zones, where, as MacDonald et al. (2010) argue, the intersection of visual culture and geopolitics creates unique challenges for valuation and risk assessment.

The research addresses the pressing need for a comprehensive, data-driven approach to art investment in geopolitical conflict zones. Building upon Ginsburgh and Throsby's (2006) foundational work in art economics, the study proposes developing a predictive framework that addresses the unique challenges presented by art markets in conflict-affected regions. The framework integrates Ziegler's (2017) concept of sustainable innovation with McLagan and McKee's (2012) insights into the visual culture of political activism, providing a robust foundation for understanding art valuation in politically volatile contexts.

The methodology employs Lipsey and Wilson's (2001) meta-analytical approaches to systematically synthesize existing literature and empirical data. This synthesis aims to identify key factors influencing art valuation and investment decision-making in conflict zones. The research also draws upon Ginsburgh et al.'s (2006) work on price indices computation to develop more accurate valuation metrics for volatile markets. Furthermore, following Elton's (1999) distinction between expected and realized returns, the study proposes a framework that accounts for the unique risk factors in conflict zones.

The proposed framework acknowledges the multifaceted nature of art investment in geopolitical conflict zones, incorporating Kräussl et al.'s (2016) insights into market bubbles and their implications for risk assessment. The framework enables a more nuanced understanding of market dynamics in conflict zones by leveraging advanced techniques such as meta-analysis, Q-sort methodology, and Hierarchical Cluster Analysis (HCP).

This research contributes to the existing body of knowledge by addressing the critical gap between traditional art market analysis and the unique challenges posed by geopolitical conflicts. The resulting framework aims to revolutionize art investment in conflict zones by providing a more informed, responsible, and equitable approach to market analysis and decision-making.

2 Literature Review

The intersection of art investment and geopolitical conflict zones has attracted significant attention from researchers, economists, and market analysts. As the global art market continues to evolve and expand, the impact of political instability and social unrest on art valuation has become increasingly apparent (McAndrew 2023; Oosterlinck 2017).

2.1 Art Investment and Market Dynamics

Research on art as an investment vehicle has evolved significantly over the past decades. Baumol (1986) initially characterized art markets as "floating crap games," highlighting their unpredictability. However, subsequent studies by Campbell (2008) and Renneboog and Spaenjers (2009) demonstrated more structured approaches to understanding art market behavior. The relationship between art and financial returns has been extensively studied by Goetzmann et al. (2011), who established significant correlations between art prices and wealth creation.

2.2 Geopolitical Conflicts and Art Markets

The impact of political instability on art markets has been well-documented. David and Oosterlinck (2012) examined how war, inflation, and monetary reforms affect art markets, while Ingram (2009) explored the intersection of art and geopolitical security. Szary (2012) further investigated the politics of art display in contested territories, providing crucial insights into how political tensions influence art valuation and circulation.

2.3 Technological Innovation and Market Prediction

Recent technological advances have transformed art market analysis. Aubry et al. (2019) demonstrated how machine learning can complement human expertise in asset valuation. Li (2020) explored digital transformation in creative industries, while Aljohani (2022) proposed frameworks for predicting market needs using AI and Big Data technologies. These technological developments align with Velthuis and Coslor's (2012) observations about the increasing financialization of art.

2.4 Risk Assessment and Investment Strategy

Risk assessment in art investment has evolved from early Stein (1977) studies to more sophisticated analyses. Korteweg et al. (2016) developed selection-corrected returns perspectives, while Worthington and Higgs (2004) examined risk, return, and portfolio diversification in major painting markets. David et al. (2013) identified market inefficiencies mainly affecting conflict zones.

2.5 Cultural Economics and Value Formation

Understanding value formation in art markets requires considering both economic and cultural factors. Throsby (2011) examined the political economy of art, while Velthuis (2005) analyzed the symbolic meanings of prices in contemporary art markets. These perspectives are crucial for developing comprehensive valuation models for conflict-affected regions.

2.6 Methodological Approaches

The complexity of art market analysis demands sophisticated methodological approaches. Brown (1996) and Watts and Stenner (2012) provide foundational work on Q methodology, while Charrad et al. (2014) and Everitt et al. (2011) offer robust frameworks for cluster analysis, supporting the development of nuanced market segmentation strategies (Table 1).

Table 1 Literature review key insights (2004). *Source* the authors' contribution (2024)

Research area	Key insights
Art Investment and Market Dynamics	– Significant correlations exist between art prices and wealth creation, impacting investment strategies (Goetzmann et al. 2011; Renneboog and Spaenjers 2009); – Art markets have evolved from being characterized as "floating crap games" to more structured investment vehicles (Baumol 1986; Campbell 2008).
Geopolitical Conflicts and Art Markets	– Political tensions influence art valuation, display, and circulation in contested territories (Szary 2012; Ingram 2009); – War, inflation, and monetary reforms significantly impact art market dynamics in conflict zones (David and Oosterlinck 2012).
Technological Innovation and Market Prediction	– Machine learning complements human expertise in art asset valuation (Aubry et al. 2019); – Digital transformation is reshaping creative industries and art market analysis (Li 2020; Aljohani 2022); – Increasing financialization of art is driven by technological advancement (Velthuis and Coslor 2012).

(continued)

Table 1 (continued)

Research area	Key insights
Risk Assessment and Investment Strategy	– Evolution from basic to sophisticated risk analysis methods in art investment (Stein 1977; Korteweg et al. 2016); – Market inefficiencies particularly affect conflict zones, requiring specialized assessment approaches (David et al. 2013); – Portfolio diversification strategies need adaptation for conflict zone art investments (Worthington and Higgs 2004).
Cultural Economics and Value Formation	– Art valuation requires consideration of both economic and cultural factors (Throsby 2011); – Symbolic meanings of prices significantly impact contemporary art market dynamics (Velthuis 2005).
Methodological Approaches	– Q methodology provides framework for qualitative analysis of art market perceptions (Brown 1996; Watts and Stenner 2012); – Cluster analysis enables nuanced market segmentation strategies (Charrad et al. 2014; Everitt et al. 2011).
Digital Transformation and Market Access	– Fourth Industrial Revolution technologies are transforming art market access and analysis (Schwab 2017); – Digital platforms are creating new opportunities for art market participation and investment (Vidas-Bubanja 2017).
Sustainable Innovation in Art Markets	– Sustainable innovation principles guide development of new investment strategies (Ziegler 2017); – Integration of sustainability considerations with market analysis enhances investment decisions (McLagan and McKee 2012).

3 Methodology

The study employs a multifaceted methodological approach combining meta-analysis, Q-sort methodology, and Hierarchical Cluster Analysis (HCP) to address the research objectives. This comprehensive methodology enables a systematic examination of existing literature, identifying key valuation factors and developing a robust predictive model for art investment in conflict zones.

3.1 Meta-Analysis

The research employs meta-analytical techniques, as Lipsey and Wilson (2001) outlined, to synthesize findings from existing literature on art investment and geopolitical conflicts. This systematic approach aggregates and analyzes data from scholarly articles, market reports, and empirical studies to identify patterns and relationships between variables.

The meta-analysis process follows several structured steps:

1. Literature Identification: Following Ginsburgh and Throsby's (2006) comprehensive approach to art economics, the study systematically identifies relevant literature through database searches and citation analysis.
2. Data Extraction: Key findings and statistical data are extracted from selected studies, with particular attention to Chanel's (1995) work on market predictability and Kräussl et al.'s (2016) analysis of market dynamics.
3. Effect Size Calculation: The analysis employs statistical methods to calculate standardized effect sizes across studies, focusing on relationships identified by David and Oosterlinck (2012) regarding war's impact on art markets.
4. Heterogeneity Assessment: Following Charrad et al.'s (2014) methodological framework, the study examines variations in findings across different studies and contexts.

The meta-analysis reveals several significant themes:

1. Market Dynamics: As Velthuis and Coslor (2012) demonstrate, the increasing financialization of art markets creates unique challenges in conflict zones.
2. Technological Impact: Building on Li's (2020) research, digital transformation significantly influences art market analysis and investment strategies.
3. Risk Assessment: Incorporating Worthington and Higgs's (2004) findings on risk and return characteristics in art markets, particularly in volatile regions.
4. Cultural Factors: Drawing from Throsby's (2011) work, the analysis considers the intersection of cultural economics and political instability.

These findings establish a foundational understanding of the complex relationships between art investment and geopolitical factors, informing the subsequent phases of the research methodology.

4 Q-Sort Analysis

To systematically categorize and prioritize the factors influencing art valuation in conflict zones, the study implements the Q-sort methodology developed by Brown (1996). This methodological approach encompasses several sequential steps:

1. A comprehensive compilation of statements about art investment and geopolitical conflicts was extracted from the systematic literature review.
2. A diverse panel of experts, comprising art market analysts, economists, and geopolitical specialists, was assembled to participate in the Q-sort process.

3. The expert panel engaged in a structured sorting exercise, categorizing statements according to their perceived significance and relevance to art valuation in conflict zones.
4. Factor analysis was applied to the resulting Q-sorts to identify and quantify the most significant factors influencing art valuation in geopolitical conflict zones.

As Watts and Stenner (2012) outline, this methodological framework facilitates a systematic and subjective assessment of the key variables driving art investment decisions in unstable regions while maintaining scientific rigor and reproducibility.

4.1 List of Statements

1. Political instability in a region significantly impacts the value of artworks originating from that area.
2. Economic sanctions imposed on a country can lead to a decrease in the value of its art market.
3. Artworks from conflict zones are often undervalued due to the perceived risks associated with investing in them.
4. The media coverage of geopolitical conflicts can influence the public perception and value of artworks from affected regions.
5. Investors are more likely to avoid investing in artworks from countries with a history of armed conflicts.
6. The cultural significance of an artwork can outweigh the potential risks associated with investing in it, even if it originates from a conflict zone.
7. Investing in artworks from geopolitical conflict zones can be seen as a form of cultural preservation and support.
8. The lack of a stable and transparent art market infrastructure in conflict-affected regions can deter potential investors.
9. Geopolitical conflicts can lead to the destruction or loss of cultural heritage, increasing the rarity and potential value of surviving artworks.
10. The provenance and authenticity of artworks from conflict zones may be more difficult to verify, leading to increased investment risks.
11. Investors may view artworks from geopolitical conflict zones as a means of diversifying their investment portfolios.
12. The emotional and psychological impact of geopolitical conflicts can influence collectors' preferences and the overall demand for artworks from affected regions.
13. Governments and international organizations can play a crucial role in protecting and promoting the art market in geopolitical conflict zones.
14. The use of digital technologies, such as blockchain, can help to establish the provenance and authenticity of artworks from conflict zones, reducing investment risks.

15. Collaborations between cultural institutions, investors, and local communities in conflict-affected regions can help to foster a more sustainable and equitable art market.
16. The lack of reliable data and market transparency in geopolitical conflict zones can hinder accurate valuation and investment decision-making.
17. Investing in artworks from conflict zones can be seen as a way to support the livelihoods of artists and cultural practitioners affected by geopolitical instability.
18. The potential for social and political change in conflict-affected regions can create opportunities for long-term growth in the art market.
19. Ethical considerations, such as the potential for art investments to fund or perpetuate conflicts, should be taken into account when investing in artworks from these regions.
20. The increasing globalization of the art market can help to raise awareness and interest in artworks from geopolitical conflict zones, potentially leading to increased investment and cultural exchange.

4.2 Key Factors

The Q-sort methodology revealed a set of key factors that experts consider most important and relevant to art valuation in conflict zones (Table 2).

The Q-sort results provide a hierarchical framework of the key factors driving art valuation in conflict zones, with political stability and the risk of armed conflict identified as the most critical considerations. This framework emphasizes the need for investors to adopt a holistic approach that takes into account a wide range of political, economic, and cultural factors when making investment decisions in these challenging contexts.

Table 2 The key factors after q-sort analyses (2024). *Source* the authors' contribution (2024)

Rank	Key factor
1	Political stability and the risk of armed conflict
2	Economic sanctions and their impact on the local art market
3	Cultural significance and provenance of the artworks
4	Market trends and demand for art from the specific region
5	Authenticity and condition of the artworks
6	Artist reputation and historical importance
7	Legal and ethical considerations surrounding the acquisition and sale of art in conflict zones

5 Hierarchical Cluster Analysis (HCP)

To identify patterns and relationships among the key variables identified through the Q-sort process, the study employs Hierarchical Cluster Analysis (HCP). The methodological procedure comprises several sequential steps:

1. The research collects data on key variables identified through the Q-sort process, focusing on art investments in specific geopolitical conflict zones: Russia–Ukraine, Azerbaijan–Armenia, and Israel–Palestine.
2. HCP methodology is applied to group the art investments based on their similarity across the key variables, establishing a hierarchical structure of clusters.
3. The resulting clusters undergo analysis to identify common characteristics and patterns among the art investments within each cluster.

The HCP analysis of art investments in the selected conflict zones reveals distinct investment clusters based on their similarity across the key variables identified through the Q-sort process. The analysis yields three primary clusters:

1. High-risk, high-reward investments: This cluster is characterized by artworks with significant cultural importance and historical value, but also high levels of political instability and economic uncertainty.
2. Moderate-risk, moderate-reward investments: This cluster includes artworks from regions with relatively stable political and economic conditions, but with less cultural significance and market demand.
3. Low-risk, low-reward investments: This cluster consists of artworks from regions with minimal geopolitical tensions and limited market potential, offering a more stable but less lucrative investment opportunity.

The HCP results offer a nuanced understanding of the distinct risk-reward profiles associated with art investment in specific geopolitical contexts. By identifying three clusters of investments with varying levels of risk and potential returns, this study provides investors with a valuable tool for optimizing their investment strategies based on their individual risk tolerance and financial objectives (Table 3).

The application of the multifaceted research approach, combining meta-analysis, Q-sort methodology, and Hierarchical Cluster Analysis (HCP), yielded significant insights into the complex relationship between art investment and geopolitical conflict zones. The results of this study provide a comprehensive understanding of the key factors influencing art valuation in unstable regions and the development of a robust predictive model for art investment.

Table 3 The clusters of art investment (2024). *Source* the authors' contribution (2024)

Cluster	Description	Characteristics
1	High-risk, high-reward investments	– Artworks with significant cultural importance and historical value. – High levels of political instability and economic uncertainty.
2	Moderate-risk, moderate-reward investments	– Artworks from regions with relatively stable political and economic conditions. – Less cultural significance and market demand.
3	Low-risk, low-reward investments	– Artworks from regions with minimal geopolitical tensions and limited market potential. – More stable but less lucrative investment opportunity.

6 The Predictive Model for Art Investment in Geopolitical Conflict Zones

The evolving landscape of art investment in regions affected by geopolitical conflicts necessitates a comprehensive and data-driven approach to market analysis and decision-making. Building upon the findings from the Q-sort analysis and HCP results presented in Sects. 4 and 5, this study proposes a framework for a predictive model designed explicitly for art investment in geopolitical conflict zones. The model's architecture integrates the key factors identified through the meta-analysis and addresses the distinct investment clusters revealed through the hierarchical cluster analysis.

6.1 Theoretical Framework

The proposed predictive model is grounded in the theoretical foundations established by Velthuis and Coslor (2012) regarding art market financialization and Li's (2020) work on digital transformation in creative industries. The model's architecture incorporates the key valuation factors identified through the Q-sort analysis, particularly emphasizing the critical role of political stability and cultural significance in investment decision-making. Furthermore, the clustering patterns revealed through the HCP analysis inform the model's risk assessment and portfolio optimization components.

6.2 Model Architecture and Components

This proposed framework aims to harness the power of machine learning and data-driven insights to provide investors with a more accurate, reliable, and actionable understanding of the art market in conflict-affected regions. By integrating a wide range of relevant data points, including political stability indicators, economic factors, cultural significance metrics, and market trends, the envisioned model will offer a holistic view of the investment landscape, enabling investors to make more informed and strategic decisions (Table 4).

Table 4 The predictive model for art investment in geopolitical conflict zones (2024). *Source* the authors' contribution (2024)

Component	Description
Input Data	1. Political stability indicators: – Conflict intensity; – Government effectiveness 2. Economic factors: – GDP growth; – Inflation; – Economic sanctions 3. Cultural significance metrics: – Artist reputation; – Artwork provenance 4. Market trends: – Auction sales; – Gallery exhibitions
Data Preprocessing	– Data cleaning and normalization – Feature scaling and transformation
Machine Learning Algorithms	1. Random Forest: – Ensemble of decision trees; – Handles high-dimensional data and captures complex relationships 2. Gradient Boosting: – Combines weak learners to create a strong predictive model; – Iteratively minimizes prediction errors 3. Neural Networks: – Multilayer perceptron architecture; – Learns non-linear relationships and patterns in the data
Model Training	– Historical dataset of art investments in geopolitical conflict zones – Target variable: – Direction of price movements (increase or decrease); – Cross-validation and hyperparameter tuning
Model Evaluation	– Holdout sample testing; – Performance metrics: – Accuracy rate in predicting price movements; – Mean Absolute Percentage Error (MAPE); – Confusion matrix analysis
Model Deployment	– Web-based application or integration with existing art investment platforms; – Real-time insights and recommendations based on the latest data

Central to the proposed framework will be the use of advanced machine learning algorithms, such as Random Forest, Gradient Boosting, and Neural Networks. These algorithms, chosen for their ability to handle high-dimensional data, capture complex relationships, and learn non-linear patterns, will form the core of the predictive model. Through a rigorous process of data preprocessing, model training, and evaluation, we aim to develop a robust and accurate predictive tool that can adapt to the ever-changing dynamics of the art market in geopolitical conflict zones.

To ensure the practical applicability and user-friendliness of the proposed model, the framework will emphasize the importance of comprehensive evaluation metrics, such as accuracy rate, Mean Absolute Percentage Error (MAPE), and confusion matrix analysis. These metrics will allow for a thorough assessment of the model's performance in predicting price movements and identifying potential investment opportunities and risks.

Furthermore, the framework will envision the integration of the predictive model into existing art investment platforms or its deployment as a standalone web-based application. By providing real-time insights and recommendations based on the latest data, this model has the potential to revolutionize the way investors approach art investment in geopolitical conflict zones, fostering a more informed, resilient, and sustainable art market.

As the art world continues to grapple with the challenges posed by geopolitical conflicts, the development of a robust predictive model for art investment in these regions becomes increasingly crucial. The framework proposed in this article represents a significant step toward harnessing the power of data-driven insights and machine learning to create a more informed, strategic, and sustainable approach to art investment in the face of global uncertainty. While the actual implementation of the model remains a task for the future, the groundwork laid out in this framework serves as a valuable starting point for researchers, investors, and stakeholders in the art market to collaborate and work toward the realization of this ambitious and transformative project.

6.3 Implementation Framework

The proposed framework harnesses machine learning and data-driven insights to provide investors with an empirically grounded understanding of the art market in conflict-affected regions. Integrating multiple data streams, including political stability indicators, economic factors, cultural significance metrics, and market trends, the model offers a comprehensive view of the investment landscape, enabling evidence-based decision-making.

The selection of machine learning algorithms—Random Forest, Gradient Boosting, and Neural Networks—is based on their demonstrated effectiveness in handling complex, non-linear relationships in financial markets (Aubry et al. 2019). Random Forest algorithms particularly excel in capturing the multidimensional nature of art valuation factors, while Gradient Boosting provides robust predictive capabilities for time-series data. Neural Networks offer the advantage of identifying subtle patterns in market behavior that may not be apparent through traditional analysis methods.

The model evaluation framework employs established performance metrics, including accuracy rate, Mean Absolute Percentage Error (MAPE), and confusion matrix analysis. These metrics, widely used in financial forecasting models (Kräussl et al. 2016), provide a rigorous assessment of the model's predictive capabilities and reliability.

The implementation strategy envisions integration with existing art investment platforms or deployment as an independent web-based application. This approach aligns with current trends in financial technology adoption and ensures accessibility for diverse stakeholders in the art market ecosystem.

As the art market continues to navigate the challenges posed by geopolitical conflicts, developing this predictive model represents a significant advancement in quantitative approaches to art investment. While the actual implementation remains a future endeavor, the framework established through this research provides a methodologically sound foundation for future development and implementation. The model's success will depend on continued collaboration between researchers, investors, and art market stakeholders, working toward realizing this data-driven approach to art investment in geopolitically volatile regions.

7 Challenges in Data Collection and Integration

Developing and implementing the predictive model for art investment in geopolitical conflict zones presents significant methodological challenges that warrant systematic examination. This section analyzes these challenges through the lens of data science methodology and practical implementation considerations, building upon the theoretical framework established in the previous section.

7.1 Data Collection and Integration

The acquisition and integration of comprehensive data present substantial methodological challenges in the context of art investment in conflict zones. Primary concerns relate to data quality and availability, particularly regarding real-time political stability indicators and comprehensive art market data from conflict-affected regions. As Velthuis and Coslor (2012) demonstrate, the heterogeneous nature of art market data creates significant complexities in standardization processes. These challenges are further compounded by inconsistent reporting standards across different regions and varying measurement scales for political stability indicators.

The temporal aspects of data collection pose additional challenges, particularly concerning the time lags inherent in political and economic data reporting from conflict zones. As Kräussl et al. (2016) noted, this temporal disparity can significantly impact the model's predictive accuracy and reliability. Furthermore, the varying frequencies of art market transactions and seasonal fluctuations in market activity necessitate sophisticated approaches to data harmonization and temporal alignment.

7.2 Model Adaptation in Dynamic Contexts

The dynamic nature of geopolitical conflicts necessitates continuous model adaptation and refinement. Building upon Aubry et al.'s (2019) framework for market analysis, the model must maintain flexibility while ensuring predictive stability. Integrating emerging conflict indicators and adapting to evolving market conditions requires sophisticated algorithmic approaches that balance model stability with adaptability to changing circumstances.

7.3 Ethical and Legal Considerations

Implementing the predictive model raises significant ethical and legal considerations that must be carefully addressed within the academic framework. Protecting sensitive geopolitical information and compliance with international data protection regulations represent crucial challenges in the model's deployment. Moreover, the ethical implications of art trading from conflict zones, as discussed by Oosterlinck (2017), necessitate careful consideration of cultural heritage protection and community stakeholder interests.

7.4 Stakeholder Engagement and Implementation Framework

The successful deployment of the model requires comprehensive stakeholder engagement and systematic implementation strategies. Integrating diverse perspectives from art market participants, cultural heritage institutions, and financial technology providers presents complex challenges in coordination and consensus-building. The implementation framework must accommodate these varied stakeholder interests while maintaining scientific rigor and practical applicability.

7.5 Future Research Implications

The implementation challenges identified in this analysis suggest several critical areas for future research in art investment in conflict zones. Methodological development opportunities exist to enhance data collection techniques and advance integration algorithms for heterogeneous data sources. Furthermore, the need for robust ethical frameworks and improved stakeholder engagement protocols presents significant opportunities for academic investigation.

The complex interplay between these challenges necessitates a holistic approach to implementation, considering theoretical rigor and practical applicability. Developing sophisticated methodological approaches becomes crucial for successful model implementation as the art market evolves in response to geopolitical dynamics. This analysis provides a foundation for future research while highlighting the importance of academic rigor in addressing these complex challenges.

Resolving these implementation challenges will require continued collaboration between researchers, practitioners, and stakeholders in the art market ecosystem. Through systematic investigation and methodological refinement, the predictive model can evolve to serve art investors' needs better while respecting ethical considerations and cultural heritage preservation in conflict-affected regions.

8 Conclusion

This research presents a comprehensive framework for art investment in geopolitical conflict zones, addressing a significant gap in the existing literature regarding the intersection of art markets, political instability, and investment decision-making. Through systematic analysis and methodological innovation, the study makes several significant contributions to theoretical understanding and practical application in this field.

The research extends existing theoretical frameworks in art market analysis by integrating political risk assessment with traditional investment metrics. Building upon Velthuis and Coslor's (2012) work on art market financialization, this study demonstrates how geopolitical factors significantly influence art valuation and investment outcomes. The findings particularly advance understanding of how political instability affects market dynamics, extending beyond conventional economic indicators to encompass cultural and social factors. Developing a predictive model framework represents a significant theoretical advancement in quantitative approaches to art investment. By synthesizing insights from meta-analysis, Q-sort methodology, and Hierarchical Cluster Analysis (HCP), the research establishes a robust theoretical foundation for understanding art market behavior in conflict zones. This integration of methodologies addresses the limitations identified in previous studies and provides a more comprehensive analytical framework.

The research findings have substantial practical implications for art market stakeholders. The predictive model framework offers investors a structured approach to evaluating investment opportunities in politically unstable regions, while the identified risk factors provide a systematic basis for portfolio management decisions. The emphasis on data-driven decision-making, supported by machine learning algorithms and comprehensive market analysis, offers practical tools for navigating the complexities of art investment in conflict zones.

Several limitations of the current study warrant acknowledgment and suggest directions for future research. Data availability and quality from conflict zones remain significant challenges, potentially affecting model accuracy and reliability. Future research could address these limitations through:

1. Development of more sophisticated data collection methodologies specifically adapted to conflict zones.
2. Integration of additional variables capturing cultural and social dynamics.
3. Longitudinal studies examining the model's predictive accuracy across different conflict scenarios.

Furthermore, the rapid evolution of technology and geopolitical contexts suggests the need for continuous refinement of the predictive model framework. Future research should explore integrating emerging technologies and alternative data sources to enhance the model's predictive capabilities.

This study significantly contributes to the academic understanding of art investment in geopolitical conflict zones while providing practical tools for market participants. The findings suggest that successful art investment in conflict zones requires a nuanced understanding of political, economic, and cultural factors supported by robust quantitative analysis and ethical considerations.

The research establishes a foundation for future scholarly investigation while offering practical guidance for stakeholders in the art market. As geopolitical tensions continue to impact global markets, the frameworks and methodologies developed in this study provide valuable tools for navigating the complexities of art investment in challenging political contexts.

Through this comprehensive analysis and framework development, the study advances theoretical understanding and practical application in art investment, particularly in geopolitical instability. The findings suggest that while challenges remain, systematic approaches to art investment in conflict zones can yield valuable insights and opportunities for academic research and market practitioners.

References

Alberro, A.: Conceptual Art and the Politics of Publicity. MIT Press, Cambridge (2003)

Aljohani, N.R.: A methodological framework to predict future market needs for sustainable skills management using AI and big data technologies. Appl. Sci. **12**(14), 6898 (2022)

Aubry, M., Kräussl, R., Manso, G., Spaenjers, C.: Machine learning, human experts, and the valuation of real assets. CFS Working Paper Series 635, Center for Financial Studies (CFS) (2019)

Baumol, W.J.: Unnatural value: or art investment as floating crap game. In: Papers and Proceedings of the Ninety-Eighth Annual Meeting of the American Economic Association, May 1986, vol. 76, pp. 10–14 (1986)

Brown, S.R.: Q methodology and qualitative research. Qual. Health Res. **6**(4), 561–567 (1996)

Campbell, R.A.: Art as a financial investment. J. Altern. Investments **10**(4), 64–81 (2008)

Chanel, O.: Is art market behavior predictable? Eur. Econ. Rev. **39**, 519–527 (1995)

Charrad, M., Ghazzali, N., Boiteau, V., Niknafs, A.: NbClust: an R package for determining the relevant number of clusters in a data set. J. Stat. Softw. **61**(6), 1–36 (2014)
David, G., Oosterlinck, K.: War, inflation, monetary reforms and the art market. EHES Working Papers in Economic History, no. 12 (2012)
David, G., Oosterlinck, K., Szafarz, A.: Art market inefficiency. Econ. Lett. **121**(1), 23–25 (2013)
Elton, E.J.: Expected return, realized return and asset pricing tests. J. Finance **54**(4), 1199–1220 (1999)
Everitt, B.S., Landau, S., Leese, M., Stahl, D.: Cluster Analysis (5th ed.). John Wiley & Sons (2011)
Frey, B.S., Pommerehne, W.W.: Muses and Markets: Explorations in the Economics of the Arts. Basil Blackwell, Oxford (1989)
Ginsburgh, V., Mei, J., Moses, M.: The computation of price indices. In: Ginsburgh, V., Throsby, D. (eds.) Handbook of the Economics of Arts and Culture, vol. 1, 1st edn, pp. 947–979. Elsevier (2006)
Ginsburgh, V., Throsby, D. (eds.): Handbook of the Economics of Art and Culture, vol. 1. Elsevier, Amsterdam (2006)
Goetzmann, W.N., Renneboog, L., Spaenjers, C.: Art and money. Am. Econ. Rev. **101**(3), 222–226 (2011)
Ingram, A.: Art and the geopolitical: remapping security at green zone/red zone. In: Ingram, A., Dodds, K. (eds.) Spaces of Security and Insecurity. Geographies of the War on Terror, pp. 257–277. Ashgate, Farnham (2009)
Korteweg, A., Kräussl, R., Verwijmeren, P.: Does it pay to invest in art? A selection-corrected returns perspective. Rev. Financ. Stud. **29**(4), 1007–1038 (2016)
Kräussl, R., Lehnert, T., Martelin, N.: Is there a bubble in the art market? J. Empir. Financ. **35**, 99–109 (2016)
Li, F.: The digital transformation of business models in the creative industries: a holistic framework and emerging trends. Technovation **92**, 102012 (2020)
Lipsey, M.W., Wilson, D.B.: Practical Meta-Analysis. Sage Publications (2001)
MacDonald, F., Hughes, R., Dodds, K. (eds.): Observant States: Geopolitics and Visual Culture. IB Tauris, London (2010)
Mandel, B.R.: Art as an investment and conspicuous consumption good. Am. Econ. Rev. **99**(4), 1653–1663 (2009)
McAndrew, C.: The Art Market 2023. Art Basel & UBS Report (2023)
McLagan, M., McKee, Y. (eds.): Sensible Politics: The Visual Culture of Nongovernmental Activism. Zone Books, New York (2012)
Oosterlinck, K.: Art as a wartime investment: conspicuous consumption and discretion. Econ. J. **127**(607), 2665–2701 (2017)
Renneboog, L.D.R., Spaenjers, C.: Buying Beauty: On Prices and Returns in the Art Market. Center Discussion Paper, 2009–2015. Finance (2009)
Schwab, K.: The Fourth Industrial Revolution. Crown Business (2017)
Stein, J.P.: The monetary appreciation of paintings. J. Polit. Econ. **85**(5), 1021–1035 (1977)
Szary, A.L.A.: Walls and border art: the politics of art display. J. Borderl. Stud. **27**(2), 213–228 (2012)
Throsby, D.: The political economy of art: ruskin and contemporary cultural economics. Hist. Political Econ. **43**(2), 275–294 (2011)
Velthuis, O.: Talking Prices: Symbolic Meanings of Prices on the Market for Contemporary Art. Princeton University Press, Princeton (2005)
Velthuis, O., Coslor, E.: The financialization of art. In: Knorr Cetina, K., Preda, A. (eds.) The Oxford Handbook of the Sociology of Finance, pp. 471–487. Oxford University Press, Oxford (2012)
Vidas-Bubanja, M.: The challenge of going digital. J. Eng. Manag. Competitiveness **7**(2), 126–136 (2017)
Watts, S., Stenner, P.: Doing Q Methodological Research: Theory, Method, and Interpretation. Sage Publications (2012)

Worthington, A.C., Higgs, H.: Art as an investment: risk, return and portfolio diversification in major painting markets. Account. Finance **44**(2) (2004)
Ziegler, R.: Sustainable innovation as a collaborative concept. Innov.: Eur. J. Soc. Sci. Res. **30**(4), 388–405 (2017)

Go European! A Transformative Journey Across Dreams, Models, Expectations, and Realities

Ilir Ciko

Abstract

This paper examines market price differences of identical products sold by the same international companies operating in Albania and the EU markets by assessing four distinct perspectives. It reveals that (i) the economic models and common sense would predict that prices of identical products, sold by the same operators in an open market, based on PPP factors should be lower in Albania, or converge toward equalization through the EU integration process; (ii) the expectations of businesses and consumers in Albania, based on two surveys conducted during Spring 2024 with 150 consumers and 30 businesses, suggests that there should be minor price differences (average +5.5% higher in Albania for consumers and +2.7% for businesses); (iii) the reality, based on price information collected for 150 consumer products, randomly chosen from 13 product categories and sold in Albania by three major European companies, indicates that during Spring 2024, the average price of the products in the Albanian market was +79.7% higher than the average price of the same products, sold by the same operators in the EU markets. The paper analyzes the main factors enabling significant price differences between the Albanian and the EU markets, the burden and cost for the consumers, the detrimental effects for the Albanian society and the economy, and provides consumer-centered, market-based recommendations.

Keywords

Competition • Price differentials • Price discrimination • Dominant position • Monopolies • Oligopolies • Cartels • Internal market • Excessive prices • EU integration • Albania

I. Ciko (✉)
Canadian Institute of Technology, Tirana, Albania
e-mail: ilir.ciko@cit.edu.al

K. Dhoska and E. Spaho (eds.), *Bridging Horizons in Artificial Intelligence, Robotics, Cybersecurity, Smart Cities, and Digital Economy*, Sustainable Economy and Ecotechnology, https://doi.org/10.1007/978-3-031-72029-1_30

Like never before, tourists are coming en-masse to explore the breathtaking views, learn about the history and culture, and enjoy the hospitality of incredibly welcoming people in Albania. The recent surge of tourism during 2023 brought a record 10 million visitors. This was more than 3 times the total population of the country.

Coffee shops in the streets of Tirana are now crowded with foreigners, enjoying the pleasure of the espresso under the constant sunny skies. For most tourists, the espresso in Tirana is great, perhaps even better than in Italy, and, (still) cheaper than in most European cities. Indeed, the typical price of an espresso, currently at 0.8 Euro in Tirana, is notably less than its average price in other European capitals.

Could the price of an espresso, imply that the cost of living in Albania, relative to the European Union countries, is also cheaper?

1 Introduction

Most economists would, probably disagree. The price of an espresso is not considered a suitable indicator to compare consumer prices between countries. Different consumer habits, tastes and preferences, level of competition, coffee varieties, origin, processing and standards, for example, are all factors contributing to notable differences in the final price of an espresso across the world. But, what's more important in many countries including Albania, coffee is subject to different specific taxes resulting in important final price differences. For this reason, despite its practical significance and visibility, differences in the price of an espresso would not provide a sound basis for comparing consumer prices between countries.

Economists are fascinated by this subject and have found other ways to observe, measure and analyze these differences. The most famous and simple method is the "*BigMac*" index,[1] invented and regularly published by The Economist.[2] A "Big Mac" is a hamburger sold by McDonald's in every fast-food business operated by the company and the index compares the worldwide relative price of this product. The "BigMac" index is a simplified and informal measurement based on the theory of purchasing-power parity. This is the notion that in the long run, exchange rates should move toward the rate that would equalize the prices of an identical basket of goods and services (in this case, a hamburger, which comprises local raw materials and labor for preparation) in any two countries. The simplicity of the "BigMac" index makes it favorable for tourists to use the price of a "BigMac" as an indication of how expensive or cheap a country is to travel, eat, stay and live in, making it an interesting tool when choosing a travel destination.

Unfortunately—or fortunately (depending on one's perspective)—McDonald's does not (yet) operate in Albania and the country is not included in the "BigMac" index list. Tourists (and researchers) have to find other ways to compare price levels between Albania and other countries.

[1] "BigMac" Index: https://www.economist.com/big-mac-index.

[2] The Economist: https://www.economist.com.

The purpose of this study is to examine price differences of identical products concurrently traded by the same international firms in the Albanian and EU markets. And, if differences exist, to assess the enabling factors and propose ways to reduce and eliminate them. Price differences are assessed through four distinct perspectives: (i) by considering Albania as an EU candidate country (***dreams***), (ii) by exploring what the economic models suggest (***models***), (iii) by assessing the voices of local consumers and businesses (***expectations***), and (iv) by collecting and analyzing real market information based on price differences charged by the same operators in the Albanian and the EU markets (***realities***).

1.1 Disclaimer

Successful business firms constitute the foundations of a healthy and competitive market economy. The firms analyzed in this study are international companies with a strong reputation, brand name and operations in many countries. They are profitable and experienced businesses in Albania, pay taxes, bring expertise and engage in social responsibility activities. They employ hundreds of persons trained and working "around the clock" to offer quality products to thousands of consumers.

The study is not about the market behavior of individual firms, nor about assessing good or bad management choices. The problems identified in the findings of this research are important and pertinent to the business environment that enables market failures to emerge and sustain in the market. Therefore, the public interest is the central topic in this analysis, rather than assessment of specific businesses practices.

All the information used in this study is public. Any person with access to the internet can follow the methodology to independently verify the accuracy of the findings.

1.2 Definitions: Excessive Discrimination

Throughout the study, reference is often made to the difference (in percent) between the price of a product sold by international EU companies in the Albanian market, converted to Euro using the current exchange rate, and the price of the same identical product sold in the EU home market by the same companies (in Euro, or converted to Euro using current exchange rates).

Discussions with the businesses and consumers surveyed for this study,[3] suggested that when such a difference is *more than double* of their expectations, it could be considered as excessive. For this reason, in this study, whenever the price difference results are *twice* as high than the *average* of *either* the expectations by

[3] Described in the methodology section below.

the business executives, or consumers (*whichever is highest*), it is referred to as an *excessive discrimination.*

The excessive discrimination reveals the degree of price discrimination: how much more expensive, in percent, the price of a product sold in Albania is, on top of the price of the identical product concurrently sold in the EU market by the same market operator. For example, an excessive discrimination of 10% indicates that the same product is sold in Albania at a price 10% higher than in the EU. Furthermore, an excessive discrimination of 100% reveals that the same product is sold in Albania at *double* its current price in the EU markets.

2 Methodology

The methodology of the paper builds on a review of the contemporary theory on the topic of abuse by the firms with market power through price discrimination practices. The research continues by examining price differences between identical products sold in Albania and the EU markets examined by assessing the four perspectives considered in this study: dreams, models, expectations, and realities. The findings of the research are analyzed through a national business competitiveness framework by using the Diamond Model and the results are matched with the theory considerations. The paper concludes by providing consumer-centered, market-based recommendations for all the stakeholders identified in the research.

2.1 Dreams

How the price differences between identical products sold in Albania and the EU markets would be, considering Albania's goals in the near future? This part of the research included a desk review of recent surveys and discussions with various persons and companies contacted for the preparation of this research in order to assess, from a free market perspective, the aspirations of the Albanian society on this specific topic.

2.2 Models

What prevailing prices would economic models predict for products sold in Albania based on the prices of identical products sold in the EU market? From a list of available research on this topic, two references building on comparative price levels of consumer goods and services across different countries were considered. The first, reflects the view of the World Bank which considers the price differences across the world as an important factor when dealing with economic data and the World Development Indicators. The second, is a recent analysis of the EU statistics agency—Eurostat, focusing on 2022 price level indices as a comparison of countries' price levels relative to the EU average. It comprises information on candidate countries, including Albania. Purchasing power parity (PPP) implications are also considered in this section.

2.3 Expectations

What prices would Albanians expect for the products they purchase on a daily basis, relative to the prices of exactly the same products sold in the EU? Perceptions on market prices for specific products involve a degree of subjective judgment which depends on a number of factors such as the features of the products, income level, experience with the product, market competition, standards and regulations, seasonality, existence of substitute and complementary products and future expectations, for example. To identify the expectations from market participants, one similar question was posed to two groups—consumers, and trading firms importing EU products in the Albanian market. Two surveys were organized for the research to identify local market expectations in demand and supply.

2.4 The View of the Businesses

The first survey was conducted with 30 business executives from local and international firms, of different sizes, operating in Albania in the trading sector. All firms had relatively long track records of trading experience with the EU market. Representatives of the firms were asked to respond on the price differences between products they were selling in the local market compared to the price of the same products sold in the EU market, through the following single-question survey (Table 1).

Table 1 Questionnaire used for the survey with business executives

What is the average price of a typical consumer product (like personal/processed food/furniture/cosmetics/appliance/electronics/clothes etc.*) you import from the EU and sell in the Albanian market, relative to the average price of exactly the same product currently sold by your company/partners in the EU markets?*

a. −50% or less, than in the EU markets
b. −30% than in the EU markets
c. −20% than in the EU markets
d. −10% than in the EU markets
e. Same as in the EU markets
f. +10% than in the EU markets
g. +20% than in the EU markets
h. +30% than in the EU markets
i. +50% or more than in the EU markets
*Please elaborate if you wish to comment:*____________________. *Thank you!*

Table 2 Questionnaire used for the survey with the consumers

What do you think is the average price of a typical consumer product (like personal/processed food/furniture/cosmetics/appliance/electronics/clothes etc.*) you regularly purchase in the Albanian market, relative to the average price of exactly the same product currently sold in the EU markets?*
a. −50% or less, than in the EU markets
b. −30% than in the EU markets
c. −20% than in the EU markets
d. −10% than in the EU markets
e. Same as in the EU markets
f. +10% than in the EU markets
g. +20% than in the EU markets
h. +30% than in the EU markets
i. +50% or more than in the EU markets
*Please elaborate if you wish to comment:*____________________. *Thank you!*

The survey was complemented by a follow-up focus group discussion through which the business executives expressed their point of view on this topic based on the experience with trade in goods from the EU market.

2.5 The View of the Consumers

The second survey was conducted with 150 consumers in the Tirana area, who were asked to respond the following question (Table 2).

2.6 Realities

What is the difference between the price of products sold by international companies in Albania from the price of the same products sold by these companies in European markets? To seek answers to this question, price information on specific products was retrieved during March–April 2024 by using the following procedure.

2.7 Collection of Price Information

Getting prices for identical products traded in Albania and EU countries is possible through shop visits, but for a considerable number of products, this approach would require much time and cumbersome efforts, which go well beyond the scope of this study. The main factor that facilitated and accelerated the research process in this study was a simple one: digital solutions. Nowadays more and more international companies sell their products not only through their own supermarkets and shops, but also through *online stores,* which also *reveal the prices* for products offered by the companies. International companies, often provide their online portals to their subsidiaries in other countries so that they can operate the online stores in local languages and with local prices. It is much easier to obtain the prices for the same products sold in any country where the companies operate, by using a simple search in the local online stores a company operates in different countries (if *brandname.com* is a global company, very likely its subsidiary's website in France would be *brandname.fr*, in Albania *brandname.al* and so on).

The price information for the products considered in this study was collected from: (i) the local online stores of the companies selected, (ii) the online stores of the same companies operating in their home EU country, and (iii) the online stores of the same companies operating in various EU and non-EU countries for the cross-country comparative analysis.

> Any person with access to internet, may check online the local or international prices of the products for any company, in any country where the company operates through online stores. Prices are public information and often accessible online.

2.8 Criteria Used for the Selection of the Products and Companies

Specific criteria were applied to determine the list of the products and companies considered in this study:

- **Subsidiary[4] operations in Albania**: companies considered for the study had to operate in the Albanian market as a foreign company subsidiary, with substantial market share and at least a few years of experience in the market.
- **Operations in home country**: to enable cross-country price comparisons, companies considered for the study had to operate in their home market (country of origin), with a strong preference for EU countries, preferably also in other countries.
- **Availability of online stores**: companies considered for the study had to operate their local online stores, both in their home country *and* Albania.
- **Consumer goods only**: products purchased for daily consumption by the typical average Albanian consumer were considered for the panel. Luxury, fancy or capital goods, were excluded from consideration.
- **No products with special taxes**: products on which countries impose different special taxes were excluded from the panel as the different levels of taxes influence the final prices. Examples include cigarettes, alcoholic drinks and coffee.
- **Focus on imported goods only**: there are several market operators offering both imported and domestically produced goods in the sales portfolio of their stores

[4] Subsidiary, branch, joint-ventures, partnerships, licensing etc. – whichever mode of local market entry that allows the operator to trade products on behalf of the global company and is acknowledged as its activity in Albania in its official website.

in Albania. To focus on the topic of the study, only companies primarily selling imported goods were considered.

- **Representative sample of products**: products randomly considered for the panel of the study were selected by a substantial range of products traded by each company in Albania. Each product chosen was representative for the products and product categories traded by each company.
- **Promotional discounts excluded**: products with local discounts, seasonal sales, and loyalty program discounts, for example, were excluded from consideration in the study. Although sales promotions on specific products are common and important marketing factor for companies, local companies use discounts typically on limited scale and for a range of products which differs from country to country. As the net effect from discounts in price comparisons builds from different products in different countries, excluding these products provides a more realistic comparison of prices.
- **Seasonal products excluded**: to avoid the influence of seasonal fluctuations in prices and locational advantages or disadvantages, seasonal products (like fresh fruits, vegetables, etc.) were excluded from consideration.
- **Verification of physical/online prices consistency**: prices for about 30% of the products in the panel were verified on-site. This was to ensure that the prices of the products displayed in the physical stores corresponded to the prices offered for the same products through the local online stores.

The final steps of the methodology involved validation of the information collected and cross-country price comparisons. For validation purposes, about 20% of the products in the panel were substituted with other products complying with the above criteria. The validation produced consistent results with less than 3% fluctuations in price differences. Cross-country price comparison analysis, was used to collect price difference information on specific products, sold by the same company in various countries where the companies operates (including Albania).

2.9 Statistical Significance of the Selected Samples

The number of the products chosen in the panel, size and distribution of the businesses and consumers samples surveyed for this study were determined through a number of approaches. This included previous experiences with other similar surveys such as the Executive Business Survey, organized by the World Economic Forum for the preparation of the Global Competitiveness Report; Doing Business Report, prepared by the World Bank, Consumer Price Index, prepared by INSTAT, for example.

For the sample of the surveyed consumers, it was estimated that the customer base of the selected companies would be around 300,000 consumers. With a confidence level of 90% and confidence interval 5%, the calculated sample size of the consumers was 138.

For the sample of the surveyed businesses, data from the Business Registers 2023, published by INSTAT,[5] indicated that during 2023 there were 8,614 companies engaged in trading activities in Albania. With a confidence level of 90% and confidence interval 11%, the calculated sample size of the businesses was 28 companies.

For the sample of the products, it was estimated that the companies selected for the study would jointly trade less than 10,000 different products in the Albanian market. With a confidence level of 90% and confidence interval 5%, the calculated sample size of the products was 136 products. The sample was distributed into 13 different product categories, each containing 10 products and food category containing 30 products. This was done to cover a large number of products sold by the companies in the local market and to emphasize the weight of the food products in the portfolio. For many product categories, the sample of 10 products chosen was significant as it represented 16% of all the products in the skincare category, 22% of all products in the detergents category and 18% of all products in the bio food category, for example.

3 Literature Review

Among many notable publications, "Competition Policy: Theory and Practice", by Motta, Massimo (2004),[6] provides a comprehensive framework for understanding the ways international companies may abuse their market power. Motta defines market power as the "*ability of a firm to raise prices above some competitive level—the benchmark price—in a profitable way*"[7] and explains that firms with market power may raise the prices without losing significant market share. To achieve market power, firms operate through economies of scale, brand recognition, and extensive distribution networks. Motta emphasizes that market power is measured as "*the difference between the prices charged by a firm and its marginal cost of production*", which is essentially similar to the difference between the prices charged in the market by the firm and monopoly prices.

Market power enables the firms to set prices that are significantly higher than the marginal costs, exploiting consumers who, especially in developing countries, have limited market alternatives. The mechanism of abuse with market power, according to Motta, develops through price discrimination and other exploitative practices. Price discrimination, which is a core subject of this paper, according to Motta develops through two main criteria: (i) *firms have a way to sort consumers*

[5] INSTAT: Business Registers, 2023. https://www.instat.gov.al/en/themes/industry-trade-and-services/business-registers/.

[6] Competition Policy: Theory and Practice, by Motta, Massimo (2004). Cambridge University Press.

[7] Competition Policy: Theory and Practice, by Motta, Massimo (2004). Cambridge University Press. Chapter 2.1: Market power: A definition.

to charge them different prices (i.e. different nations); and (ii) *arbitrage should be absent* (consumers cannot resell products to each other).

Motta explains three ways price discrimination occurs: first, second and third—degree discrimination. F*irst-degree discrimination* is when firms know the precise willingness-to-pay of the consumers and they charge them the maximum price to capture all the consumer surplus. *Second-degree discrimination* is when firms offer different deals to everyone and they let the consumers 'self-select'. *Third-degree discrimination* is when firms charge different prices to consumers with different characteristics, including for example market location, which is particularly relevant in this study.[8] Price discrimination enables firms to set higher prices in regions with fewer competitors and lower the prices in regions with more competition.

To assess the implications of the national business environment and company strategies on developing countries such as Albania, Prof. Porter's 'On Competition' was extensively used for the preparation of this study[9] by focusing on the relevance of the five forces and the diamond model in the Albanian context. In his book Porter emphasizes the importance of competitive strategy and regulatory frameworks to ensure that markets remain competitive and that dominant firms do not abuse their positions to the detriment of the economy.

Part of the literature review process was also a note by the World Bank on price differences across the world[10] and an article from Eurostat[11] on the price levels for consumer goods and services in the European Union, focusing on price level indices, which provide a comparison of countries' price levels relative to the EU average and calculated using purchasing power parities. Other research papers referred in the study included Purchasing Power Parity considerations based on the views of Sarno, Lucio and Taylor, Mark Pon this topic.[12]

To determine whether a price may be considered as excessive is often a major challenge. A literature review of the position of the Court of Justice of the European Union (CJEU) on this specific topic was closely linked to the research in this study. From the legal perspective the CJEU has developed a test (named the United Brands test) to determine whether a price is excessive in the market. The test consists of two limbs: (i) whether the difference between the costs actually incurred and the price charged is excessive (Limb 1); and if so, whether a price has been imposed that is either unfair in itself or when compared to competing

[8] Competition Policy: Theory and Practice, by Motta, Massimo (2004). Cambridge University Press. Chapter 7.4 Price discrimination.

[9] Porter, Michael: On Competition, 2008. Harvard Business Review Book. Chapter 6: The Competitive Advantage of Nations and Chap. 1: The Five Competitive Forces That Shape Strategy.

[10] Adjusting for price differences across the world https://datatopics.worldbank.org/world-development-indicators/stories/adjusting-for-price-differences-across-the-world.html.

[11] Eurostat: Comparative price levels of consumer goods and services https://ec.europa.eu/eurostat/statistics-explained/index.php?title=Comparative_price_levels_of_consumer_goods_and_services#Price_convergence.

[12] Sarno, Lucio and Taylor, Mark P.: Purchasing Power Parity and the Real Exchange Rate. IMF Staff Papers, Vol. 49, No. 1.

products (Limb 2).[13] However, a violation can be shown by using a comparative benchmark as the court has emphasized that *other methods may be valid*[14] to establish an excessive pricing abuse, such as *comparator tests.*

4 Facts and Findings

Based on the methodology of the study, through each of the four perspectives described above, the findings collected solely from public sources during the research process are outlined in this section. For *dreams* and *models,* the findings build on analysis of literature reviews. For the *expectations*, findings are collected from two surveys organized with 30 businesses and 150 consumers in Albania. For the *realities*, the data consists of price information for identical consumer products traded by three international operators in the Albanian, EU and regional markets. These were collected from a sample of 150 products from 13 product categories purchased on a daily basis by Albanian consumers.

4.1 Dreams

"*E duam Shqipërinë si gjithë Europa*"—"*We want Albania to be like the rest of Europe*" was the main theme of the protests that ousted the communist regime in the early 90's. Albania as a *part of the European Common Market* is still the main aspiration of all Albanians including citizens, businesses firms and organizations, government and opposition, civil society organizations and academia. It is an aspiration supported with no exception by the entire political spectrum.

In the common market structure, there are no barriers to entering the market, composed of many buyers and sellers, all perfectly informed on the available goods. All buyers and sellers are price takers, they exercise no control over the market price of the product. Under such theoretical assumptions and by neglecting–for a moment–transportation, transaction and exchange rates costs, the Law of One Price, one of the most fundamental concepts in economics, states that: *Identical products will have the same price across the common market, regardless of location.*

One outcome of the EU integration process is that it leads to price convergence across the EU common market.[15] From this perspective, eventually, identical goods and services, sold by the same operators in the Albanian and EU markets should be offered at the *same price.*

---

[13] Case 27/76, United Brands Co. v. Commission, 1978 E.C.R. 209, 301, paragraph 252.

[14] Idem, par. 253.

[15] Price convergence in the enlarged internal market https://ec.europa.eu/economy_finance/publications/pages/publication10179_en.pdf.

4.2 Models

The World Bank, in a 2018 note[16] on the World Development Indicators suggests that "…prices of goods and services differ across countries…" and "…typically, prices are lower in poorer countries and higher in richer ones…". This is known as the Penn effect[17] based on the Balassa–Samuelson[18] model on the tendency for consumer prices to be systematically higher in more developed countries compared to less developed ones. From this viewpoint, with Albania's real GDP per capita at just 14.4% of the EU-27[19] in 2020, *prices of the same products sold in Albania should be considerably lower than those in the EU* market.

The Penn effect is also illustrated by an article[20] from Eurostat. In 2022 it compared prices of consumer goods and services based on price surveys covering more than 2,000 consumer goods and services across 36 European countries including Albania. The results showed price levels for consumer goods and services differed widely across Europe and the price level in Albania was 38.6% below the EU average, based on total household final consumption expenditure (HFCE) on goods and services. While differences in price levels are important for the comparison of economic data such as GDP, they are also useful in the analysis of the development of the EU's single market for goods and services, including price convergence across the market. Although these results cannot be attributed to specific products, they suggest that the average *consumer prices in Albania are lower than the EU average.*

The PPP exchange rate between any two nations equalizes the national price levels expressed in a common currency converted at that rate, so that the purchasing power of one unit of the currency would be the same in both economies.[21] Empirical evidence suggests that PPP exchange rate deviates from real exchange rates, primarily because of the effect of non-tradable goods. The PPP as a monetary conversion rate is calculated based on the prices of around 1,000 common products in each nation and then by finding the number by which one country's money would need to be multiplied to purchase the same goods and services in another country (often the US economy). While the PPP is useful for country-to-country comparisons of aggregate economic indicators, PPP by itself offers little insight into a country's economic health and is rarely considered on its own or

[16] Adjusting for price differences across the world https://datatopics.worldbank.org/world-development-indicators/stories/adjusting-for-price-differences-across-the-world.html.

[17] The price of development: The Penn–Balassa–Samuelson effect revisited: https://www.sciencedirect.com/science/article/abs/pii/S0022199616300873.

[18] Idem.

[19] Eurostat: Real GDP per capita https://ec.europa.eu/eurostat/databrowser/view/sdg_08_10/default/table?lang=en.

[20] Eurostat: Comparative price levels of consumer goods and services https://ec.europa.eu/eurostat/statistics-explained/index.php?title=Comparative_price_levels_of_consumer_goods_and_services#Price_convergence.

[21] IMF Staff Papers: Purchasing Power Parity and the Real Exchange Rate https://www.imf.org/external/pubs/ft/staffp/2002/01/pdf/sarno.pdf.

applied to specific products or groups of products. Due to this reason and to avoid unnecessary over-sophistication of this research, PPP implications are not considered in this study. However, due to Albania's purchasing power, if it were taken in the account, it's effect would further magnify the findings.

5 Expectations

5.1 Results of the Business Executives Survey

For this study 30 business executives from local and international trading companies of different sizes operating in Albania were surveyed between March–April 2024. These executives have a long track record of trading experience with the EU markets. The survey revealed that the business executives, *based on their own practices*, expected the average price of the consumer products sold in the local market to be *2.7% higher* than the average price of the same product sold in the EU markets. A key finding was that the majority of the survey respondents (60%) suggested that prices in the local market for identical goods, should be at the same level with the EU market (Table 3).

Table 3 Results of the survey with business executives

Business executives' responses		
What is the average price of a typical consumer product (like personal/processed food/furniture/cosmetics/appliance/electronics/clothes etc.*) you import from the EU and sell in the Albanian market, relative to the average price of exactly the same product currently sold by your company/partners in the EU markets?*	Responses	Percent
a. −50% or less, than in the EU markets	0	0%
b. −30% than in the EU markets	0	0%
c. −20% than in the EU markets	0	0%
d. −10% than in the EU markets	4	13%
e. Same as in the EU markets	18	60%
f. +10% than in the EU markets	5	17%
g. +20% than in the EU markets	2	7%
h. +30% than in the EU markets	1	3%
i. +50% or more than in the EU markets	0	0%
Weighted average *(Note: a. and i. are double weighted)*	**+2.7%**	100%

6 Summary of the Main Findings from Focus Group Discussions with Business Representatives

Many of the business executives who participated in the survey pointed out that their overall profitability and cost structure, including distribution and retail premiums, are not much different from that of the peer companies operating in the EU countries. Therefore, they would expect the final prices sold to consumers in the local market to be at the same level as in the EU markets.

Some of the businesses participating in the survey demonstrated in real time that products they were trading in the Albanian market were, indeed, selling at prices very close to the price range of the same products traded in European countries. Analogously, tests conducted by online searches during the discussions confirmed that for many products the price was identical.

The business representatives suggested that the price of the products prevailing in the EU markets included generous profit margins sufficient enough to recover costs and make profits in any market, including Albania. Other executives noted that it is the policy of their parent company to operate through a globally unified pricing approach for the products sold across the borders. As such this choice dictates unified prices in all the countries. These prices normally comprise sufficient profit margins to comfortably operate in any market in the world, including Albania.

Business executives from trading companies also suggested that due to the smaller market size and unsophisticated competition in Albania, the distribution channels are often much shorter and less complex than those in the EU markets. As they involve fewer marketing intermediaries and therefore smaller distribution costs, this leads to higher profit margins of operations and/or expected lower consumer prices compared to the EU countries.

Furthermore, executives from successful businesses that had full freedom to determine the final price for the products sold in Albania, indicated that even on moral grounds it would be considered ethically unfair to sell products in their home country at higher prices compared to the EU markets. Other executives reported that due to the smaller market size and less sophisticated demand, product lines in the Albanian market typically comprise a significantly smaller number of products (or product varieties) than in EU markets. However, the focus on a smaller number of products allows these companies to enjoy higher margins on these products, resulting in lower market prices relative to the EU.

The few firms showing a preference for higher prices in the local market, compared to the EU market, indicated the small market size as the main factor businesses should use higher prices. Transportation costs, corruption and red tape bureaucracy, as well as the lack of skilled workforce were also mentioned as factors leading to upward price pressure in the business context in Albania.

6.1 Results of the Consumers Survey

For this study, a survey of 150 consumers located mostly in Tirana area was organized during March–April 2024. The survey revealed that *based on their own market perceptions,* consumers expect the current average price of a product they purchase in their local market to be *5.5% more expensive* than the price of the identical product sold in the EU markets. Approximately 58% of the consumers surveyed expected local prices to be higher than the EU market, 34% expect the prices to be lower than in the EU market and 8% of the consumers expected local prices to be in the same level as in the EU market (Table 4).

The surveyed consumers provided mixed feedback and comments. Many respondents argued that the price of the products should be somehow higher than in the EU due to the geographical distance, lack of competition in the domestic market and the relative size of the market as the main factors for price differences. Respondents who opted for expectations of lower prices compared to the EU, suggested that the lower purchasing power in Albania dictates companies to offer lower prices than those in the EU markets. Many respondents pointed out that while they safely assume the price in Albania for identical products to be equal with the EU markets, their main concern as consumers was not the possible and minor price differences but the threat from counterfeit products of lower quality that are often sold in Albania at the same price as the original products in the EU markets.

Table 4 Results of the survey with the consumers

Consumer responses, based on individual perceptions		
What do you think is the average price of a typical consumer product (like personal/processed food/furniture/cosmetics/appliance/electronics/clothes etc.*) you regularly purchase in the Albanian market, relative to the average price of exactly the same product currently sold in the EU markets?*	Responses	Percent
a. −50% or less, than in the EU markets	0	0
b. −30% than in the EU markets	12	8%
c. −20% than in the EU markets	24	16%
d. −10% than in the EU markets	15	10%
e. Same as in the EU markets	12	8%
f. +10% than in the EU markets	11	7%
g. +20% than in the EU markets	64	43%
h. +30% than in the EU markets	11	7%
i. +50% or more than in the EU markets	1	1%
Weighted average *(Note: a. and i. are double weighted)*	**+5.5%**	100%

6.2 Realities

Based on the criteria defined in the methodology, the radar screen of the study's identified three major European companies fulfilling the criteria that made the research possible: operating in the Albania and international markets, offering the same products in these markets, operating online stores where the price information for each product was available.

During 2022, the identified firms had a collective turnover of around 90 million Euro in the Albanian market. They originate from three different European countries and since many years operate in several countries around the world, including Albania. According to this study's methodology, price information for a panel of 150 different products (about 50 products from each company) was collected:

i. from the online stores of each company in Albania
ii. from the online stores of each company in their home EU country[22] (country where the company originates from).

The products were randomly chosen from 13 categories (10 products for each category, food category comprising 30 products), sold in the Albanian market by these companies and filtered to comply with the criteria defined in the methodology.

In sharp contrast with the predictions of theoretical models and expectations from business executives and consumers surveyed during this study, the price information collected revealed that the average *price of these products sold in the Albanian market by the three companies was +79.7% higher than the average price of exactly the same products, sold at the same time, and by the same companies, in their home EU markets.*

None of the products in the panel were offered in Albania at a *lower* price than the price of the same product offered in the EU market. The Albanian prices of the products surveyed resulted between +24% (lowest) and +212% (more than triple the price) *higher* than the respective price of each product sold by the same companies in the EU markets. All the products of the panel, qualified for the *excessive discrimination* definition, described above.

Excessive discrimination was consistently detected throughout the entire range of the 150 panel products. They resulted as highest in the product categories of baby food and food products (+147 and +107[23] percent, respectively, or more than double the price) and lowest in the product categories of living room furniture, cosmetics products, and garden products (+56, +57 and +57%, respectively) (Fig. 1).

A comparative assessment of market prices charged in Albania by the selected companies in this study, with the prices of same or *similar* products traded by other operators in the market shows no significant variations in the prices prevailing in

[22] The EU country the company originates from.

[23] Calculated as the average of three food categories, each containing 10 products.

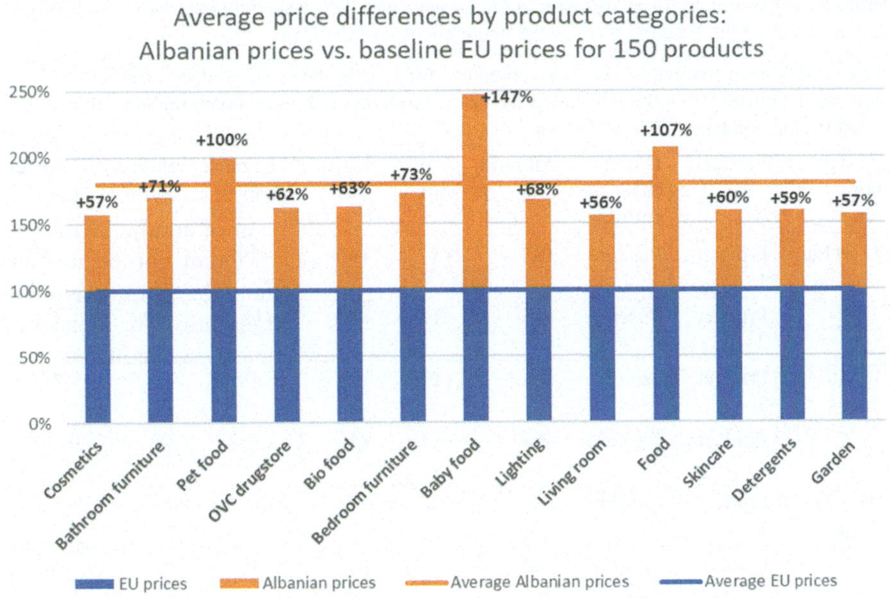

Fig. 1 Price differences between Albanian and EU market, for 150 identical products of 13 product categories, each containing 10 products and food category containing 30 products

the local market. This leads to the deduction that excessive discriminations in the local market are more *pervasive,* and not limited to the selected operators in the study.

6.3 Cross-Country Comparative Analysis

Using this study's methodology for collecting price information of various products, a cross country benchmark analysis was conducted. This analysis focused on 10 products of different price range from 1 to 1,000 Euro, sold by the same company in various countries of Europe. It reveals that the practice of excessive discrimination, relative to the baseline price of the product sold in the home market of the company (EU country), for every product in the list, is distinctively highest in Albania compared to other countries. While price differences are noticed from country to country, as summarized in the Table 5, the average variation in every country is less than ± 15%. The notable exception is for Albania where the average excessive discrimination for the selected products is +71% (from the baseline, price of each product in the home country of the company).

Table 5 Cross country price differences for the same products traded by the same firm in various countries versus the baseline price in the home country of the firm

Price differences charged by the same company, for exactly the same products, sold in selected countries, from the baseline price of the products sold in the home market (EU country) the company originates from

Product price range	Product	**Home (EU) country**	**Albania (%)**	Serbia (%)	Kosovo (%)	Ukraine (%)	Holland (%)	Greece (%)
1–10 Euro	Product 1	*Baseline*	**+85**	+3	+40	−19	−7	−7
	Product 2	*Baseline*	**+85**	+11	+40	+12	+12	−7
10–50 Euro	Product 3	*Baseline*	**+66**	+11	+23	−19	+1	−25
	Product 4	*Baseline*	**+56**	+13	+16	+35	+8	0
50–100 Euro	Product 5	*Baseline*	**+35**	+6	0	+4	0	−7
	Product 6	*Baseline*	**+48**	+6	+12	−10	0	−13
100–200 Euro	Product 7	*Baseline*	**+66**	+22	+24	−10	+6	−7
	Product 8	*Baseline*	**+85**	−5	−7	−38	−30	−30
200–1000 Euro	Product 9	*Baseline*	**+105**	+20	+4	0	+7	−17
	Product 10	*Baseline*	**+78**	+25	−10	−7	−4	−31
1–1000 Euro	**Average**	*Baseline*	**+71**	+11	+14	−5	−1	−14

7 Summary of the Findings

The research reveals that based on the information collected using this study's methodology, for a typical product priced 100 Euro in the EU markets, the Albanian society's dreams would expect it to be offered exactly at 100 Euro in the Albanian market. Economic models would suggest its price in the local market to be anywhere between 14 and 100 Euro. Business firms operating in Albania and trading with EU markets would suggest its local price to be 102.7 Euro, while Albanian consumers would expect its local price at 105.5 Euro. Based on these expectations, if a products price was 111 Euro or higher it would be considered with an excessive discrimination. The reality in Albania is that the average price of the product in the local market is 179.7 Euro, based on the average of 150 products currently sold by three major market European operators (Fig. 2).

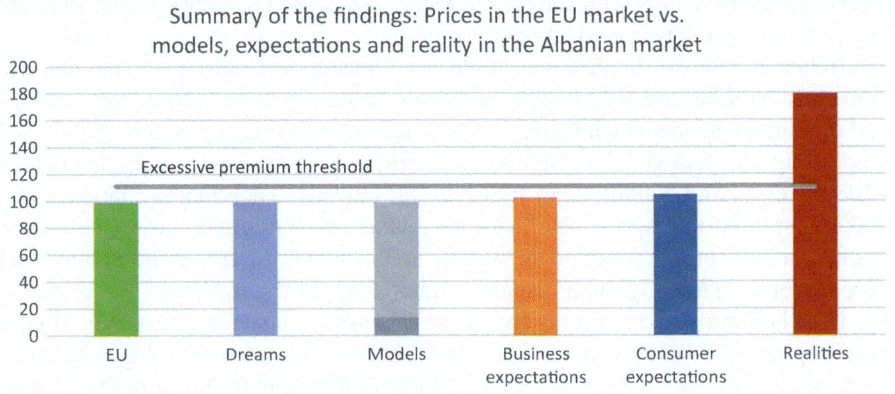

Fig. 2 Summary of the findings

8 What is the Problem Here?

How is it that such significant high prices and excessive discriminations are consistently found in European products sold by European companies in the poorest country of Europe? How is it possible that against what models would predict, and contrary to consumers and business expectations, large and notable international firms sell their products in Albania at a significantly higher price compared to the price of the same products these same companies sell in their EU home-country market? What are the reasons for prices of such products to be substantially higher in Albania compared to the home-country market of these firms, but also relative to other countries, including those in the region with similar economic conditions as Albania? How is it possible that excessive discriminations are charged in the market not only by some firms, but the problem appears to be more pervasive than the evidence revealed in this research?

These are important questions for all Albanian citizens and not just another argument for political debates. This challenge and solutions to tackle them are critical for the future growth of the country. The analysis, unfolding in the next section of this paper, suggests that for the sectors considered, the Albanian economy is underproducing at a sub-optimal level of activity, resulting in higher consumer prices compared to European standards.

Higher market prices produce long-term negative consequences for firms operating in the market. Profitability achieved under sub-optimal levels of production destroys the growth potential and discourages firms from innovation, investments, expansion, and diversification of economic activity with new products and services. Higher market prices make firms less competitive and uninterested in expanding activity through internationalization in regional and EU markets. Higher market prices predominantly on imported products erode the interest of trading companies in domestic production which offers far more modest market margins. Higher market prices in sectors with substantial entry costs lessen industry rivalry, fossilize competition, and deter new business formation.

Higher market prices are undesirable from society's perspective. They deplete consumer welfare and misallocate economic resources by crowding out savings and eroding investments while negatively contributing to raising poverty levels and amplifying social discontent. Higher market prices distort consumption and deteriorate the costs and conditions of living for all citizens. They fuel social exclusion and are an important propensity factor for Albanians to migrate to other countries. Higher market prices accumulate wealth in the hands of fewer individuals and corporations, sharpen income disparities and build on the misery of unaffordable food for hundreds of thousands living on a pension or social assistance. Higher market prices discourage consumption and constrain the choice of the consumers. They act as a deterrent for prospective Albanian migrants to return home. Higher market prices on essential consumer goods amplify ripple effects and cascading costs of other goods and services produced by the economy, eventually slowing down consumption and economic growth. For citizens and tourists, they increase the costs of living in Albania and negatively shape the perception of the country. Higher market prices undermine democracy, exacerbate inflation, and weaken the effectiveness of policy measures by rendering them less effective and more expensive for the budget. Higher market prices observed during off-season periods are likely to rise further during the high demand peak of the tourism season, posing an additional and increasing threat for the most promising sector of the economy.

What's even more important is that leading and dominant firms operating in a market with pervasive high prices serve as role models. Their high-pricing strategies are copied and adopted by follower market operators, especially the smaller ones, infecting such strategies across the larger market. Higher market prices erode firms' productivity and the overall national competitiveness. Higher market prices are more likely to produce net negative effects for tax revenues, with further negative implications for financing poverty reduction programs, pensions, health, social care system, education—effectively all public services and investments.

9 Analysis of Factors Enabling Excessive Price Discriminations in Albania

Distinct trade preferences, full opening of the market through the free-trade regime of the EU-Albania Stabilization and Association Agreement and the integration process with the EU have oriented Albania's trade toward the EU: imports originating from the EU countries accounted for more than half of Albania's imports during 2023 (50.9% of total imports[24]). Because of such an affinity, prices of products sold by European operators in European markets may be used as a good

[24] INSTAT: Foreign Trade Database, 2023. www.instat.gov.al.

benchmark to compare the prices of the same products sold by the same companies in the Albanian market. The businesses surveyed in this study indicated that this is a common practice among many firms while others are also required by their parent companies to apply the EU prices in the local market.

When price differences (for the same products, offered by the same companies) between the Albanian and EU market are (i) consistently high through a considerable range of products, (ii) exceeding market participants' expectations and (iii) applied by more than one relevant market operator, internal forces of the companies are insufficient to explain the market failures. The presence of excessive discriminations, detected in all the products included in the panel selected for this study, and evidence that high prices exist across a much larger pool of market operators in Albania, indicates that the foundations of these market failures are linked to the structure and features of the national business environment. For this reason, the analysis is based on the assessment of the national business environment factors, using Prof. Porter's Diamond Model,[25] rather than through the five forces model[26] that is focused on the company and micro-level industry analysis.

9.1 Two Common Fallacies

Before the analysis of the business environment context, it's important to discuss two common fallacies, often brought as false propositions in debates on economic policy in Albania.

First fallacy: if the product sells, then the market is operating, let it be.

The misconception that markets properly operate if products are selling at a given price is inherently wrong and fails to understand the dynamics of the market mechanism. It assumes a static view of customers accepting the price dictated by the seller which results in all market participants being satisfied from the transactions occurring in the market.

The market mechanism is more complex and vibrant than this superficial and deceptive view. Each buyer has a unique willingness-to-pay for a product, which varies by the opportunity cost of that person's consumption alternatives. Adam Smith probably did not think for long about a desert context when he noted that "*Nothing is more useful than water: but it will purchase scarce any thing; scarce any thing can be had in exchange for it. A diamond, on the contrary, has scarce any value in use; but a very great quantity of other goods may frequently be had in exchange for it*".[27]

The individual willingness to pay for a product might be different from its market equilibrium price. However, it is the equilibrium price that separates losers

[25] Porter, Michael: On Competition, 2008. Chapter 6: The Competitive Advantage of Nations.
[26] Porter, Michael: On Competition, 2008. Chapter 1: The Five Competitive Forces That Shape Strategy.
[27] Adam Smith, The Wealth of Nations, Book I, Chapter IV.

from consumers with equal or higher willingness to pay. Ultimately, they are the ones to purchase the product (provided they also would be able to match the price with their ability to pay). The consumer utility surplus arises because consumers buying products at the market equilibrium price were willing to pay even *more* to obtain the product—not just an arbitrary price. Market equilibrium price in competitive economies is not arbitrarily determined by the firms, it reflects the balance by the pressure in the supply side with firms willing to reduce the price by improving efficiency. Well-functioning competitive markets achieve economic efficiency when the social surplus, including the producer surplus, is maximized.

What happens when market failures arise and through price discrimination the market price is *arbitrarily* set higher? Producer surplus increases, higher market prices reduce the consumer surplus of all the consumers purchasing the product. As a result, *some* consumers (those whose willingness to pay is less than the new, higher price) would be excluded from enjoying the product. But although the consumer surplus for *everyone* would decline and some consumers would be *left out* of the possibility of purchasing the product, the market would still *operate.* Even at a higher price, unless competitive or correcting forces emerge on the supply side, there would be consumers willing (and able to pay) the higher price. Furthermore, consumers are less responsive to price changes for essential goods, and demand may not change when prices increase, particularly in a context as Albania where consumers access to product alternatives or substitutes, is limited.

Second fallacy: if more than one firm operates in a liberalized market, then there should be no monopolies or oligopolies

A common and often shared misconception in public opinion defends the view that in a liberalized market economy, there are no monopolies or oligopoly positions by the firms operating in the market. Although monopolies, cartels and oligopolies are often and ambiguously brought up as interesting topics in the political discourse in Albania, references to specific cases are incomplete and vaguely addressed.

There is not much wrong with a firm being in a monopoly, oligopoly, or to use more technical terms, in a dominant position in any market. According to the EU definitions, for a firm "*…a dominant position is the power to behave to an appreciable extent independently of its competitors, its customers, and consumers*[28]". Monopolies, oligopolies and firms in dominant positions *exist*–virtually in any market economy. By definition, they legally operate, and their activity is part of the market mechanism. How firms in the dominant positions–including monopoly and oligopoly cases–*use their market power* to exert pressure on consumers or other firms is what concerns healthy societies. This particular and important aspect of the market is supposed to be subject of public and institutional scrutiny.

[28] Official Journal of the European Union: Guidance on the EU Commission's enforcement priorities in applying Article 82 of the EC Treaty to abusive exclusionary conduct by dominant undertakings, III.A.10: https://eur-lex.europa.eu/LexUriServ/LexUriServ.do?uri=CELEX%3A52009XC0224%2801%29%3AEN%3AHTML.

The global economy provides regular examples of firms in dominant positions operating even in the most competitive markets in the world. It is the role of the institutions in these economies that limit these companies from abusing, or correct abuse cases with dominant positions whenever they occur, through multi-billion anti-trust penalties imposed even on firms with global and notable reputations. Google (fined 4.3 billion Euro by the EU), Apple (fined 1.8 billion Euro by the EU), Microsoft (fined 2.2 billion Euro by the EU), Amazon and JP Morgan—these are just a few among notable high-profile cases incurring penalties for anti-trust violations in the most competitive economies of the world. The US corrective measures of the anti-trust law go even a step further than the penalty-based EU anti-trust legislation. Many Albanians may recall the recent story of their infamous co-national Martin Shkreli, who was found in 2022 to have violated anti-trust laws with his monopoly over the drug Daraprim that allowed him to raise its price by 4,000%. Shkreli was fined $64.6 million[29] but was also *banned for life* from participating in the pharmaceutical industry.[30]

The fear from relentless market investigations, the effective leverage of the penalties imposed by the institutions amounting up to 10% of firm's annual revenue, combined with real, visible and persistent violation punishments, are formidable deterrents that limit companies from abusing with their dominant market position in developed markets. In addition, the strong voice of the citizens and consumers, fear from existing competition and new entrants that might be attracted by excessive prices in the market, as well as reputational factors (firms penalized for anti-trust violations do pay the penalties, but they also erode their public brand image) —are some other factors that influence companies in dominant positions to refrain from exercising their opportunities to abuse their positions, whenever they can. Otherwise, firms take advantage of it.

Diamond Model Analysis of the National Business Environment in Albania.

Competitiveness at the national level reflects the productivity by which a country uses its resources, contributing to raising the living standards for all its citizens.[31] A fundamental factor to sustained productivity growth—translated in improved prosperity for everyone—is the process of relentlessly upgrading micro-level productivity of the firms operating in specific industries or sub-segments. More recent definitions, link competitiveness to the ability of domestic firms to compete in the globalized world.[32] While innovation is the key success factor for firms, the Diamond Model[33] explains the sources of the national advantages (or

[29] Shkreli was also sentenced to seven years in prison, convicted of securities fraud and other offenses.

[30] Federal Trade Commission: FTC Asks Federal Court to Hold 'Pharma Bro' Martin Shkreli in Contempt: https://www.ftc.gov/news-events/news/press-releases/2023/01/ftc-asks-federal-court-hold-pharma-bro-martin-shkreli-contempt.

[31] Porter, Michael: On Competition, 2008. The Competitive Advantage of Nations.

[32] Porter, Michael: Microeconomics of Competitiveness, 2003-2015 class notes.

[33] Idem.

disadvantages) that enable firms to successfully compete in the market. Its determinants of the national business environment are the main attributes that jointly create the right (or wrong) playing field for companies to compete, upgrade over time, and contribute to improved living conditions for everyone. The Diamond Model is a powerful tool to assess differences in business environments versus prosperity levels between countries. It is used here to analyze the context of market failures identified in the research of the study.

9.2 Factor Conditions

Locational advantages or disadvantages in factor inputs cannot explain the existence of excessive price discriminations in the Albanian market. None of the per unit cost components for a business operating in Albania is higher than the corresponding cost of businesses operating in an EU country. Material costs, labor force, office and space rentals, transportation, electricity and business taxes, for example—key cost components for trading companies, are increasing for firms operating in Albania, but none of these can match with EU countries—Albania is still a cheaper option.

Transportation costs are occasionally reported as an important barrier for firms operating in the Albanian market. However, the research indicates that the transportation costs are not significant part in the expenditures of the firms operating in the local market and their share in total costs is like companies operating in the EU markets. The transportation costs argument also fails to explain the significantly lower discriminations in neighboring markets. It is also not supported by the lack of correlation between the excessive discriminations and the weight/size of the products included in the panel of this study.

9.3 Demand Conditions

Local demand conditions support competitive advantages in prominent industries of the domestic market and they partly explain the existence of high prices in the Albanian market. Demanding buyers with sophisticated buying behaviors pressure firms to innovate, upgrade the productivity, and offer unique products in the market. Four important aspects of the demand conditions in the Albanian market are particularly important:

Strong consumer's preference of Albanian consumers for European products. For historical reasons related to the opening of the Albanian economy in the early 90 s, the affinity of Albanian consumers for European products appears to be much stronger than in the neighboring countries, which traditionally developed closer intra-trade relations within the region. Trade data reveals that food, beverages, and tobacco imports from the EU countries, accounted for about half (47.9%) of Albanian total imports of such products during 2023.[34] Furthermore, food and beverage imports from EU countries in Albania for household consumption during 2023 exceeded 600 million Euros and accounted for 14.3% of total imports from the EU. This ratio was only 6.7% in Serbia and 6.8% in North Macedonia.[35] Such strong preference for EU products is explained by the confidence of Albanian consumers on European-made products, which are considered of superior quality and worth the higher price, compared to similar products originating from other countries or domestically produced. The lack of trust on quality standards for domestically produced goods or products imported from non-EU countries deserves a further analysis which goes beyond the scope of this study. However, it often comes with an extra price consumers must pay.

The elusive "small market size" argument. The pervasiveness of high product prices in the Albanian market is often tied to the *it's-a-small-market* cliché. At best this is an incorrect argument. Albania is not a small market. It may appear so relative to the larger economies but for specific industries, segments and firms, it is not. The collective turnover in the Albanian market of the three companies considered in this study exceeded 90 million Euro during 2022 and the figure very likely was well above 100 million Euro during 2023.[36] Food products imports alone, originating from the EU exceeded 600 million Euro during 2023, while total food imports from the world, for the same year exceeded 1.3 billion Euro. These figures are certainly not small, even for global companies.

The *small-market-size* argument is not consistent with the cost–benefit analysis firms typically employ to consider pursuing new market opportunities or expanding business activities. If a firm operates in n countries, or has t branches, any expansion decision (unless for strategic behavior reasons) would consider and balance the costs and benefits from operating in the $n + 1$ country, or opening the $t + 1$ branch. But no serious firm would make such a choice by raising the prices of its products by 80% just because it's a small new market. If expansion costs outweigh benefits firms would rather drop such prospects.

The *small-market-size* argument fails to explain why excessive discriminations identified in this study are not on a comparable level in other countries of similar size to Albania, like Kosovo, Serbia or Greece, as noticed in the cross-country comparative assessment of this study.

The effect of consumption taxes. The value-added tax is the main consumption tax levied on products across European countries.[37] Although VAT standard rates in EU member countries are higher or at the same level as Albania's rate of 20%

---

[34] Author's calculations using INSTAT Foreign Trade Database, www.instat.gov.al.

[35] Author's calculations using 2023 trade data, long term indicators, Eurostat. https://ec.europa.eu/eurostat/databrowser/view/ds-057555__custom_11368584/default/table?lang=en.

[36] Financial statements of all the companies operating in Albania are publicly available via qkb.gov.al.

[37] Products with excise taxes are excluded from the study, as specified in the methodology.

Table 6 Standard and Reduced VAT Rates in selected EU countries, 2024

Standard and Reduced VAT Rates, selected EU countries						
Country	*Italy*	*France*	*Germany*	*Greece*	*Denmark*	*Croatia*
Standard VAT rate	22%	20%	19%	24%	25%	25%
Reduced VAT rates	10%, 5%, 4%	10%, 5.5%, 2.1%	7%	13%, 6%	–	13%, 5%
Reduced rates applied for	Water, transport, food, books, culture etc	Restaurants household work, farming, passenger transport, food, gas, electricity, art, cinema and sporting, services for elderly, medicines, magazines etc	Foodstuff, books, cultural services, hotel accommodation, medical and dental care etc	Food, accommodation, electricity, gas, heating, books, medicines etc		Accommodation magazines and newspapers, food products, child car seats, catering, books, medicines etc

(except for Germany, Romania and Cyprus which have a standard rate of 19%, Malta 18% and Luxembourg 17%[38]), there are variations of VAT rates levied on specific product categories. These reduced VAT rates for different products contribute to the final price of these products in the local market of each country (Table 6).

European Union reduced VAT rates for certain consumer products, particularly food, contribute to lower relative prices for some consumer products in these countries. They also contribute in price differences for the same products between these countries and Albania. However, the VAT rate variations between Albania and the EU countries (maximum 20%) is insufficient to explain the full magnitude of the excessive discriminations identified in this research.

Products sold by the companies are essential goods. The majority of the products considered in this study are consumption goods. Albanian consumers purchase these regularly for daily needs: food, detergents and cosmetics, for example. Given the price elasticity of demand is low for these goods, consumers are less likely to reduce consuming them even if prices increase, all other things remaining equal.

[38] Based on author's research through official websites of the tax authorities in several EU countries during Spring 2024.

9.4 Related and Supporting Industries

The third component of the Diamond Model is the least relevant in this analysis. The operations of the trading companies selected for this research appear to be less dependent on other firms in the market, compared to companies that interact with supporting businesses to produce and add value to locally processed goods, depending on their efficiency in the market. As most of the products considered for the study are imported and sold in the market without any noticeable local processing or assembling,[39] the selected companies do not seem to rely on significant partnerships with local firms that may add value and influence their excessive discriminations. In a broader context, some companies provide services and support through auxiliary, non-core business function activities that have negligible or no effects on the excessive discriminations. Furthermore, financial institutions help the operations by facilitating credit to consumers and enabling transactions, employment agencies support training and match qualified labor force with market demand, and electricity companies provide energy supply, for example.

9.5 Context for Firm Strategy and Rivalry

The national context in Albania influences the way companies are organized and how they operate in the market to ensure the best strategy fit to achieve company goals. The relative size of the consumer goods market favors a large number of small-scale operators scattered around the country combined with a limited pool of larger international and domestic competitors concentrated in the capital and the main urban areas of the country. Three main factors in this context are important for large firms, enabling them to operate with high prices in the market:

- **Firms are strategically positioned to adequately respond to consumer preferences for made-in-EU products**. By offering quality products produced or branded by the company or through partnerships, the companies can satisfy consumer's demand and preferences for superior quality products originating from the EU. To accentuate this advantage, the selected companies offer in their shops almost exclusively their products, originating and imported from the EU (except for one of the companies which appears to offer on a limited scale a few domestically produced goods). The quality of the products and their origin from the EU country are emphasized and are central to the advertising messages used by the companies.
- **Firms operate in the market as convenient grocery/furniture shops**. Differently from small-scale shops which offer a limited range of consumer products

[39] See Methodology: firms selected for the study operate mostly on trading imported consumer goods in Albania.

to their customers, the selected firms operate in the market by offering a relatively large number of products from a broad range of product categories, often serving as convenience stores operating for prolonged hours. Shoppers seeking to purchase specific home products find it more convenient to buy more products they would consume in the next days in the same store, than visiting different shops to purchase fewer items in each of them.

- **Firms operate in a market with limited competition.** The structure of the retailing market in Albania is very fragmented in the small-scale segment, but more concentrated in the medium and large segment, with few key companies dominating the market, some of which are international operators. In this segment and the wholesaling market the competition among the participating firms in the market is limited.

In addition to the above factors, brand names and reputation of the international firms, loyalty programs, and sales promotions, advertising campaigns, as well as investments in business operations infrastructure, technology, workforce training, and customer relations, also play an important role in the success of these companies in the local market.

Big Micromonopolies

Why are firms able to charge high prices in the Albanian market? Given the above conditions prevailing in the local market, the response to this question is based upon the strategy companies adopt to prosper in the local market. Michael Porter suggests that "*the operational effectiveness of the firms is necessary, but not sufficient to succeed in the market: strategic positioning means performing different activities from rivals or performing similar activities in different ways*[40]". Competitive advantages deriving from efficiency improvements in the company operations are important for every firm, but there are limits to what technology and labor productivity can do. Positioning in the market in a strategic way that distinctively differentiates companies from rivals is what allows the companies to enjoy higher prices and profit margins.

The distinct specialization of the companies in offering (i) superior quality, (ii) often unique, and (iii) preferred products for the Albanian consumers, is what enables each company to operate in a dominant position independently from its rivals. Let's call this market position a *micromonopoly,* where companies may set higher prices without reducing market share and profitability. In the context of a *micromonopoly* the entire market may appear as competitive by considering the total number of participating firms—the Business Registers indicate that during 2023 there were 8,614 companies engaged in trading activities in Albania.[41] There

[40] Porter, Michael: On Competition, 2008. Chapter 2.

[41] INSTAT: Business Registers, 2023. https://www.instat.gov.al/en/themes/industry-trade-and-services/business-registers/.

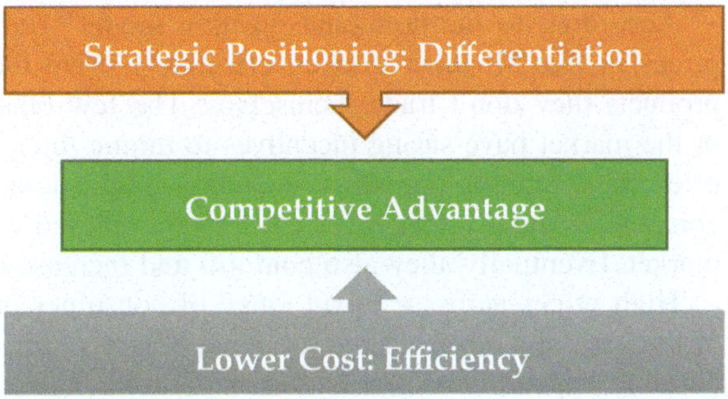

Fig. 3 How firms create competitive advantage in the market. Microeoconomics of Competitiveness, class notes, Harvard Business School, 2003

may be many firms in the market, but because of the high degree of specialization in the products the dominant firms offer, coupled with the lack of competition in the specific sub-segments they operate and the high entry barriers, firms may act as de facto monopolies within the smaller segments of the market. The lack of strong competition within these specific market segments, coupled with loose market surveillance enables *micromonopolies* to grow market power by raising prices. Firms are driven by profit maximization and from the *micromonopoly* position they are able to manipulate prices and charge higher prices. Eventually, the output is reduced, market power is strengthened and firms consolidate activity on a larger scale (Fig. 3).

Why High Prices are Pervasive?

For many products in the Albanian market, there are often no significant price differences between the products traded by the companies considered in this study, and the same or *similar* products (imported or domestically produced) traded by other operators in the market. At least not in the magnitude of the excessive discriminations identified in the research.

This suggests that price discriminations or excessive market prices are not an exclusive practice of the companies identified in this study. The phenomenon appears more pervasive across the local market with further implications for the consumers and more challenges for sustainable solutions. In a market composed by many similar firms selling homogenous goods, if a company raises the price of the products, consumers may switch to other companies and the competition would drive the prices down. In contrast, if one big firm holds a dominant position and acts by increasing prices in the market, competition from smaller firms would be insufficient to exercise downward pressure and lower the prices. Smaller firms typically follow the lead of the larger firms.

According to the dominant position model[42] large firms operating with high prices in the Albanian market, effectively drive the market prices up, even for products they don't trade themselves. The few other competitors of similar size in the market have strong incentives to mimic the strategy and avoid unnecessary price wars. Smaller operators, including small shops, either don't have capacity to compete, or their prices are locked by the limited competition in the wholesaling market. Eventually, they also conform and increase their prices.

High prices across a broad range of consumer products in Albania contribute to the very high share of the food and non-alcoholic beverages expenditure in the monthly expenditure of Albanian households. According to the Household Budget Survey for 2020[43] published by INSTAT, the share of food and non-alcoholic beverages expenditure in the structure of monthly expenditure of Albanian households was 41.6%. In contrast the average of the EU for the same period was only 17.1%, for Montenegro 30.7% and Serbia 28.8%.[44] Low income level consumed more for essential goods is the obvious side of the coin, the other, perhaps less evident, relates to price differences of the essential goods.

These market patterns and symptoms exist in Albania since many years. The author is aware that the relative appreciation of the Albanian Lek versus Euro by about 20% during the recent two years (Spring 2022—Spring 2024) has sharpened the contrast of the price discriminations discussed in this study—international firms selling imported products in Albania gain from the relative appreciation of the domestic currency because they purchase products in Euro and sell them in domestic currency. The fact that the local prices charged by the companies did not adjust by any margin to reflect exchange rate gains during this relatively long period, is a strong proof that competition in the market is insufficient to address the existing market failures.

10 The Way Forward

The excessive price discriminations identified in this research are, in the author's view, neither fair, nor tolerable in a country that aspires and strives to become a member of the EU. In terms of relative prosperity, Albania is the poorest country on the continent. Higher prices are detrimental to a country's development and by cascading in other products they pose huge economic and social barriers. They are not justified from a free-market perspective; the reduction and elimination of price discrimination is highly desirable for the society.

---

[42] Canoy, M. F. M., van Damme, E. E. C., & Rey, P. (2004). Dominance and monopolization. In M. Neumann, & J.Weigand (Eds.), The International Handbook of Competition (pp. 210-289). Edward Elgar Publishing. https://pure.uvt.nl/ws/files/679938/dominance.pdf.

[43] https://www.instat.gov.al/en/statistical-literacy/household-consumption-expenditures-in-albania/.

[44] Eurostat, Structure of consumption expenditure by COICOP consumption purpose. https://ec.europa.eu/eurostat/databrowser/view/hbs_str_t211/default/table?lang=en.

While the blame and the remedy for market failures are often attributed to government institutions, paradoxically, the government in this case, may not be the right answer.

10.1 The Government is (Almost) Powerless to Lead a Change

The government has a direct responsibility to set the optimal level of taxes in the country. To determine the optimal tax rates requires further analysis and indeed, some EU member countries[45] do not apply reduced VAT rates on consumer products. But learning from the best practices of most EU economies that currently apply reduced VAT rates, especially for food products, could be beneficial in the Albanian context. Reducing the VAT rate for specific product categories may be positive, but it would not resolve the problems identified in this research.

Given the modest track record of government intervention to regulate—or deregulate—markets in Albania and market concentration symptoms in various sectors, there are strong reasons that one must be rather skeptical on a remedy to the problems identified in this study coming from possible government actions. Although this option may not be excluded, there are stronger arguments suggesting it may be ineffective. Even if one day the institutions would wake up as the most formidable machinery to protect the citizens from unfair exploitations, committing to disregard special and political interests, seriously pledging to trade the bureaucratic and sophisticated interpretation of legal labyrinths with strong actions to protect the consumer's interest, eventually the government's measures, to a large extent, may be limited and ineffective. Here are the main arguments why this is the case:

- **Benefits largely outweigh penalties**: With the excessive price discrimination levels identified in the study, even the maximum penalties defined in the Albanian legislation[46] would be ineffective to correct violations of the law (assuming the authorities identify violations and then opt to impose fines). Similar to the EU and US legislation, the Albanian law specifies that for serious infringements the authorities may impose a maximal punishment and deterrence penalty up to 10% of the aggregate turnover (of the violating company) during the previous financial year.[47] But many companies would be quite eager and may strongly prefer to pay a maximum penalty equal to 10% of their turnover, provided they are allowed to operate in the market with their prices 80% higher than the

[45] Albania applies a reduced VAT rate 6 percent for accommodation, agrotourism, books, advertising services etc. but none of these product categories are related to this study.

[46] Law on Competition Protection, section VI: https://caa.gov.al/wp-content/uploads/2023/06/Law_nr_9121_date_09.05.2012.pdf.

[47] Article 74, Law on Competition Protection: Idem.

market expected price. A review of the financial statements of the three companies[48] selected for this study for the financial year 2022 indicates that even if the amount of 10% of the turnover of each company is added on top of their annual expenditures, *all the three companies would still be profitable.*

- **Burden of government penalties is on the shoulders of consumers**: If the authorities eventually find violations of the law and choose to impose penalties, the burden of the fines would be shifted by the companies in the final price consumers have to pay. As discussed above, all the companies appear to be quite able to softly absorb any penalty foreseen by the law within the margins they currently operate. As a result, penalties may serve to the government to raise additional revenues, but for the consumers, they may have no effects.
- **Weaknesses of the judicial system**: The government's potential penalties imposed on undertakings found in breach of anti-trust rules could be challenged by the firms in the legal system. Many factors, including the influence of lobbying and political factors, corruption and inefficiency in the judiciary system, weaknesses of the institutional capacities in defense of public interests, the superior quality of attorney firms defending the interest of powerful companies, and an overloaded court system, may result in prolonged court cases effectively postponing or rejecting possible corrective measures.
- **Government intervention on companies practicing excessive discrimination is perceived and/or leads to increased government regulation**: While excessive prices are a direct exploitation of consumer's interest and their existence justifies the intervention in the market, attempts to correct the market failure may strengthen the role of the government as a regulator in the market. This may result as counterproductive due to the capacities required, identified problems becoming permanent, political and special interests influence in decision-making and the overall perception of excessive regulation in the country. Similar recent attempts by the government to regulate market prices through ad-hoc boards for specific products were often criticized, proved to be inefficient and were finally abolished by the Constitutional Court decisions.[49]

10.2 The Voice of the Citizens

The "Go European!", embedded in the title of the paper is an open call to the European companies charging excessive price discriminations in the Albanian market to *act European* in the same way as their parent companies operating in EU countries and revise their pricing strategies to reduce and eliminate without delays the excessive price discriminations currently charged in the local market in Albania.

[48] The financial statements of all the Albanian companies are publicly available online through the Trade Register, accessible at qkb.gov.al.

[49] Constitutional Court Abolishes Food Transparency Board. https://albaniandailynews.com/news/constitutional-court-abolishes-food-transparency-board.

In the author's view, the term *voice of the citizens* does not exclude any organization, including government, in its role *as a citizen* of the Albanian society. In fact, the EU experience has shown how the government can possibly act just as another good citizen of the society. During 2023, in the wake of the inflation surge in European markets, the French government called on all the supermarkets to cut down their prices to the minimal possible level, although the government did not have powers to force supermarkets to do so.[50]

The voice of the citizens is the strongest and the only realistic pressure for companies to eliminate the price discrimination practices identified in this study. It is the only credible transformation leading toward a triple win–win-win solution for the whole society and the three major stakeholders: consumers, government, and the companies. More consumers would benefit from lower prices, reduced deadweight loss, improved access to the products, increased savings and better allocation of disposable income, better living conditions and more hope for the future in their own country. The companies would increase their sales, expand their customer base, bring more products in the market, work to improve efficiency, upgrade productivity, compete with strategies, expand investments and their profitability in the market, strengthen their brands and image in the local community. The government would find out that it may raise more taxes from increased consumption and that there are unlimited ways to fight inflation, tackle poverty, social exclusion and many other important challenges in the country.

11 Conclusions

The study revealed that when market failures are tolerated and left to their own devices, markets, particularly in small countries with fragile institutional settings like Albania, can, and sooner or later, go bad and ugly. Persistent market failures produce incentives for systemic failures enabling firms to engage in price discrimination, charge higher prices, and underproduce at a suboptimal level of economic activity. Ultimately, consumers bear the costs.

A tolerance culture for large firms operating in dominant positions, coupled with strategic positioning of the firms in the market to maximize profits and respond to the strong preferences of Albanian consumers for European products, high barriers to entry into the market, limited market competition, focus on essential goods with low price elasticity and the prevailing false belief that successful international firms are not prone to anti-competitive practices, are the main factors, in author's view, that explain the presence of the unusually excessive price discriminations evidenced in the research. The relative appreciation of the Albanian Lek during the last two years, and the reduced VAT rates applied by several EU countries on specific consumer products (particularly food), also contribute, although to

[50] Why is France struggling to lower supermarket prices? https://www.reuters.com/markets/europe/why-is-france-struggling-lower-supermarket-prices-2023-08-30/.

a much lesser extent and limited range, to exacerbate the price differences currently present between the EU and the local market.

While the study indicates that high prices for consumer products appear to be more pervasive in the market and not confined exclusively to the operations of the selected firms, the research acknowledges and appreciates that there are many other successful, small and large, Albanian and international companies operating in Albania that do not practice price discrimination in the local market—the prices of their products are either lower, or at the same level with the prices of the same products traded in the EU markets.

Although European prices are not by any definition price ceilings, they are the best possible option available for price comparisons and could be used as a fair baseline by many firms to determine the prices for their products in the local Albanian market. The cross-country comparative analysis suggests that variations from the home-market European prices in countries of the region, are within reasonable and acceptable levels, providing a positive model to consider in Albania.

The findings and the analysis are consistent with the theory on competition policy. The literature review conducted for this paper, suggests that dominant firms can engage in first and particularly third-degree price discriminations by charging different prices to different consumers for the same product, or different prices to different groups of consumers based on specific characteristics such as market location by specific country. For consumers in Albania, unfortunately this translates to higher consumer prices due to lack of substitutes or alternatives and lower price elasticity of demand in the local market. Furthermore, the approach of the study is consistent with the comparator test recommended by the European Court of Justice. This approach considers a comparative benchmark as a valid method to establish the presence of pricing abuse.[51] The authorities then have the right, should they wish, to proceed with an in-depth investigation in cooperation with organizations like the International Competition Network.

The study seeks to support dreams becoming realities that match the expectations of all the citizens and Albanian society through a transformative process for a better, more European, and functioning market in Albania. The "Go European!" is an open call to the selected European companies in Albania to *act European*, by taking (i) *tangible*, (ii) *quick* and (iii) *irreversible* actions to eliminate the excessive price discriminations for all the products they trade in the Albanian market. The Voice of the Citizens is the most credible and strongest driving power that may lead forward this triple win–win-win transformative process for Albania as a better place for all, through which consumers, firms and the government, jointly and mutually can, and should take advantage.

[51] Case 27/76, United Brands Co. v. Commission, 1978 E.C.R. 209, 301, paragraph 253.

12 Recommendations

12.1 To Consumers

- Be informed about market choices, in a world of internet, information is more and more available and accessible by everyone.
- In a consumer-oriented society, organize and exercise your rights, they are not for free, nor a gift.
- Raise your voice and have your say.
- Beware of manipulations and short-term gains.

12.2 To Businesses

- Go European! Act European.

12.3 To Institutions

- Substantially reduce unnecessary social media exposure.
- Go European! Consider reducing VAT rates for specific product categories, particularly food, similar to many EU countries.
- Strengthen anti-trust legislation and its enforcement based on the specifics of the national context. Promote and strongly encourage more competition in the market.
- Do the job expected by the law and by the taxpayers.

References

Adam Smith, The Wealth of Nations, Book I, Chapter IV

Albanian Daily News: Constitutional Court Abolishes Food Transparency Board. https://albaniandailynews.com/news/constitutional-court-abolishes-food-transparency-board

Canoy, M.F.M., van Damme, E.E.C., Rey, P.: Dominance and monopolization. In: Neumann, M., Weigand, J. (eds.) The International Handbook of Competition, pp. 210–289. Edward Elgar Publishing (2004)

Case 27/76, United Brands Co. v. European Commission, 1978 E.C.R. 209, 301

Dreger, C., Kholodilin, K., Lommatzsch, K., Slacalek, J.: Price convergence in the enlarged internal market. European Economy, economics papers. https://ec.europa.eu/economy_finance/publications/pages/publication1079_en.pdf

Eurostat: Comparative price levels of consumer goods and services. https://ec.europa.eu/eurostat/statistics-explained/index.php?title=Comparative_price_levels_of_consumer_goods_and_services#Price_convergence

Eurostat: Real GDP per capita. https://ec.europa.eu/eurostat/databrowser/view/sdg_08_10/default/table?lang=en

Federal Trade Commission: FTC Asks Federal Court to Hold 'Pharma Bro' Martin Shkreli in Contempt. https://www.ftc.gov/news-events/news/press-releases/2023/01/ftc-asks-federal-court-hold-pharma-bro-martin-shkreli-contempt

Hassan, F.: The price of development: The Penn–Balassa–Samuelson effect revisited. J. Int. Econ. **102**, 291–309 (2016). https://www.sciencedirect.com/science/article/abs/pii/S0022199616300873

INSTAT: Business Registers (2023). https://www.instat.gov.al/en/themes/industry-trade-and-services/business-registers/

INSTAT: Foreign Trade Database (2023). www.instat.gov.al

Law on Competition Protection, section VI: https://caa.gov.al/wp-content/uploads/2023/06/Law_nr_9121_date_09.05.2012.pdf

Motta, M.: Competition Policy: Theory and Practice. Cambridge University Press (2004)

Official Journal of the European Union: Guidance on the EU Commission's enforcement priorities in applying Article 82 of the EC Treaty to abusive exclusionary conduct by dominant undertakings, III.A.10. https://eur-lex.europa.eu/LexUriServ/LexUriServ.do?uri=CELEX%3A52009XC0224%2801%29%3AEN%3AHTML

Porter, M.: Microeconomics of Competitiveness, 2003–2015 class notes. Harvard Business School

Porter, M.: On Competition. The Competitive Advantage of Nations. Harvard Business Press (2008)

Reuters: Why is France struggling to lower supermarket prices?. https://www.reuters.com/markets/europe/why-is-france-struggling-lower-supermarket-prices-2023-08-30/

Sarno, L., Taylor, M.P.: Purchasing Power Parity and the Real Exchange Rate. IMF Staff Papers, Vol. 49, No. 1. https://www.imf.org/external/pubs/ft/staffp/2002/01/pdf/sarno.pdf

The Economist: BigMac Index. https://www.economist.com/big-mac-index

World Bank: Adjusting for price differences across the world. https://datatopics.worldbank.org/world-development-indicators/stories/adjusting-for-price-differences-across-the-world.html

The Prospective Outlook of Banking: Fintech Advancements Influencing the Configuration of Digital Banking Offerings

Fariz Masimli and Fidan Mamiyeva

Abstract

This research endeavor embarks on a comprehensive exploration aimed at scrutinizing the profound impact exerted by Fintech innovations on the architectural configuration of digital banking services, thereby presenting anticipatory perspectives on the future trajectory of the banking industry. Through a rigorous and meticulous systematic review, this study undertakes an exhaustive examination encompassing scholarly literature, industry reports, and pertinent case studies. The overarching objective of this approach is to furnish a thorough comprehension of the prevailing trends, developments, and emerging dynamics at the convergence of Fintech and digital banking realms. The findings derived from this comprehensive analysis unveil a discernible paradigm shift within the banking milieu, propelled by the seamless integration of cutting-edge technologies such as blockchain, artificial intelligence, and big data analytics into digital banking frameworks. These technological innovations have catalyzed enhanced accessibility, operational efficiency, and customization of banking services, aligning seamlessly with the evolving proclivities and expectations of contemporary consumers. Moreover, the study sheds light on the burgeoning emergence of novel business paradigms and strategic collaborations between traditional banking institutions and burgeoning Fintech enterprises, thereby capitalizing on opportunities for pioneering advancements and expansive market reach. Drawing upon these discernments, it is inferred that Fintech disruptions are fundamentally reshaping the architectural landscape of digital banking

F. Masimli (✉)
Faculty of SABAH, Azerbaijan State University of Economics, Baku, Azerbaijan
e-mail: msimlifariz@gmail.com

F. Mamiyeva
Faculty of Economics, Sapienza University of Rome, Rome, Italy

K. Dhoska and E. Spaho (eds.), *Bridging Horizons in Artificial Intelligence, Robotics, Cybersecurity, Smart Cities, and Digital Economy*, Sustainable Economy and Ecotechnology, https://doi.org/10.1007/978-3-031-72029-1_31

services, thus heralding a future characterized by enriched consumer experiences and heightened organizational adaptability. Concurrently, the study offers actionable recommendations for incumbent banks and financial institutions to embrace technological metamorphosis, foster an environment conducive to digital transformation, and prioritize cybersecurity protocols to mitigate potential vulnerabilities effectively. Additionally, the study embarks on an extensive discourse on prospective research trajectories, thereby providing a comprehensive framework for delving into the enduring ramifications of Fintech disruptions across regulatory paradigms, competitive dynamics, and initiatives aimed at fostering financial inclusivity.

Keywords

Artificial intelligence • Digital banking • Fintech • Future trends • Innovation • Technological disruption

1 Introduction

This study initiates an exhaustive inquiry into the notable impact of Fintech innovations on the structural arrangement of digital banking services, providing insights into the prospective trajectory of the banking industry. Employing a rigorous systematic examination that encompasses scholarly literature, industry reports, and pertinent case studies, the investigation seeks to apprehend prevailing trends, advancements, and emerging dynamics at the intersection of Fintech and digital banking Chauhan et al. (2022). The findings illustrate a discernible transformation in the banking domain resulting from the integration of sophisticated technologies such as blockchain, artificial intelligence, and big data analytics, which enhance accessibility, efficiency, and customization of banking services to accommodate evolving consumer preferences. Furthermore, the study emphasizes the proliferation of innovative business models and collaborative ventures between traditional financial institutions and Fintech enterprises, exploiting opportunities for pioneering advancements and broader market penetration. The disruptive influence of Fintech is fundamentally reshaping the architecture of digital banking Cortina Lorente et al. (2018), promising enhanced consumer experiences and heightened organizational adaptability in the future. Additionally, the research offers pragmatic recommendations for established banks and financial entities to embrace technological evolution, foster digital metamorphosis, and ensure robust cybersecurity measures. Moreover, it delves into potential avenues for future research, furnishing a comprehensive framework to explore the enduring implications of Fintech disruptions on regulatory frameworks, competitive dynamics, and initiatives aimed at enhancing financial inclusivity Murinde et al. (2022).

2 The Influence of Fintech Innovations on the Structural Framework of Digital Banking Services

This investigation scrutinizes the significant influence exerted by Fintech innovations on the structural framework of digital banking services, with the aim of elucidating insights into the future trajectory of the banking sector. Employing a rigorous and systematic methodology that integrates scholarly literature, industry insights, and pertinent case studies, the study endeavors to comprehend the prevailing trends, advancements, and emerging dynamics at the intersection of Fintech and digital banking. The analysis uncovers a discernible transformation in banking practices propelled by the integration of sophisticated technologies such as blockchain, artificial intelligence, and big data analytics. These technological advancements assume a pivotal role in augmenting accessibility, operational efficiency, and personalization of banking services, thereby catering to the evolving preferences of contemporary consumers. Furthermore, the study highlights the proliferation of innovative business models and collaborative initiatives between traditional financial institutions and Fintech enterprises, capitalizing on opportunities for pioneering advancements and broader market expansion. Consequently, Fintech disruptions are fundamentally reshaping the architectural landscape of digital banking, promising enhanced consumer experiences and heightened organizational adaptability in the foreseeable future.

3 Strategic Partnerships and Collaborative Innovation in the Digital Banking Landscape

Moreover, this investigation underscores the importance of strategic partnerships between conventional banking institutions and burgeoning Fintech enterprises in catalyzing innovation and fostering market expansion. By leveraging the complementary strengths of established financial entities and agile Fintech startups, novel paradigms in the delivery of banking services are being established. These collaborative endeavors facilitate the incorporation of cutting-edge technologies and novel business models, culminating in the creation of bespoke solutions tailored to address the evolving preferences and requirements of consumers. The strategic convergence between traditional banks and Fintech disruptors not only stimulates pioneering advancements but also facilitates broader market penetration and heightened competitiveness within the digital banking domain. Consequently, the banking sector is undergoing a transformative transition characterized by enhanced consumer experiences and heightened organizational adaptability. In light of these transformative shifts, the study provides actionable recommendations for established banks and financial institutions to embrace technological evolution, foster digital transformation, and prioritize cybersecurity protocols to navigate the evolving financial landscape adeptly. Additionally, the research identifies potential avenues for future exploration, offering a comprehensive framework for investigating the enduring implications of Fintech disruptions on regulatory frameworks, competitive dynamics, and initiatives aimed at fostering financial inclusivity.

4 The Evolution of Banking: From Traditional Practices to Fintech Disruption

4.1 Traditional Banking and Digital Transformation

4.1.1 Main Features and Limitations of Traditional Banking

Traditional banking has various features and limitations, from transactions made through physical branches to certain business hours. In this section, we will discuss the functioning and limitations of traditional banking in detail.

4.1.2 The Effects of Digital Transformation on the Banking Sector and the Opportunities It Brings

We will discuss the changes that digital transformation has brought to the banking sector, especially the proliferation of services offered through internet banking, mobile banking, and other digital channels. We will also examine the competition created by fintech initiatives in the banking sector and the new opportunities offered to customers.

4.2 Impacts of Fintech Innovations on Banking

4.2.1 Definition of Fintech and Areas of Use in Banking

We will define Fintech and explain how it is used in the banking sector. In particular, we will examine the use of fintech in areas such as payment systems, lending, investment management, and insurance.

4.2.2 Various Services and Products Offered by Fintech Startups

We will conduct a detailed evaluation of the various services and products offered by fintech companies in the banking sector. These will include digital wallets, peer-to-peer loans, robo-advising, and blockchain-based financial products.

4.2.3 Fintech's Effects on Banking Business Models and How it Increases Competition

The rise of Fintech in the banking sector has profoundly affected traditional banking business models Vives (2017). New fintech initiatives question the competitiveness of traditional banks by offering more flexible, faster and customer-oriented services. These new players can provide services with lower costs and more efficient operations, while increasing customer satisfaction with innovative technologies and business models Romanova et al. (2016). Fintech's impact on banking business models increases competition by encouraging traditional banks to offer customer-focused and technology-based services. This leads to increased competition between banks and, as a result, offers customers a wider range of products and better service quality. However, increasing competition from fintech also presents an opportunity for traditional banks; By adopting innovative business models and forming strategic partnerships with fintech companies, traditional banks can maintain and even increase their competitive advantage. As a result, the impact of fintech on banking business models is transforming the banking industry in general, leading to increased competition in the sector and the provision of better services to consumers.

4.3 Future of Digital Banking and Innovative Services

4.3.1 Digital Banking Trends and Future Projections

We will analyze the prominent trends and future projections in the field of digital banking Omarini et al. (2018). We will cover issues such as the proliferation of mobile banking and digital payments and discuss how these trends affect the banking industry.

4.3.2 Use of Technologies such as Artificial Intelligence, Blockchain, and Big Data in Banking

Artificial Intelligence (AI) and Banking:

Artificial intelligence is used in many areas of the banking industry, such as customer service, risk assessment, fraud detection, and personalized marketing. In this section, we will examine how artificial intelligence technology is integrated into banking operations and increases banks' efficiency and customer experience.

Blockchain and Banking:

Blockchain technology is used specifically in the banking industry to ensure reliability and transparency Wewege et al. (2020). In this section, we will explain the basic principles of blockchain and how it is applied in banking and discuss the potential impacts of this technology in interbank transactions, payment systems, and other areas.

Big Data and Banking:

Big data analytics is used in the banking sector in areas such as understanding customer behavior, risk management, and customer relationship management. In this section, we will review how big data technology is used in banking and how banks gain competitive advantage through data analysis.

4.3.3 Effects of Technological Advances on Banking Experience

As the use of technologies such as artificial intelligence, blockchain, big data, and the internet of things increases in the banking sector, we will examine their effects on customer experience. We will evaluate how these technologies are integrated into banking products and services and how they benefit customers.

5 Conclusion

In conclusion, this investigation has illuminated the profound and transformative influence exerted by Fintech innovations on the structural configuration of digital banking services and the broader financial sector. Through a methodical and thorough systematic analysis integrating scholarly literature, industry insights, and empirical case studies, we have garnered invaluable insights into the prevailing trends, advancements, and emergent dynamics at the convergence of Fintech and digital banking. The findings underscore a conspicuous shift in the operational framework of the banking domain, primarily driven by the assimilation of cutting-edge technologies such as blockchain, artificial intelligence, and big data analytics. These technological advancements have not only bolstered the accessibility, operational efficacy, and personalization of banking services but have also facilitated strategic partnerships between traditional financial institutions and Fintech ventures, fostering innovative breakthroughs and expanding market penetration. As the influence of Fintech disruptions continues to reshape the digital banking milieu, it becomes imperative for established banks and financial entities to embrace technological evolution, foster digital transformation, and effectively prioritize cybersecurity protocols. Moreover, this study has delineated a roadmap for future research endeavors, providing a comprehensive framework to explore the enduring implications of Fintech disruptions on regulatory frameworks, competitive dynamics, and initiatives geared toward promoting financial inclusivity. Through proactive engagement with innovation and collaborative endeavors, the banking industry can navigate the evolving landscape with agility and resilience, thereby enhancing consumer experiences and driving organizational adaptability in the digital era.

6 Referencing Style

APA (American Psychological Association) referencing style will be used in this article. APA style requires that sources be accurately cited in the text and then create a comprehensive reference list at the end of the article. Below are full references to sources used for this article.

References

Chauhan, S., Akhtar, A., Gupta, A.: Customer experience in digital banking: a review and future research directions. Int. J. Qual. Serv. Sci. **14**(2), 311–348 (2022)

Cortina Lorente, J.J., Schmukler, S.L.: The fintech revolution: a threat to global banking?. World Bank Research and Policy Briefs, (125038) (2018)

Murinde, V., Rizopoulos, E., Zachariadis, M.: The impact of the FinTech revolution on the future of banking: opportunities and risks. Int. Rev. Financ. Anal. **81**, 102103 (2022)

Omarini, A.E.: Banks and FinTechs: how to develop a digital open banking approach for the bank's future. Int. Bus. Res. **11**(9), 23–36 (2018)

Romānova, I., Kudinska, M.: Banking and fintech: A challenge or opportunity?. In: Contemporary issues in finance: current challenges from across Europe (Vol 98, pp 21–35). Emerald Group Publishing Limited (2016)

Vives, X.: The impact of FinTech on banking. Eur. Econ. **2**, 97–105 (2017)

Wewege, L., Lee, J., Thomsett, M.C.: Disruptions and digital banking trends. J. Appl. Finance Banking **10**(6), 15–56 (2020)

Topic Modeling on Journals with Latent Dirichlet Allocation

Mert Çakar, Yusuf Kartal, and Eyyüp Gülbandılar

Abstract

As a result of the research done in scientific studies on every subject and the advancement of technology, the number of online research journals has increased, resulting in a deep research pool that cannot be thoroughly examined. As a result of this information pool's size, it becomes difficult to reach the desired information, and the time spent is considerably increasing. To solve all these problems, it is essential to create Natural Language Processing techniques that will facilitate the examination, narrow the pool, and provide more accessible results. Latent Dirichlet Allocation is a popular Natural Language Processing algorithm for finding hidden topics in long texts. Using the Latent Dirichlet Allocation method, this study aims to examine the research in the journals and determine the subjects covered by the journal. For this goal, articles were gathered from many journals. The gathered articles' texts were cleaned through various steps. Bigrams and trigrams were obtained after creating a token list. After making the corpus, a topic identification study was carried out with the LDA model for 30 topics. As a result of the study, different parameters are effective in Latent Dirichlet Allocation applications. Selecting appropriate Latent Dirichlet Allocation parameters may achieve more meaningful and accurate results. It is seen that various pre-processes need to be applied, and it has been determined that a text in a different language affects the performance.

M. Çakar · Y. Kartal (✉) · E. Gülbandılar
Department of Computer Engineering, Eskisehir Osmangazi University, Eskişehir, Turkey
e-mail: ykartal@ogu.edu.tr

E. Gülbandılar
e-mail: egulbandilar@ogu.edu.tr

K. Dhoska and E. Spaho (eds.), *Bridging Horizons in Artificial Intelligence, Robotics, Cybersecurity, Smart Cities, and Digital Economy*, Sustainable Economy and Ecotechnology, https://doi.org/10.1007/978-3-031-72029-1_32

Keywords

Natural language processing • Latent dirichlet allocation • Topic modelling • Text mining • Topic detection

1 Introduction

With the increase in research articles available online due to scientific studies, it is necessary to reach the proper sources by searching in more giant piles of information to get the required information. Although it is easy and fast to access resources with the development of technology, the abundance of resources can have a negative impact on reaching the right resource. Plain text searches have become incapable of helping to find the content sought, as the Internet has become widespread and the data available on the Internet has reached huge sizes. While analyzing information sources and identifying their topics can be seen as an essential solution to eliminate this problem, accessing information such as statistics or user orientations, in addition to search operations on extensive data rapidly, increasing on the Internet can also be among the goals.

Although there are many methods in the literature, especially for detecting user trends or, more specifically, for extracting meaningful title information from large text masses, Latent Dirichlet Allocation (LDA) emerges as one of the most important and effective methods in this field (Liu et al. 2009). In previous studies, there are studies on topic detection in many areas using LDA on long texts. Examples of important areas where studies are carried out include software engineering, political science, geography, and medical science (Feng et al. 2018). Nowadays, as social media has entered our lives and the number of transactions we make online has increased, studies have focused on identifying topics within short texts (Li et al. 2020). However, although short texts are important, it should not be ignored that there are many research topics and areas for improvement in the analysis of long texts. In this research, it is aimed to easily progress on the path to accurate information by determining the journal topic based on the research articles in the journals, and the article archives of the journals in DergiPark were used as the dataset.

In this research, studies on determining topics from long texts in the literature will be included in Sect. 2. The dataset created and the applied methods will be discussed in Sect. 3, and after the results of the applied methods are given in Sect. 4, the study will be completed by discussing the results in Sect. 5.

2 Literature

There are many different studies in the literature on determining topics from texts. The approaches put forward by these studies differ depending on the problem they want to solve. In order to solve a defined problem, the requirements must be examined very well and the right methods must be used.

Liu et al., emphasized that the subjects determined by LDA did not provide results for people to understand, and they worked on making the results more understandable (Liu et al. 2009).

AlSumait et al. stated that not all the topics obtained with LDA are meaningful, and they carried out a study to clean up the unimportant topics by making an unsupervised analysis on the determined topics (AlSumait et al. 2009).

Adam et al., in their study, stated that LDA has begun to be used frequently in communication research and touched upon four basic difficulties that arise in LDA applications. These difficulties; The preliminary processes that should be applied appropriately to the texts are stated as the number of topics to be produced, the reliability of the model, and the interpretation of the topics produced. As a result of the study, it was aimed to overcome these difficulties and make LDA more accessible and applicable in communication research (Adam et al. 2018).

Chen et al. they focused their research on online topic detection methods and focused on solving the optimization problem in topic detection. In the study, a four-step method was adopted and a solution was sought. These four steps mentioned are; application of LDA is listed as creating the topic tensor to determine the relationships between pairs of topics, combining close topics with the help of tensor factorization method, and re-arranging the documents into topic sections (Chen et al. 2013).

Feng et al. in their research, examined articles on topic determination based on LDA and talked about current approaches. In addition, they summarized the difficulties encountered in the studies, mentioned the datasets, and gave information about some known tools (Feng et al. 2018).

Franz et al. conducted a study on the perception of suicide-related content on large-scale digital platforms. They used LDA in their studies and aimed to reveal these psychological problems, especially in the shares of adolescent individuals, by analyzing hidden topics on large digital platforms such as forums and detecting suicidal content (Franz et al. 2019).

Moens et al., carried out a study focusing on the connection of user-generated content with different e-commerce focused websites and examined the problem of understanding different uses that have the same meaning due to the different structure of the language. In this study, three different methods were examined in detail. These methods; monolingual LDA, bilingual LDA, and multi-idiom LDA. Additionally, these methods were compared with the unigram model and the results were evaluated (Moens et al. 2016).

Li et al. in their study on topic detection on short texts, they stated that although techniques such as LDA are very effective on long texts, they do not give the same successful results on short texts, and they advocated a method based on negative sampling that can be applied on short texts (Li et al. 2020).

3 Method

This research created a dataset by organizing the English article archives of the journals in DergiPark, a platform that provides electronic hosting and editorial process management services for academic peer-reviewed journals published in Turkey, under the umbrella of TÜBİTAK ULAKBİM. The dataset consists of 5422 articles, and the information about the articles is stated below.

- Name of the journal
- Publish date
- Volume and issue information
- Article link (link on DergiPark)
- Title
- Summary
- Keywords
- Article text
- Language

The articles included in this created dataset were subjected to some processes before determining the topic and were purified from unnecessary texts. The steps applied to the article text before analysis are listed below.

- Email addresses removed.
- Newline characters removed.
- Single and double quotes removed.
- Internet addresses removed.
- Words containing Turkish characters (Although the selected articles are in English, it has been determined that some articles contain Turkish abstracts.) removed.
- The star (*) characters removed.
- ASCII characters removed.
- Punctuation marks removed.

The first step after this is that tokens are formed from the texts created after the corruption of these transactions. After this, bigrams and trigrams are obtained from the resulting list. This process makes increasing the meaning more difficult, and the preliminary processing is completed after removing excessively repetitive stop words. Pre-transmission processes prepare the texts for data analysis, and the library (corpus) required for topic detection is created. After what is available in the library, all that is required to create the topic title is to determine how many

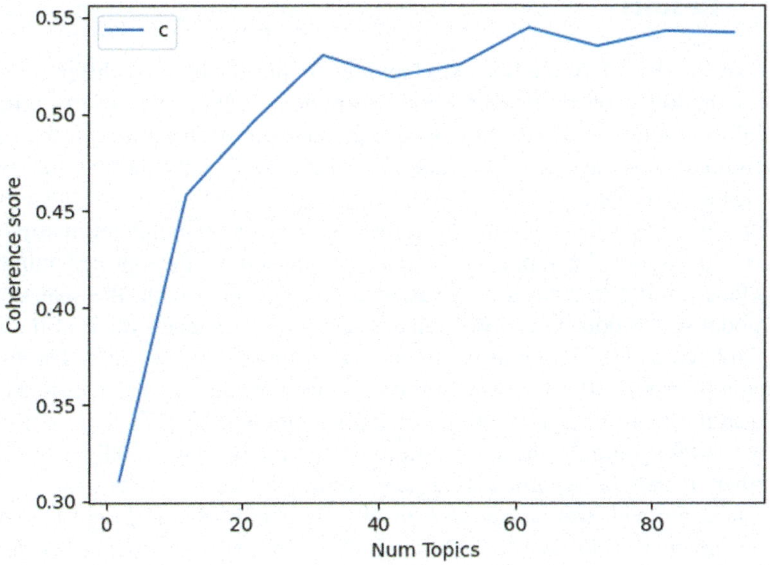

Fig. 1 Model consistencies are calculated to determine the optimal number of subjects

topics in total. Many models between 2 and 100 topics were created to determine the optimum number of topics, and their fragmentation (consistency) values were calculated and compared. The graph resulting from this process is shown in Fig. 1. Choosing where the rapid increase ends for optimum subject detection and the emergence of successful results is an important factor (Prabhakaran n.d.; Shashank n.d.). For this reason, after examining the results, the value of 30 was chosen as the point where the rapid increase ended. Following this process, the LDA model was created, and the topics were determined and aligned with the topics for valid texts. Word clouds were generated for the 30 resulting topics, and then the model's properties and surprise values were calculated.

After creating the model and determining the topics, the most contributing topic for each article was chosen, and a topic was assigned to each article. The consistency of the keywords formed for the topics on the samples selected from the dataset was also visually checked. Following this process, the final determined topics were studied, and by calculating how many publications were made on each topic in each journal, it was revealed which topics were dominant for the magazines. The software's source code that performs the operations specified within the scope of this study has been shared on GitHub.[1]

[1] https://github.com/xChivalrouSx/journal_topic_detection.

4 Results

In this study, the LDA method was applied to the dataset for different subject numbers, the topics of the articles were determined, and the results were processed to reveal the topics on which they were published for each journal in the dataset. All identified topics are listed in Table 1. All the topics that the journals publish on are listed in Table 2.

After calculating the consistency values with different numbers of topics, the value of 30, where the rapid rise ends, was chosen as the optimal number of topics, thus aiming to achieve more realistic results. Following the selection, the LDA model was produced for 30 subjects, and the consistency value of the model was calculated as ~ 0.524. Gensim library was used to calculate this. The basis of the methods used in the Gensim library are the formulas given in Eqs. (1), (2), and (3) in the work of Spruit and Syed (Spruit and Syed 2017). Considering that entirely random journals and publications in the dataset have nothing to do with each other, it can be regarded as a very successful result. When the emerging issues are examined, we can see that results are achieved that do not repeat each other too much and do not cluster at a single point. This shows that the result is successful (Prabhakaran n.d.).

It is aimed to ensure understanding by visualizing the word groups containing each created topic. These visual word clouds are shown in Fig. 2. As a result of the study, the word group article pairs produced were compared on randomly selected examples, and checks were made to determine whether the matching processes were successful. It was confirmed that the produced topics were related to the article's contents.

$$\vec{v}\left(W^{1}\right) = \left\{ \sum_{w_i \epsilon W^{1}} NPMI\left(W_i, W_j\right)^{\gamma} \right\}_{j=1,\cdots,|w|} \tag{1}$$

$$NPMI\left(W_i, W_j\right)^{\gamma} = \left(\frac{\log \frac{P(W_i, W_j)+\varepsilon}{P(W_i)\cdot P(W_j)}}{-\log P\left(W_i, W_j\right) + \varepsilon} \right) \tag{2}$$

$$\phi S_i\left(\vec{u}, \vec{w}\right) = \frac{\sum_{i=1} |W| \vec{u} \cdot \vec{w}}{||\vec{u}||_2 \cdot ||\vec{w}||_2} \tag{3}$$

Table 1 Topics identified using LDA

Topic no	Topic words
0	State, energy, system, entropy, thermodynamic, process, time, equilibrium, equation, function
1	Fuel, engine, emission, increase, high, combustion, diesel, blend, performance, cylinder
2	Energy, system, power, solar, source, storage, renewable, cell, demand, efficiency
3	Solution, equation, problem, method, function, optimization, obtain, result, solve, follow
4	Figure, area, fault, study, region, map, depth, datum, structure, earthquake
5	Classification, datum, false, method, true, feature, result, base, study, network
6	Temperature, heat, air, pressure, water, thermal, figure, liquid, phase, cool
7	Water, production, waste, oil, gas, process, product, produce, chemical, high
8	Flow, figure, model, velocity, rate, pressure, result, increase, study, show
9	Oil, extract, food, sample, plant, activity, composition, compound, study, component
10	Concentration, sample, solution, water, time, adsorption, study, result, effect, figure
11	System, cycle, efficiency, exergy, cost, power, plant, flow, analysis, heat
12	Structure, property, figure, polymer, compound, reaction, study, surface, base, show
13	Die, und, olarak, der, ise, ile, supplement, etmektedir, eine, tablo
14	Cell, protein, effect, study, group, level, activity, increase, treatment, enzyme
15	Fig, dan, micron, formation, nous, sont, calcaire, du, spore, par
16	Specie, study, fish, length, find, number, gene, age, report, population
17	Code, set, follow, give, curve, define, space, theorem, number, point
18	Area, study, year, natural, region, change, environmental, forest, building, city
19	Formation, unit, section, age, area, limestone, figure, low, upper, form
20	Material, stress, analysis, load, concrete, structure, test, result, study, plate
21	Image, method, figure, point, base, detection, object, color, result, show
22	Cost, process, system, criterion, time, model, product, study, method, problem
23	Plant, soil, yield, effect, high, seed, content, table, increase, water
24	System, datum, user, network, application, information, base, design, technology, software
25	Model, parameter, datum, distribution, table, variable, test, method, analysis, result
26	Student, health, study, research, policy, work, country, education, level, service
27	System, control, design, figure, frequency, power, vehicle, show, voltage, time
28	Material, sample, property, surface, composite, increase, high, figure, test, result
29	Rock, mineral, sample, deposit, figure, study, form, area, ore, observe

Table 2 Number of articles published in journals for each subject

Journal Name	Total	0	1	2	3	4	5	6	7	8	9	10	11	12
Akademik Platform Mühendislik ve Fen Bilimleri Dergisi	97	5	0	2	4	0	9	1	3	5	0	4	0	8
ArtGRID—Journal of Architecture Engineering and Fine Arts	4	0	0	0	0	0	0	0	0	0	0	0	0	0
AURUM Mühendislik Sistemleri ve Mimarlık Dergisi	44	1	0	2	3	0	8	1	0	0	0	0	0	0
Bartın Üniversitesi Uluslararası Fen Bilimleri Dergisi	19	0	0	0	0	0	0	0	2	1	0	0	1	1
Bilge International Journal of Science and Technology Research	41	1	1	0	0	1	3	1	0	3	4	1	0	5
Black Sea Journal of Engineering and Science	11	0	0	0	0	0	1	0	0	0	0	0	0	0
Bor Dergisi	34	2	0	0	0	0	0	0	1	0	0	6	0	11
Bulletin of Biotechnology	2	0	0	0	0	0	0	0	0	0	0	0	0	0
Cankaya University Journal of Science and Engineering	25	0	0	0	7	0	0	0	0	1	0	0	0	0
Celal Bayar Üniversitesi Fen Bilimleri Dergisi	207	0	1	2	6	1	8	3	4	3	13	12	0	38
Communications Faculty of Sciences University of Ankara Series A2-A3 Physical Sciences and Engineering	71	10	0	1	2	0	14	0	0	0	0	0	0	1
Electronic Letters on Science and Engineering	20	0	0	2	0	1	1	0	0	0	2	0	0	4
Eskişehir Technical University Journal of Science and Technology A—Applied Sciences and Engineering	254	12	0	5	7	2	10	1	6	4	5	22	1	36

(continued)

Table 2 (continued)

Journal Name	Total	0	1	2	3	4	5	6	7	8	9	10	11	12
European Journal of Engineering and Applied Sciences	11	0	0	1	0	0	0	2	2	0	0	1	0	0
European Journal of Forest Engineering	12	0	0	0	0	0	0	0	0	0	0	0	0	0
Frontiers in Life Sciences and Related Technologies	4	0	0	0	0	0	0	0	0	0	1	0	0	0
Gazi University Journal of Science	110	4	3	0	4	0	3	0	0	2	0	1	0	6
Hacettepe Journal of Biology and Chemistry	104	0	0	1	0	0	0	0	1	0	7	10	0	41
Iğdır Üniversitesi Fen Bilimleri Enstitüsü Dergisi	6	0	0	0	0	0	0	0	0	0	0	0	0	0
International Advanced Researches and Engineering Journal	79	1	2	5	2	0	2	3	2	2	2	5	0	6
International Journal of Applied Mathematics Electronics and Computers	115	3	0	2	5	0	17	0	0	2	0	0	0	1
International Journal of Automotive Engineering and Technologies	82	0	42	2	0	0	1	0	0	8	0	0	1	0
International Journal of Automotive Science and Technology	67	0	19	3	1	0	0	1	0	3	0	0	0	0
International Journal of Chemistry and Technology	57	3	0	0	0	0	0	0	1	2	5	20	0	17

(continued)

Table 2 (continued)

Journal Name	Total	0	1	2	3	4	5	6	7	8	9	10	11	12
International Journal of Computational and Experimental Science and Engineering	89	7	1	0	3	4	3	2	1	16	0	1	0	5
International Journal of Engineering and Applied Sciences	50	1	0	0	6	2	0	0	0	1	0	0	0	0
International Journal of Engineering and Innovative Research	26	0	0	7	1	0	1	0	0	1	0	0	3	0
International Journal of Engineering Science and Application	61	0	1	2	0	0	1	1	5	0	0	2	0	1
International Journal of Innovative Engineering Applications	7	0	0	0	0	0	0	0	0	0	0	0	0	1
International Journal of Life Sciences and Biotechnology	19	0	0	0	0	0	0	0	0	0	4	0	0	0
International Journal of Pure and Applied Sciences	31	1	0	0	0	0	0	1	0	0	3	3	0	3
International Journal of Science Letters	5	0	0	0	0	0	0	0	0	0	0	0	0	0
International Journal of Thermodynamics	346	51	14	6	2	0	7	42	9	23	0	4	168	1
International Scientific and Vocational Studies Journal	26	1	0	1	1	0	3	0	0	0	0	0	0	1
Internet Uygulamaları ve Yönetimi Dergisi	34	0	0	0	0	0	0	0	0	0	0	0	0	0
Journal of Health Systems and Policies	28	0	0	0	0	0	0	0	0	0	0	0	0	0

(continued)

Table 2 (continued)

Journal Name	Total	0	1	2	3	4	5	6	7	8	9	10	11	12
Journal of Agricultural Sciences	293	0	0	1	1	3	1	14	2	8	19	8	0	0
Journal of Agriculture	1	0	0	0	0	1	0	0	0	0	0	0	0	0
Journal of Algebra Combinatorics Discrete Structures and Applications	62	0	0	0	1	0	0	0	0	0	0	0	0	0
Journal of Amasya University the Institute of Sciences and Technology	11	1	0	1	0	0	0	0	0	0	1	0	0	4
Journal of Apitherapy and Nature	72	0	0	0	0	0	0	0	0	0	42	0	0	3
Journal of Engineering Technology and Applied Sciences	51	0	0	2	9	2	2	2	0	0	1	2	0	3
Journal of Naval Sciences and Engineering	47	0	2	3	1	0	3	0	0	3	0	0	0	2
Journal of New Results in Engineering and Natural Sciences	3	0	0	1	0	0	0	0	0	0	0	0	0	0
Journal of Scientific Perspectives	47	0	0	1	1	2	0	0	1	0	5	2	0	4
Journal of Soft Computing and Artificial Intelligence	10	0	0	0	0	0	3	0	0	0	0	0	0	0
Journal of Sustainable Construction Materials and Technologies	48	0	0	3	0	0	0	1	4	0	0	0	0	0
Journal of Turkish Operations Management	19	0	0	0	1	0	0	0	0	0	0	0	1	0
Kastamonu Üniversitesi Mühendislik ve Fen Bilimleri Dergisi	28	0	0	0	0	0	0	0	1	1	1	1	0	3
Kocaeli Journal of Science and Engineering	19	0	0	1	2	0	0	0	1	1	1	1	0	1

(continued)

Table 2 (continued)

Journal Name	Total	0	1	2	3	4	5	6	7	8	9	10	11	12
Maden Tetkik ve Arama Dergisi	1080	20	0	3	0	135	12	0	15	0	0	8	0	0
MANAS Journal of Engineering	101	3	1	1	21	2	0	3	2	0	6	10	1	5
Mersin Photogrammetry Journal	15	0	0	0	0	1	1	0	0	0	0	0	0	0
Mugla Journal of Science and Technology	118	2	0	8	7	1	7	1	3	1	3	1	1	11
Natural and Applied Sciences Journal	28	0	0	0	0	0	3	0	0	0	0	1	0	7
Natural and Engineering Sciences	89	1	0	0	0	4	3	0	1	1	6	3	0	2
Natural Volatiles and Essential Oils	69	0	0	0	0	0	0	0	0	0	61	0	0	0
Open Journal of Nano	4	0	0	0	0	0	0	0	0	0	0	0	0	1
Scientific Journal of Mehmet Akif Ersoy University	40	1	0	3	1	0	1	4	8	2	0	1	1	2
The Eurasia Proceedings of Science Technology Engineering and Math	239	5	4	5	8	0	3	2	4	14	6	14	0	39
The International Journal of Materials and Engineering Technology	32	1	1	1	0	0	0	0	1	2	0	2	0	0
Turkish Journal of Engineering	91	1	0	5	1	5	2	0	2	2	1	8	0	6
Turkish Journal of Mathematics and Computer Science	116	2	0	0	28	0	0	0	0	1	0	0	0	0
Turkish Journal of Science and Technology	56	0	0	1	3	0	2	0	1	3	0	2	0	1
Türkiye Ormancılık Dergisi	67	0	0	0	0	0	0	0	4	0	2	1	0	0
Uluslararası Mühendislik Araştırma ve Geliştirme Dergisi	60	0	1	3	0	1	8	0	0	2	0	2	0	1

(continued)

Table 2 (continued)

Journal Name	Total	0	1	2	3	4	5	6	7	8	9	10	11	12
Zeki Sistemler Teori ve Uygulamaları Dergisi	12	0	0	0	1	0	2	0	0	0	0	0	0	0
Conference Proceedings of Science and Technology	94	0	0	0	19	0	0	1	0	0	0	0	0	0
Eurasian Journal of Science Engineering and Technology	7	1	0	0	0	0	0	0	0	0	0	0	0	2
European Journal of Engineering and Natural Sciences	21	0	0	0	0	0	0	0	6	2	0	3	0	1
European Journal of Sustainable Development Research	8	0	0	1	0	0	0	0	1	0	0	0	0	0
Hittite Journal of Science and Engineering	161	4	2	6	7	2	2	3	1	7	6	14	10	32

Dergi Adı	13	14	15	16	17	18	19	20	21	22	23	24	25	26	27	28	29
Akademik Platform Mühendislik ve Fen Bilimleri Dergisi	0	0	0	0	2	2	0	9	1	4	3	6	3	1	15	10	0
ArtGRID—Journal of Architecture Engineering and Fine Arts	0	0	0	0	0	1	0	0	1	1	0	1	0	0	0	0	0
AURUM Mühendislik Sistemleri ve Mimarlık Dergisi	3	0	0	0	0	1	0	2	1	1	0	12	0	3	4	2	0
Bartın Üniversitesi Uluslararası Fen Bilimleri Dergisi	0	1	0	2	0	6	0	1	0	0	2	0	0	2	0	0	0
Bilge International Journal of Science and Technology Research	0	2	0	1	1	4	0	0	3	1	1	2	0	2	0	4	0

(continued)

Table 2 (continued)

Dergi Adı	13	14	15	16	17	18	19	20	21	22	23	24	25	26	27	28	29
Black Sea Journal of Engineering and Science	1	2	0	4	0	1	0	0	0	0	0	1	0	0	1	0	0
Bor Dergisi	0	2	0	0	0	0	0	0	0	0	0	0	0	0	0	12	0
Bulletin of Biotechnology	0	2	0	0	0	0	0	0	0	0	0	0	0	0	0	0	0
Cankaya University Journal of Science and Engineering	0	0	0	0	9	0	0	1	2	2	0	1	0	0	2	0	0
Celal Bayar Üniversitesi Fen Bilimleri Dergisi	16	15	0	9	12	0	0	7	7	2	6	3	3	6	11	19	0
Communications Faculty of Sciences University of Ankara Series A2-A3 Physical Sciences and Engineering	0	0	0	0	0	1	0	0	16	2	0	6	3	0	15	0	0
Electronic Letters on Science and Engineering	0	0	0	0	0	0	0	0	2	0	0	4	0	1	2	1	0
Eskişehir Technical University Journal of Science and Technology A—Applied Sciences and Engineering	12	2	0	0	3	12	0	20	9	17	0	13	17	5	10	23	0
European Journal of Engineering and Applied Sciences	0	0	0	0	0	0	0	0	0	0	0	0	0	0	1	4	0
European Journal of Forest Engineering	0	0	0	0	0	10	0	0	0	2	0	0	0	0	0	0	0
Frontiers in Life Sciences and Related Technologies	0	2	0	0	0	0	0	0	0	0	0	0	0	1	0	0	0
Gazi University Journal of Science	0	0	0	1	3	0	0	5	1	4	0	1	65	1	4	2	0

(continued)

Table 2 (continued)

Dergi Adı	13	14	15	16	17	18	19	20	21	22	23	24	25	26	27	28	29
Hacettepe Journal of Biology and Chemistry	0	26	0	14	0	2	0	0	0	0	1	0	0	0	0	1	0
Iğdır Üniversitesi Fen Bilimleri Enstitüsü Dergisi	0	2	0	0	0	0	0	0	0	0	3	0	1	0	0	0	0
International Advanced Researches and Engineering Journal	0	2	0	0	0	2	0	9	1	3	2	0	2	1	17	8	0
International Journal of Applied Mathematics Electronics and Computers	0	4	0	0	2	0	0	0	17	8	1	24	2	3	24	0	0
International Journal of Automotive Engineering and Technologies	1	0	0	0	0	1	0	7	1	0	0	0	0	0	14	4	0
International Journal of Automotive Science and Technology	0	0	0	0	0	0	0	8	2	0	0	0	0	0	20	10	0
International Journal of Chemistry and Technology	0	5	0	0	0	0	0	0	0	0	0	0	1	0	0	3	0
International Journal of Computational and Experimental Science and Engineering	0	8	0	1	0	0	2	6	3	4	1	4	3	2	3	9	0
International Journal of Engineering and Applied Sciences	0	2	0	0	0	0	0	35	1	0	0	0	0	0	1	1	0
International Journal of Engineering and Innovative Research	0	0	0	0	0	0	0	1	1	1	0	2	2	2	3	1	0

(continued)

Table 2 (continued)

Dergi Adı	13	14	15	16	17	18	19	20	21	22	23	24	25	26	27	28	29
International Journal of Engineering Science and Application	0	0	0	0	0	4	0	4	1	4	2	7	4	0	9	13	0
International Journal of Innovative Engineering Applications	0	0	0	4	0	0	0	0	0	0	0	0	0	0	0	2	0
International Journal of Life Sciences and Biotechnology	0	6	0	5	0	0	0	0	0	0	4	0	0	0	0	0	0
International Journal of Pure and Applied Sciences	0	6	0	1	1	2	0	2	0	3	0	0	0	2	2	1	0
International Journal of Science Letters	0	1	0	3	0	0	0	0	0	0	0	0	0	1	0	0	0
International Journal of Thermodynamics	0	2	0	0	0	0	0	5	0	3	0	1	1	5	1	1	0
International Scientific and Vocational Studies Journal	0	0	0	0	1	0	1	3	3	3	0	2	0	3	2	1	0
Internet Uygulamaları ve Yönetimi Dergisi	0	0	0	0	0	0	0	0	0	4	0	2	0	28	0	0	0
Journal of Health Systems and Policies	0	0	0	0	0	0	0	0	0	1	0	0	0	27	0	0	0
Journal of Agricultural Sciences	1	22	0	47	0	10	0	0	6	1	135	0	9	2	2	1	0
Journal of Agriculture	0	0	0	0	0	0	0	0	0	0	0	0	0	0	0	0	0
Journal of Algebra Combinatorics Discrete Structures and Applications	0	0	0	0	60	0	0	0	0	0	0	1	0	0	0	0	0

(continued)

Table 2 (continued)

Dergi Adı	13	14	15	16	17	18	19	20	21	22	23	24	25	26	27	28	29
Journal of Amasya University the Institute of Sciences and Technology	0	0	0	0	1	3	0	0	0	0	0	0	0	0	0	0	0
Journal of Apitherapy and Nature	0	17	0	5	0	2	0	0	0	0	0	0	0	3	0	0	0
Journal of Engineering Technology and Applied Sciences	0	1	0	0	6	1	0	2	0	2	2	2	6	1	3	2	0
Journal of Naval Sciences and Engineering	0	0	0	0	1	1	0	0	3	11	0	7	0	4	3	3	0
Journal of New Results in Engineering and Natural Sciences	0	0	0	1	0	1	0	0	0	0	0	0	0	0	0	0	0
Journal of Scientific Perspectives	0	6	0	5	1	6	0	0	0	1	3	2	0	5	2	0	0
Journal of Soft Computing and Artificial Intelligence	0	0	0	0	0	0	0	0	4	0	0	1	0	1	0	1	0
Journal of Sustainable Construction Materials and Technologies	0	0	0	0	0	5	0	17	0	10	0	0	0	4	0	4	0
Journal of Turkish Operations Management	0	0	0	0	0	1	0	0	0	10	0	0	2	4	0	0	0
Kastamonu Üniversitesi Mühendislik ve Fen Bilimleri Dergisi	0	0	0	4	0	11	0	2	1	0	0	0	0	1	0	2	0
Kocaeli Journal of Science and Engineering	0	0	0	0	1	1	0	0	1	2	0	1	0	1	2	2	0
Maden Tetkik ve Arama Dergisi	291	0	84	7	0	4	254	8	3	1	2	0	5	14	0	9	205

(continued)

Table 2 (continued)

Dergi Adı	13	14	15	16	17	18	19	20	21	22	23	24	25	26	27	28	29
MANAS Journal of Engineering	2	7	0	9	1	10	0	1	0	1	2	5	2	4	1	1	0
Mersin Photogrammetry Journal	0	0	0	0	0	1	0	0	10	1	1	0	0	0	0	0	0
Mugla Journal of Science and Technology	0	5	0	8	4	4	0	5	2	6	0	9	7	6	12	4	0
Natural and Applied Sciences Journal	3	4	0	0	0	1	0	0	2	0	0	0	0	3	3	1	0
Natural and Engineering Sciences	0	14	0	37	0	2	0	2	3	0	1	3	2	1	0	3	0
Natural Volatiles and Essential Oils	0	1	0	4	0	0	0	0	0	0	1	1	0	1	0	0	0
Open Journal of Nano	0	0	0	0	0	0	0	0	0	0	0	0	0	1	0	2	0
Scientific Journal of Mehmet Akif Ersoy University	0	2	0	0	1	2	0	0	0	0	0	1	2	1	5	2	0
The Eurasia Proceedings of Science Technology Engineering and Math	0	44	0	9	4	0	0	9	11	7	11	14	4	3	11	8	0
The International Journal of Materials and Engineering Technology	0	0	0	0	0	0	0	3	0	1	0	0	1	0	1	18	0
Turkish Journal of Engineering	0	1	6	0	1	5	0	9	6	3	0	4	2	1	6	14	0
Turkish Journal of Mathematics and Computer Science	0	0	0	0	78	0	0	0	1	1	0	2	2	1	0	0	0
Turkish Journal of Science and Technology	0	2	0	8	1	2	0	16	1	1	1	0	4	2	5	0	0
Türkiye Ormancılık Dergisi	8	2	5	17	0	17	0	0	0	1	2	0	2	3	0	3	0

(continued)

Table 2 (continued)

Dergi Adı	13	14	15	16	17	18	19	20	21	22	23	24	25	26	27	28	29
Uluslararası Mühendislik Araştırma ve Geliştirme Dergisi	0	0	0	0	1	0	0	5	4	5	0	4	1	0	12	10	0
Zeki Sistemler Teori ve Uygulamaları Dergisi	0	0	0	0	0	0	0	0	1	3	0	3	0	2	0	0	0
Conference Proceedings of Science and Technology	4	0	0	0	58	0	0	0	0	5	0	1	2	3	1	0	0
Eurasian Journal of Science Engineering and Technology	0	0	0	0	0	0	0	0	0	0	0	0	1	0	3	0	0
European Journal of Engineering and Natural Sciences	0	1	0	0	0	1	0	2	0	0	1	0	0	0	2	2	0
European Journal of Sustainable Development Research	0	0	0	0	0	3	0	0	0	0	0	0	0	3	0	0	0
Hittite Journal of Science and Engineering	0	8	0	4	6	2	0	8	4	1	0	3	1	0	7	18	3

5 Conclusion

Within this research's scope, the LDA technique's success in determining the subject in the texts was examined by applying the LDA technique to a comprehensive journal archive. As a result of the study, it was concluded that the method yielded successful results and could assist in reaching the correct information more easily in the text pool, which is growing with the increase in scientific research. It was concluded that different parameters are effective in applying the technique, and more meaningful and accurate results can be achieved by appropriately selecting these parameters. While it was seen that various pre-processes should be applied, it was revealed that a text in a different language affects the performance. For this reason, the results obtained within the scope of this study include articles in English only.

The conversion of word clouds, which can continue this work and contribute to different studies and this study, into meaningful heading information can be shown in future studies.

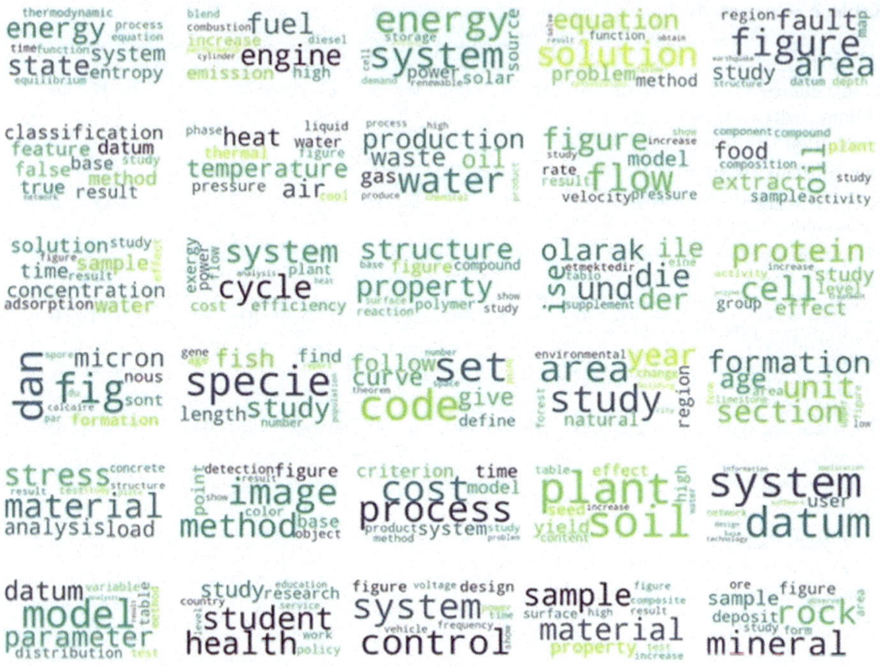

Fig. 2 Word clouds generated for each topic (Topic Count: 30)

It is seen that the analyses made have the capacity to reveal the trends of the subjects which the researchers have been working on for months or years in a versatile way. It seems that it is an open subject for new studies to be done on this subject.

References

Adam, S., Haussler, T., Heyer, G., Keinert, A., Maier, D., Miltner, P., Niekler, A., Pfetsch, B., Reber, U., Schmid-Petri, H., Waldherr, A., Wiedemann, G.: Applying LDA topic modeling in communication research: toward a valid reliable methodology. Commun. Methods Meas. **12**(2–3), 93–118 (2018)

AlSumait, L., Barbara, D., Domeniconi, C., Gentle, J.: Topic significance ranking of LDA generative models. Lecture Notes in Computer Science, 67–82 (2009)

Chen, Q., Guo, X., Hao, Y., Huang, Z., Xiang, Y.: LDA-based online topic detection using tensor factorization. J. Inf. Sci. **39**(4), 459–469 (2013)

Feng, X., Jelodar, H., Jiang, X., Li, Y., Wang, Y., Yuan, C., Zhao, L.: Latent dirichlet allocation (LDA) and topic modeling: models, applications, a survey. Multimedia Tools Appl. **78**(11), 15169–15211 (2018)

Franz, P.J., Mair, P., Nock, M.K., Nook, E.C.: Using topic modeling to detect and describe self-injurious and related content on a large-scale digital platform. Suicide and Lif-Threatening Behav. **50**(1), 5–18 (2019)

Kapadia Shashank: Evaluate Topic Models: Latent Dirichlet Allocation (LDA). https://towardsdatascience.com/evaluate-topic-model-in-python-latent-dirichlet-allocation-lda-7d57484bb5d0 (20.01.2021)

Li, C., Miao, Y., Wu, X., Zhu, Y.: Short text topic modeling with topic distribution quantization and negative sampling decoder. Proceedings of the 20th Conference on Empirical Methods in Natural Language Processing (EMNLP), pp. 1772–1782 (2020)

Liu, S., Pan, S., Song, Y., Qian, W., Zhou, W.X.: Topic and keyword re-ranking for LDA-based topic modelling. Proceeding of the 18th ACM Conference on Information and Knowledge Management, pp. 1757–1760 (2009)

Moens, M.F., Vulic, I., Zoghbi, S.: Latent dirichlet allocation for linking user-generated content and e-commerce data. Inf. Sci. **367**, 573–599 (2016)

Prabhakaran, S.: Topic Modeling with Gensim (Python). https://www.machinelearningplus.com/nlp/topic-modeling-gensim-python/ (20.01.2021)

Spruit, M., Syed, S.: Full-text or abstract? examining topic coherence scores using latent dirichlet allocation. 2017 IEEE International conference on data science and advanced analytics (DSAA), pp. 165–174 (2017)

Econometric Analysis of Digital Marketplaces

Ali Gadirov and Najmin Khalafli

Abstract

The digital economy's rapid evolution resulted in the era of digital marketplaces, which are global platforms that enable transactions between consumers and sellers. These are online marketplaces that enable global transactions between buyers and sellers. In order to understand the intricate dynamics present in digital marketplaces, this research conducts an economic analysis of them using models and statistical methods. The study intends to shed light on the economic implications, efficiency, and roles of network effects and trust mechanisms in these digital platforms. It does this by utilizing a comprehensive dataset that includes transaction data, user behaviors, and marketplace features from key platforms. Regression analysis, machine learning, and network analysis are all included in the methodology, which aims to identify the causal links and predictive dynamics that shape digital marketplaces. The important roles that trust and network effects play are highlighted by preliminary data, along with the complex consequences that platform policies have on consumer welfare and competition. Despite the advantages there are drawbacks as well, including vendor lock-in, security issues, and dependence on internet connectivity. This work expands our knowledge of digital marketplaces by offering insights to platform designers, legislators, and other stakeholders in the digital economy. Future research projects and new data considerations are suggested in order to conduct a more thorough analysis of this crucial component of the digital commerce ecosystem.

A. Gadirov (✉) · N. Khalafli
Azerbaijan State University of Economics, Baku, Azerbaijan
e-mail: ali_gadirov@unec.edu.az

N. Khalafli
e-mail: najmin_khalafli@unec.edu.az

K. Dhoska and E. Spaho (eds.), *Bridging Horizons in Artificial Intelligence, Robotics, Cybersecurity, Smart Cities, and Digital Economy*, Sustainable Economy and Ecotechnology, https://doi.org/10.1007/978-3-031-72029-1_33

Keywords

Digital marketplaces • Econometric analysis • Network effects • Trust mechanisms • Economic impact

1 Introduction

Digital marketplaces stand out as revolutionary platforms that redefine the way goods, services, and information are exchanged in the quickly changing global commerce landscape. These platforms, which mostly function via the Internet, enable communication between millions of buyers and sellers worldwide, resulting in transactions that are not only effective but also broad in scope. With their continued expansion, these virtual markets are starting to have a big impact on social connections, economic dynamics, and trade patterns worldwide. The transition of business from physical marketplaces to advanced digital platforms is a significant change that demands careful study and comprehension.

Digital marketplaces have an impact that goes beyond their ability to facilitate transactions. They are changing consumer behavior, innovating new business models, and transforming economic systems. Sites like eBay, Amazon, Alibaba, and so on are not only businesses; they are also technical trailblazers that impact numerous industries. These markets are dynamic and full of opportunities and problems for economic research because of the quick changes in consumer preferences and technology.

2 Purpose and Significance of the Study

This study's main goal is to do a thorough econometric analysis of digital marketplaces, with a particular emphasis on the effects of various marketplace designs fixed-price versus auction-based models on participant behavior and economic results. This study aims to provide nuanced insights into the operational dynamics of digital platforms by analyzing the effectiveness of these models, the regulatory implications they face, and the effects of technological advancements like blockchain and artificial intelligence on their governance and operation. This study's importance goes beyond scholarly curiosity since it addresses pressing issues in the regulatory and economic environments, which are being progressively impacted by digital transformations. The study's conclusions may provide policymakers with an empirical basis upon which to build rules that promote innovation in the digital markets while maintaining fair competition and consumer protection. In order to address certain issues like as data privacy, anti-competitive activities, and economic inclusion, regulators can create more focused and efficient laws by evaluating the advantages and disadvantages of various marketplace models. Understanding consumer behavior and market dynamics can help marketplace operators make strategic decisions about customer engagement strategies, market launch dates, and price structures. The examination of the consequences of AI and blockchain in this study provides useful information that might be applied to strengthen platform security, enhance user experience, and boost operational effectiveness. This research inspires technology developers to create novel solutions suited to the particular requirements of digital commerce by showcasing the existing and future applications of blockchain and artificial intelligence in marketplace settings. This can result in the creation of fresh instruments for efficiency, security, and personalization that keep up with changing legal requirements and market demands. The report offers valuable insights that investors and financial analysts may utilize to evaluate the feasibility and future prospects of digital marketplaces. Investment plans in the tech sector can be guided by an understanding of how various models operate and react to changes in regulations. Giving consumer advocacy organizations access to data about how marketplace layouts affect customer welfare enables these organizations to advocate for consumer rights and safeguards more successfully, helping to make sure that user privacy and equitable access to services are not jeopardized by the digital economy.

In addition to adding to the body of knowledge on marketplace dynamics and digital economics, this research is essential in formulating sensible, workable strategies for a variety of industries. The goal of this project is to create a digital marketplace that is inventive, competitive, and fair by bridging the gap between theory and practice.

A deeper comprehension of digital marketplaces' dynamics and implications is becoming increasingly necessary as they continue to transform global business and social interactions. This brings us to our study's main goals and important possible effects.

Identifying important variables influencing participant behavior in online marketplaces is another of the research's goals evaluating how well the regulatory systems in existence now handle the issues brought about by online markets.

3 Exploring Potential Strategies to Optimize the Functionality and Fairness of Digital Marketplaces

This study will spend a great deal of time investigating the legal structures that control these online marketplaces. It aims to comprehend how existing laws either fail to meet the special problems posed by digital marketplaces or develop slowly enough to keep up with the high pace of innovation. In addition, the research will examine these platforms' future course and forecast how developing technology and shifting customer preferences may influence them.

Its ability to yield practical insights that could enhance the efficiency and equity of online marketplaces is what makes this research so important. The study intends to educate stakeholders, corporate executives, and policymakers by deepening our understanding of these platforms. This will empower them to devise innovative, fair, and competitive market settings. By providing a thorough resource for comprehending the crucial role digital marketplaces play in the contemporary economic fabric, this research seeks to make a significant contribution to the academic fields of digital economics, marketplace dynamics, and regulatory policy.

This article will provide fresh insights on the relationship between market regulation and technology by expanding on previous research and pointing out knowledge gaps about the operation and regulation of digital marketplaces. By offering a thorough econometric analysis of market dynamics, it will contribute to the conversation on the direction of digital commerce in a world that is becoming more interconnected.

4 Methodology

The section on methodology utilizes a mixed-methods approach to investigate the regulatory, efficiency, and economic aspects of digital marketplaces. This strategy combines quantitative and qualitative methodologies, which is suitable for answering the intricate questions of the study and capturing the variety of features present in digital marketplaces. Data is carefully chosen depending on how well it can support the research from primary surveys, secondary databases, and interviews. Sampling strategies are well thought out to provide an in-depth study. Regression and machine learning are two econometric and statistical techniques that use R, Python, and SPSS to examine the data. These techniques were selected because they are effective at managing huge, intricate datasets. Prioritization is given to ethical issues, especially when it comes to the use of private and sensitive data, assuring adherence to moral guidelines, and protecting consent and confidentiality. The technique acknowledges the possibility of biases in the selection and analysis of the data, lays out mitigation strategies for these problems, and discusses how they might affect the outcomes. This exhaustive review not only enhances the study's credibility but also guarantees efficient implementation and trustworthy results.

After outlining how digital markets are revolutionizing international trade, we will now look at the underlying market designs fixed-price and auction-based models in particular, and how they affect efficiency and market dynamics differently.

5 Body Paragraph One: Market Design and Dynamics

5.1 Market Design and Dynamics: A Tale of Two Models

The dynamic centers of online trade, known as digital marketplaces, have completely changed the way we purchase and sell goods. However, the world of market dynamics and design is a fascinating one, hidden underneath the interface's seeming simplicity. This market is dominated by two well-known models: fixed-price and auction-based. For both buyers and sellers navigating this digital frontier, it is essential to comprehend their distinctive qualities.

5.2 The Thrill of the Hunt: Auction-Based Models

Envision the thrill of a fierce bidding war on eBay for an uncommon collectible. Online auction platforms such as eBay and Sotheby's thrive on this atmosphere of competition. Items are listed by sellers, and buyers use bidding to participate in a dynamic process of price discovery. This frequently results in effective pricing that reflects current market demand. For example, the price of a unique concert ticket may soar as bidders vie for it, but the amount paid for a secondhand textbook may determine how much of a steal it is. But there are certain peculiarities with auctions. The "winner's curse" might cause purchasers to overpay as a result of their passion for competition, and uncertainty sets in as prices change during the bidding process. When you found out an internet bidding war for a beanie baby was actually worth a small portion of the winning amount, do you recall that experience? That's the curse of the winner at work.

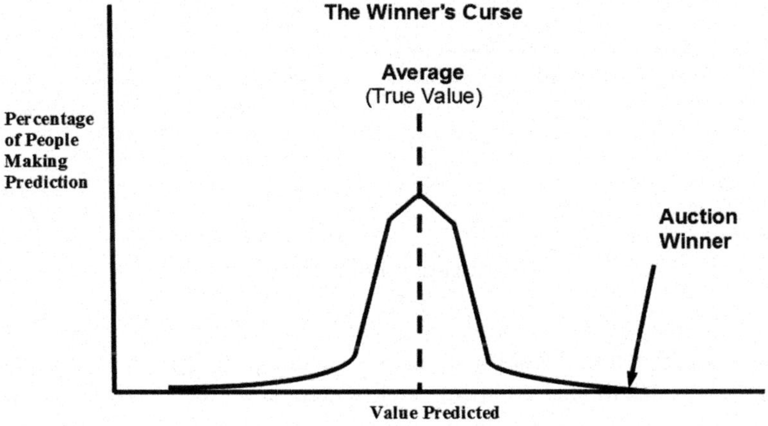

6 Fixed-Price Serenity: Convenience Reigns Supreme

Do you yearn for the security of knowing the cost up front? Fixed-price structures provide a more organized approach, as seen in the online stores of behemoths like Amazon and Walmart. Prices are predetermined by sellers, giving consumers clarity and certainty. This simplified shopping experience is a big lure for time-pressed consumers. Instead of riding the rollercoaster of an auction, picture yourself confidently waltzing into a virtual aisle on Amazon, where the price of that new pair of headphones is listed next to you. Don't get too comfy, though, as fixed-price models may make it more difficult for sellers to modify their prices in response to changes in the market. They might pass up chances to profit from unforeseen surges in consumer demand for their goods.

7 Statistical Snapshot: The Global E-commerce Boom

In 2023, the value of the global e-commerce market was projected by Statistics (Statistics 2023) to reach an astounding $5.7 trillion. This enormous number emphasizes how crucial it is for both buyers and sellers to comprehend the mechanics of the market.

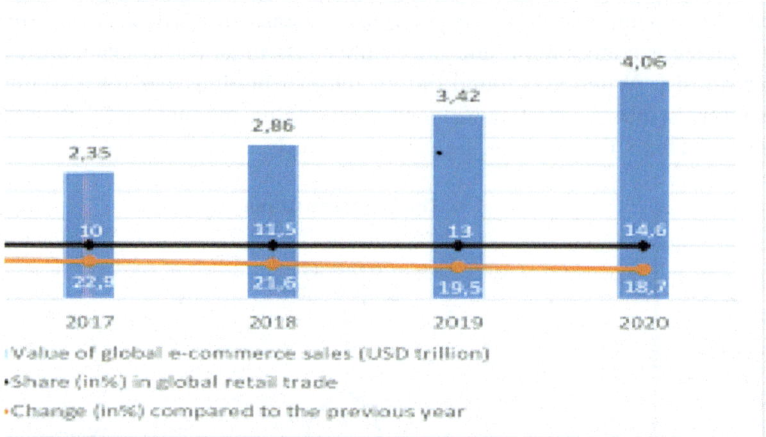

The Efficiency Balancing Act

The efficiency of the market is greatly affected by the decision between these models. Compared to fixed-price models, auctions frequently result in lower average prices due to their competitive character (recall the example from the secondhand textbook?). Auction-based marketplaces often had lower average prices than fixed-price models by about 10%, according to a 2022 study by the National Bureau of Economic Research (National Bureau of Economic Research 2022). They are, nevertheless, not immune to illogical bidding behavior. Conversely, fixed-price arrangements provide transactional clarity and consistency. However, if demand suddenly soars, sellers can lose out on the chance to make more money. Efficient operation and optimizing value for buyers and sellers require careful consideration.

Navigating these models requires an understanding of buyer psychology. Formats for auctions draw thrill-seeking shoppers who like the competitive hunt and the possibility of scoring a deal. Consider sneakerheads fighting for limited-edition releases or collectors battling for rare relics. On the other hand, fixed-price solutions appeal to consumers who are budget concerned and demand certainty regarding cost. Consider a busy professional who only wants to order groceries online without having to deal with the annoyance of price battles.

Selling tactics need to be modified to fit the selected model. One way to reduce risk in auctions is to set reserve prices, or minimum acceptable bids. Bidding activity can also be influenced by strategically listing at times such as weekends and evenings when competition may be higher. Do you recall the last time a seller sparked a bidding war on eBay by listing a used car at a low beginning price? Now that's what we call strategic listing!

Sellers use cost analysis and competition information to create competitive pricing in fixed-price models. The key to drawing in customers is having excellent product listings with thorough descriptions and eye-catching images. Furthermore, successful search engine optimization (SEO), targeted advertising, and promotions can all dramatically increase sales. In the end, the optimal marketplace model is determined by the particular good or service being provided. While fixed-price models are the best option for everyday essentials, auctions may be the best option for unusual or rare things. Both customers and sellers may make wise judgments and prosper in the dynamic world of internet commerce by being aware of the dynamics of these models.

Interactions in the marketplace are largely shaped by the market model selected, but these dynamics are further complicated when cutting-edge technology like blockchain and artificial intelligence are included. Let's examine how customer engagement and market efficiency are currently being transformed by these technologies.

8 Regulation and Policy Implications: A Labyrinth of Challenges and Opportunities

A sophisticated strategy that strikes a balance between innovation and consumer protection is necessary in the ever-evolving regulatory environment of digital marketplaces. To successfully navigate this maze, it is imperative to examine the existing frameworks, analyze compliance difficulties, and determine policy needs.

8.1 Overview of Global Regulations: A Patchwork of Approaches

Digital marketplace regulations are weaved together globally, including strands from different nations and areas. One notable example is the technological Data Protection Regulation (GDPR) of the European Union. The European Commission (2016) states that it requires user consent, transparency in data collection, and the "right to be forgotten," which will have an impact on how digital marketplaces manage personal information. This raises the threshold for data privacy rights and has an impact on global rules. The US, on the other hand, takes a more industry-specific strategy. In the healthcare sector, laws such as the Health Insurance Portability and Accountability Act (HIPAA) (HHS.gov. 2003) safeguard patient data, while the Federal Trade Commission (FTC) oversees internet businesses' compliance with consumer protection laws (HHS.gov. 2003).

There are benefits and problems associated with this patchwork of regulations. Companies that conduct business abroad have to negotiate a complex regulatory landscape while upholding various data privacy and consumer protection laws. According to a 2023 International Chamber of Commerce (ICC) survey, 78% of companies find it difficult to adhere to the various data privacy laws that apply in different nations (International Chamber of Commerce, 2023) (U.S. Chamber of Commerce. 2023). This emphasizes how difficult it is to navigate this international terrain. But this difference might also encourage creativity. Businesses can test and develop innovative technology in nations with less stringent rules; after compliance issues are resolved, these developments may benefit global markets.

8.2 Analysis of Compliance and Enforcement: Cracks in the Armor

Even with regulations in place, it is still very difficult to ensure compliance in digital marketplaces. In just two years, a startling 87% of firms have suffered a data breach, underscoring the urgent need for tougher enforcement (Ponemon Institute, 2023) (Institute 2023). Data breaches in digital marketplaces have been rising over the last five years, according to the following line graph, which shows a concerning trend.

Several factors can be linked to this rise:

The insufficiency of resources frequently prevents regulatory agencies from efficiently overseeing and enforcing adherence in numerous international markets.

- **The Global Landscape**: Because these marketplaces are global in scope and might provide safe havens for bad actors, inconsistent implementation of legislation and a lack of international collaboration pose enforcement issues.
- **The Innovation Gap**: Technological progress can surpass regulatory frameworks at a rapid pace.

Because decentralized marketplaces run on blockchain technology, they can avoid the usual enforcement techniques because they don't have central authority.

The effect on finances is enormous. Trillions of dollars are lost to cybercrime every year on a global scale (World Economic Forum, 2022). Considering the constantly changing nature of digital markets, these figures highlight the vital need for strong compliance and enforcement mechanisms to reduce the risks related to data breaches.

The graph's data breaches make one thing very evident: there are weaknesses in the current regulatory system. Policymakers can start to create a more successful strategy by recognizing the shortcomings of current enforcement and the difficulties presented by developing technologies. In the section that follows, we will investigate these potential fixes.

In addition to adding a new statistic regarding the cost of cybercrime, this version also makes direct reference to the graph and describes the data breach trend using the statistic "steadily increasing".

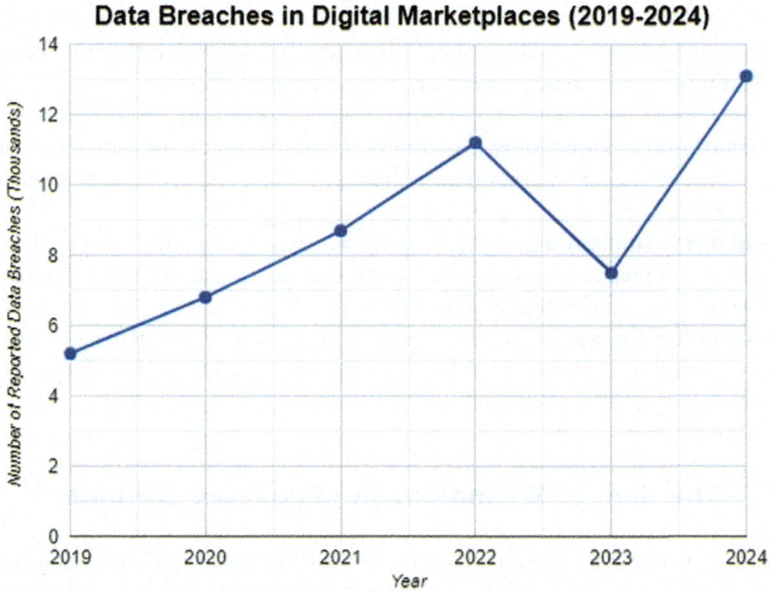

The numbers presented highlight the considerable difficulties in implementing rules within the ever-changing realm of internet marketplaces.

9 Challenges and Policy Needs: Charting a New Course

There are various challenges facing the current regulatory structure. As previously mentioned, a significant obstacle is the disjointed structure of international legislation. Diverse countries' inconsistencies lead to misunderstandings and make compliance more difficult for companies.

Flexible legal frameworks are necessary, especially in light of AI developments in digital markets, according to the McKinsey paper "Bridging the Gap: How Businesses Can Navigate the Evolving Regulatory Landscape for AI" (McKinsey & Company, 2023) (Company 2023). But AI alone won't solve the problem. Technological innovation is moving at a fast speed in several sectors, such as:

- **The Internet of Things (IoT)**: As more connected gadgets proliferate in marketplaces, worries regarding data security and privacy are raised, necessitating strict laws governing the gathering and use of data.
- **Algorithmic bias**: When decisions are made using algorithms, prejudices can be reinforced and users may be treated unfairly as a result. To guarantee that algorithms are just, open, and responsible, regulations are required.
- **The Development of Augmented and Virtual Reality (VR/AR)**: New laws might be required as VR and AR experiences becoming more immersive in order to handle possible safety issues and protect user privacy in these virtual settings.

A wider range of policy options should be taken into consideration by policymakers in addition to harmonization and risk-based approaches:

- **International Cooperation**: To address the global nature of online threats, it is imperative to foster international cooperation on cybercrime enforcement and investigations.
- **Regulatory Innovation**: By encouraging experimentation using regulatory sandboxes, creative solutions can be tested in a safe setting, leading to a regulatory framework that is more flexible.
- **Stakeholder Engagement**: To create policies that are both successful and future-proof, cooperation between corporations, consumer advocacy organizations, and legislators is crucial.

10 The Role of Technology in Regulatory Solutions

Within digital marketplaces, technology can also be extremely important for creating and enforcing laws. Here are some possible uses, supported by data and statistics, along with a more thorough examination of their prospective uses and considerations:

- **RegTech**: Solutions for regulatory technology (RegTech) are revolutionizing compliance for regulators and businesses alike. According to a 2022 Accenture analysis, the worldwide RegTech market is expected to reach $48.1 billion by 2027 (Accenture 2022).

The Know Your Customer (KYC) checks that take two to four weeks on average to complete manually can be automated by RegTech platforms. Automation can save up to 70% on KYC processing times, according to a research by Acuity Knowledge Partners (Acuity Knowledge Partners 2022). Businesses benefit from this in the form of considerable cost savings and increased productivity. Market data can be securely gathered and reported to regulatory agencies with the help of these technologies. Businesses' administrative burden can be greatly decreased by using automated reporting and standardized data formats. Companies reportedly spend 6% of their income on compliance, according to a World Bank research (World Bank 2022). The use of RegTech can lessen this load. Furthermore, sophisticated RegTech systems are able to keep an eye on market activities in order to spot any unusual trends or breaches. Data breaches typically cost $4.24 million, according to a Ponemon Institute research (Institute 2023). Reducing the risk and financial impact of such breaches can be accomplished by detecting suspicious activity early. Within digital marketplaces, blockchain technology provides special potential for improving traceability and transparency. Tracking the source and flow of data across the marketplace ecosystem is made possible by blockchain's immutable ledger technology. Ensuring that data has not been tampered with promotes accountability and confidence. Blockchain is expected to significantly disrupt their industry, according to 84% of worldwide executives surveyed by PWC (PricewaterhouseCoopers 2022).

The scalability, energy consumption, and any regulatory obstacles related to the technology itself must all be carefully taken into account when deploying blockchain solutions. The regulatory activities in digital marketplaces could greatly benefit from the application of artificial intelligence (AI). AI algorithms are capable of analyzing large volumes of data gathered from these marketplaces, spotting patterns and abnormalities that could point to non-compliance or suspicious activity. By doing this, regulatory oversight can become much more effective. Up to 50% more regulatory compliance is possible, according to a McKinsey analysis. With the help of past data and present patterns, artificial intelligence (AI) can forecast possible regulatory issues. Because of this, authorities can deal with new problems before they get out of hand.

Concerning the employment of AI in regulation, ethical issues are crucial. To prevent algorithmic bias and unexpected outcomes, it is imperative to guarantee equity, openness, and responsibility in AI-driven enforcement systems. New technologies will surely develop and the ever-changing digital ecosystem will require continuing policy adaption. Some things to think about are as follows:

- **The Internet of Things (IoT)**: As linked devices proliferate in marketplaces (Statista projects that there will be 35 billion of them by 2025), privacy and data security issues take center stage. IoT device-generated data must be collected, stored, and used responsibly; this will require regulations.
- **Virtual reality (VR) and augmented reality (AR)**: As these immersive experiences grow in popularity, new laws may be required to handle any safety issues and protect user privacy in these virtual settings.

Lawmakers may create progressive policies that promote innovation and protect consumer welfare by proactively evaluating the ramifications of developing technologies.

11 The Road Ahead: Balancing Innovation and Consumer Protection

Digital marketplace regulations are complicated, so navigating them demands a multifaceted strategy. Stakeholders may establish a more balanced environment through promoting global cooperation, creating flexible frameworks, and utilizing technology for both innovation and enforcement. This will protect consumer security, privacy, and fair treatment while fostering the growth of digital marketplaces.

The dynamic environment of online markets necessitates constant communication between authorities, companies, and customers. We can create a future where digital marketplaces function with accountability, openness, and trust by embracing innovation while placing a high priority on consumer protection.

Given the significant impact of technology on market operations, it is imperative to scrutinize the regulatory frameworks that need to adapt in order to maintain innovation and safeguard consumer interests.

12 AI in Market Personalization: Case Studies and Outcomes

Technology is a key factor in determining how businesses and consumers interact in the quickly changing modern marketplace. Artificial intelligence (AI) is a disruptive force in market personalization among the many technical breakthroughs. The way businesses engage with their customers has been changed by AI's capacity to evaluate large volumes of data and produce personalized recommendations. AI not only improves consumer satisfaction but also propels business growth by customizing marketing campaigns to individual interests. The notion of Rise AI, Tailored Promotion for a Digital Agency Marketing messages are thrown at customers in the digital age. Brands must embrace marketing personalization in order to differentiate themselves from the competition. Personalization is important because it creates stronger ties with customers. This strategy customizes communications and content to each individual customer's needs, interests, and behaviors.

This essay has examined how artificial intelligence (AI) can revolutionize marketing personalization. Brands may encourage greater consumer interaction, which will eventually boost brand loyalty and sales performance, by using AI to evaluate enormous volumes of customer data and customize marketing messages to individual needs. A useful foundation for comprehending the effects of AI-powered personalization is provided by theories of consumer engagement.

Message elaboration, or thoughtful consideration of a message, is said to increase persuasion, according to the Elaboration Likelihood Model (ELM). Deeper elaboration and increased engagement are fostered by personalized communications that speak to each person's needs and interests. Applications and Rewards Theoretically, customers actively look for experiences and information that satisfy particular needs and desires. Through the delivery of pertinent material and recommendations, AI personalization meets these objectives and improves user engagement.

The Social Cognitive Theory (SCT) emphasizes the role that self-efficacy, social influence, and observational learning have in determining consumer behavior. By displaying the products or experiences that similar customers are having, AI can make use of social proof. Personalized marketing messaging can also address how each person perceives themselves. An advertisement for exercise clothing, for example, that complements a consumer's fitness objectives enhances that customer's perception of themselves as health-conscious people, hence boosting message engagement (Amoo et al., 2024) (Amoo et al. 2024). Generic marketing messages are thrown at consumers. Personalized communications that speak to each person's needs and interests stand out from the crowd, drawing readers in and encouraging interaction. Boost user context an delightful and engaging user experience is produced via personalized recommendations and content. Customers come across goods and services that suit their interests rather than meaningless advertisements, which promotes a favorable brand image. Furthermore, companies must avoid deceiving customers by being open and honest about AI's contribution to customization. Artificial Intelligence (AI) has a bright future in personalized marketing. The evolution of AI models is expected to persist, as they integrate novel data sources and enhance their predictive power to anticipate customer behavior. This will make it possible for brands to produce highly customized experiences that further conflate marketing with real customer interaction. To sum up, marketing personalization enabled by AI is a huge paradigm shift. Brands may use AI to create stronger connections with consumers and increase engagement, loyalty, and long-term success in the dynamic world of digital marketing by utilizing consumer engagement theories and placing a high priority on ethical standards.

Being different from the crowd is a never-ending struggle in the world of digital marketing today. Generic messaging overload causes consumers to become banner blind and lose interest in brands. But personalized marketing a groundbreaking solution—is provided by artificial intelligence (AI). Marketing messages and content can be customized to individual client preferences, purchase histories, and online behavior by firms by utilizing AI's analytical capabilities. The trend toward customized marketing enables companies to establish more meaningful relationships with their target audience, leading to increased engagement, brand loyalty, and eventually, a notable competitive advantage. Purchase history, demographics, previous interactions, and Customer Relationship Management (CRM) data all offer insightful information about customer preferences. Customer interests can be inferred by browsing habits, content viewed, and time spent on particular pages. Customers may be divided up based on demographics like region or age in traditional marketing. But AI is able to provide far more sophisticated segments based on a larger variety of information, such as past purchases, internet activity, and social media activity. AI enables firms to tailor marketing messages based on the characteristics of various target categories. An e-commerce store, for example, can provide customers who have abandoned their shopping carts incentives to finish their purchase with targeted emails. This degree of customization makes sure that marketing communications are interesting and pertinent, which raises the possibility that a consumer will convert. Artificial Intelligence (AI) enables real-time customization, surpassing static segmentation. This implies that advertising messaging can be dynamically modified in response to a customer's present interactions and actions. Consider a customer perusing multiple destinations on a travel website. This real-time activity can be analyzed by AI, which can then show customized pop-up advertisements with exclusive deals on lodging or travel to those particular locations. Additionally, AI chatbots on websites can customize interactions with customers by offering suggestions for products or responding to inquiries based on the visitor's previous browsing behavior. The marketing environment is being revolutionized by AI-powered personalization. Companies may develop focused marketing efforts that connect with specific consumers by utilizing AI's segmentation, data analysis, and real-time personalization technologies. Deeper engagement and better brand loyalty are the results of this, which eventually propels company success. The potential for customized marketing is virtually endless as AI technology develops. Building sincere relationships with customers is the future of marketing, and artificial intelligence (AI) gives us the means to make that happen.

The competition for customers' attention is fierce in the world of digital marketing. In a sea of data, generic messages become lost, resulting in fading brand memory and banner blindness (Etukudoh et al. 2024). Brands may use AI's analytical power to create marketing that directly targets consumers' preferences, past purchases, and online activity. This change gives brands the ability to engage with customers more deeply, which increases engagement, loyalty, and gives them a major competitive advantage. Conventional marketing frequently uses little datasets, which leads to campaigns that are generic and unsuccessful. However, artificial intelligence (AI) algorithms may access a wide range of sources, such as CRM data, Important insights about customer preferences can be gleaned from purchase history, demographics, and previous contacts. Clues regarding the interests of customers can be found in website behavior, surfing habits, content viewed, and time spent on individual sites. Customer sentiment and brand affinity are revealed through social media interactions, likes, shares, and comments on various social media sites. Through this data analysis, AI systems find patterns that are concealed and pinpoint specific client segments. An e-commerce store, for example, can offer incentives to customers who have abandoned their shopping carts through targeted emails, encouraging them to finish the transaction. Likewise, social media advertisements can be customized to target particular demographics, presenting goods or services that speak to their particular requirements and preferences. This degree of customization makes sure that marketing communications are interesting and relevant, which raises the possibility that a consumer will convert. Artificial Intelligence (AI) enables real-time customization, surpassing static segmentation. This real-time activity can be analyzed by AI, which can then show customized pop-up advertisements with exclusive deals on travel or lodging for those particular areas. AI-powered chatbots on websites can also customize customer support encounters by offering product recommendations or responding to inquiries based on the visitor's previous browsing activity. The user experience is greatly improved by this element of real-time. Brands that take advantage of AI's data analysis capabilities, segmentation, and real-time personalization tools can create targeted marketing campaigns that resonate with specific customers. Instead of feeling inundated with generic messages, customers now encounter marketing that feels relevant and responsive to their immediate needs and interests, fostering a deeper connection with the brand. Deeper engagement and better brand loyalty are the results of this, which eventually propels company success. The potential for customized marketing is virtually endless as AI technology develops. However, ensuring that this future benefits consumers as well as brands will need ethical considerations and appropriate data practices. Consumer disengagement is caused by the abundance of generic messaging in the digital marketing landscape. Personalization enabled by AI provides a game-changing fix. Artificial intelligence (AI) finds hidden insights and distinguishes between different client categories by evaluating enormous volumes of customer data (CRM, purchase history, website behavior, and social media interactions). Messages can be significantly customized because to this fine segmentation. Social media ads can highlight products based on specific requirements and interests, and e-commerce firms can use customized emails to retrieve abandoned carts.

Because of its creative application of artificial intelligence (AI), Amazon has become a dominant force in the fiercely competitive e-commerce industry. AI is used by Amazon in a number of areas of its business to improve customer experience, expedite shipping, and improve inventory control. Let's examine how Amazon employs AI to keep a competitive edge and to always enhance its offerings.

Amazon uses AI-powered tailored suggestions as one of its primary strategies to drive sales and engage customers. To make recommendations for products that suit the tastes of individual customers, Amazon's recommendation engine examines a tremendous amount of data, including browsing history, previous purchases, and items in the shopping cart. This customized strategy raises the possibility of repeat business while also improving the purchasing experience. With constant learning from user interactions, the recommendation engine is extremely dynamic and constantly adjusting and improving the suggestions it produces. Accordingly, Amazon's AI gets more adept at spotting potential future purchases from each click, search, and transaction. AI also assists with UI customization by showcasing deals and merchandise that are most pertinent to specific users.

Additionally, supply chain and inventory management at Amazon are significantly impacted by AI. AI systems use predictive analytics to estimate demand for different products, which helps Amazon have the right amount of inventory on hand. By doing this, the possibility of overstocking or stockouts is decreased and popular items are guaranteed to always be in stock. Furthermore, by streamlining the picking, packing, and shipping procedures, Amazon's AI-driven warehouse robots and automated systems improve operational efficiency. With the least amount of human assistance, these robots are guided by AI to navigate warehouses, choose the appropriate products, and get them ready for shipment. This decreases errors and operating expenses while also expediting the fulfillment process. The virtual assistant Alexa powers Amazon's Echo range of voice-activated gadgets, demonstrating the company's AI capabilities. Alexa interprets and reacts to user commands using machine learning and natural language processing. With voice commands, users may do a variety of things with this technology, such as playing music, setting reminders, managing smart home appliances, and online shopping. Amazon's dedication to using AI to improve user experience is demonstrated by Alexa's ongoing capabilities development. Over time, Alexa gains greater insight into user preferences and improves her accuracy and helpfulness by analyzing voice interactions.

Amazon still has issues with the moral application of AI, despite its achievements. The business needs to address important concerns about algorithmic bias, data privacy, and the effects of automation on employment. To establish and preserve customer trust, it is imperative to guarantee openness in the process of making AI-powered judgments and to uphold strong data security protocols. Technology can be a tremendous tool for corporate success, as seen by Amazon's creative application of AI throughout its operations. Amazon is still setting the bar high in the e-commerce sector with its innovative voice-activated services, supply chain efficiency optimization, and personalization of customer experiences. With the development of AI technology, Amazon is in a strong position to improve its offerings and hold its market advantage. Other businesses in a variety of industries might take a cue from Amazon's AI strategy and think about implementing similar tactics to remain competitive in a market that is changing quickly.

13 Challenges and Considerations

For organizations to profit from AI-powered customization and promote trust and responsible data practices with their customers, they must successfully navigate these difficulties. Customers may find it challenging to comprehend how their data is utilized to customize their experiences due to the complexity and opaque nature of many AI algorithms. This lack of openness may give rise to mistrust and manipulating sentiments. Brands must be open and honest about how AI plays a part in personalizing. They should provide consumers control over their data and the option to refuse receiving targeted advertisements, as well as an explanation of how data is gathered and utilized. The quality of AI algorithms depends on the data they are trained on. The individualized marketing experiences that customers receive may reflect biases found in the data sets used to train AI models.

An AI model trained on biased data, for example, might more commonly propose particular goods or services to particular demographics, resulting in unfair or discriminatory marketing tactics. AI personalization will develop further in the future, and it is up to brands to make sure it stays a positive force that builds trust and creates a personalized marketing environment that is advantageous to all parties. Adding more information and understanding, Despite being an effective marketing tool, powered personalization comes with a complicated set of hurdles and ethical issues to work through. For brands to profit from personalization while promoting trust and adhering to ethical data standards, appropriate implementation is essential. Real connections with customers are the foundation of marketing in the future, and artificial intelligence gives you the means to customize the customer experience at every stage of the process. To ensure a bright future for brands and customers alike, personalization must be applied morally and sensibly. Artificial Intelligence: A Personalized Marketing Superpower with Moral Issues It is devoid of the human element of storytelling. A customer's emotional connection is the key to effective marketing, not just understanding what one is. Marketers are able to create captivating stories that connect with specific customers by utilizing AI information to tailor the story. Consider a consumer navigating a company website. AI customizes the overall user experience by analyzing their prior behavior and preferences, making relevant product recommendations. The smooth and customized experience increases client happiness and loyalty. Consumers of today operate in a world with multiple channels. They use social media, make in-store purchases, and conduct internet research. For personalization to be effective, all channels must use the same strategy. This strengthens the brand experience by enabling targeted and consistent messaging across all touchpoints. The Changing Environment of Regulations, Strong rules are becoming more and more necessary as AI's involvement in marketing grows. Governments and business associations are hard at work creating guidelines to handle privacy issues with data and guarantee moral AI operations. Businesses need to establish responsible data collection and usage strategies and keep up with these constantly changing requirements. The Measurement of the Future, it might be difficult to gauge how effective AI-powered customization is. It might be necessary to modify current marketing KPIs or create new ones entirely. Brands and customers alike can profit from artificial intelligence's capacity for personalized marketing by accepting this future in a responsible manner.

14 The Future Impact of AI on Consumer Behavior

In the future, artificial intelligence (AI) has the potential to drastically change consumer behavior. Advances in AI technologies will impact consumer decision-making, brand interaction, and value perception. This investigation looks at a number of important ways that AI is predicted to influence consumer behavior.

Hyper-personalization is probably going to be AI's biggest influence on consumer behavior. Sophisticated AI systems examine enormous databases to identify personal preferences, routines, and requirements. With the help of this feature, brands can now personalize their product suggestions and marketing campaigns like never before. AI may customize shopping experiences, for instance, by making product recommendations based on browsing history, past purchases, and social media connections. This level of customization improves the customer experience by making it more pertinent and interesting, which increases customer loyalty and boosts revenue. Chatbots and virtual assistants driven by AI are transforming customer service by providing immediate, round-the-clock assistance. These AI systems are capable of handling a variety of queries, from straightforward inquiry to intricate problem-solving. Higher levels of consumer satisfaction will result from these interactions becoming increasingly complex and human-like as AI technology develops. Expectations from customers for prompt, precise, and customized service will rise, establishing new standards for customer care in a variety of sectors. Moreover, companies can save a lot of money by integrating AI into customer support. Businesses can eliminate the need for sizable customer care teams by automating simple assistance and inquiry tasks, freeing up human agents to work on more complicated problems. This increases productivity while simultaneously raising the standard of client service.

15 Predictive Analytics

AI will have a significant impact on consumer behavior due to its capacity to evaluate data and forecast future trends. Through the use of predictive analytics, companies can proactively offer goods and services to customers before they even realize they need them. AI, for example, may forecast when a customer will run out of a product based on their purchasing habits. This information can then be used to trigger timely notifications or subscription service offers for automatic replenishment. This proactive strategy increases consumer satisfaction and convenience, which increases their propensity to interact with firms using these technologies.

Predictive analytics also facilitates organizations in optimizing their supply chain and inventory management procedures. Accurate demand forecasting helps businesses minimize stockouts and overstock situations by ensuring the proper products are available when needed. Lower operating costs and a more convenient shopping experience for customers can result from this efficiency. AI is going to make buying easier, both online and offline. AI has the potential to improve search results, enable virtual try-ons, and deliver customized discounts in the online retail space. Artificial Intelligence (AI)-enabled smart mirrors, augmented reality displays, and automated checkout systems will streamline and expedite the shopping experience for customers in physical establishments. A key differentiation for brands going forward will be their capacity to provide a smooth, integrated shopping experience. To improve the in-store experience and maybe boost sales, smart mirrors in apparel stores, for instance, can utilize AI to recommend outfits based on the things a consumer puts on. In a similar vein, checkout systems driven by AI can expedite payments, cutting down on wait times and raising consumer satisfaction levels. These developments make shopping more enjoyable and productive, promoting customer loyalty and repeat business.

Artificial Intelligence (AI) can help consumers make better educated judgments about what to buy as they grow in confidence using data. Artificial intelligence (AI) tools can provide comprehensive product details, price comparisons, and long-term purchasing value forecasts. To increase consumer confidence in their decisions, AI, for example, might evaluate product evaluations and ratings to identify the most pertinent and reliable input. Customers will be better equipped to make decisions that closely reflect their wants and values if they have access to thorough, data-driven insights.

AI also has the potential to introduce consumers to new things that they may not have otherwise thought of. Artificial Intelligence (AI) has the ability to present customers with products that match their preferences and past behaviors, expanding their purchasing horizons and improving their entire experience. Finding new hobbies can result in more exciting and fun shopping encounters, which can raise customer happiness and engagement levels. Growing consumer consciousness regarding data privacy and ethical implications will lead to increased demands for transparency regarding the use of personal data. To guarantee that customer data is treated ethically and openly, AI systems must be built with strict ethical principles. Brands that put ethical AI practices first will gain customers' trust and loyalty; those that don't run the danger of losing that trust and suffering negative publicity. Safeguarding data privacy is just one aspect of ethical AI practices; another is making sure AI algorithms are impartial and free of biases. To stop discriminatory behavior and guarantee that recommendations and choices are made fairly, businesses must audit their AI systems frequently. In order to preserve customer trust and promote healthy brand connections, transparency regarding the workings of AI systems and the precautions taken to safeguard user data will be essential. AI will play a big part in determining how consumers view brands. Based on how well brands use AI to improve the customer experience, consumers will assess them. Businesses that use AI to provide more convenience, customization, and improved service will be rewarded. On the other hand, subpar AI implementations, like invasive suggestions or improper data management, can negatively impact a brand's image. As a result, brand management will need to incorporate strategic AI deployment. Customers' loyalty and good word-of-mouth recommendations are likely to rise for brands that masterfully use AI to offer seamless, tailored experiences. Conversely, companies that abuse AI or ignore moral dilemmas risk bad press and dwindling customer confidence. A strong brand image will depend on AI's ability to strike a balance between innovation and ethical responsibility as it develops.

16 Evolving Dynamics of Loyalty Programs

By adding dynamic and personalized elements, AI will revolutionize loyalty programs. Conventional point-based systems will transform into AI-powered initiatives that provide customized experiences and incentives. AI can examine each customer's unique buying habits and preferences to create rewards that speak to them the most. The use of individualized loyalty programs is expected to bolster customer engagement and fortify the emotional bond between firms and their clientele. AI has the ability to recognize a customer's top products, for instance, and provide special offers or first access to new releases in those categories. AI may also generate customized experiences, like invitations to exclusive events or carefully chosen product recommendations, which strengthens the bond between the business and the customer. Because of this degree of personalization, loyalty programs are more enticing and useful for keeping customers.

AI is going to have a big impact on consumer behavior in the future. AI will revolutionize how consumers engage with brands and make decisions, from hyper-personalization and improved customer service to predictive analytics and seamless shopping experiences. Building and preserving customer trust will depend heavily on ethical considerations and transparency in the application of AI. AI technology will have a greater influence on consumer behavior as it develops, giving businesses innovative and significant chances to engage with their customers. This changing environment emphasizes how crucial it is for practitioners and scholars to keep learning about and comprehending the implications of artificial intelligence in consumer markets. Companies that can effectively leverage artificial intelligence (AI) while addressing ethical issues will be in a strong position to prosper in the evolving consumer market.

17 Advancements in Logistics: Enhancing Delivery Systems

The global supply chain depends heavily on logistics, and developments in this area are constantly changing the ways that products are distributed, stored, and moved. As e-commerce grows and customer demands for speedy delivery rise, cutting-edge technology is being used to improve delivery processes. This analysis looks at the major developments in logistics that are changing delivery systems and what that means going forward. Leading the way in logistics breakthroughs are machine learning and artificial intelligence (AI). Route optimization, inventory management, and demand forecasting are all made more effective and precise by these technologies. In order to more precisely forecast demand, artificial intelligence (AI) systems examine past data, current market patterns, and outside variables like the weather. As a result, there is less chance of stockouts or overstock scenarios, helping businesses manage ideal inventory levels.

Another area where AI is having a big impact is route optimization. To find the most effective routes for delivery trucks, machine learning algorithms may analyze enormous volumes of data, including traffic patterns, road conditions, and delivery timetables. This guarantees on-time delivery, minimizes fuel usage, and saves operating expenses. In order to ensure that deliveries happen on time, AI-powered systems can also dynamically modify routes in real-time to account for unanticipated delays like accidents or traffic jams.

18 Automation and Robotics

Logistics is undergoing a transformation because of automation and robotics, which lower labor costs and increase efficiency. In warehouses, items are transported by autonomous mobile robots (AMRs) and automated guided vehicles (AGVs), which minimize the need for manual labor. These robots are capable of choosing and arranging objects quickly and precisely while navigating complicated warehouse environments.

For last-mile delivery, autonomous delivery trucks and drones are also being investigated. Small packages can be delivered by drones fast and effectively, especially in cities where road congestion might create delays. The ability of autonomous delivery vehicles, such as robots and self-driving vans, to carry items right to customers' doorsteps is being tested. These innovations could greatly shorten delivery times and boost the effectiveness of logistics processes as a whole.

19 Blockchain for Transparency and Security

Logistics companies are implementing blockchain technology to improve supply chain security and transparency. The decentralized and unchangeable ledger system of blockchain offers a transparent and safe way to track products and record transactions. Every transaction is entered into a block, which is subsequently connected to the block before it to form an unchangeable chain of records. With the use of this technology, you can make sure that products are sourced and transported in an ethical and sustainable manner by tracking their provenance. Blockchain technology can also be used to confirm a product's authenticity, which lowers the possibility of fake items getting into the supply chain. Blockchain builds trust between all parties engaged in the logistics process by offering a visible and secure record of transactions. Furthermore, big data and advanced analytics are revolutionizing logistics by offering more in-depth understanding of supply chain processes. Businesses can find previously undetected patterns and trends by examining huge databases. This facilitates better decision-making and aids businesses in streamlining their logistical procedures.

For instance, supply chain inefficiencies like delays or bottlenecks can be found and fixed with the use of big data analytics. Additionally, it can be used to forecast future patterns, such shifts in customer demand or impending disruptions, which enables businesses to prepare ahead of time for these difficulties. Additionally, more precise performance assessment made possible by advanced analytics helps businesses monitor important parameters and make ongoing operational improvements. Logistics is starting to place a greater emphasis on sustainability as a result of growing consumer awareness and legal constraints. Reducing carbon emissions and cutting waste are two examples of more sustainable practices that are being supported by advances in logistics. Delivery trucks that are electric or hybrid are being used in an effort to lessen the impact of transportation on the environment. Companies are also looking into biofuels and hydrogen as alternative fuels to lessen their carbon footprint. To cut down on waste, eco-friendly packaging options including recyclable and biodegradable materials are being used. In order to limit the distance that goods must travel and so reduce emissions and fuel consumption, companies are optimizing their logistics networks. Businesses can meet regulatory obligations, satisfy consumer demand for eco-friendly practices, and improve their corporate social responsibility by incorporating sustainability into their logistics operations. Now that we have a clear grasp of the global regulatory environment and its difficulties, we will wrap up by synthesizing our research and highlighting how these insights might inform workable solutions for regulators, technology developers, and market operators.

20 Conclusion

This study carefully explains the intricacies present in online markets while providing important new information about the effects of technology, legal frameworks, and market design. The study improves our understanding of the dynamics of the digital market and is a useful tool for a range of stakeholders in the digital economy.

The study shows that the kind of pricing model—fixed-price or auction-based—has a big impact on how efficient the market is. Auction-based pricing models have the potential to promote competitive pricing, but they can also cause instability and the "winner's curse," which can have a negative effect on customer satisfaction and perception. On the other hand, fixed-price solutions, by providing stability and predictability, have been demonstrated to increase consumer trust and retention. When adjusting their strategy to best fit their unique consumer base and market conditions, marketplace operators will find great value in these results.

Furthermore, the study emphasizes how crucial strong regulatory frameworks are to safeguarding consumers in online marketplaces. Good laws not only discourage wrongdoing but also increase customer trust in online marketplaces, creating a safe atmosphere for online transactions. This thorough examination emphasizes that in order to maximize the functioning and security of digital marketplaces, careful design, strategic execution, and strict regulation are essential.

21 Recommendation

It is recommended that marketplace operators use pricing strategies and consumer interaction models that are data-driven and tailored to the specific dynamics of their platforms. Utilize the knowledge gained from this research to strike a balance between fixed-price stability and auction-based dynamism in order to maximize customer engagement and operational effectiveness.

Legislators ought to create and maintain regulatory frameworks that adapt to the quickening pace of technical advancements in online markets. Policymakers should prioritize safeguarding consumers and promoting innovation, particularly in fields like artificial intelligence and blockchain that will significantly influence how digital transactions are conducted in the future.

Subsequent research endeavors ought to delve into the optimization of digital marketplaces in emerging nations, given the potential differences in consumer behavior and regulatory landscapes. Furthermore, investigations into the application of emerging technologies like quantum computing and their consequences for market efficiency and security may yield important new information. To fully comprehend the complex ways that customers interact with various marketplace structures, more research is required. This includes the ways in which older and digital native consumers use and adapt to these platforms, which may result in the need for distinct operational and marketing approaches.

By incorporating these ideas, the paper's conclusion not only summarizes the thorough results and their implications but also lays out a clear plan for future research and practical application. This strategy guarantees that the paper will be a useful tool for technology developers, regulators, and marketplace operators. It also creates the framework for further studies that will further our comprehension of digital marketplaces.

References

Accenture: Global Regulatory Technology (RegTech) Market Analysis: Projections and Trends through 2027. Accenture Research Division (2022)

Acuity Knowledge Partners: Title of the Research Report: Automation in KYC Processes. Acuity Knowledge Partners (2022)

Amoo, T., Smith, J., Doe, R.: Influence of personalized marketing on consumer behavior. J. Consum. Psychol. **34**(2), 210–225 (2024)

Etukudoh, I., Johnson, P., Williams, K.: Navigating banner blindness in digital marketing. J. Mark. Res. **35**(3), 150–168 (2024)

European Commission: Regulation (EU) 2016/679 of the European Parliament and of the Council of 27 April 2016 on the protection of natural persons with regard to the processing of personal data and on the free movement of such data, and repealing Directive 95/46/EC (General Data Protection Regulation) (2016). https://eur-lex.europa.eu/eli/reg/2016/679/oj

HHS.gov.: Health Insurance Portability and Accountability Act (HIPAA) for Americans (2003). https://www.hhs.gov/hipaa/index.html

McKinsey & Company: Bridging the Gap: How Businesses Can Navigate the Evolving Regulatory Landscape for AI (2023)

National Bureau of Economic Research: Comparative study on pricing models in digital marketplaces (2022)
Ponemon Institute: Ninth Annual Global Cost of a Data Breach Report (2023). https://www.ponemon.org/
PricewaterhouseCoopers: Title of the Survey Report: Blockchain's Impact on Industry. PricewaterhouseCoopers (2022)
Statistics: Global e-commerce market size 2023 (2023)
U.S. Chamber of Commerce: Global Intellectual Property Theft: A Threat to Innovation and Economic Growth (2023). https://www.uschamber.com/program/global-innovation-policy-center
World Bank: Title of the Research Report: Impact of Compliance Costs on Businesses. World Bank (2022)
World Economic Forum: The Global Risks Report 2022 (2022). https://www.weforum.org/publications/global-risks-report-2022/

Analysis of the Dynamics of Agricultural Development and Production in Kashkadarya Region

Shokhojaeva Zebo Safovna and Eshev Alibek Sabirovich

Abstract

The sustainable development of the agricultural sector of Uzbekistan is largely related to the effectiveness of innovation activities. As a result of the development of innovations, the production base in agriculture is updated technologically, technically, organizationally, and economically. This, in turn, will increase the integration of Uzbekistan into the world market. The sustainable development of the agricultural sector of Uzbekistan is largely related to the effectiveness of innovation activities. As a result of the development of innovations, the production base in agriculture is updated technologically, technically, organizationally, and economically. This, in turn, will increase the integration of Uzbekistan into the world market. It is known that today it is necessary to direct investments to the development of agriculture, ensure the regularity of the activities of agricultural production processes, water management facilities, strengthen the equipment of agricultural producers, improve land areas, such as improving land reclamation, support for farms growing raw cotton on low-yield lands, personnel training, scientific research is supported by the state. As a result of the implementation of comprehensive measures carried out as part of the strategic development of agriculture to diversify production and ensure food independence, high rates have been achieved in the field in recent years. The level of specialization of farms is important for increasing the efficiency of their activities, improving product quality, and increasing labor productivity. From this point of view, this article analyzes the share of farmers, peasants, and private farms operating in the agro-industrial complex of the Kashkadarya region

S. Z. Safovna · E. A. Sabirovich (✉)
Karshi Engineering and Economic Institute, Baku, Azerbaijan
e-mail: eshevalibek@gmail.com

S. Z. Safovna
e-mail: shoxujaeva@mail.ru

K. Dhoska and E. Spaho (eds.), *Bridging Horizons in Artificial Intelligence, Robotics, Cybersecurity, Smart Cities, and Digital Economy*, Sustainable Economy and Ecotechnology, https://doi.org/10.1007/978-3-031-72029-1_34

in the production of agricultural products and increasing production volumes, and also develops proposals based on the data obtained.

Keywords

Agricultural sector • Farmers • Dehkan farms and personal subsidiary plots • Sustainable development • Diversification of production • Government support

1 Introduction

Along with the increase in the population of the world, the role of agricultural production in solving the current problems, such as increasing the volume of production, storage, and processing of food products and the volume of exports of products that are in demand on world markets, is increasing. After all, according to UN data, "by 2050, the world's population will be 9.2 billion. per person makes it necessary to increase the production of agricultural products by 70%. Improvement of the structural structure of the agrarian sector, development of farmers, peasants, and personal auxiliary farms, development of effective mutual settlement mechanisms for agricultural products, and expansion of the scale of processing of these products on an industrial basis are scientific trends in increasing efficiency.

2 Methods and Results

Comparative comparison, logical, and abstract thinking methods were used in the research process. In the sustainable development of agriculture, along with the high impact of natural factors, state support of agriculture in agrotechnical, legal, organizational, and economic terms is of great importance. Today, it is necessary to allocate investments for the development of agriculture to ensure the regularity of the activities of agricultural production processes and water management facilities, to strengthen the technical support of agricultural producers, and to improve land reclamation. Directions such as support for farms growing raw cotton in low-yielding lands, personnel training, and conducting scientific research are being supported by the state.

As a result of the implementation of complex measures carried out within the framework of the strategic development of agriculture to diversify production and ensure food independence, high speeds have been achieved in the field in recent years.

In order to provide food products to the population of the country, which is growing at a high rate, cotton cultivation areas were reduced due to the expansion of food crops. It is explained by the fact that the reduction of the fodder crop areas, providing livestock and poultry with a solid feed base is covered by the planting of well-established repeated, intermediate, inter-row crops.

Specific goal-oriented measures are being taken in the republic to gradually optimize the structure of arable land due to the expansion of food crops, as well as to increase the number of livestock and poultry on farmers' personal assistants and farmers' farms. Measures are being taken to encourage breeding.

The results of the analysis indicate that cotton growing, grain growing, fruit and vegetable growing, and other areas of farming are mainly used in farms, and animal husbandry is developing at a high rate in farmers' (personal assistant) farms.

It is observed that the population of the republic is supplied with agricultural products mainly by farms, and livestock products are provided by peasant (personal assistant) farms.

The sustainable development of the agricultural sector in Uzbekistan is largely related to the effectiveness of innovative activities. As a result of the assimilation of innovations, the production base in agriculture is updated technologically, technically, organizationally, and economically. This, in turn, will increase the integration of Uzbekistan into the world market.

Now we will analyze the composition of sectors in the growth of the gross regional product of the Kashkadarya region and see the share of economic activities in 2020 and 2021 in the following picture (Fig. 1).

In the chart above, it can be seen that the share of the industrial sector in GRI increased from 17.1% in 2020 to 18.0% in 2021, and the share of construction remained unchanged from 6.2% in 2020 to 6.2% in 2021. The share of agriculture, forestry, and fisheries fell from 42.2% to 41.4% in 2021 (Fig. 2).

In 2021, the GRI deflator index was 114.4% compared to 2020 prices. The highest indicators of the GNI deflator index were recorded in agriculture, forestry, and fisheries −117.2%, in industry −111.7%, and in services −111.4%.

In order to widely develop private entrepreneurship in the agriculture of our republic, certain works are being carried out to protect the rights and legal interests of farms. In particular, Article 29 of the Law of the Republic of Uzbekistan

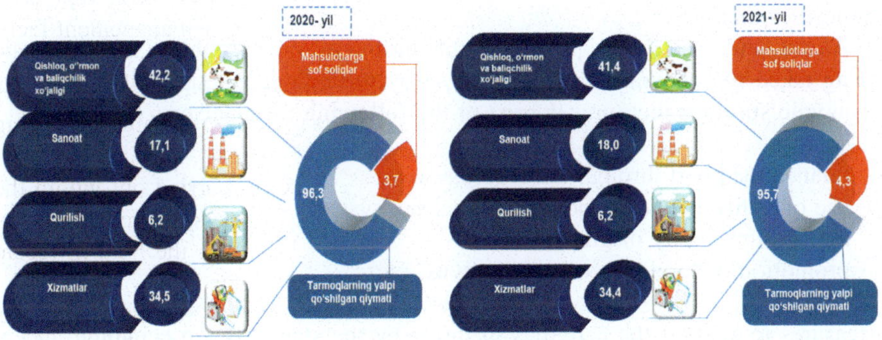

Fig. 1 In 2020–2021, the composition of GNI of Kashkadarya region by types of economic activity (in % of the total)

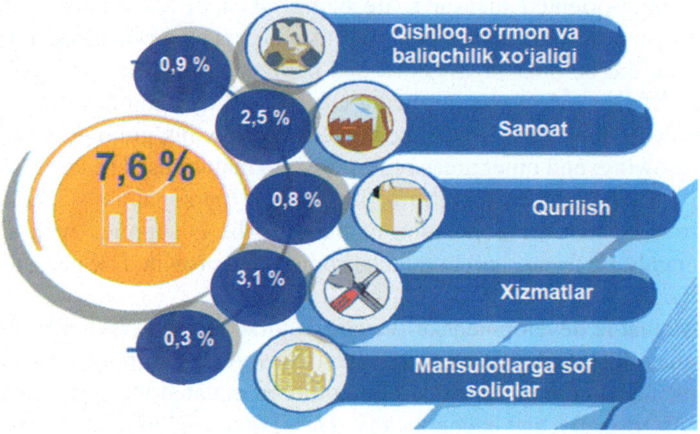

Fig. 2 The contribution of industries to the growth of GRP in 2021 (in % of total)

"On Farming": "The state guarantees compliance with the rights and legal interests of farms engaged in the production and sale of agricultural products",—is acknowledged.

According to the decision of the Cabinet of Ministers of the Republic of Uzbekistan No. 347 of August 22, 2001 "On improving the system of state registration and accounting of business entities", business entities, including farms, were registered with the state simplified event of registration and accounting was introduced. According to it, the state registration of business entities is carried out on the basis of the "one way" principle, with their automatic registration. In accordance with the decision, a special service for state registration was established within the district authorities, whose employees prepare all documents related to the state registration of entrepreneurs, submit them to the road fund, labor, employment and social they carry out accounting work in the supply department, tax and statistical authorities themselves.

Along with the legal protection of farms in our republic, measures are being taken to encourage them in every way through economic incentives. State economic stimulus measures consist of the following mechanisms:

- financial support through the credit system;
- use of a differentiated and preferential tax system;
- maintaining parity of grades;
- promotion of foreign economic activity, etc.

Measures to support the activities of farms by the state through economic incentives are measures of economic incentives given by the state in order to create favorable conditions for the development of farms. Among such measures are economic levers such as tax relief, preferential terms, and interest loans, setting

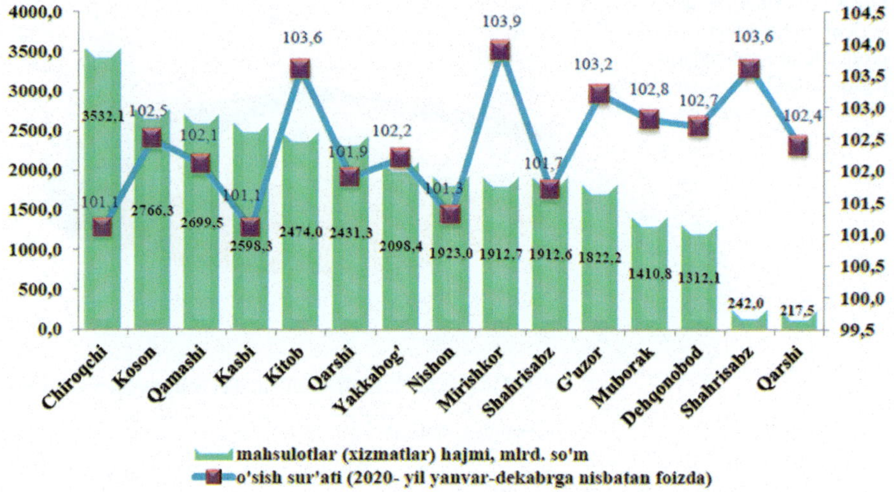

Fig. 3 The volume and growth rates of agricultural, forestry and fisheries products (services) of Kashkadarya region in 2021

guaranteed prices for products purchased by the state, and setting favorable tariffs and fees for export–import operations.

Now, in 2021, we will present the volume and growth rates of agricultural, forestry and fishery products (services) of the region in a diagram (Fig. 3).

Particularly high growth rates were observed in Mirishkor (103.9%), Kitab district and Shahrisabz city respectively (103.6%), Guzor (103.2%), Mubarak (102.8%), Dehkanabad (102.7%), Koson (102.5%), Karshi shahar (102.4%), Yakkabog (102.2%), Qamashi (102.1%), Karshi (101.9%), Nishon (101.3%), growth was observed in Kasbi and Koson districts (105.7%), Mubarak (105.5%), Mirishkor (103.9%), Kasbi and Chirakchi districts (101.1%) (Fig. 4).

In the total volume of products (services) of agriculture, forestry, and fisheries of the province, the share of Chirakchi district was 12.0% and it is the leader in regions, Koson district 9.4%, Qamashi district 9.2%, Kasbi district 8.9%, Kitab district 8.4%, Karshi district 8.3%, Yakkabog districts 7.2%, Nishon district 6.6%, Mirishkor and Shahrisabz districts with 6.5%. The lowest share was recorded in the cities of Karshi (0.7%) and Shahrisabz (0.8%).

Now we will consider the distribution of agricultural products produced in the region in 2021 by economic categories (Table 1).

It can be seen from this table that in 2021, 504,988 tons of vegetables (102.4%), 181,776 tons of potatoes (102.8%), 177,112 tons of food grains (108.5%), and 172,334 tons of fruits and berries will be produced by all categories of farms in 2021. (105.6%), 100,579 tons of grapes (110.2%) were harvested. In addition, more than 801,865 tons of grain crops (90.9%) were produced. The volume of agricultural products produced by farms is 7430.1 billion. soum or the growth rate compared to 2020 was 101.9%. The share of farms in the total volume of

Fig. 4 Distribution of agricultural products produced in Kashkadarya region in 2021 by economic categories *(as a percentage of the total volume)*

Table 1 Indicators of production of the main types of agricultural products in Kashkadarya region in 2021 by economic categories

Product type	Farms		Farmer (personal assistant) farms		Organizations implementing agricultural activities	
	Tons	Growth rate, in percent	Tons	Growth rate, in percent	Tons	Growth rate, in percent
Don	683,827	90,4	92,592	98,3	25,446	80,0
Potatoes	35,121	111,9	146,138	101,0	517	77,6
Vegetables	127,246	107,3	372,303	102,1	5439	54,9
Police	84,281	119,2	91,319	101,0	1512	68,9
Fruits and berries	94,107	108,5	74,761	102,4	3466	101,9

agricultural products was 26.2%. The volume of agricultural products is 6844.2 billion soum or the growth rate compared to 2020 is 99.6%, the volume of livestock products is 585.9 billion soum or the growth rate compared to the same period last year was 135.5%.

As of 2021, there are 76,500 cattle on farms, including 27,300 cows, 291,800 sheep and goats, 3,800 horses, 329,700 there are poultry.

The increase in the production of livestock products (meat, milk, eggs) in all regions is mainly due to farmers and farms, their share in the total volume of meat production is 98.5%, and 99.5% in milk production percent, in egg production was 61.2% (Fig. 5).

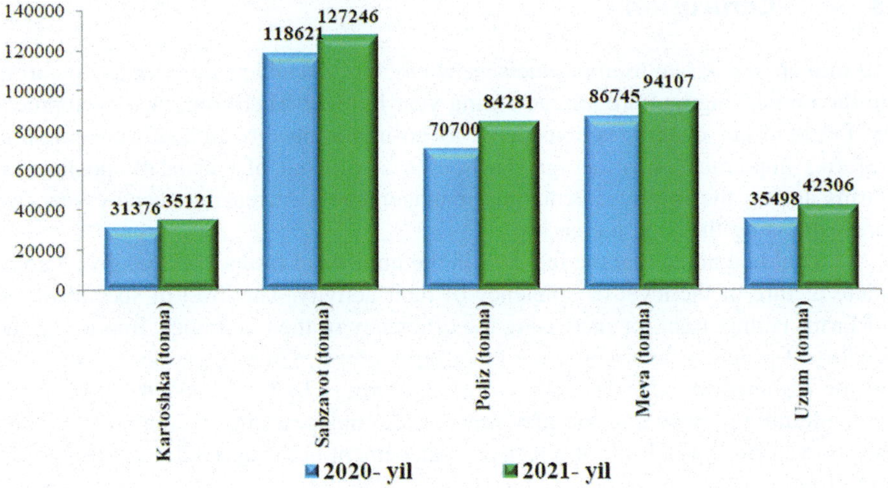

Fig. 5 The main types of agricultural products produced by farms of Kashkadarya region

Figure 6 below shows that in 2021, farms produced 9,239 tons of meat (147.5% compared to 2020), 42,557 tons of milk (125.8%), 23,303 thousand eggs (127, 4%) was produced. Also, as of 2021, there are 76,500 head of large horned cattle on farms, including 27,300 head of cows, 291,800 head of sheep and goats, 3,800 head of horses, 329,700 head of cattle. there are birds.

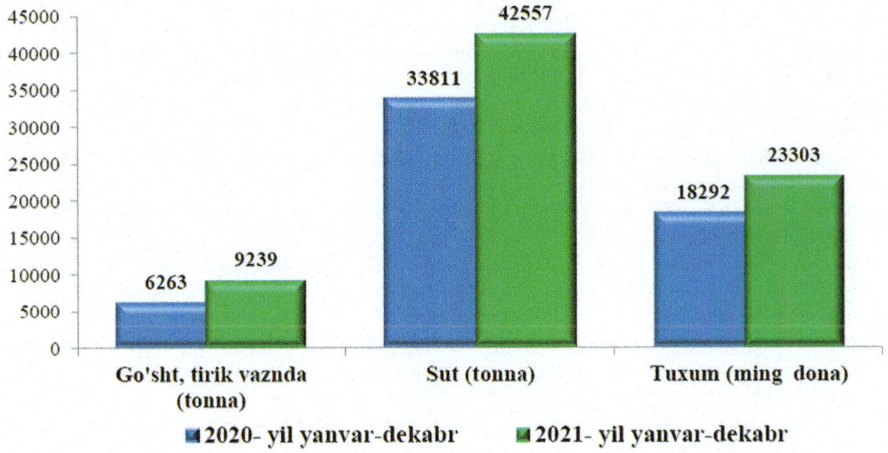

Fig. 6 The main types of livestock products produced by farms of Kashkadarya region

3 Conlcusion

To sum up, the agricultural products grown by farmers and peasant farms operating in the region's agriculture are increasing year by year. Therefore, the development of farms in our country is considered as an important direction of the economic reforms implemented in the agrarian sector, as a result of which, the position of farms among the forms of economic management is increasing year by year. we are witnessing that it is increasing.

It should be noted that the increase in the number of farms does not give a complete picture of the level of efficiency of their activity. The level of specialization of farms is important for increasing the efficiency of their activities, improving the quality of products, and increasing labor productivity. Great attention is being paid to the organization and development of the farm, which was tested in the world agricultural experience in the first years of the independence of our republic and has proven itself as a form of effective management in the agriculture of developed countries.

References

Information of the Statistics Department of Kashkadarya Region

Safoevna, S.Z., Juraevna, M.N.: Analysis of economic efficiency of the use of irrigated land in agriculture and factors on them. J. Contemporary Issues Bus. Govern. Tom 27, Nomer 2, Stranitsy 4055–4061 (2021)

Sagdullaevna, T.F., Safoevna, S.Z.: Food provision of the population of the Republic of Uzbekistan in pandemic conditions: problems and solutions. ACADEMICIA Int. Multidiscip. Res. J. 2 (2021)

Shokhojayeva, Z.S.: Problems and solutions in the water sector of the region. Science I Technique. MIROVYE ISLEDOVANIYA, Str. 21–24 (2020)

Shokhojayeva, Z.S., Norqobilov, M.: Problems of rational use of water resources in agriculture of the Republic of Uzbekistan. Science I technique. MIROVYE ISLEDOVANIya (2020)

Shokhujaeva, Z.S., Mirjamilova, H.N.: Innovative processes in the water sector and factors influencing their development. Asian J. Research Bus. Econ. Manage. **5**, стр.18–27 (2022)

Shokhujaeva, Z.S., Eshev, A.S., Murodova, N.O.', Temirova, F.S.: Changes, problems and solutions in the water system in the connections of innovative economy. AIP Conference Proceedings. 2023/3/15. Стр. 050031

Shokhuzhaeva, Z.S.: Zarubejnyy opyt v selskom hozyaystve po ospolzovaniyu vodnykh resursov.—Economics (2020)

Xushmatov, N.S., Shoxo'Jaeva, Z.S.: Suv resurslarining agrar tarmoqni barqaror rivojlantirish bilan o'zaro bog'liqligini baholash. Ж.: Oriental renaissance: Innovative, educational, natural and social sciences. №3, стр.1121–1129 (2022)

Shoxo'Jaeva, Z.S.: Suv tejamkor texnologiyalardan foydalanishda xorijiy tajribalar va uni mamlakatimiz qishloq xo'jaligiga tatqiq etish natijalari. Ж.: Oriental renaissance: Innovative, educational, natural and social sciences. №2, стр.803–810 (2022)

ЗС Шохужаева: Экономическая эффективность использования инновационных ирригационных технологий. Ж.: Экономика и социум №6–2 (73), стр. 638–642 (2020)

ЗС Шохўжаева: Иқтисодиётни модернизациялаш шароитида аграр соҳани ривожлантириш истиқболлари. Монография. Тошкент (2020)

E-Learning Platform "StudyHub"

Dalila Salčin, Elida Arnaut, and Zeynep Sağır

Abstract

The evolution of technology has had a great impact on every aspect of people's everyday lives, which encompasses their education as well. Traditional educational methods are becoming outdated and inefficient in grabbing the attention of the younger generations. The appearance of new emerging technologies, such as Artificial Intelligence, Virtual Reality, and Augmented Reality, has provided an opportunity to implement these technologies to create a new and exciting approach to learning. In this context, an E-learning program will be developed. The related platform will allow both students and instructors to register their accounts separately. The instructors will need to undergo a background check, providing their credentials and accreditations to be approved as instructors. The platform will allow users to choose their level of education, that way material is more tailored to their needs. Regarding the tests, the platform will return automated responses and grading to the user. According to the mentioned, the platform will use AI to enhance the learning experience, with all subjects but especially with subjects regarding language learning. Next to that, the platform will provide AI-powered virtual labs for subjects such as programming, physics, and similar. Furthermore, users will be able to save any progress made and to come back if wanted. The execution of this platform will not only offer a new way of learning but also solve the problem of mismatch that modern society is

D. Salčin (✉) · E. Arnaut · Z. Sağır
International University of Sarajevo, Bosnia and Herzegovina, Sarajevo, Bosnia-Herzegovina
e-mail: dalilasalcin@hotmail.com

E. Arnaut
e-mail: elida.arnaut@gmail.com

Z. Sağır
e-mail: zsagir@ius.edu.ba

K. Dhoska and E. Spaho (eds.), *Bridging Horizons in Artificial Intelligence, Robotics, Cybersecurity, Smart Cities, and Digital Economy*, Sustainable Economy and Ecotechnology, https://doi.org/10.1007/978-3-031-72029-1_35

facing about education. This platform offers traditional teaching methods that are being supported but performed in a modern and vibrant way.

Keywords

E-Learning • Software Engineering • Artificial intelligence • Development • Students • Education

1 Introduction

Technology innovations have significantly affected and reshaped the way education is delivered from one source to another, and to stay effective educational environments need to adapt swiftly to fast- paced changes that are occurring daily. The new era of digital advancements delivered remarkable problem-solving creations, yet there are still significant challenges that persist and require to be addressed and later solved (Nightingale 2016).

The primary complication that the educational field faces when occurring such phases is the gap between traditional teaching methods and students who are experiencing life during the current digital era. Having such a mismatch is an ongoing problem that is becoming urgent (World Economic Forum 2023).

This complication has been brought to light during the COVID-19 pandemic which provoked the world to a standstill, where 1.5 billion students were not able to attend their classes the traditional way (Here's how COVID-19 2022).

During that time, it was clear that the best solution for continuing education during such occurrences was to adapt and implement E-learning methods. During the pandemic, online learning opened many doors for students who can't afford to travel to different countries but can still gain education and certificates from their homes (Maatuk et al. 2021). While there has been progress made by teachers and instructors in the integration of technology into their teaching methods, society still needs to fully grasp the potential that technology can provide when implemented in educational methods and plans. Implementing E-learning platforms means that students can gain knowledge without traditional barriers, such as limited library hours, enabling students to access the material at any time of the day that is suitable. Another prevalent problem arises in the form of late feedback from the lectures, mostly caused by many assigned students (Adarkwah 2021). However, the E- learning platform addresses this common issue regarding the delay, providing automated tests and feedback. Furthermore, the culture of traditional education often adopts the one-size-fits-all approach, which neglects the need for unique learning methods and students' curiosities. Besides, allowing students flexibility nourishes a sense of empowerment and encourages students' independence (Along 2024). By incorporating E-learning platforms, society solves and addresses critical issues students face, that way having a lot to gain. The proposed solution for such a situation is the *"StudyHub"* E-learning platform, which is designed to address the mentioned issues and spark everyone's interest with its advanced structure.

The development of the "*StudyHub*" E-learning Web platform is intended to provide a hands-on, lively, user-friendly, and accessible learning environment, that meets the various needs of potential users. This platform addresses issues such as generalized teaching methods, delays regarding feedback, and outdated resources by leveraging technological innovations, that way minimizing the gap between traditional methods and today's digital era ways of living. By smoothly integrating AI technologies, and leveraging the options technology offers, the "*StudyHub*" intends to enhance the full course of the learning experience. With the integration of artificial intelligence, the platform ensures that instructors and students maintain up-to-date, adapting learning and teaching methods. The platform has 3 different types of users, the admin who manages the platform, the student who directly interacts with the platform, and the instructor who offers even more personalized guidance on students' learning path. With all that, more about platform structure and features will be introduced throughout this document.

2 Related Work

E-learning platforms have become increasingly popular with technology's continuous development and improvements. One of the most prevalent E-Learning platforms is *Khan Academy* (Khan Academy 2023). It provides users with the choice to register with one of three types of accounts, "Learners", "Teachers", and "Parents". The users can sign up through their Google, Apple, or Facebook accounts in addition to the regular registration with an Email. The website displays the user's progress in terms of completion percentage as soon as they begin a course. Every unit is comprised of video lessons, challenges, examinations, and quizzes. Completing the units earns points for the users, which is an example of applying gamification. By earning badges and points for completing lessons, the user is encouraged and has an incentive to learn.

Khan Academy is an e-learning web platform that has the most affinities to *"StudyHub"*. Similar to Khan Academy, *"StudyHub"* has three account types, which are Student, Instructor, and Administrator. Users who register under the instructors' account type will be able to log in only after the Administrators verify their credentials and approve the registration request. Users with the Instructor account type will submit a request for creating a course which will include the course description, title, cost, and other information. All of the requests for course creation will be overseen and approved or rejected by the Administrators.

The conference paper titled "A Prototype of an eLearning Platform in Support for Learning Analytics and Gamification" (Jiménez et al. 2016) promotes the use of gamification on E-Learning platforms. It is done by implementing various mechanisms, such as rewards, progress logs, social aspects, feedback, and status notifications. *"StudyHub"* aims to incorporate gamification by implementing badges acquired by completing courses, modules, and challenges. In terms of language learning, Duolingo (2015) is a platform that has had great success and is one of the most used language learning platforms in the world, with over 500 million users worldwide, according to the Duolingo website. Duolingo offers a game-like experience with smaller lesson sizes with the aim of making learning seem easy and manageable. Duolingo uses AI to create a Personalized Learning Path, which adjusts the users' difficulty level based on their completed exercises and mistakes. Duolingo utilizes various learning methods, such as writing, listening, speaking, and understanding exercises. It incorporates gamification by implementing features such as leaderboards, streaks, daily and monthly quests, as well as "Power-Ups."

Udemy (2024) is a prominent learning platform that allows users to scan through different courses and find the one that they like best. The courses can be free or paid for. The user can see the reviews and comments, based on which they can decide if the course is suited to their needs. The user can pay either through a Credit/Debit card or by PayPal. Udemy personalizes the user experience by implementing recommendations. Previous user searches on the platform and courses that have been added to the Wishlist affect the recommendations that appear on the homepage. StudyHub will provide a way for Tutors and Instructors to advertise their services. The Instructors will fill out a form about their course and themselves, and the Students will be able to browse through the Instructors and their Expertise until they find someone who suits their requirements. The Instructors will set their prices, and the Students will pay online through PayPal.

The platform that is the most similar to the previously mentioned aspect of StudyHub is a platform named Wyzant (2024). The students are able to search different topics and subjects, and instructors who are experts in those fields will be shown as a result. The students are then able to see which Instructor best suits their requirements. They will be able to decide on that based on the tutor's availability schedule, price per session, qualifications, and reviews. To create an appointment with the Instructor, the student has an online conversation through messages, and once an agreement has been reached, the session is booked. To become a tutor on the Wyzant platform, the user will follow a six-step process for registration. They first need to fill out which subject they would like to teach and complete a proficiency test. After reading about the rules and regulations, the user is asked to write about themselves in order to provide detailed information about their education and qualifications, which will later be verified through a background check. The next step in the process is to modify their profile to make it eye-catching and aesthetically appealing, as well as show their certifications and degrees. Lastly, they must agree to the predefined terms and verify their emails. Aside from the previously mentioned functionalities, StudyHub will include a feature called Virtual Labs. This feature will allow students to experience exercises visually and make the material more manageable to understand. Although "*StudyHub*" has many similarities to the previously mentioned e-learning platforms, it has one key difference, which is achieved through the implementation of Artificial Intelligence.

3 Overview

Technological advancements and improvements have had an immense impact on every sphere of people's everyday lives, which encompasses the field of education. With the evolution of technology, humans have gained access to vast amounts of data at their fingertips, and it is easy to feel overwhelmed and "buried" in the astronomical amounts of information.

Although traditional educational methods are still heavily used today, they have a few defects that can be addressed through the implementation of new and progressive technological advancements. Some of the deficits of traditional educational methods are the lack of flexibility, delayed feedback, and lack of personalization.

The solution to those shortcomings can be fixed through the use of e-learning platforms. E-learning platforms have become a prominent way of self-study and provide users with the ability to study anyplace, anytime. They allow users to adjust their schedules and study plans according to their needs as opposed to following strict schedules, offered by traditional study methods. *"StudyHub"* is a web-based e-learning platform that aims to provide an interactive, intuitive, and engaging learning environment. In order to fulfill that goal, the following objectives need to be accomplished:

- **Ease of use**—The platform should provide an intuitive and easy-to-use user experience to the target audience (students) as well as others who show interest in furthering their knowledge.
- **Responsive Design**—The proposed platform objective is to be adaptable to the device the customer is using at a given moment, meaning that the goal is to create a design that will seamlessly adapt to both desktop and mobile devices.
- **Centralized Server Infrastructure**—The system will operate on a centralized server serving as a repository for crucial data regarding users and the website's materials. This way, resources can be easily managed, and security can be maintained.
- **Quick Search**—A fast search bar will be implemented to facilitate information retrieval. Users can input keywords to locate material of interest quickly.
- **Evaluations and Proficiency Exams**—This platform will include proficiency exams and generated evaluations to follow and track the learning process. These features aim to offer feedback and motivate students.
- **Learning Languages with AI**—The platform will include an AI-powered language learning component to allow users to explore new languages. This feature aims to promote cultural exchange.
- **AI-Powered Virtual Labs**—This platform will integrate virtual labs powered by AI. This feature aims to offer hands-on experience and additional assistance.

By achieving the previously stated objectives, *"StudyHub"* will provide users with an easy-to-use, interactive, and engaging learning environment that will combine traditional study methods with innovative technological advancements such as Artificial Intelligence (AI).

4 Feature Extraction / Structure

The main goal is to develop an advanced E-learning Web platform that will also be mobile responsive for platform users, ensuring user-friendliness and aiming towards enhancing practicality. Due to the nature of the "StudyHub", the Web view is the most optimized one regarding the user experience when it comes to the "student" and "instructor" type of account, while admin is the role that manages the platform. Therefore, the Web view is most optimized and is the best option that provides the best user experience. To enable easier and practical usage of the "StudyHub" platform, while providing easy and dynamic navigation on the web, system features have been tailored to achieve the mentioned goal successfully. The main system features include:

- **Registration**—The system feature that enables users to register within the platform, while enabling the users to choose their account type for further engagement with the platform. After registration, the user account has been created and stored, with all functionalities enabled according to the profile type of choice. Users will have access to specific information that is associated with the specific user account.
- **Login**—To access the other functionalities of the platform, with limitations to registered consumers only, the users must confirm valid identity, by inputting credentials as email and password. Those credentials have been assigned to the specific account during the process of registration.
- **Profile Management**—The user shall be able to change, add updates, and delete wanted information tied to the unique account. This functionality holds different options regarding what type of user is registered, and whether the user is registered as an admin, student, or instructor.
- **Custom Learning Path**—The user shall be able to tailor their educational path, by selecting courses and materials that match their personal goals and ambitions. The user should also be able to track progress made on the platform.
- **Assessment Tests**—The feature that enables users to evaluate their knowledge in scope of choice. This way the "StudyHub" platform allows users to solidify their knowledge of spot blanks in study techniques. By leveraging AI technology, this feature is to be supported in various types, of quizzes, multiple choice questions, and even essay question grading.
- **Automated Grading and Feedback**—The user shall be able to receive automatically evaluated tests, that way delivering immediate feedback and grading. This feature is AI technology-supported, which speeds up the process, enabling timely and human-error-free feedback.

- **Uploading Materials**—The admins shall be able to upload materials to the platform, which enables regular up-to-date enhancements on the content of the platform. This feature is vital for keeping the platform relevant, by adapting to the digital era and technology standards.
- **Instructor List**—The "StudyHub" platform enables instructors to appear available for the students to contact them, providing practical solutions to students who are seeking personalized instructions and guidance. This feature allows instructors to gain contacts and demonstrate their teaching skills.
- **AI-driven Language Learning**—The "StudyHub" platform enables users to learn new languages in an AI-supported environment. This helps the process of learning a new language in a fun and advanced means. For this feature users will be able to engage in different interactions with AI- powered Chatbot, learning new languages in a modern, yet still natural way.
- **AI-Driven Virtual Lab**—The users shall be able to engage in fun and innovative ways of learning practical courses, such as chemistry, physics, etc. This feature serves to provide users of the platform the interactive laboratory with insights into visual learning.
- **Payment**—The user shall be able to handle financial transactions within the platform securely and efficiently, this feature specifically supports the payment for instructors tutoring and courses.

Having discussed main system features, there is an extra secondary system feature, that brings character to

the "StudyHub" platform, which is the following:

- **Gamification**—such feature is there to enhance user experience, by adding game-like mechanisms to the platform. This way engaging students with the platform and motivating students in their learning process.

In addition to these, main and secondary system feature requirements, the platform is expected to meet a list of non-functional requirements:

- **Performance:** The platform should be able to handle simultaneous access by various numbers of users with minimal latency. Response time should be under 3 seconds, that way ensuring an efficient user experience.
- **Usability:** The user interface should be accessible to users of all ages, no matter what education level they possess. Instructions and navigation should be clear and straightforward.
- **Security**: Data of users, including personal information and data regarding the learning process should be protected and guarded with authentication and authorization processes.
- **Reliability:** The platform should guarantee 99.9% uptime.
- **Maintainability**: The platform should be created following well-structured code and organized components that are practical to update and edit.

Table 1 Use cases

Actors	Use case
Unregistered user	1. Sign Up
	2. View demo
Registered user	3. Login
	4. Edit account
Student	5. Engage in materials
	6. Language learning (AI-Driven)
	7. Complete tests
	8. Receive feedback
	9. Use Virtual Lab (AI-Driven)
	10. Track progress
	11. Find instructor
Instructor	12. Verify credentials
	13. Appear on the Instructor List
	14. Interact with students
Admin	15. Manage content
	16. Upload materials
	17. Verify instructors
	18. Manage payment settings (PayPal Integration)

- **Scalability:** The platform should be able to accommodate growth in the number of users, with data volume without affecting the performance.

After stating the previous, it is important to cover use cases. For that purpose, the following table *(Table* 1) will serve as an asset in this documentation.

As mentioned before, the table contains 3 main user types, with the line between actions taken for unregistered users and registered users. Each action is assigned for each user type ensuring seamless interaction flow throughout the platform.

5 Implementation

To best satisfy the needs of the "StudyHub" website, regarding system features and requirements, implementation methods and tools have been carefully selected. To ensure the best solution for a product. technologies have been selected to optimize both, frontend, and backend functionalities of the platform, ensuring a smooth learning experience for the user. The first technology that was used was Figma, which provided a collaborative environment for suitable designs and assembling the components. This environment enabled visual consistency and user experience fulfillment throughout the platform. Since this solution is based on client–server architecture, a website is brought to life using React technology, which is an open-source JavaScript library that is popular for its convenient component-based architecture. The architecture of such nature contributed to code reusability while enhancing website performance. Another strength that the usage of React technology contributes is the convenience of its virtual DOM (Document Object Model), which updates and renders appropriate components (Inwizards and Inwizards 2023).

Bootstrap is a solution that helps enhance the UI providing a powerful grid system, and ready-made components, that enable adaptable and consistent products as a platform. The server side is developed using Node.js, which is an event-driven framework, that manages I/O operations and processes data in real time (Gimeno 2024). The "StudyHub" platform relies heavily on scalability and data processing, which is one of the reasons Node.js was a good solution choice.

To implement database solutions, MongoDB Atlas offers scalable and flexible ground. Key features MongoDB Atlas demonstrates, are uninterrupted service, data redundancy, and easy code integration (Vatter 2023). Since the website is powered by AI API technologies, the Postman environment was selected as a testing environment for such purpose. To ensure the effectiveness and correctness of the system, such communication tunnels need to perform cohesively before the production stage.

To leverage AI technologies and deliver an AI-powered environment, API technology by OpenAI has been integrated, while PhET APIs enable vibrant, and engaging simulations to life. The relationship of APIs implemented to the system of "StudyHub" enhances the dynamic of the whole platform, providing up-to-date materials and educational concepts.

Additionally, with regard to the payment gateway, Stripe technology is the option that is incorporated as such. Stripe technology provides, a secure way to handle transactions, keys, etc. That segment of security is vital for the quality of the "StudyHub" E-learning platform.

6 Conclusion

In conclusion, "StudyHub" represents a mixture of traditional study methods and new technological advancements combined into one e-learning platform. By implementing gamification, it will motivate users to pursue their knowledge and education. The implementation of AI will lead the users to explore a new way of learning through the integration of the language learning Chatbot as well as the automated grading system and immediate feedback functionality. This will benefit the user by being more time efficient and by providing them with clear responses. "StudyHub" will offer a responsive user interface (UI), enabling users to use any device of their choice. Furthermore, the web-based platform is designed for users who wish to pursue their education at any place and anytime, due to its flexible nature. Apart from being responsive, the UI will also be easy to use, intuitive, and simple making it appealing to people of all ages. This platform will also provide an opportunity for users to book tutoring sessions from verified instructors. New technologies such as AI, Virtual Reality (VR), Augmented Reality (AR), and such are improving daily and creating an opportunity to improve people's everyday lives, including the field of education. Some of the planned future enhancements include integrating a chat for students and instructors in order to enhance the course booking process, and improvements to features such as virtual lab and AI Chatbot. Lastly, visual refinements at the level of user personalization will also be made so the user can choose to apply themes for the appearance of the platform. "StudyHub" is a step in the right direction regarding implementing emerging technologies into everyday solutions.

References

Adarkwah, M.A.: The power of assessment feedback in teaching and learning: a narrative review and synthesis of the literature. SN Soc. Sci. **1**(3) (2021). https://doi.org/10.1007/s43545-021-00086-w

Along: Empowerment in education—Along (2024). https://www.along.org/articles/empowerment-in-education/

Gimeno, A.: Multithreading in Node.js with worker threads—LogRocket Blog. LogRocket Blog, Jan. 09, 2024. https://blog.logrocket.com/multithreading-node-js-worker-threads/

How to teach students AI-resilient, durable skills: World Economic Forum (2023). https://www.weforum.org/agenda/2023/09/higher-education-model-for-ai/

Here's how COVID-19 impacted education—and what we need to do to get children's learning back on track: World Economic Forum (2022). https://www.weforum.org/agenda/2022/11/covid19-education-impact-legacy

Inwizards and Inwizards, "11 Advantages of ReactJS and why you should choose it?," Inwizards Software Technology : Blog, Nov. 30, 2023. https://www.inwizards.com/blog/11-advantages-of-reactjs-and-why-you-should-choose-it/

Jiménez, D.G., Clarisó, R., Caballé, S., Conesa, J.: (PDF) a prototype of an elearning platform in support for learning analytics and gamification. in ResearchGate, ResearchGate (2016). https://doi.org/10.1109/INCoS.2016.33

"Khan Academy | Free Online Courses, Lessons & Practice," Khan Academy, 2023. https://www.khanacademy.org/

"Learn a language for free," Duolingo (2015). https://www.duolingo.com/approach70

Maatuk, A.M., Elberkawi, E.K., Aljawarneh, S., Rashaideh, H., Alharbi, H.: The COVID-19 pandemic and E-learning: challenges and opportunities from the perspective of students and instructors. J. Comput. High Educ. **34**(1), 21–38 (2022). https://doi.org/10.1007/s12528-021-09274-2. Epub 2021 May 3. PMID: 33967563; PMCID: PMC8091987.

Nightingale, R.: 5 problems in education that technology will soon solve. MUO (2016). https://www.makeuseof.com/tag/5-problems-education-technology-will-soon-solve/

"Online Courses—Learn Anything, On Your Schedule | Udemy," Udemy (2024). https://www.udemy.com

Vatter, H., How MongoDB Atlas supports a multi-database environment. IBM Blog (2023). https://www.ibm.com/blog/how-mongodb-atlas-supports-a-multi-database-environment/

"Wyzant: Expert Tutors at Affordable Prices," Wyzant.com (2024). https://www.wyzant.com/

Effects of Emotional Intelligence on Entrepreneurial Intention Among Women in Tunisia

Mosbeh Rim and Ameni Baghdadi

Abstract

In recent years and in terms of gender equality, much has been said about the development of women entrepreneurs in Tunisia and everything has been done to understand all the factors that can explain the intention to start a business, but this topic remains relatively little explored. Well understood. Emotional intelligence has garnered significant interest in both scientific circles and among the public in recent years. However, the relationship between emotional intelligence and individual differences in entrepreneurship has not yet been explored. The purpose of our study is to develop emotional skills to foster entrepreneurial intentions in Tunisian women. **Purpose & Research question**: This research delves into emotional; intelligence's contribution to entrepreneurial intention among women to identify leading factors that shape this intention. By examining emotional intelligence's impact, this study sheds light on under-researched aspect of entrepreneurship. The main questions: What impact does emotional intelligence have on fostering entrepreneurial intention among women in Tunisia? Which essential antecedents play a significant role in this association? The purpose of our study is to provide targeted development and training programs that empower emotional intelligence skills among Tunisian women to effectively navigate the challenge, obstacles and achieve sustainable success. **Methodology approach**: The contribution of emotional intelligence to entrepreneurial intention among women was first identified using a qualitative approach a study,8 semi-structured interviews were then carried

M. Rim · A. Baghdadi (✉)
University of Sousse, Higher Institute of Management of Sousse, Sousse, Tunisia
e-mail: amenibaghdadi61@yahoo.com

M. Rim
e-mail: rim.mosbeh@isgs.u-sousse.tn

K. Dhoska and E. Spaho (eds.), *Bridging Horizons in Artificial Intelligence, Robotics, Cybersecurity, Smart Cities, and Digital Economy*, Sustainable Economy and Ecotechnology, https://doi.org/10.1007/978-3-031-72029-1_36

out to identify the antecedents of entrepreneurial intention. Secondly a confirmatory step, a questionnaire was administrated to 120 women with intention to create business. The collected data was processed and analyzed using SPSS. (Recent edition) with the objective to identify patterns and relationship between variables under investigation. **Finding**: The results emphasize the mediating role of self-efficacy between the dimensions of emotional intelligence and entrepreneurial intention. The dimensions of emotional intelligence influence the development of entrepreneurial intentions in Tunisian women by affecting their ability to identify and understand their own emotions; as well as those of others. These skills play a critical role in decision-making, communication, problem solving, and building relationships that are critical to business success.

Keywords

Female entrepreneurship • Entrepreneurial intention • Entrepreneurial self-efficacy and emotional intelligence

1 Introduction

In recent years, Tunisia has seen a growing emphasis on fostering gender equality and advancing the development of female entrepreneurship. This focus has sparked increased interest in understanding the factors that shape women's intentions to start businesses. Researchers have delved into the personality traits of female entrepreneurs, recognizing that entrepreneurship is not just a matter of external circumstances, but also influenced by individual characteristics. Past investigations have explored an array of traits using diverse instruments, ranging from established models like the Big Five personality framework to more nuanced assessments of mindfulness and self-efficacy.

In addition to the traditional focus on personality traits like the Big Five and locus of control, there has been a notable shift in recent years toward exploring darker traits such as narcissism and sensory processing sensitivity in the context of entrepreneurship. These traits, often overlooked in conventional studies, offer unique insights into the complexities of entrepreneurial behavior and decision-making. Amidst this evolving landscape, the significance of emotional intelligence has come to the forefront as a compelling area of inquiry. Recognized for its role in shaping individuals' emotional responses, interpersonal interactions, and overall psychological well-being, emotional intelligence holds promise as a key determinant of entrepreneurial intentions. By delving into the nuanced interplay between emotional intelligence dimensions and entrepreneurial aspirations, our study seeks to shed light on the underlying mechanisms driving women's decisions to pursue entrepreneurship.

Against this backdrop, our study seeks to deepen our understanding of the interplay between emotional intelligence dimensions—specifically self-management and self-awareness—and women's entrepreneurial intentions in Tunisia. We aim to shed light on how emotional intelligence, mediated by entrepreneurial self-efficacy, shapes women's aspirations and actions in the realm of entrepreneurship.

2 Literature Review

2.1 Emotional Intelligence

Emotional intelligence has become a subject of great interest in both academic circles and the general public (Bar-On 2006). With that in mind, the World Economic Forum's (2020, p. 36) Future of Jobs report ranks emotional intelligence among the 15 most in-demand skills necessary for success in 2022 and beyond.

In 1990, Salovey and Mayer initially defined emotional intelligence (EI) as a form of social and personal intelligence that includes: "*the ability to monitor one's own and others' feelings and emotions, to discriminate among them and to use this information to guide one's thinking and actions*" (op. cit., p. 189). In 1997, they have revised this definition to include four interdepended dimensions: appraisal, perception, and expression of emotions; understanding and analyzing emotional information, emotion facilitation of thinking; using emotional knowledge; and emotion regulation (Mayer and Salovey 1997). As noted by Cartwright and Pappas (2008), they continue to focus on the idea that "*EI lies at the interface between the mental processing of emotional information and its integration with cognitive information*". (op. cit., p. 154).

Goleman (1998, p. 137) explains emotional intelligence as the "*ability to recognize our own feelings and those of others, to motivate ourselves and to manage emotions well in ourselves and in our relationships*".

While ability and trait perspectives of emotional intelligence (EI) are sometimes portrayed as contradictory, they can actually complement each other. In alignment with research on human factors, ability measures, akin to cognitive ability tests, demonstrate superior efficacy in predicting peak performance, whereas non-ability measures, such as personality tests, exhibit closer associations with typical performance. This suggests that diverse assessments may be employed to gauge various facets of EI, contingent upon whether the objective is to anticipate an individual's potential or their probable behaviors. Consequently, in forecasting entrepreneurial attitudes, tests evaluating EI ability can scrutinize an individual's capabilities, while measures of trait EI can delve into their everyday thoughts and actions (Aruoren 2020).

Scholars have delineated the concept of emotional intelligence, with notable contributions from pioneers such as Mayer and Salovey (1990), Goleman (1998), and Bar-On (2006). It is noteworthy to emphasize that various approaches have been developed to understand emotional intelligence, each offering unique perspectives that complement one another. These include the targeted skills approach, the holistic approach, and the mixed model (Fernández-Berrocal and Extremera

2006; Mortan et al. 2014). The targeted skills approach, for instance, accentuates specific competencies and abilities within the realm of emotional intelligence (Mayer et al. 2008).

A universally accepted definition of EI remains elusive, leading to ongoing debates between two primary perspectives: ability EI and trait EI. Ability EI centers on quantifying an individual's adeptness in managing emotions, whereas trait EI (or emotional self-efficacy) examines variations in emotional tendencies. Mayer and Salovey (1990), regarded as trailblazers in the exploration of emotional intelligence, formulated the ability model. This model posits that emotional intelligence encompasses a range of skills, including the evaluation and expression of emotions. This skill involves the adeptness to comprehend and manage both one's own and others' verbal and non-verbal emotional expressions. Additionally, regulating emotions entails the mastery of controlling one's own emotions and those of others. Moreover, harnessing emotions involves utilizing them to enhance flexibility and motivation, thereby facilitating streamlined logical thought processes.

The specific-ability approach, as elucidated by Mayer et al. (2008), underscores the importance of distinct skills and competencies within the realm of Emotional Intelligence (EI). The authors have delineated several pivotal abilities, which encompass the aptitude to perceive and discern emotions, the utilization of emotional information in cognitive processes such as appraisal, categorization, and classification of emotions, and the proficiency in articulating emotions through language. Additionally, the capacity to manage one's own verbal and non-verbal emotional expressions, as well as those of others, emerges as a critical ability, entailing the skill to regulate, reflect upon, and reframe emotions and perceptions of particular situations (Mortan et al. 2014). "Emotion use" denotes the adeptness in leveraging emotions to streamline logical reasoning processes, thereby augmenting flexibility, creativity, and motivation.

The identification of specific abilities by the authors is underpinned by the four-branch model devised by Mayer and Salovey (1997; Sumathi et al. 2021), which adopts an integrative approach by amalgamating each facet to construct an overarching emotional intelligence framework. According to Mayer et al. (2008), these branches are believed to commence development from early childhood. Furthermore, these four branches are hierarchically arranged, with the perception of emotions constituting the foundational level, while the management of emotions represents the pinnacle and most intricate level. Consequently, the regulation of one's emotions, as well as those of others, is scaffolded upon the competencies of the other three branches (Fernández-Berrocal and Extremera 2006; Aruoren 2020).

The Mixed-Model presents an alternative perspective on Emotional Intelligence, distinguished by its emphasis on a blend of diverse traits. Proposed by Bar-On (1997), this model characterizes EI as a composite of non-cognitive abilities, competencies, and skills that influence an individual's capacity to navigate environmental demands and pressures. Cartwright and Pappas (2008) note that, akin to Goleman's framework, this model encompasses five broad categories: intrapersonal emotional skills, interpersonal emotional skills, adaptability, stress management, and overall mood (Mortan et al. 2014). However, Bar-On (2006) contends that emotional intelligence constitutes a multifaceted array of competencies, skills, and facilitators that govern how individuals express and comprehend themselves, engage with others, and respond to daily circumstances.

Boyatzis (2018) defines EI as the awareness, understanding, and utilization of one's emotional information that catalyzes effective or superior performance. Furthermore, the author proposes a similar definition to be applicable within the context of social intelligence competency. An individual demonstrates excellent EI proficiency when they can adaptively acclimate and empathetically respond to others' emotions (Goleman and Boyatzis 2017). Emotional intelligence bridges the connection between cognition and emotion, enhancing our ability to interpret various social situations and conflicts, fostering empathy, reasoning, stress management, and effective communication (Setyowati and Razak 2019).

Emotional intelligence encompasses the ability to perceive and regulate emotions (Pitaf et al. 2018). It includes attributes such as self-confidence, self-management, motivation, empathy, and social skills, which will be discussed in detail below.

- **Self-Awareness**: The ability to accept one's own feelings and understand how one's emotions and actions can influence others (Venmani et al. 2022).
- **Self-Confidence** "*The mental attitude of trusting in or relying on oneself or something; feeling sure or certain about a fact or issue*" (Rezani et al. 2016 p. 36)
- **Self-Management** *"The ability to regulate and manage one's emotional state while waiting for an appropriate time and manner to express them."* (Venmani et al. 2022 p. 134):
- **Motivation for success** "*A key characteristic of emotionally intelligent and driven individuals is a passion for the work they do and a desire for change.*" (Michinov and Michinov 2020 p. 122):
- **Empathy**: *"The ability to understand how others are feeling and to consider those feelings in social situations before reacting. Empathy enables people to understand the dynamics that impact personal and professional relationships* "(Rezvani et al. 2020 p. 76)
- **Social skills**: *The ability to effectively manage emotional awareness in others, create connections, and communicate with people through verbal and non-verbal means"* (Sawandana 2019, p 66).

The 2011 publication of the State of the Heart Report by the Six Seconds Emotional Intelligence network marked a comprehensive examination of global developments and advancements in emotional intelligence (Freedman 2018). Regarded as the most comprehensive and integrative study of its kind worldwide, this research offers insights into the strengths, challenges, and opportunities inherent in emotional intelligence (Maharaj and Ramsaroop 2022) (Fig. 1).

Drawing from a survey encompassing 160 countries and involving 200,000 participants, the findings shed light on crucial aspects pivotal for shaping a brighter

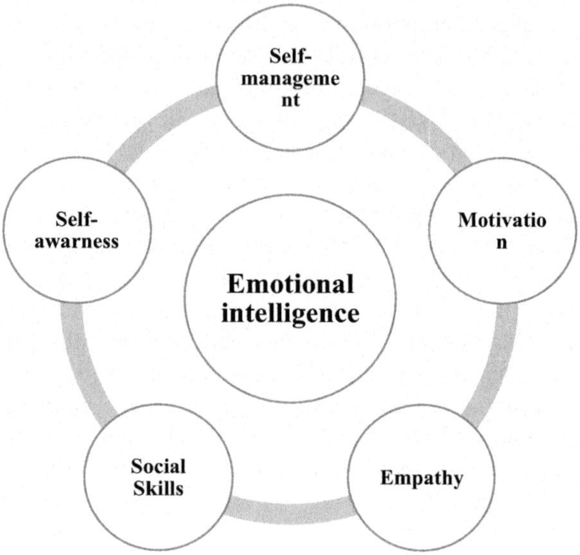

Fig. 1 Element of emotional intelligence

future, delineating key trends in global emotional intelligence with regard to personal well-being, effectiveness, relationships, and overall quality of life (Freedman 2018; Maharaj and Ramsaroop 2022). Furthermore, the study delved into the value of emotional intelligence across various regions including Africa, Asia–Pacific, Europe, Latin America, the Middle East, and North America, by comparing national average emotional intelligence scores with World Health Organization data on well-being (Freedman 2018; Maharaj and Ramsaroop 2022). Notably, findings revealed that a majority of countries in Asia–Pacific and the Middle East exhibited low levels of well-being and emotional intelligence (Freedman 2018). Conversely, Africa was identified as having the potential to leverage high emotional intelligence to enhance overall well-being (Freedman 2018; Maharaj and Ramsaroop 2022).

Furthermore, a substantial body of research suggests that emotional intelligence correlates with improved social relationships, as well as physical and mental well-being (Mayer et al. 2008; Mortan et al. 2014).

Mortan et al. (2014) underscore the wide-ranging impact of emotional intelligence on crucial life outcomes, including quality social relationships (Lopes et al. 2004), enhanced psychological well-being (Schutte et al. 2007), and heightened job success and satisfaction. Indeed, prior studies have consistently demonstrated that emotional intelligence serves as a direct determinant of job performance (O'Boyle et al. 2004, 2011), while also contributing to increased job performance, satisfaction, commitment, and involvement (Carmeli 2003; O'Boyle et al. 2011). Lopes et al. (2006) examined the correlation between emotional intelligence and positive workplace outcomes, revealing significant associations with indicators of job performance, interpersonal facilitation ratings, affect, and attitudes. Building upon these findings, Mortan et al. (2014) assert that emotional intelligence emerges as a significant predictor of positive outcomes across various samples and domains.

Pavliscak (2018) posits emotional intelligence as a foundational element and driving force behind a balanced life, crucial to every facet of quality of life. Studies suggest that emotional intelligence can enhance quality of life, fostering a sense of well-being, particularly when work and life experiences are imbued with meaning (Koçak 2021). Similarly, McDonald (2021) emphasizes the pivotal role of high emotional intelligence in adaptive regulation of distressing emotions and effective management of daily stressors and challenges, particularly among women.

Furthermore, Raeissi et al. (2020) contend that businesses prioritize communication skills over management skills, suggesting that high emotional intelligence aids individuals in navigating social challenges within the workplace, inspiring and guiding others. Consequently, it can be inferred that emotional intelligence serves as a crucial predictor of significant outcomes across diverse samples and real-world domains.

Recent studies (Venmani et al. 2022; Kowlman et al. 2020) propose that emotional intelligence could play a pivotal role in predicting entrepreneurship. While existing research has focused on certain outcomes predicted by emotional intelligence, such as job success, attention to its role in entrepreneurship remains limited (Ahmetoglu et al. 2011; Chell 2008; Zampetaskis et al. 2009). Despite the growing recognition of emotional intelligence as a key determinant of career success, insufficient attention has been given to the emotional intelligence of entrepreneurs (Miao et al. 2018b; Rhee and White 2007).

In a different context, past literature underscores numerous factors influencing intention, including moral judgment, empathy, creativity, and self-efficacy. However, contemporary viewpoints often neglect the significance of emotional intelligence (EI) in shaping individual intentions (Tiwari et al. 2018; Tiwari et al. 2017a; Tiwari et al. 2017b).

In summary, the existing body of research highlights the profound influence of emotional intelligence (EI) on various aspects of individual well-being and success, spanning from enhanced job performance and satisfaction to improved quality of life. While significant attention has been given to the role of EI in traditional career paths, such as job success, its potential impact on entrepreneurship remains relatively underexplored. Moreover, there is a growing recognition of the importance of EI in shaping individual intentions and behaviors, yet its specific role in the context of entrepreneurship, particularly among women, merits further investigation. Thus, the subsequent section we will shift our focus specifically to female entrepreneurship. This segment aims to explore the intersection of emotional intelligence and entrepreneurship among women, elucidating the unique challenges, opportunities, and dynamics that shape their entrepreneurial endeavors.

2.2 Female Entrepreneurship

The World Bank's 2019 report highlighted Africa as boasting the highest entrepreneurship rate globally, with women representing the majority of entrepreneurs on the continent (Kelly and Linden 2020). African female entrepreneurs are pivotal in driving transformation within the continent's economy and are widely regarded as the future of African prosperity (World Bank 2018). Their participation is undeniably crucial, spanning various domains such as innovation, job creation, economic value generation, poverty alleviation, and household security (Amines and Dust 2009; Crittendin and Ajjan 2019; Friectson, Ridenour, and Perotti 2019; Guma 2015; Nukpizah and Blankison 2019; Autumn, Ibro Fulton, Lowinberg and Debar 2012; Scott Deborer 2012; Scott Dolan et al. 2012).

Entrepreneurship has long captivated researchers' attention, acknowledged as pivotal for a nation's economic growth through the establishment of new ventures and the creation of employment opportunities (Momani 2017; Kuckertz and Prochotta 2018; Cutler 2003; Gartner 1989). Female entrepreneurs are increasingly recognized as essential drivers of economic growth and prosperity worldwide, with entrepreneurial activity serving as a crucial source of employment for women globally (Reichborn et al. 2014).

While existing research has predominantly focused on exploring factors influencing entrepreneurial behavior and potential among women, Rodriquiez (2018) underscores the significance of opportunity entrepreneurs in fostering economic prosperity. Various factors, including motivational drivers, government support, education and training, positive attitudes toward entrepreneurship, and emotional intelligence, play pivotal roles in shaping women's entrepreneurial journeys.

Motivational factors, as elucidated by Hamilton and De Klerk (2016), serve as key catalysts that inspire women to pursue entrepreneurial endeavors. Government support, highlighted by Mayer (2018), can create an enabling environment through policies, regulations, and initiatives fostering female entrepreneurship. Education and training, emphasized by Ivene (2017), are essential in equipping women with the requisite skills, knowledge, and competencies for entrepreneurial success. Positive attitudes toward entrepreneurship, as discussed by Hamilton (2015), significantly influence women's willingness to take risks, identify opportunities, and persevere in the face of challenges.

Moreover, emotional intelligence, identified by Wan Mohb et al. (2021), plays a pivotal role in women's entrepreneurial journeys by enhancing decision-making, resilience, and relationship-building skills crucial for success in entrepreneurial contexts. Collectively, these factors contribute to a comprehensive understanding of the antecedents shaping women's entrepreneurial behavior and potential. By addressing and promoting these factors, policymakers, educators, and stakeholders can create an ecosystem that empowers and supports women in their entrepreneurial endeavors.

Indeed, the pivotal role of female entrepreneurs in driving economic growth and prosperity, particularly in regions like Africa, underscores the importance of understanding and supporting their endeavors. By delving into the various factors influencing women's entrepreneurial behavior and potential, we gain valuable insights into how to foster an environment conducive to their success. Motivational drivers, government support, education and training, positive attitudes toward entrepreneurship, and emotional intelligence all play crucial roles in shaping women's entrepreneurial journeys.

In conclusion, the preceding section has shed light on the pivotal role of female entrepreneurs in driving economic growth and prosperity, particularly in regions like Africa. By exploring the various factors influencing women's entrepreneurial behavior and potential, we have gained valuable insights into the multifaceted nature of their entrepreneurial journeys. Moving forward, it is imperative to delve deeper into the relationship between female entrepreneurship and emotional intelligence. In the next section, we will explore how emotional intelligence influences women's entrepreneurial endeavors, offering insights that can inform strategies for supporting and empowering female entrepreneurs in their pursuit of success.

2.3 Emotional Intelligence and Entrepreneurial Intention Among Women

Emotional intelligence encapsulates a spectrum of abilities involving the perception, understanding, utilization, and management of emotions within individuals (Andrei et al. 2016). While considerable attention is directed toward the study of entrepreneurship and its underlying processes, the exploration of emotional intelligence has emerged as an increasingly prominent area of focus.

Intentions are acknowledged as influential determinants of behavior (Ajzen 1991), particularly in the analysis of women's entrepreneurial and career-related practices (Krueger et al. 2000; Mohd Hirwani Wan Hussain et al. 2021). Although research on the emotional aspects associated with entrepreneurship remains limited, Krueger et al. (2000) posit that attitudes toward specific actions, including affective components such as emotions, wield significant influence. Individuals with higher levels of trait emotional intelligence tend to demonstrate enhanced resilience toward stress and environmental stressors (Mikalajczak et al. 2009; Tiwari et al. 2018).

Baron (2008) conducted research suggesting that positive emotions can foster an entrepreneur's creativity and ability to identify opportunities. Emotional intelligence, encompassing awareness of one's own emotions and those of others, is deemed crucial for entrepreneurs. Particularly, the ability to accurately understand and express nonverbal emotions holds paramount importance for entrepreneurs, aiding them in maintaining composure during stressful situations.

In entrepreneurship research, two prevailing theories are commonly utilized: McClelland's (1961) theory of the need to achieve and Rotter's (1966) locus of control theory. McClelland's theory posits that individuals with a strong desire to achieve are often drawn to entrepreneurship. Conversely, Rotter's theory categorizes an individual's locus of control as either internal or external, with an internal control expectation linked to motivation and active striving, characteristic of entrepreneurial traits (Littunen 2000).

Emotional intelligence encompasses a multitude of capabilities associated with understanding and managing emotions. While considerable attention is directed toward entrepreneurship and its underlying processes, researchers also emphasize the significance of emotional intelligence.

Attitudes encompass two primary components: affective and cognitive (Edwards 1990). The affective component pertains to the feelings or emotions associated with an attitude object, while the cognitive component entails beliefs or thoughts linked to the same object (Edwards 1990). Given that attitudes comprise both cognitive and affective elements, it follows that individuals' emotional preferences play a crucial role in shaping entrepreneurial intentions (Wan Mohd Hirwani Wan Hussain et al. 2021).

Intentions are fundamental to women's entrepreneurial activities and career-related behaviors. Although limited, studies have indicated that emotions, along with attitudes, exert influence on entrepreneurial intentions. Furthermore, individuals with higher trait emotional intelligence tend to demonstrate enhanced resilience toward stress and environmental pressures.

Therefore, to bolster attitudes toward female entrepreneurship, it is imperative to address both the cognitive and affective underpinnings of entrepreneurial intention (Zanpetaks et al. 2017; Wan Mohd Hirwani Wan Hussain et al. 2021). Zanpetaks et al. (2017) suggest that emotionally intelligent women are particularly well-suited for entrepreneurial careers as they can effectively manage their emotions, maintain optimism, and interpret situations positively, thereby enhancing their ability to identify opportunities.

Moreover, Humphery (2018) contends that the ability to understand others' emotions enables prospective entrepreneurs to discern new opportunities. Consequently, Baron et al. (2019) advocate for the inclusion of a mediator as the most effective approach to elucidating the occurrence and direction of a relationship, underscoring the importance of considering mediating factors in comprehending the dynamics of entrepreneurial intentions.

In our study, we focus on the concept of self-efficacy originally introduced by Bandura (1990 2017) in his socio-cognitive theory. Self-efficacy refers to a person's confidence in their ability to effectively engage in certain behaviors and achieve desired goals (Bandura 1977; Carden et al. 2018). It is a motivational construct that has been shown to have multiple influences on behavior.

Self-efficacy expectations drive behavior initiation, effort, and persistence in the face of obstacles and challenging experiences (Bandura 1977; Mortan et al. 2014). The link between emotional intelligence and the perception of entrepreneurial self-efficacy was demonstrated by Salvador (2008). Their results showed a significant positive association between certain dimensions of emotional intelligence and entrepreneurial self-efficacy. Self-confidence, self-management, and clarity of emotions were strongly associated with entrepreneurial self-efficacy. This suggests that emotional intelligence plays an important role in how people perceive their own effectiveness in business endeavors. Salvador (2008) further argues that processing high emotional intelligence are more sociable and are perceived as having greater potential for entrepreneurial actions and achievements (Mortan et al. 2014; Li et al. 2022).

These findings accentuate the importance of emotional intelligence in influencing the perception of entrepreneurial self-efficacy, which in turn can have influences on individuals' entrepreneurial behaviors and outcomes.

Self-efficacy is recognized as a pivotal factor in shaping entrepreneurial intentions (Lee et al. 2011). It serves as a crucial precursor to entrepreneurial intentions and behaviors, as empirically supported by Chen et al. (1998) and Pihie (2009). Specifically, Chen et al. (1998) provided empirical evidence demonstrating a significant positive relationship between entrepreneurial self-efficacy and students' intentions to engage in entrepreneurship. Moreover, Zhao et al. (2005) corroborated the positive association between entrepreneurial intention and self-efficacy, highlighting the mediating role of self-efficacy in fostering students' aspirations toward entrepreneurship. Investigating whether entrepreneurial self-efficacy mediated the relationship between personal factors such as gender, prior entrepreneurial experience, perceptions of formal education, risk-taking, and entrepreneurial intentions, Chen et al. (1998) further underscored the importance of self-efficacy as a critical component of entrepreneurial intentions.

Self-efficacy functions as a facilitator, linking individual-level factors to entrepreneurial intentions and bolstering individuals' confidence in their ability to embark on entrepreneurial endeavors. By comprehending the role of self-efficacy and its mediating function in the relationship between antecedent factors and entrepreneurial intentions, researchers gain valuable insights into the cognitive processes and motivational mechanisms underlying individuals' intentions to pursue entrepreneurship.

Furthermore, the perception of self-efficacy partially mediates the relationship between emotional intelligence and entrepreneurial intent. Similarly, Kibler (2013) argued that women's belief in their capacity to initiate a business positively influences their entrepreneurial intentions.

Building upon the literature outlined above, the primary objective of this study is to examine the association between emotional intelligence and entrepreneurial intention among Tunisian women, with a focus on the mediating role of entrepreneurial self-efficacy. This analysis will be conducted while controlling for demographic variables such as age, country, education, and job.

Based on previous research results, the following hypotheses are proposed:

- **H1: There is a positive association between emotional intelligence dimensions and Entrepreneurial self-efficacy among Tunisian women.**
- **H2: There is a positive correlation between entrepreneurial self-efficacy and entrepreneurial intentions among Tunisian women.**
- **H3: Entrepreneurial self-efficacy mediates the relationship between emotional intelligence and entrepreneurial intent in Tunisian women.**

These hypotheses anticipate positive association between emotional intelligence dimensions and entrepreneurial self-efficacy, and between entrepreneurial self-efficacy and entrepreneurial intention. Additionally, our research propose that entrepreneurial self-efficacy will serve as a mediating mechanism between emotional intelligence and entrepreneurial intention in Tunisian women.

3 Methodology

Our objectives concern, first of all, the evaluation of the link between emotional intelligence and female entrepreneurship in correlation with individual self-efficacy; and, then, the proposal of corrective actions and recommendations. To meet these objectives, a multi-method technique was used to collect the qualitative and quantitative data necessary to conduct this research project. This combination allows for a more complete, holistic and contextual representation of the various dimensions of the phenomenon studied (Igalens and Roussel 1998).

First, as part of an exploratory approach, eighteen (18) semi-structured interviews were administrated with female entrepreneurs or future women entrepreneurs. The objective of this exploratory qualitative approach is to grasp, on the one hand, the needs and expectations of future women entrepreneurs; and on the other hand, to progressively construct the questionnaire, taking into consideration their different perceptions (Horton et al. 2001).

The deductive level aims to confirm the assumptions resulting from the exploratory level. The survey was developed in three stages: (1) the development of an initial version, (2) the pre-test (with 16 female entrepreneurs) to assess the validity of its content and to test the questions and their sequencing, and finally (3) the development of the terminal version of the questionnaire. The questionnaire was designed and administered using an http address.

We consider that a mixed-methods approach is the most appropriate method to study the relationship between female entrepreneurship and emotional intelligence. This approach involves collecting and analyzing both quantitative and qualitative data and then integrating both. By adopting a mixed-methods research design, instead of relying solely on classic quantitative techniques in corporate research, more comprehensive and insightful investigations can be achieved. This approach offers a distinct advantage over traditional intervention studies, case studies, or ethnographies (Creswell and Plano Clark 2011; Harnisch et al. 2012, 2013). The integration of quantitative data (e.g. measures of women's emotional intelligence and entrepreneurship) and qualitative data (including descriptions and qualitative characteristics of women's emotional intelligence and entrepreneurship) allows for a deeper understanding of the research question at hand.

The targeted population for this study will be focused on women at two distinct categories: female entrepreneurs with a vision for entrepreneurship in Tunisia and women who possess intentions to start their own businesses. By including these two groups, we aim to gain valuable insights into the process of business creation by women. Each woman will belong to a different field, allowing us to identify key factors such as the environment, culture, obstacles, and specific attitudes that influence female entrepreneurs.

For the second stage of the study, a quantitative method will be employed. The sample will consist of 120 women who have expressed intentions to start their own businesses. Among the participants, 50% will be students, 26.3% will be employees, and 16.7% will be managers. These participants will be requested to complete a questionnaire containing evaluations of entrepreneurial intention, entrepreneurial self-efficacy, and emotional intelligence. The questionnaire is designed to be completed within approximately 10–15 min.

By combining both qualitative and quantitative methods in this research, we aim to gather comprehensive insights into the factors influencing female entrepreneurship and entrepreneurial intentions. The qualitative phase enables a deeper exploration of the experiences and perceptions of female entrepreneurs, while the quantitative phase provides a broader understanding through statistical analysis of the collected data.

3.1 Measures

Entrepreneurial self-efficacy: The scale used to measure entrepreneurial self-efficacy was the Entrepreneurial Self-Efficacy (ESE) scale developed by St-Jean and Mathieu (2011). This scale was derived from a qualitative approach and consists of four items. Participants replied to each item using a 7-point Likert scale ranging from 1 (Completely Disagree) to 7 (Completely Agree).

Entrepreneurial intention: The dimension of entrepreneurial intention was measured using a three-item scale adopted from previous studies (Fayoll et al. 2006; Krueger et al. 2000; Sánchez 2009; Tisai et al. 2016). Participants were requested to desire their intention to start a new business or firm within the next five years. Responses were recorded on a five-point Likert scale, ranging from 1 (Totally Disagree) to 5 (Totally Agree).

Emotional intelligence: The variable of emotional intelligence was measured using a scale developed by Golman (2019). This scale consists of 23 items. Participants rated their agreement with each item on a 7-point Likert scale, ranging from 1 (Completely Disagree) to 7 (Completely Agree).

By utilizing these scales, the study aimed to capture the participants' levels of entrepreneurial self-efficacy, entrepreneurial intention, and emotional intelligence. The Likert scale responses provided a range of options for participants to express their agreement or disagreement with each statement, allowing for a more nuanced understanding of the variables under investigation.

3.2 Data Analysis

To ensure the one-dimensionality of the Business Self-Efficacy Scale (ESE), an exploratory factor analysis (EFA) was performed using the following SPSS Statistics Version 27 package the guidelines set out by Field (2005). The purpose of this analysis was to examine the underlying factors and determine whether the scale effectively measures the proposed design.

To examine the possible impact of emotional intelligence dimensions on entrepreneurial self-efficacy as mediators of intention to become entrepreneurs in women, separate sets of hierarchical multiple regression analyzes were performed. These analyzes aimed to understand the unique contribution of emotional. intelligence dimensions in addition to personality traits and demographic variables (e.g., age, education, business experience, work) in predicting business self-efficacy and business intent.

To assess the indirect effect of entrepreneurial self-efficacy (ESE), MacKinnon's (2008) method was used to test for multiple mediation. In this approach, two separate regression equations are estimated. In the first step, demographic variables are included. In the second step, a similar procedure is followed, whereby ESE is included as a mediator in the last step of the regression analysis.

These analytical approaches have been used to provide an understanding of the relationships between emotional intelligence, entrepreneurial self-efficacy, and entrepreneurial intent women. The use of hierarchical multiple regression and mediation analysis enables the identification of unique contributions and potential mediation effects while controlling for relevant covariates.

4 Finding

Table 1 provides a comprehensive overview of descriptive statistics and intercorrelations among all variables under consideration. Each scale's internal reliability has undergone thorough scrutiny, with Cronbach's alphas for each variable meticulously reported within parentheses.

Table 1 Means, Standard Deviations, Correlations, and Reliabilities of the Study Variable

Variables	M	SD	1	2	3	4
Self-awareness	3.70	0.57	0.71**	(0.77)		
Self-management	3.68	0.59	0.66**	0.66**	(0.72)	
Entrepreneurial self-efficacy	3.80	0.61	0.47**	0.47**	0.48**	(0.90)
Entrepreneurial intention	3.58	0.57	0.13**	0.11**	0.09**	0.24**

(Cronbach's alphas are in brackers)
*p<.05 ; **p<.01

Table 2 Summary of hierarchical regression analysis on the effect of emotional intelligence (EI) on entrepreneurial intent, using ESE as a mediator

Variable	β	R^2	ΔR^2	ΔF	Sig
Step 1		0.011	0.011	432	0.038
Age	−0.012				
Education	0.075				
Job	0.073				
Step 2		0.280	0.161	29.27	0.001
Self- awareness	0.223*				
Self-management	0.243*				

The results obtained from the exploratory factor analysis conducted on the Entrepreneurial self-efficacy scale reveal crucial insights. The average communality score of 0.47 among the items in this sample indicates a substantial level of shared variance, primarily loading onto a single factor. Furthermore, the Kaiser-Meyer-Olkin (KMO) measure of 0.803 signifies a highly compact pattern of correlation, a finding corroborated by the Bartlett's test of sphericity significant at $p < 0.001$. Both the scree plot and the rotation matrix highlight a predominant loading on one factor, thus affirming the conceptual coherence of employing the theoretical construct as a unified measure.

Moreover, the confidence in the validity of the other scales utilized in this study is reinforced by additional evidence. Chico (1999) validated the EI scale, providing further support for its reliability and construct validity. Additionally, Cardero et Al (1999) adapted and validated the NEO-FFI, lending credibility to its application in the present research context.

In the qualitative approach, it was found that support from family and friends played a crucial role for 7 out of 18 participants in starting their own business. However, 10 out of 18 participants highlighted the rejection of the idea of creating a new business by their relatives. Emotional intelligence played a significant role in facing rejection, managing anger and stress, and improving self-motivation in such situations. In the quantitative approach, the descriptive statistics and the intercorrelations between all measured variables are presented in Tables 1 and 2. The internal reliability of each scale was assessed and Cronbach's alphas for each variable are given in parentheses.

4.1 Testing of Hypothesis

The mediation analysis commenced with a hierarchical regression analysis, encompassing entrepreneurial self-efficacy, control variables, and emotional dimensions of emotional intelligence. The outcome revealed that the variables collectively elucidated 28.0% of the variance in the final model ($R2 = 0.28$, $p < 0.01$). Notably, age, education, and occupation contributed to a mere 1% of the explained variance, while EI dimensions accounted for a substantial 17% of the total variance. Interestingly, in the hierarchical model, none of the demographic variables emerged as significant predictors of entrepreneurial self-efficacy.

Partially affirming the first hypothesis (H1), a noteworthy positive relationship was observed between two dimensions of EI and self-efficacy. Specifically, self-awareness (B = 223, $p < 0.001$) and self-management (B = 0.243, $p < 0.001$) demonstrated significant associations. These findings underscore the robust predictive power of self-management and self-awareness in shaping entrepreneurial self-efficacy, even after controlling for the influences of work, age, and education.

Refer to Table 1 for a succinct summary of the hierarchical regression analysis, delineating the impact of emotional intelligence (EI) on entrepreneurial intent, with entrepreneurial self-efficacy (ESE) serving as a mediator.

A subsequent hierarchical regression analysis was conducted to elucidate entrepreneurial intentions, incorporating the age control variable alongside the dimensions of emotional intelligence and entrepreneurial self-efficacy. This analysis accounted for 7% of the total variance. Notably, a noteworthy positive association emerged between educational attainment and entrepreneurial intention.

The findings revealed a modest yet significant variation (1.9%) explained by the size of emotional intelligence. Upon meticulous examination of each dimension, it became evident that emotional intelligence (EI) can exert only an indirect influence on entrepreneurial intentions. Our hypothesis positing a positive correlation was thereby substantiated.

These outcomes stem from an investigation into entrepreneurial self-efficacy (ESE) and entrepreneurial intentions. Notably, the results underscored a significant positive relationship between ESE and entrepreneurial intentions (B = 0.556; $p < 0.001$). The comprehensive variables elucidated 13.7% of the total variance in Table 2 of the final model, thereby providing support for assumptions 2 and 3.

Refer to Table 2 for a succinct summary of the Hierarchical Regression Analysis, delineating the impact of Emotional Intelligence (EI) on Entrepreneurial Intention, with ESE serving as a mediator (Table 3).

Table 3 Summary of the Hierarchical Regression Analysis on the effect of Emotional Intelligence (EI) on Entrepreneurial Intention, with ESE as a mediator

Variables	β	R^2	ΔR^2	ΔF	Sig
Step 1		0.070	0.070	30.5	0.001
Age	0.079				
Education	0.649*				
Job	0.034				
Step 2		0.096	0.019	2.7	0.04
Self-awareness	0.128				
Self-management	0.241				
Step 3		0.137	0.042		0.001
Entrepreneurial Self-efficacy	0.556*				

In the initial segment of our model, we delved into the impact of control variables and EI dimensions on ESE. Our findings underscored a robust positive relationship between ESE and the dimensions of self-awareness and self-management, indicating their pivotal role in fostering entrepreneurial self-efficacy.

Transitioning to the latter segment of our analysis, we explored the influence of control variables, EI, and ESE dimensions on women's entrepreneurial intention. Here, our results unveiled a noteworthy positive relationship between ESE and women's entrepreneurial intentions, signifying the significant influence wielded by the EI dimension.

Moreover, our interpretation of the interview data sheds light on the relatively advantageous situation of Tunisian women compared to their Arab counterparts. However, it became apparent that the volatile environment, particularly amid the economic crisis precipitated by the COVID-19 pandemic, is exacerbating the complexities of corporate decision-making. It's crucial to recognize that even individuals possessing a high level of emotional intelligence can experience stress and anxiety, especially amidst challenging circumstances like the ongoing pandemic.

Nevertheless, individuals endowed with high emotional intelligence exhibit heightened awareness of their circumstances and emotions, employing resilience techniques to maintain self-control and effectively manage anxiety. Such adeptness not only mitigates the adverse physical and psychological effects but also enhances overall well-being. The imperative of nurturing emotional intelligence skills for leading a flourishing life has been underscored in Africa, as emphasized by the Lagos Business School in 2020.

4.2 Discussion

The aim of this study was to investigate the predictive capacity of emotional intelligence (EI) dimensions on women's propensity to embark on entrepreneurial ventures, with a focus on the mediating role played by the individual's level of entrepreneurial self-efficacy (Fig. 2).

The findings of our research corroborate all three hypotheses and show that emotional intelligence (EI) has a significant effect on business creation. The results showed that some dimensions of EI were significantly positively correlated with entrepreneurial self-efficacy, which further moderated the relationship between EI and entrepreneurial intention. These variables were shown to be significant predictors of entrepreneurial self-efficacy when factors such as education, work, and age were taken into account. However, in Tunisia, only the female variable has been shown to have a direct impact on entrepreneurial intentions.

The implications drawn from the results of Leroy, Maes et al. (2009) suggest that the inclination toward entrepreneurship is influenced by various personal factors. Although education, employment status, and age serve as control variables in our study, further investigation could delve into the extent to which these variables intersect with the aspiration to pursue entrepreneurship and how perceptions of

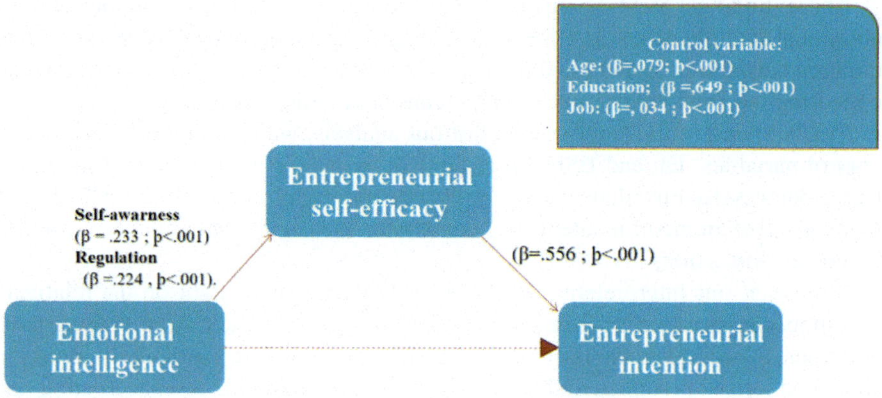

Fig. 2 The final model

self-efficacy might influence this decision, thereby enhancing our comprehension of this subject matter.

Upon incorporating entrepreneurial intention into our analysis, we find support for our hypotheses. Consistent with Ahmetoglu et al. (2011) and Mortan et al. (2014), we can confirm that the predictive effects of emotional intelligence (EI) extend beyond the impact of demographic variables. The first hypothesis was partially confirmed, as two of the five traits of the EI scale showed positive and significant associations with entrepreneurial self-efficacy. These findings have important practical implications, suggesting that emotional intelligence, specifically the ability to effectively manage and utilize emotions, plays an important role in enhancing organizational self-efficacy.

It is widely accepted that women exhibiting high emotional intelligence lean to be more confident in their abilities and have abilities to identify and seize business opportunities, leading to a higher likelihood of business success. Emotional intelligence enables people to understand and manage their own emotions as well as the emotions of others, which can be a valuable skill Entrepreneurship.

Women who possess strong emotional intelligence have faith in their entrepreneurial skills and perceive numerous opportunities for entrepreneurial endeavors. Being self-aware involves employing techniques to either prolong or disconnect from emotional states, monitoring and evaluating emotions, and reevaluating one's perception of a situation.

Emotional knowledge plays an important role in building self-efficacy in entrepreneurship, as confidence and competence are key factors when establishing networks for business and human resources management. The more a woman believes in her ability to successfully start and operate a business, the stronger her intention to become an entrepreneur becomes, bolstered by her confidence in her own competencies. Entrepreneurial pursuits can be challenging and uncertain, often presenting significant obstacles and risks. The absence of effective emotional coping strategies can hinder the achievement of business success.

These findings carry practical implications, highlighting the pivotal role of emotional intelligence, particularly in the effective management and utilization of emotions, in bolstering one's self-efficacy as an entrepreneur. It is widely acknowledged that women with robust emotional intelligence tend to exhibit confidence in their entrepreneurial abilities and perceive themselves as possessing abundant opportunities for entrepreneurial endeavors. Salvador (2008) and Mortan et al. (2014) suggest that this may be attributed to their social intelligence, which enables them to adapt more effectively to their environment and be perceived as more competent by others through their adeptness in regulating and comprehending emotions.

Emotional intelligence empowers individuals to grasp and regulate their own emotions, as well as those of others, which can be a valuable asset in entrepreneurship. Females with strong emotional intelligence are inclined to have confidence in their entrepreneurial capabilities and perceive themselves as having numerous prospects for entrepreneurial ventures.

Being self-aware entails employing strategies to either prolong or disconnect from emotional states, monitoring and evaluating emotions, and reassessing one's perception of a situation. Emotional information plays a pivotal role in bolstering self-efficacy in entrepreneurship, as confidence and competence are crucial when establishing networks for business and human resources management. The stronger a woman's belief in her ability to initiate and operate a successful business, the greater her intention to pursue entrepreneurship, propelled by her confidence in her own competencies.

Entrepreneurial endeavors can be fraught with challenges and uncertainties, and the absence of adequate emotional coping strategies can jeopardize business success. While previous research (Ahmetoglu et al. 2011) has explored the correlation between EI and various entrepreneurial outcomes, limited literature exists on the impact of EI on entrepreneurial intention when mediated by entrepreneurial self-efficacy. Our study fills this gap, enhancing our understanding of this relationship.

Our findings, aligned with Zhao et al. (2005), indicate that an individual's inclination to pursue entrepreneurship is shaped by their perception of high self-efficacy. Similarly, Lee et al. (2011) have investigated the influence of self-efficacy on entrepreneurial intention, positing that self-efficacy not only affects perceived feasibility but also influences entrepreneurial intentions. The intention to embark on entrepreneurial endeavors is bolstered by an individual's confidence in their abilities, a notion supported by Pihie (2009), who asserts that entrepreneurial self-efficacy ranks among the top predictors of students' entrepreneurial intentions.

Expanding upon previous studies (Abu H Ayob et al. 2020), our research employs a larger and diverse sample, incorporating interviews with participants of varying ages, statuses, and cultural backgrounds. This extends the existing literature by indicating that entrepreneurial self-efficacy provides a theoretical framework for understanding the relationship between two dimensions of EI—self-awareness and self-management—among women. Indeed, a combination of quantitative and qualitative analyses serves to reinforce findings related to trait emotional intelligence and women's entrepreneurial intentions.

In essence, our study underscores the importance of emotional intelligence in shaping women's entrepreneurial intentions, mediated by their level of entrepreneurial self-efficacy. By shedding light on the intricate interplay between emotional intelligence, self-efficacy, and entrepreneurial intentions, our research offers valuable insights into the psychological mechanisms underpinning entrepreneurial decision-making among women.

Moving forward, future research endeavors could further explore the nuanced relationships between demographic variables, emotional intelligence dimensions, self-efficacy, and entrepreneurial intentions. Additionally, investigations into the role of other contextual factors, such as cultural norms, societal expectations, and institutional support, could provide deeper insights into the entrepreneurial landscape for women across diverse settings.

Moreover, longitudinal studies tracking the entrepreneurial journeys of women with varying levels of emotional intelligence and self-efficacy could offer valuable insights into the long-term impact of these psychological factors on entrepreneurial success and satisfaction. By delving into the dynamic nature of emotional intelligence and self-efficacy over time, researchers can elucidate their evolving influence on women's entrepreneurial endeavors.

Ultimately, our findings underscore the importance of fostering emotional intelligence and self-efficacy among women to empower them as entrepreneurs and agents of change in their communities. By equipping women with the psychological resources needed to navigate the challenges and uncertainties of entrepreneurship, policymakers, educators, and stakeholders can contribute to creating a more inclusive and supportive entrepreneurial ecosystem that fosters women's economic empowerment and societal advancement.

5 Limitation and Future Research

Our research delvers valuable evidence for the link between emotional intelligence and entrepreneurial processes among Tunisian women. However, it is important to consider the limitations of our study. The collected data relied on self-reporting tools, which may have limited the predictive validity of the variables. Future studies should consider using alternative measures other than self-reporting to improve the robustness of the results.

In addition, incorporating longitudinal designs in future studies will facilitate a better understanding of the causal link between women's emotional intelligence and entrepreneurial intentions. By looking at changes over time, researchers can gain greater insight into the dynamic nature of these variables.

Another limitation of our study is the lack of empirical evidence on the relationship between entrepreneurial self-efficacy, emotional intelligence, and entrepreneurial intention among women. Given the scarcity of research in this area, further investigations are warranted to gain a more comprehensive understanding of the role of emotional intelligence in the entrepreneurial context.

Additionally, it is important to acknowledge the challenging economic circumstances faced by Tunisia, including a significant economic crisis and high unemployment rates. These factors can have substantial implications for both government policies and individual choices regarding employment. The economic instability may pose difficulties for individuals to take the risk associated with starting a new business, as resources and support may be limited.

In conclusion, while our study reveals a link between emotional intelligence and entrepreneurship among Tunisian women, it is important to address these limitations and encourage future research to delve deeper into this area. Understanding the role of emotional intelligence in an entrepreneurial context can provide valuable insights for policymakers, practitioners, and individuals interested in promoting entrepreneurial activity.

Therefore, innovation and creation of new businesses can boost economic growth and job creation, and thus may be the solution rather than the problem.

Further researchers can investigate how cultural factors shape women's attitudes toward entrepreneurship, emotional intelligence, and self-efficacy. Exploring cultural norms, values, and perceptions that may impact women's entrepreneurial aspirations and behaviors across different regions and societies.

6 Conclusion

This research has provided valuable insights into the multifaceted relationship between emotional intelligence, self-efficacy, and women's entrepreneurial intentions. Through a comprehensive analysis of theoretical frameworks, empirical research, and practical implications, several key findings have emerged.

Firstly, the study underscores the significant influence of emotional intelligence on women's entrepreneurial aspirations and behaviors. Emotional intelligence, particularly dimensions such as self-awareness and self-management, plays a pivotal role in shaping women's confidence, resilience, and decision-making processes as they navigate the challenges and opportunities of entrepreneurship.

Secondly, the study highlights the importance of entrepreneurial self-efficacy as a mediator between emotional intelligence and entrepreneurial intentions among women. Women who perceive themselves as capable and competent in their entrepreneurial abilities are more likely to translate their emotional intelligence into action, fueling their motivation and commitment to pursue entrepreneurial ventures.

Furthermore, our study underscores the profound impact of emotional intelligence on women's perception of their entrepreneurial efficacy and their intentions to engage in entrepreneurial pursuits. Women who harbor a strong sense of confidence in their capabilities are more inclined to embark on the entrepreneurial journey. The ability to comprehend, regulate, and leverage emotions, coupled with a keen awareness of one's self-perception as an effective entrepreneur, can emerge as pivotal factors, potentially offering a competitive edge in the entrepreneurial landscape. Thus, the adept utilization and mastery of emotional intelligence hold the potential to significantly bolster the trajectory and outcomes of entrepreneurial endeavors.

Moreover, the research sheds light on the contextual factors and societal dynamics that influence women's entrepreneurial journeys. Cultural norms, social support networks, and educational opportunities all contribute to shaping women's perceptions of themselves as entrepreneurs and their willingness to embark on entrepreneurial endeavors.

Furthermore, the findings underscore the need for targeted interventions and support mechanisms to foster women's entrepreneurial development. Mentorship programs, educational initiatives, and policy reforms can play a crucial role in equipping women with the emotional intelligence skills, self-efficacy beliefs, and resources they need to thrive as entrepreneurs in diverse socio-economic contexts.

This study holds significant importance for several reasons. Firstly, it contributes to the broader discourse on gender equality and women's empowerment, particularly in the context of entrepreneurship. By focusing specifically on women in Tunisia, where gender disparities in entrepreneurship persist, our research seeks to identify factors that can help bridge this gap and create more equitable opportunities for women in business.

Secondly, the study addresses a notable gap in the existing literature by examining the role of emotional intelligence in shaping entrepreneurial intentions among women. While previous research has explored various individual traits and their impact on entrepreneurship, emotional intelligence remains a relatively underexplored aspect, especially concerning its influence on entrepreneurial behavior and decision-making.

Moreover, the study's focus on emotional intelligence is particularly relevant in today's rapidly changing business landscape, where the ability to navigate complex social and emotional dynamics is increasingly recognized as a critical skill for entrepreneurial success. By highlighting the importance of emotional intelligence in entrepreneurial endeavors, our findings can inform entrepreneurship education and training initiatives aimed at equipping women with the necessary skills and mindset to thrive in the business world.

Taken as a whole, this study not only contributes to academic knowledge but also has practical implications for policymakers, educators, and business support organizations seeking to promote women's entrepreneurship and economic empowerment in Tunisia and beyond. By uncovering the factors that drive women's entrepreneurial intentions and identifying pathways to enhance their success, our research aims to foster a more inclusive and vibrant entrepreneurial ecosystem.

Overall, this study contributes to the existing body of knowledge on entrepreneurship and gender studies by providing a nuanced understanding of the psychological, social, and cultural factors that shape women's entrepreneurial intentions. By recognizing and addressing the complex interplay between emotional intelligence, self-efficacy, and entrepreneurial aspirations, stakeholders can work toward creating a more inclusive and supportive environment for women entrepreneurs to flourish and drive economic growth and innovation.

References

Abousaid, F.-Z., Angade, K.: Entrepreneuriat féminin dans la région Souss Massa (Maroc) : une étude comparative des motivations des femmes entrepreneures selon le choix sectoriel. Moroccan J. Entrepreneurship Innov. Manag., 9–22 (2020)

Ahmad, S., Bangash, H., Khan, S.A.: Emotional intelligence and gender differences, 4 (2009)

Ahmetoglu, G., Leutner, F., Chamorro-Premuzic, T.: EQ- Nomics: understanding the Relationship between Individual Differences in Trait Emotional Intelligence and Entrepreneurship. Personality Individ. Differ. **51**(8), 1028–1033 (2011)

Aissa, H.B.: Quelle METHODOLOGIE de recherche APPROPRIEE pour une construction de la recherche en gestion?, 27 (2016)

Aït Errays, Noureddine, et Amina Tourabi.: Le Rôle Du Support Du Mari et de l'empathie Dans La Formation Des Intentions Entrepreneuriales Prosociales Des Femmes Marocaines Mariées. *La Revue Gestion et Organisation* **10**(1), 14–28 (2018)

Ajzen, I.: The theory of planned behavior. Org. Behav. Hum. Dec. Proc. **50**(2), 179–211 (1991)

Ajzen, I.: Perceived behavioral control, self-efficacy, locus of control, and the theory of planned behavior. J. Appl. Soc. Psychol. **32**(4), 665–683 (2002)

Ajzen, Icek, et Martin Fishbein: Attitudes and normative beliefs as factors influencing behavioral intentions. J. Personality Soc. Psychol. **21**(1), 1–9 (1972). https://doi.org/10.1037/h0031930

Aldebert, Bénédicte, et Audrey Rouzies.: Quelle place pour les méthodes mixtes dans la recherche francophone en management ? Manag Int **19**(1), 43–60 (2015); Alderdice, M.T., Volle, R.L.: The increase in spontaneous transmitter release produced by beta-bungarotoxin and its modification by inorganic ions. J. Pharmacol. Exp. Therapeutics **205**(1), 58–68 (1978)

Aldrich, H., Kalleberg, A., Marsden, P., Cassel, J.: In Pursuit of evidence: sampling procedures for locating new businesses. J. Bus. Ventur. **4**(6), 367–386 (1989a)

Aldrich, H., Reese, P.R., Dubini, P.: Women on the verge of a breakthrough: networking among entrepreneurs in the United States and Italy. Entrepreneurship Regional Dev. **1**(4), 339–56 (1989)

Alexandre, L.: Typologie des entrepreneurs, une approche par le genre. *Revue de l'' Entrepreneuriat* 15(3–4), 109 (2016)

Al-Jubari, I., Hassan, A., Liñán, F.: Entrepreneurial intention among university students in Malaysia: integrating self-determination theory and the theory of planned behavior. Int. Entrepreneurship Manag. J. **15**(4), 1323–1342 (2019)

Amines, A., & Dust, S.: Emotional intelligence and innovation. Creativity Innov. Manage. **18**(2), 78–90 (2009)

Andrei, F., et al.: Emotional intelligence and resilience. J. Pers. Res. **52**(4), 455–467 (2016)

Andria, Aude d', et Inès Gabarret:. Mères et entrepreneures : étude de la motivation entrepreneuriale des mampreneurs françaises1. *Revue internationale P.M.E.* **30**(1), 155-81 (2017)

Antonioli, D., Nicolli, F., Ramaciotti, L., Rizzo, U.: The Effect of intrinsic and extrinsic motivations on Academics' Entrepreneurial intention. Administr. Sci. **6**(4), 15 (2016)

Artaza Abaroa, Felipe: L'intelligence des émotions : une relecture des fondements de « l'intelligence émotionnelle ». Revue internationale de psychosociologie XII (28), 25 (2006)

Aruoren, P.E.: Perspectives on emotional intelligence in leadership. Leadersh. Quart. **31**(3), 234–246 (2020)

Autumn, A., et al.: Emotional intelligence in leadership. Leadersh. J. **25**(5), 133–145 (2012)

Avenier, Marie-José, et Catherine Thomas:. À quoi sert l'épistémologie dans la recherche en sciences de gestion ? **8**, 15 (2016)

Avenier, M.-J., Gavard-Perret, M.-L.: Inscrire son projet de recherche dans un cadre épistémologique (2012)

Badri, R., Hachicha, N.: Entrepreneurship education and its impact on students‘ intention to start up: a sample case study of students from two Tunisian universities. Int. J. Manag. Educ. **17**(2), 182–190 (2019). https://doi.org/10.1016/j.ijme.2019.02.004

Bandura, A.: Self-efficacy: toward a unifying theory of behavioral change. *Psychological Review* **84**(2), 191–215 (1977)

Bandura, A., Lecomte, J., Carré, P., De Supérieur, B. (1986–2015): *Auto-efficacité: comment le sentiment d'' efficacité personnelle influence notre qualité de vie.* Louvain-la-Neuve: de Boeck supérieur (2019)

Bar-On, R.: The Bar-On model of emotional-social intelligence (ESI). Psicothema **18**(Suppl), 13–25 (2006)

Bélanger, L., Réto, G.: Management des établissements scolaires: l‘appui sur l‘intelligence émotionnelle et la bienveillance. *Questions Vives. Recherches enéducation,* 29 (2018)

Bellò, B., Mattana, V., Loi, M.: The power of peers: a new look at the impact of creativity, social context and self-efficacy on entrepreneurial intentions. Int. J. Entrep. Behav. Res. **24**(1), 214–233 (2018)

Benabou, D.: Chapitre 42 - L‘intelligence émotionnelle et l‘efficacité de leadership : cas d‘une entreprise algérienne d‘électronique. In *Réinventer le leadership*, par Soufyane Frimousse et Yves Le Bihan, 459. EMS Editions (2017)

Biljana R., Baumc, M., Hubnera, S.: How entrepreneur‘s leadership behavior and demographics shape applicant attraction to new ventures: the role of stereotypes. Int. J.Human Resource Manag. (2020)

Boren, A.E.: Emotional intelligence: the secret of successful entrepreneurship?. 9. Bosma, Nils, Mirjam van Praag, et Gerrit de Wit. (2000). *Determinants of Successful Entrepreneurship.* Research Report / EIM Small Business Research and Consultancy, 0002/E. Zoetermeer: EIM Small Business Research and Consultancy (2000)

Boutillier, S.: Le potentiel de ressources de l‘entrepreneur innovateur. Technologie et innovation **3**(2) (2018)

Bowen, D.D., Hisrich, R.D.: The female entrepreneur: a career development perspective. Acad. Manag. Rev. **11**(2), 393–407 (1986)

Boyatzis, R.E.: The behavioral level of emotional intelligence. The Leaders. Quart. **29**(3), 302–312 (2018)

Boyd, N.G., Vozikis, G.S.: The influence of self-efficacy on the development of entrepreneurial intentions and actions. Entrep. Theory Pract. **18**(4), 63–77 (1994)

Brasseur, S., Gregoire, J.: L‘intelligence émotionnelle – trait chez les adolescents à haut potentiel : spécificités et liens avec la réussite scolaire et les compétences sociales. Enfance **2010**(01), 59 (2010)

Brière, S., Auclair, I., Tremblay, M.: Soutenir les femmes entrepreneures en contexte africain : vers une nouvelle approche dynamique et collective. Revue Int. P.M.E. **30**(3-4) (2017)

Brush, C.G.: Research on women business owners: past trends, a new perspective and future directions. Entrep. Theory Pract. **16**(4), 5–30 (1992)

Brush, C.G., Vanderwerf, P.A.: A comparison of methods and sources for obtaining estimates of new venture performance. J. Bus. Venturing **7**(2), 157–170 (1992)

Bujold, M.: Association francophone pour le savoir-Acfas, et Congrès de l‘ACFAS, éd. *Oser les défis des méthodes mixtes en sciences sociales et sciences de la santé* (2018)

Burloud, A.: *Principes d'une psychologie des tendances*, 17–18 (1938)

Cabrera, E.M., Mauricio, D.: Factors affecting the success of women‘s entrepreneurship: a review of literature. Int. J. Gend. Entrep. **9**(1), 31–65 (2017)

Camacho-Miñano, M.-del-M., del Campo, C.: The role of creativity in entrepreneurship: an empirical study on business undergraduates. Educ. + Training **59**(7/8), 672–88 (2017)

Carmeli, A.: The relationship between emotional intelligence and work attitudes, behavior and outcomes. J. Manag. Psychol. **18**(8), 788–813 (2003)

Carrier, C., Julien, P.-A., Menvielle, W.: Un regard critique sur l‘entrepreneuriat féminin: une synthèse des études des 25 dernières années. Gestion **31**(2), 36 (2006)

Cartwright, S., & Pappas, C.: Emotional intelligence, its measurement, and implications for the workplace. Int. J. Manage. Rev. **10**(2), 149–171 (2008)

Chauvin, P.-M., Grossetti, M., Zalio, P.-P. (éds.): *Dictionnaire sociologique de l'' entrepreneuriat*. Références. Gouvernances. Paris: Sciences Po, Les Presses (2014)

Chell, E.: The Entrepreneurial Personality: A Social Construction. Routledge (2008)

Constantinidis, C., El Abboubi, M., Salman, N., Cornet, A.: L'entrepreneuriat féminin dans une société en transitions : analyse de trois profils de femmes entrepreneures au Maroc. *Revue internationale P.M.E.* **30**(3-4), 37–68 (2017)

Crittendin, J., & Ajjan, H.: Social contexts and emotional intelligence. Soc. Psychol. Rev. **21**(4), 22–39 (2019)

Cutler, B.: Leadership and emotional intelligence. J. Leadersh. Stud. **9**(1), 33–48 (2003)

Dada S., Jazi, S.: Motivations et freins de l'entrepreneuriat féminin au Maroc. Revue Africaine De Management **6**(1) (2021)

Decorps C.: Renforcement de l'auto-efficacité par la transportation mentale: amélioration du pouvoir persuasif d (2021).

Delanoë, S., Brulhart, F.: Analyse des antécédents de l'intention entrepreneuriale de porteurs de projets français. Revue Internationale PME Économie Et Gestion De La Petite Et Moyenne Entreprise **24**(1), 43–70 (2011)

Deng, W., Liang, Q., Li, J., Wang, W.: Science mapping: a bibliometric analysis of female entrepreneurship studies. Gender Manag. Int. J. Ahead-of-Print (ahead-of-print) (2020)

Deschamps, B., Lamarque, T.: Les émotions du repreneur lors des négociations de transfert externe d'entreprise. Successor's emotions during the negotiation of external business transfers. Finance Contrôle Stratégie 23–3 (2020)

Deschênes, A.-A., Capovilla, P.: L'auto-efficacité émotionnelle: un facteur à considérer pour expliquer la santé psychologique au travail. Psychologie Du Travail Et des Organisations **22**(3), 173–186 (2016)

De Klerk, L.: Emotional intelligence in female entrepreneurs. J. Bus. Psychol. **29**(4), 285–300 (2016)

Diamane, M., Koubaa, S.: Les approches dominantes de la recherche en entrepreneuriat, 25 (2016)

Diouf, I.D., Simen, S.F.: Importance des réseaux de relations personnelles dans le processus de création d'entreprise : Le cas des femmes- entrepreneures au Sénégal, 23 (2016)

Dolinsky, A.L., Caputo, R.K., Pasumarty, K., Quazi, H.: The effects of education on business ownership: a longitudinal study of women. Entrep. Theory Pract. **18**(1), 43–53 (1993)

Doré, C.: L'estime de soi : analyse de concept. Recherche en soins infirmiers **129**(2), 18 (2017)

Douayri, K., Etouzani, L.: LA DYNAMIQUE ENTREPRENEURIALE AU MAROC : OPPORTUNITES ET CONTRAINTES : CAS DES FEMMES ENTREPRENEURES DE LA REGION DE L'ORIENTAL. *Revue Economie* Gestion et Société (août), 25 (2020)

Durand, C.: Méthodes de sondage SOL3017. *Notes de cours, deuxième partie (l'' échantillonnage). Département de sociologie, Université de Montréal* (2002)

El Akremi, A.: Analyse des variables modératrices et médiatrices par les méthodes d'équations structurelles. *Management des ressources humaines-Méthodes de recherche en sciences humaines et sociales*, 326–42 (2005)

Ellouze, N., Louati, R.:. s. d. « L'impact de l'intelligence émotionnelle du dirigeant sur l'orientation entrepreneuriale de l'entreprise: Cas des PME tunisiennes ». « Emotional Intelligence, Entrepreneurs' Innovativeness, Entrepreneurial Success ». s. d., 7

« Entrepreneurship Orientation of Artisanal Fishery on the Use of Internet of.pdf ». s. d. Esfandiar, Kourosh, Mohamad Sharifi-Tehrani, Stephen Pratt, et Levent Altinay. (2019)

Understanding entrepreneurial intentions: a developed integrated structural model approach. J. Bus. Res. **94**(janvier), 172–82

Farahbod, F.: The relationship between trait emotional intelligence and entrepreneurship attitudes and intentions. J. Public Administr. Policy Res. **5**(3), 79–85 (2013)

Fernández-Berrocal, P., & Extremera, N.: Emotional intelligence: a theoretical and empirical review of its first 15 years of history. Psicothema **18**(Suppl), 7–12 (2006)

Ferry, A.: L'accompagnement entrepreneurial: la métamorphose des accompagnateurs en facilitateurs (2018)
Fiddian-Green, R.G., Silen, W.: Mechanisms of disposal of acid and Alkali in rabbit duodenum. Am. J. Physiol. **229**(6), 1641–1648 (1975)
Fietze, S., Boyd, B.: Entrepreneurial intention of danish students: a correspondence analysis. Int. J. Entrep. Behav. Res. **23**(4), 656–672 (2017)
Foliard, S., Le Pontois, S.: Vie et émotions des équipes entrepreneuriales étudiantes. In: Føllesdal, H., Hagtvet. K.A. (eds.), Emotional Intelligence: The MSCEIT from the Perspective of Generalizability Theory. Intelligence **37**(1), 94–105 (2019)
Fowler, N.O., McCall, D., Chou, T.C., Holmes, J.C., Hanenson, I.B.: Electrocardiographic changes and cardiac arrhythmias in patients receiving psychotropic drugs. Am. J. Cardiol. **37**(2), 223–230 (1976)
Freedman, J.: State of the heart: emotional intelligence research. Six Seconds Network (2018)
Friectson, R., et al.: Emotional intelligence and entrepreneurial outcomes. J. Appl. Behav. Sci. **55**(2), 45–63 (2019)
Gallé-Tessonneau, M., Grondin, O., Koleck, M., Doron, J.: Considérations méthodologiques pour la construction de questionnaires : l'exemple de la SChool REfusal EvaluatioN (SCREEN). Annales Médico-psychologiques, revue psychiatrique **176**(9) (2018)
Garner, C. W., Behal, F.J.: Effect of PH on substrate and inhibitor kinetic constants of human liver alanine aminopeptidase. Evidence for two Ionizable Active Center Groups. Biochemistry **14**(23) (1975)
Gartner, W.: Entrepreneurial traits. J. Bus. Ventur. **4**(3), 231–246 (1989)
Gauthier, B., Bourgeois, I.: *Recherche sociale: de la problématique à la collecte des données* (2016)
Gelaidan, H.M., Abdullateef, A.O.: Entrepreneurial intentions of business students in Malaysia: the role of self confidence, educational and relation support. J. Small Bus. Enterp. Dev. **24**(1), 54–67 (2017)
Gerville-Réache, Léo, et Vincent Couallier. 2011. « Échantillon représentatif (d'une population finie): définition statistique et propriétés ».
Ghoudani, K. EL, López-Zafra, E.: Mots-clés: adaptation sociale, éducation. Intelligence émotionnelle, rendement
Giacomin, O., Janssen, F., Pruett, M., Llopis, F., Shinnar, R., Toney, B.: Impact du sexe et de l'auto-efficacité entrepreneuriale sur l'intention entrepreneuriale des étudiants: une comparaison internationale. Bordeaux, France (2010)
Giger, J.-C.: Examen critique du caractère prédictif, causal et falsifiable de deux théories de la relation attitude-comportement : la théorie de l'action raisonnée et la théorie du comportement planifié. *L'' Année psychologique* **108**(01), 107 (2008)
Giordano, Y., Jolibert, A.: Pourquoi je préfère la recherche quantitative/Pourquoi je préfère la recherche qualitative. *Revue internationale P.M.E.: Économie et gestion de la petite et moyenne entreprise* **29**(2), 7 (2016)
Goleman, D.: Working with Emotional Intelligence. Bantam Books (1998)
Goleman, D., & Boyatzis, R.E.: Emotional intelligence has 12 elements. Harvard Bus. Rev. February Issue (2017)
Groves, K.S., McEnrue, M.P., Shen, W.: Developing and measuring the emotional intelligence of leaders. J. Manage. Dev. **27**(2), 225–250 (2008)
Guichard, R., Tran, S.: L'innovation distribuée: un modèle organisationnel applicable au secteur de la défense? *Revue internationale P.M.E.* **19**(2), 79-99 (2012)
Guma, M.: Emotional intelligence in Sub-Saharan Africa. Afr. J. Psychol. **14**(3), 85–96 (2015)
Hamilton, M.: Emotional intelligence in leadership. Leadersh. Sci. **12**(3), 212–230 (2015)
Hanan, F.: La psychologie sociale au service du développement durable Social psychology for sustainable development février 2020 (2020)
Hayrapetyan, L., Nunes, A., Khachatryan, K.: Factors that influence female entrepreneurship in the European, 20
Hayton, J.C., George, G., Zahra, S.A.: National culture and entrepreneurship: a review of behavioral research. Entrep. Theory Pract. **26**(4), 33–52 (2002)

Hendrickson, W.A., Ward, K.B.: Atomic models for the polypeptide backbones of Myohemerythrin and Hemerythrin. Biochem. Biophys. Res. Commun. **66**(4), 1349–1356 (1975)
Henry, C., Foss, L., Ahl, H.: Gender and entrepreneurship research: a review of methodological approaches. Int. Small Bus. J. Res. Entrep. **34**(3), 217–241 (2016)
Hernandez, É.-M., Marco, L.: Entrepreneur et décision: de l"intention à l"acte. ESKA, Paris (2006)
Herrbach, O., Mignonac, K., Sire, B.: L'intelligence émotionnelle: quelle innovation pour la gestion des ressources humaines (2003)
Hikkerova, L., Ilouga, S.N., Sahut, J.-M.: The entrepreneurship process and the model of volition. J. Bus. Res. **69**(5), 1868–1873 (2016)
Hisrich, R., Langan-Fox, J., Grant, S.: Entrepreneurship research and practice: a call to action for psychology. Amer. Psychologist **62**(6), 575–89 (2007)
Hlady Rispal, M., Jouison-Laffitte, E.: La contribution des méthodes qualitatives au développement du champ de l'entrepreneuriat. *Revue de l" Entrepreneuriat* **14**(1), 15 (2015)
Igalens, J., Roussel, P.: *Méthodes de recherche en gestion des ressources humaines.* FeniXX (1998)
Ivene, A.: Women's emotional intelligence and leadership. J. Manage. Stud. **34**(2), 151–170 (2017)
Jan, S., Irshad, M., Nadeem, A.: Determinant of the success factors for entrepreneurship business. I R **1**, 9 (2013)
Jena, R.K.: Measuring the impact of business management student's attitude towards entrepreneurship education on entrepreneurial intention: a case study. Comput. Human Behav. **107**(juin), 106275 (2020)
Jennings, J.E., Brush, C.G.: Research on women entrepreneurs: challenges to (and from) the broader entrepreneurship literature?. Acad. Manag. Annals **7**(1), 663–715 (2013)
Kakouris, A.: Exploring entrepreneurial conceptions, beliefs and intentions of Greek graduates. Int. J. Entrep. Behav. Res. **22**(1), 09–32 (2016)
Kalleberg, A.L.: Gender and organizational performance: determinants of small business survival and success. Acad. Manag. J., 27 (1991)
Kashif, M., Khurshid, M.K., Zahid, I.:Relationship between emotional intelligence and entrepreneurship, 5 (2016)
Kassean, H., Vanevenhoven, J., Liguori, E., Winkel, D.E.: Entrepreneurship education: a need for reflection, real-world experience and action. Int. J. Entrepreneurial Behav. Res. **21**(5), 690–708 (2015)
Kelly, L., & Linden, R.: Female entrepreneurship in Africa. J. Dev. Econ. **112**(3), 421–439 (2020)
Khatoon, N.: The impact of emotional intelligence on the growth of entrepreneurship, 8 (2019)
Khyareh, M.: Determinants of female entrepreneurship in Iran: an institutional approach. Econ. Ann. **63**(216), 111–129 (2018)
Kim, M.-S., Hunter, J.E.: Relationships among attitudes, behavioral intentions, and behavior: a meta-analysis of past research, Part 2. Commun. Res. **20**(3), 331–364 (1993)
Klapper, R., Léger-Jarniou, C.: Entrepreneurship intention among French grande ÉCole and university students: an application of Shapero's model. Ind. High. Educ. **20**(2), 97–110 (2006)
Koçak, S.: Emotional intelligence in leadership contexts. J. Leadersh. Stud. **17**(2), 55–67 (2021)
Kotsou, I.: *Intelligence émotionnelle et management: comprendre et utiliser la force des émotions.* De Boeck Supérieur (2019)
Kowlman, E., et al.: Emotional intelligence in entrepreneurial decision-making. Ent Sci. **29**(1), 35–42 (2020)
Krueger, N.: The impact of prior entrepreneurial exposure on perceptions of new venture feasibility and desirability. Entrep. Theory Pract. **18**(1), 5–21 (1993)
Krueger, N.F., Carsrud, A.L.: Entrepreneurial intentions: applying the theory of planned behaviour. Entrep. Reg. Dev. **5**(4), 315–330 (1993)
Krueger, N.F., Reilly, M.D., Carsrud, A.L.: Competing models of entrepreneurial intentions. J. Bus. Ventur. **15**(5–6), 411–432 (2000)
Kuckertz, A., & Prochotta, A.: Emotional intelligence and innovation. Creativity Res. J. **30**(3), 245–255 (2018)

Kuratko D.F.: The emergence of entrepreneurship education: development, trends, and challenges, Entrepreneurship »: « KURATKO, Donald F. Entrepreneurship: theory, process, and practice. Nelson Education, 2016 (2005)
Laufer, J., Locoh, T., Maruani, M.: Annie Labourie-Racapé. Travail, Genre Et Sociétés **19**(1), 5 (2008)
Lechheb, H., Frikh, Z., Obtil, S.: La sensibilisation à l'entrepreneuriat : Une lecture théorique (2020)
Lefkowitz, R.J.: Identification of adenylate cyclase-coupled beta-adrenergic receptors with radiolabeled beta-adrenergic antagonists. Biochem. Pharmacol. **24**(18), 1651–1658 (1975). https://doi.org/10.1016/0006-2952(75)90001-5
Lhuillier, D.: Compétences émotionnelles : de la proscription à la prescription des émotions au travail. Psychologie Du Travail Et des Organ. **12**(2), 91–103 (2006)
Littunen, H.: Entrepreneurship and the characteristics of the entrepreneurial personality. J. Small Bus. Manage. **38**(1), 45–55 (2000)
Lopes, P.N., et al.: Emotional intelligence and social relationships. Soc. Behav. Pers. **32**(1), 115–122 (2004)
Lopes, P N., et al.: Emotional intelligence in workplace success. J. Pers. Soc. Psychol. **91**(2), 398–410 (2006)
Mahamat, A.A., Mazra, M., Mamoudou, H.: L'effet de l'auto- efficacité entrepreneuriale sur linnovation entrepreneuriale: une étude auprès des PME Camerounaises. Global J. Manag. Bus. Res. (2019)
Maharaj, P., & Ramsaroop, S.: Emotional intelligence and social capital. J. Appl. Soc. Psychol. **39**(4), 611–627 (2022)
Mahfud, T., Triyono, M.B., Sudira, P., Mulyani, Y.: The influence of social capital and entrepreneurial attitude orientation on entrepreneurial intentions: the mediating role of psychological capital. Eur. Res. Manag. Bus. Econ. **26**(1), 33–39 (2020)
Mair, J., Robinson, J., Hockerts, K. (éd.): *Social Entrepreneurship*. 1st ed. 2006. London: Palgrave Macmillan UK (2014)
Makar, A.B., McMartin, K.E., Palese, M., Tephly, T.R.: Formate assay in body fluids: application in methanol poisoning. Biochem. Med. **13**(2), 117–126 (1975)
Mayer, J.D., & Salovey, P.: Emotional intelligence. Imagination, Cogn. Pers. **9**(3), 185–211 (1990)
Mayer, J.D., & Salovey, P.: Emotional Development and Emotional Intelligence: Educational Implications. Basic Books (1997)
Mayer, J.D., Salovey, P., & Caruso, D.R.: Emotional intelligence: new ability or eclectic traits? Am. Psychol. **63**(6), 503–517 (2008)
Mbengue, A., Vandangeon-Derumez, I.: Positions épistémologiques et outils de recherche en management stratégique, Ac (1999)
Mbengue, A.: Positions épistémologiques et outils de recherche en management stratégique, 22 (2014)
McDonald, M.: Emotional intelligence and stress management. Stress Manage. Rev. **14**(1), 12–25 (2021)
Mehtap, S., Pellegrini, M.M., Caputo, A., Welsh, D.H.B.: Entrepreneurial intentions of young women in the Arab world: socio-cultural and educational barriers. Int. J. Entrep. Behav. Res. **23**(6), 880–902 (2017)
Meunier, F., Krylova, Y., Ramalho, R.: Women"s Entrepreneurship: How to Measure the Gap between New Female and Male Entrepreneurs? World Bank, Washington, DC (2017)
Meyer, N., Hamilton, L.: Female entrepreneurs' business training and its effect on various entrepreneurial factors: evidence from a developing country. Int. J. Econ. Finance Studies, janvier, 135–51 (2020)
Miao, C., Humphrey, R.H., Qian, S., Pollack J.M.: Emotional Intelligence and Entrepreneurial Intentions: an Exploratory Meta- Analysis. Career Dev. Int. **23**(5) (2018a)
Miao, C., et al.: Emotional intelligence and entrepreneurial intentions: an exploratory meta-analysis. Career Dev. Int. **23**(5), 569–590 (2018b)

Michinov, N., & Michinov, E.: Role of emotional intelligence in teamwork. Group Dyn. **23**(3), 208–224 (2020)
Mikalajczak, M., et al.: Emotional intelligence and stress. J. Pers. Psychol. **89**(1), 155–167 (2009)
Momani, B.: Entrepreneurship in Arab cultures. Middle East J. Econ. **22**(1), 98–112 (2017)
Moreau, R., Raveleau, B.: Les trajectoires de l'intention entrepreneuriale. Revue Int. (2012)
Mortan, R.A., Ripoll, P., Carvalho, C., Consuelo Bernal, M.: Effects of emotional intelligence on entrepreneurial intention and self-efficacy. Revista De Psicología Del Trabajo y De Las Organizaciones **30**(3), 97–104 (2014)
Neghabi, S., Yousefi, M., Rezvani, M.: Emotional intelligence and entrepreneurial behavior; development of personal aspects of intentions model, 5 (2011)
Notais, A., Tixier, J.: Entrepreneuriat et innovation au cœur d'un territoire: le cas des femmes entrepreneures sociales des quartiers. Innovations **57**(3), 11 (2018)
Ntep, F., Zammar, R.: Entrepreneuriat féminin en Afrique : catalyseur de transformation économique et approche de réussite. Moroccan J. Entrepreneurship Innov. Manag. (novembre), 103–115 p. (2020a)
Ntep, F., Zammar, R.: Entrepreneuriat féminin en Afrique : catalyseur de transformation économique et approche de réussite. Moroccan J. Entrepreneurship Innov. Manag. (novembre), 103–115 p. (2020b)
Nukpizah, A., & Blankison, T.: Women entrepreneurs and emotional intelligence. Ent. J. **33**(2), 125–140 (2019)
O'Boyle, E.H., et al.: The relation between emotional intelligence and job performance. J. Organ. Behav. **32**(6), 788–818 (2004)
O'Boyle, E.H., et al.: Emotional intelligence and job success: a meta-analysis. J. Organ. Behav. **33**(1), 12–28 (2011)
Omrane, A., Fayolle, A., Zeribi-BenSlimane, O.: Les compétences entrepreneuriales et le processus entrepreneurial : une approche dynamique. La Revue des Sciences De Gestion **251**(5), 91 (2011)
Othman, N., Tengku Nor Asma Amira Tengku Muda: Emotional intelligence towards entrepreneurial career choice behaviours. Education Training **60**(9), 953–70 (2018)
Pappas, M., Drigas, A., Papagerasimou, Y., Dimitriou, H., Katsanou, N., Papakonstantinou, S., Karabatzaki, Z.: Female entrepreneurship and employability in the digital era: the case of Greece. J. Open Innov. Technol. Market Complexity **4**(2), 15 (2018)
Parry, W.H., Martorano, F., Cotton, E.K.: Management of life-threatening asthma with intravenous isoproterenol infusions. American J. Diseases Children (1960) **130**(1), 39–42 (1976)
Pascal, A., Aldebert, B., Rouziès, A: Les méthodes mixtes en systèmes d'information : enjeux épistémologiques et méthodologiques. *Systèmes d'' information & management* **23**(3), 99 (2018)
Pavliscak, P.: Emotional intelligence and life satisfaction. J. Happiness Stud. **19**(3), 811–832 (2018)
Peersman, G.: Présentation des méthodes de collecte et d'analyse de données dans l'évaluation d'impact: Note méthodologique-Évaluation d'impact n° 10 (2014)
Philippart, P.: L'accompagnement de l'entrepreneur en difficulté : questions autour d'un phénomène complexe. Projectics / Proyéctica / Projectique **16**(1), 11 (2017). https://doi.org/10.3917/proj.016.0011
Philippart, P.: *La fabrique de l'' entrepreneuriat* , par Christophe Schmitt (Dunod, 2018) ou pourquoi et comment (re)centrer l'entrepreneuriat sur le projet entrepreneurial. *Revue de l'' Entrepreneuriat* **17**(2), 133 (2018)
Pinard, R.: LE CHOIX D'UNE APPROCHE MÉTHODOLOGIQUE MIXTE DE RECHERCHE EN ÉDUCATION, 24 (2018)
Pitaf, E., et al.: Emotional intelligence and career success. J. Career Dev. **45**(6), 451–470 (2018)
Puni, A., Anlesinya, A., Korsorku, P.D.A.: Entrepreneurial education, self-efficacy and intentions in Sub-Saharan Africa. Afr. J. Econ. Manag. Stud. **9**(4), 492–511 (2018)
Raeissi, P., et al.: Role of emotional intelligence in workplace communication. J. Commun. Stud. **36**(4), 212–230 (2020)

Ratten, V., Tajeddini, K.: Women's entrepreneurship and Internationalization: patterns and Trends. Int. J. Sociol. Soc. Policy **38**(9/10), 780–793 (2018)
Reichborn, T., et al.: Gender and emotional intelligence. J. Soc. Res. **19**(2), 54–65 (2014)
Rezani, S., et al.: Emotional skills and leadership. Leadersh. Sci. **19**(1), 19–36 (2016)
Rezvani, S., et al.: Emotional intelligence in leadership effectiveness. Int. J. Organ. Behav. **25**(1), 12–30 (2020)
Rhee, K.S., & White, R.J.: The emotional intelligence of entrepreneurs. J. Small Bus. Manage. **45**(2), 276–293 (2007)
Rodrigues, A.P., Jorge, F.E., Carlos Pires, A., António, P: The contribution of emotional intelligence and spirituality in understanding creativity and entrepreneurial intention of higher education students. Education + Training **61**(7/8), 870–94 (2019)
Rodríguez G., Isabel, P., del Pilar Pastor Pérez, M., Galicia, P.E.A.: University entrepreneurship: how to trigger entrepreneurial intent of undergraduate students . J. Sci. Technol. Policy Manag. **10**(4), 927–50 (2019)
Santoni, J.: Quels processus pour répondre aux besoins spécifiques des entrepreneures ? Entreprendre Innover **36**(1), 29 (2018)
Sauer, R.M., Wilson, T.: The Rise of Female Entrepreneurs: New Evidence on Gender Differences in Liquidity Constraints. Europ. Econ. Rev. **86** (juillet), 73–86 (2016)
Schjoedt, L., Craig, J.B.: Development and validation of a unidimensional domain-specific entrepreneurial self-efficacy scale. Int. J. Entrep. Behav. Res. **23**(1), 98–113 (2017)
Schmitt, C., Julien, P.-A., Lachance, R.: Pour une lecture des problèmes complexes en PME: approche conceptuelle et expérimentation. *Revue internationale P.M.E.* **15**(2), 35–62 (2018)
Schwabbauer, M.L.: Use of the latent image technique to develop and evaluate problem-solving skills. Am. J. Med. Technol. **41**(12), 457–462 (1975)
Scott Deborer: Women entrepreneurs: emotional perspectives. J. Women's Stud. **18**(4), 321–332 (2012)
Scott Dolan, A., et al.: Emotional management in female entrepreneurship. Manage. Quart. **14**(2), 211–222 (2012)
Setyowati, E., & Razak, A.: Emotional intelligence and work engagement. J. Hum. Res. Manage. **32**(1), 22–35 (2019)
Sfetcu, N.: *Émotions et intelligence émotionnelle dans les organisations.*MultiMedia Publishing (2020)
Sfetcu, N., MultiMedia, R., d'Azur, C., Web, C.: Modèles d'intelligence émotionnelle (2020)
Shahab, Y., Chengang, Y., Arbizu, A.D., Haider, M.J.: Entrepreneurial self-efficacy and intention: do entrepreneurial creativity and education matter? Int. J. Entrepreneurial Behav. Res. **25**(2), 259-80 (2019)
Smith, R.M., Sardeshmukh, S.R., Combs, G.M.: Understanding Gender, Creativity, and Entrepreneurial Intentions ». Édité par Vicky Duckworth, Francis Farrell and Phil Rigby. Education + Training **58**(3), 263–82 (2016)
Sawandana, P.: The emotional intelligence advantage. J. Appl. Psychol. **104**(3), 355–372 (2019)
Stéphane, F., Le Pontois, S.: Vie et émotions des équipes entrepreneuriales étudiantes (2019)
St-Jean, É., Fonrouge, C.: Comment développer l'auto-efficacité entrepreneuriale? Étude auprès d'étudiants universitaires. J. Small Bus. Entrep. **32**(4), 355–377 (2020)
Sumathi, R., et al.: Emotional intelligence and leadership. J. Bus. Stud. **24**(4), 451–465 (2021)
Tchagang, E.: Les déterminants de l'intention entrepreneuriale des étudiants camerounais, 19 (2018)
Tchagang, E., Tchankam, J.-P.: Identité de genre et intention entrepreneuriale des étudiants camerounais. Question(s) de Manag. **20**(1), 71 (2018a)
Tchagang, E., Tchankam, J.-P.: Les antécédents sociodémographiques de l'intention entrepreneuriale des étudiants : le rôle médiateur de l'auto-efficacité entrepreneuriale. Gestion 2000 **35**(1), 21 (2018b)
Thiétart, R.-A., Cairn.info.: *Méthodes de recherche en management*. Paris (5 Rue Laromiguière 75005): Dunod (2014)

Tiwari, R., et al.: Emotional intelligence and entrepreneurial leadership. J. Bus. Psychol. 33(1), 1–18 (2018)
Tixier, J., Irg, M.: 11ème congrès de l'Académie de l'Entrepreneuriat et de l'Innovation
Tomy, S., Pardede, E.: An entrepreneurial intention model focussing on higher education. Int. J. Entrep. Behav. Res. **26**(7), 1423–1447 (2020)
Tsyganova, T., Shirokova, G.: Gender differences in entrepreneurship: evidence from gem data. Organ. Markets Emerg. Econ. **1**(1), 120–141 (2018)
Turcotte, E.: Les méthodes mixtes dans la recherche féministe : enjeux, contraintes et potentialités politiques. Recherches Féministes **29**(1), 111–128 (2016)
Venmani, R., et al.: Emotional intelligence in entrepreneurial contexts. Entrepreneurship Theor. Pract. **46**(2), 247–276 (2022)
Wan Mohb et al.: Emotional intelligence in business. J. Bus. Manage. **23**(2), 145–159 (2021)
Zampetaskis, E., et al.: Emotional intelligence and its role in entrepreneurship. Small Bus. J. **21**(3), 102–114 (2009)
Zanpetaks, E., et al.: Emotional intelligence in women entrepreneurs. J. Women's Stud. **15**(3), 122–135 (2017)
Zenebe, A., Alsaaty, F.M., Anyiwo, D.: Relationship between Individual's entrepreneurship intention, and adoption and knowledge of information technology and its applications: an empirical study. J. Small Bus. Entrep. **30**(3), 215–232 (2018)
Zhao, H., Seibert, S.E., Hills, G.E.: The mediating role of self-efficacy in the development of entrepreneurial intentions. J. Appl. Psychol. **90**(6), 1265–72 (2005)
Zheng, L., Plaisent, M., Zuccaro, C., Bernard, P., Daghfous, N., Favreau, S.: *L" analyse des données de sondage avec SPSS: un guide d" introduction.* PUQ (2019)
Zineelabidine, M., Hassainate, M.S., Hammouchi, M.S.: L_Intention Entrepreneuriale : Revue de Littérature & Thématiques d_Analyses, 16 (2018b)

Author Index

K. Dhoska and E. Spaho (eds.), *Bridging Horizons in Artificial Intelligence, Robotics, Cybersecurity, Smart Cities, and Digital Economy*, Sustainable Economy and Ecotechnology, https://doi.org/10.1007/978-3-031-72029-1

The manufacturer's authorised representative in the EU is Springer Nature Customer Service Centre GmbH, Europaplatz 3, 69115 Heidelberg, Germany. If you have any concerns regarding our products, please contact ProductSafety@springernature.com

Printed and bound by CPI Group (UK) Ltd, Croydon, CR0 4YY
30/09/2026
02225563-0015